Microstructure and Properties in Metals and Alloys

Microstructure and Properties in Metals and Alloys

Editors

Andrea Di Schino
Claudio Testani

Basel • Beijing • Wuhan • Barcelona • Belgrade • Novi Sad • Cluj • Manchester

Editors
Andrea Di Schino
University of Perugia
Perugia
Italy

Claudio Testani
CALEF-ENEA CR Casaccia
Rome
Italy

Editorial Office
MDPI
St. Alban-Anlage 66
4052 Basel, Switzerland

This is a reprint of articles from the Topic published online in the open access journals *Aerospace* (ISSN 2226-4310), *Alloys* (ISSN 2674-063X), *Crystals* (ISSN 2073-4352), *Materials* (ISSN 1996-1944), and *Metals* (ISSN 2075-4701) (available at: https://www.mdpi.com/topics/EY0KVP6Y85).

For citation purposes, cite each article independently as indicated on the article page online and as indicated below:

Lastname, A.A.; Lastname, B.B. Article Title. *Journal Name* **Year**, *Volume Number*, Page Range.

ISBN 978-3-0365-8496-6 (Hbk)
ISBN 978-3-0365-8497-3 (PDF)
doi.org/10.3390/books978-3-0365-8497-3

Contents

 *metals*

Editorial

Microstructure and Properties in Metals and Alloys

Andrea Di Schino [1,*] and Claudio Testani [2]

1   Dipartimento di Ingegneria, Università degli Studi di Perugia, Via G. Duranti 93, 06125 Perugia, Italy
2   CALEF-ENEA CR Casaccia, Via Anguillarese 301, Santa Maria di Galeria, 00123 Rome, Italy; claudio.testani@consorziocalef.it
*   Correspondence: andrea.dischino@unipg.it

1. Introduction and Scope

Microstructure design is key in targeting the desired material's properties. It is therefore essential to understand the relations between properties and microstructure and how they are driven via a specific process [1–5]. The following five journals contribute to this topic: *Aerospace, Alloys, Crystals, Materials, and Metals*. Contributions related to microstructure design and characterization are collected in this topic, together with their relation to the mechanics, fatigue, wear, and corrosion resistance of different kinds of metals and alloys. The goal of this Topic is to present contributions related to the relationship between the microstructure and properties of metals and alloys for different applications, including aeronautical and aerospace applications. Different process routes are considered (thermo-mechanical routes and additive manufacturing) in this topic. Welding is a mandatory issue in many applications: this is why contributions related to welding are also included in this topic.

2. Contributions

This topic includes 18 articles, two communications and a review paper. Among these, 10 papers were published in *Metals*, 4 papers in *Materials*, 1 in *Alloys*, 3 in *Crystals*, and 3 in *Aerospace*, covering several aspects concerning microstructure properties in metals and alloys.

X. Zhu et al. [6] present a method for the automatic detection of sorbite content in high-carbon steel wire rods. A semantic segmentation model of sorbite based on DeepLabv3+ is established. The sorbite structure is segmented, and the prediction results are analyzed and counted based on the metallographic images of high-carbon steel wire rods marked manually. For the problem of sample imbalance, the loss function of Dice loss + focal loss is used, and the perturbation processing of training data is added. The results show that this method can realize the automatic statistics of sorbite content. The average pixel prediction accuracy is as high as 94.28%, and the average absolute error is only 4.17%. The composite application of the loss function and the enhancement of the data perturbation significantly improve the prediction accuracy and robust performance of the model. In this method, the detection of sorbite content in a single image only takes 10 s, which is 99% faster using the manual cut-off method, which takes 10 min. On the premise of ensuring detection accuracy, the detection efficiency is significantly improved, and the labor intensity is reduced.

Z. Li et al. [7] report in situ observations of the austenite grain growth and martensite transformations in developed NM500 wear-resistant steel conducted via confocal laser scanning high-temperature microscopy. The results indicated that the size of the austenite grains increased with the quenching temperature (37.41 μm at 860 °C → 119.46 μm at 1160 °C) and austenite grains coarsened at ~3 min at a higher quenching temperature of 1160 °C. Furthermore, a large amount of finely dispersed (Fe, Cr, Mn)3C particles redissolved and broke apart at 1160 °C, resulting in many large and visible carbonitrides. The transformation kinetics of martensite were accelerated at a higher quenching temperature (13 s at 860 °C → 2.25 s at 1160 °C). In addition, selective prenucleation dominated, which

Citation: Di Schino, A.; Testani, C. Microstructure and Properties in Metals and Alloys. *Metals* **2023**, *13*, 1320. https://doi.org/10.3390/met13071320

Received: 5 July 2023
Accepted: 20 July 2023
Published: 24 July 2023

divided untransformed austenite into several regions and resulted in larger-sized fresh martensite. Martensite can not only nucleate at the parent austenite grain boundaries, but also nucleate in the preformed lath martensite and twins. Moreover, the martensitic laths presented as parallel laths (0–2°) based on the preformed laths or were distributed in triangles, parallelograms, or hexagons with angles of 60° or 120°.

Z. Wu et al. [8] propose solving the problem of poor metallurgical bonding of Cu/Al bimetallic composites caused by high-temperature oxidation of Cu, by different coating thicknesses of Ni layer on Cu rods used to fabricate the Cu/Al bimetallic composite by gravity casting. The effect of liquid–solid volume ratio and coating thickness on microstructure and properties of a Cu/Al bimetallic composite were investigated in this study. The results indicated that the transition zone width increased from 242.3 µm to 286.3 µm and shear strength increased from 17.8 MPa to 30.3 MPa with a liquid–solid volume ratio varying from 8.86 to 50. The thickness of the transition zone and shear strength increased with the coating thickness of the Ni layer varying from 1.5 µm to 3.8 µm, due to the Ni layer effectively preventing oxidation on the surface of the Cu rod and promoting the metallurgical bonding of the Cu/Al interface. The presence of a residual Ni layer in the cast material hinders the diffusion process of the Cu and Al atom. Therefore, the thickness of the transition zone and shear strength exhibited a decreasing trend as the coating thickness of the Ni layer increased from 3.8 µm to 5.9 µm. Shear fracture observation revealed that the initiation and propagation of shear cracks occurred within the transition zone of the Cu/Al bimetallic composite.

L. Li et al. [9] investigate the impact of various heat treatments on the strength and toughness of TA15 aviation titanium alloys. Five different heat treatment methods were employed in the temperature range of 810–995 °C. The microstructure of the alloy was examined using a scanning electron microscope (SEM) and X-ray diffraction (XRD), and its mechanical properties were analyzed through tensile, hardness, impact, and bending tests. The findings indicate that increasing the annealing temperature results in an increase in the phase boundary and secondary α phase, while the volume fraction of the primary α phase decreases, leading to a rise in hardness and a decrease in elongation. The tensile strength of heat-treated samples at 810 °C was notably improved, displaying high ductility at this annealing temperature. Heat treatment (810 °C/2 h/WQ) produced the highest tensile properties (ultimate tensile strength, yield strength, and elongation of 987 MPa, 886 MPa, and 17.78%, respectively). Higher heat treatment temperatures were found to enhance hardness but decrease the tensile properties, bending strength, and impact toughness. The triple heat treatment (810 °C/1 h/AC + 810 °C/1 h/AC + 810 °C/1 h/AC) resulted in the highest hardness of 601.3 MPa. These results demonstrate that various heat treatments have a substantial impact on the strength and toughness of forged TA15 titanium alloys.

S. Kusmanov et al. [10] describe how to modify the surface of austenitic stainless steel by anodic plasma electrolytic treatment. Surface treatment was carried out in aqueous electrolytes based on ammonium chloride (10%) with the addition of ammonia (5%) as a source of nitrogen (for nitriding), boric acid (3%) as a source of boron (for boriding), or glycerin (10%) as a carbon source (for carburizing). Morphology, surface roughness, phase composition, and microhardness of the diffusion layers in addition to the tribological properties were studied. The influence of physicochemical processes during the anodic treatment of the features of the formation of the modified surface and its operational properties are shown. The study revealed the smoothing of irregularities and the reduction in surface roughness during anodic plasma electrolytic treatment due to electrochemical dissolution. An increase in the hardness of the nitrided layers to 1450 HV with a thickness of up to 20–25 µm was found due to the formation of iron nitrides and iron-chromium carbides with a 3.7-fold decrease in roughness accompanied by a 2-fold increase in wear resistance. The carburizing of the steel surface leads to a smaller increase in hardness (up to 700 HV) but a greater thickness of the hardened layer (up to 80 µm) due to the formation of chromium carbides and a solid solution of carbon. The roughness and wear resistance of the carburized surface change are approximately the same values as after nitriding. As a result of the boriding of the austenitic stainless steel, there is no hardening of the surface, but, at the same time, there is a decrease in roughness and an increase in wear resistance on the surface. It has been established that frictional bonds in the friction process are destroyed after all types of processing as a result of the plastic displacement of the counter body material. The type of wear can be characterized as fatigue wear with boundary friction and plastic contact. The correlation of the friction coefficient with the Kragelsky–Kombalov criterion, a generalized dimensionless criterion of surface roughness, is shown.

Y. Sun et al. [11] study the microstructure and mechanical properties of as-homogenized Mg-xLi-3Al-2Zn-0.2Zr alloys (x = 5, 7, 8, 9, 11 wt.%). As the Li content increased from 5 wt.% to 11 wt.%, the alloy matrix changed from the α-Mg single-phase to α-Mg+β-Li dual-phase and then to the β-Li single-phase. With the increase in Li content, the alloy strength decreased while the elongation increased, and the corresponding fracture mechanism changed from cleavage fracture to microvoid coalescence fracture. This is mainly attributed to the matrix changing from α-Mg with hcp structure to β-Li with bcc structure. Additionally, the increase in the AlLi softening phase led to the reduction of Al and Zn dissolved in the alloy matrix with increasing Li content, which is one of the reasons for the decrease in alloy strength.

C. Ferro et al. [12] clarify the crucial role of Additive Manufacturing (AM) in the fourth industrial revolution. The design freedom provided by this technology is disrupting limits and rules from the past, enabling engineers to produce new products that are otherwise unfeasible. Recent developments in the field of Selective Laser Melting (SLM) have led to a renewed interest in lattice structures that can be produced non-stochastically in previously unfeasible dimensional scales. One of the primary applications is aerospace engineering where the need for light weights and performance is urgent to reduce the carbon footprint of civil transport around the globe. Of particular concern is fatigue strength. Being able to predict fatigue life in both LCF (Low Cycle Fatigue) and HCF (High Cycle Fatigue) is crucial for a safe and reliable design in aerospace systems and structures. In the present work, an experimental evaluation of compressive–compressive fatigue behavior has been performed to evaluate the fatigue curves of different cells, varying sizes, and relative densities. A Design of Experiment (DOE) approach has been adopted in order to maximize the information extractable in a reliable form.

Y. Guo et al. [13] report on the Nd-Fe-B hot-deformation magnet with high resistivity which was successfully prepared by hot-pressing and hot-deformation of Nd-Fe-B fast-quenched powder with amorphous glass fiber. After the process optimization, the resistivity of the magnet was increased from 0.383 mΩ·cm to 7.2 mΩ·cm. Therefore, the eddy current loss of magnets can be greatly reduced. The microstructure shows that the granular glass fiber forms a continuous isolation layer during hot deformation. At the same time, the boundary of Nd-Fe-B quick-quenched the flake and glass fiber from the transition layer, which improves the binding of the two, and which can effectively prevent the spalling of the isolation layer. In addition, adding glass fiber improves the orientation of the hot deformation magnet to a certain extent. The novel design concept of insulation materials provides new insights into the development and application of rare earth permanent magnet materials.

G. Stornelli et al. [14] show how the presence of micro-alloying elements in HSLA steels induces the formation of microstructural constituents, capable of improving the mechanical performance of welded joints. Following double welding thermal cycle, with second peak temperature in the range between Ac$_1$ and Ac$_3$, the IC GC HAZ undergoes a strong loss of toughness and fatigue resistance, mainly caused by the formation of residual austenite (RA). The present study aims to investigate the behavior of IC GC HAZ of a S355 steel grade, with the addition of different vanadium contents. The influence of vanadium micro-alloying on the microstructural variation, RA fraction formation and precipitation state of samples subjected to thermal cycles experienced during double-pass welding was reported. Double-pass welding thermal cycles were reproduced by heat treatment using a dilatometer at five different maximum temperatures of the secondary peak in the inter-critical area, from 720 °C to 790 °C. Although after the heat treatment it appears that the addition of V favors the formation of residual austenite, the amount of residual austenite formed is not significant for inducing detrimental effects (from the EBSD analysis the values are always less than 0.6%). Moreover, the precipitation state for the variant with 0.1 wt.% of V (high content) showed the presence of vanadium rich precipitates with size smaller than 60 nm of which, more than 50% are smaller than 15 nm.

S. Najafi et al. [15] study the influence of rare earth (RE) elements on the microstructure and mechanical performance of an extruded ZK60 Mg alloy. Two types of RE elements were added to a ZK60 material and then extruded at a ratio of 18:1. The first new alloy contained 2 wt% Y while the second one was produced using 2 wt% Ce-rich mischmetal. The microstructure, the texture, and the dislocation density in a base ZK60 alloy and two materials with RE additives were studied by scanning electron microscopy, electron backscattered diffraction, and X-ray line profile analysis, respectively. It was found that the addition of RE elements caused a finer grain size, the formation of new precipitates, and changes in the initial fiber texture. As a consequence, Y- and Ce-rich RE elements increased the strength and reduced the ductility. The addition of these two types of RE elements to the ZK60 alloy decreased the work hardening capacity and the hardening exponent mainly due to grain refinement.

C. Xia et al. [16] study coarse particles in Cu-0.39Cr-0.24Zr-0.12Ni-0.027Si alloy with scanning electron microscopy and transmission electron microscopy. Three types of coarse particles were determined: a needle-like Cu5Zr intermetallic phase, a nearly spherical Cr9.1Si0.9 intermetallic phase and (Cu, Cr, Zr, Ni, Si)-rich lath complex particles. The crystallographic orientation relationships of the needle-like and nearly spherical coarse particles were also determined. The reasons for formation and the role of the coarse phases in Cu-Cr-Zr alloys are discussed, and some suggestions are proposed to control the coarse phases in the alloys.

C. Liu et al. [17] study Ta hard coatings prepared on PCrNi1MoA steel substrates by direct current magnetron sputtering, and their growth and phase evolution could be controlled by adjusting the substrate temperature (T_{sub}) and sputtering power (P_{spu}) at various conditions (T_{sub} = 200–400 °C, P_{spu} = 100–175 W). The combined effect of T_{sub} and P_{spu} on the crystalline phase, surface morphology, and mechanical properties of the coatings was investigated. It was found that higher P_{spu} was required in order to obtain α-Ta coatings when the coatings are deposited at lower T_{sub}, and vice versa, because the deposition energy (controlled by T_{sub} and P_{spu} simultaneously) within a certain range was necessary. At the optimum T_{sub} with the corresponding P_{spu} of 200 °C-175 W, 300 °C-150 W, and 400 °C-100 W, respectively, the single-phased and homogeneous α-Ta coatings were obtained. Moreover, the α-Ta coating deposited at T_{sub}-P_{spu} of 400 °C-100 W showed a denser surface and a finer grain, and as a result exhibited higher hardness (9 GPa), better toughness, and larger adhesion (18.46 N).

M. Cohen et al. [18] analyze an assemblage of tiny provincial silver coins of the local (Judahite standard) and (Attic) obol-based denominations from the Persian and Hellenistic period Yehud and dated to the second half of the fourth century BCE to determine their material composition. Of the 50 silver coins, 32 are defined as Type 5 (Athena/Owl) of the Persian period Yehud series (ca. 350–333 BCE); 9 are Type 16 (Persian king wearing a jagged crown/Falcon in flight) (ca. 350–333); 3 are Type 24 series (Portrait/Falcon) of the Macedonian period (ca. 333–306 BCE); and 6 are Type 31 (Portrait/Falcon) (ca. 306–302/1 BCE). The coins underwent visual testing, multi-focal light microscope observation, XRF analysis, and SEM-EDS analysis. The metallurgical findings revealed that all the coins from the Type 5, 16, 24, and 31 series are made of high-purity silver with a small percentage of copper. Based on these results, it is suggested that each series was manufactured using a controlled composition of silver–copper alloy. The findings present novel information about the material culture of the southern Levant during the Late Persian period and Macedonian period, as expressed through the production and use of these silver coins.

A. Khajesarvi et al. [19] study the effects of carbon, Si, Cr, and Mn partitioning on ferrite hardening using a medium Si low alloy grade of 35CHGSA steel under ferrite–martensite/ferrite–pearlite dual-phase (DP) conditions. The experimental results illustrated that an abnormal trend of ferrite hardening had occurred with the progress of ferrite formation. At first, the ferrite microhardness decreased with increasing volume fraction of ferrite, thereby reaching the minimum value for a moderate ferrite formation, and then it surprisingly increased with subsequent increase in ferrite volume fraction. Beside a considerable influence of martensitic phase transformation induced residual compressive stresses within ferrite, these results were further rationalized in respect of the extent of carbon, Si, Cr and Mn partitioning between ferrite and prior austenite (martensite) microphases leading to the solid solution hardening effects of these elements on ferrite.

Q. Liu et al. [20] investigate the anchoring performance of a short-lapped-rebar splice with a corrugated metal duct and spiral hoops. A total of 30 specimens were designed considering the influences of the rebar diameter and the lapped length, and the tension testing of the splice was carried out. The results show that the specimens with 0.15 times the suggested length in GB 50010-2010 fail by the fracture of rebar, while the specimens with 0.1 times and 0.05 times the suggested length show the pull-out failure of rebar. The ultimate bond strength of specimens with the suggested length is higher than that of the conventional specimens. The stress of the anchored rebar in the short-lapped-rebar splices is distributed symmetrically along the longitudinal direction. The maximum bond stress of the anchored rebar reaches 35 MPa, which is approximately 1.4 times higher than in the conventional specimens. A semi-empirical model for predicting the ultimate bond strength of the short-lapped-rebar splice is proposed, and it shows good agreement with tested values; the average error estimated from the proposed model is only 4.49%.

R. Canumalla et al. [21] evaluated, ranked, and selected near-α Ti alloys from the literature for high-temperature applications in aeroengines driven by decision science by integrating multiple attribute decision making (MADM) and principal component analysis (PCA). A combination of 12 MADM methods ranked a list of 105 alloy variants based on the thermomechanical processing (TMP) conditions of 19 distinct near-α Ti alloys. PCA consolidated the ranks from various MADMs and identified top-ranked alloys for the intended applications as: Ti-6.7Al-1.9Sn-3.9Zr-4.6Mo-0.96W-0.23Si, Ti-4.8Al-2.2Sn-4.1Zr-2Mo-1.1Ge, Ti-6.6Al-1.75Sn-4.12Zr-1.91Mo-0.32W-0.1Si, Ti-4.9Al-2.3Sn-4.1Zr-2Mo-0.1Si-0.8Ge, Ti-4.8Al-2.3Sn-4.2Zr-2Mo, Ti-6.5Al-3Sn-4Hf-0.2Nb-0.4Mo-0.4Si-0.1B, Ti-5.8Al-4Sn-3.5Zr-0.7Mo-0.35Si-0.7Nb-0.06C, and Ti-6Al-3.5Sn-4.5Zr-2.0Ta-0.7Nb-0.5Mo-0.4Si. The alloys have the following metallurgical characteristics: bimodal matrix, aluminum equivalent preferably ~8, and nanocrystalline precipitates of Ti3Al, germanides, or silicides. The analyses, driven by decision science, make metallurgical sense and provide guidelines for developing next-generation commercial near-α Ti alloys. The investigation not only suggests potential replacement or substitute for existing alloys but also provides directions for improvement and development of titanium alloys over the current ones to push out some of the heavier alloys and thus help reduce the engine's weight to gain advantage.

Q. Zhao et al. [22] study the effect of the trace rare-earth element Ce on the microstructure and properties of cold-rolled medium manganese steel after ART (austenite-reverted transformation, ART) annealing was studied. The microstructure of the experimental steel was observed using SEM, and the mechanical properties were tested using a universal tensile testing machine. The volume fraction of the retained austenite and the texture of the steel were measured using XRD. The results showed that the original austenite grain size of the experimental steel was smaller after adding the trace rare-earth element Ce. After ART annealing, the grain size distribution of the experimental steel with rare-earth Ce was more uniform, and the comprehensive mechanical properties were better. Under the conditions of quenching at 800 °C for 5 min and annealing at 645 °C for 15 min, the maximum product of tensile strength and elongation was 28.47 GPa%.

W. Wang et al. [23] report on the dendritic growth and physical properties of broad-temperature-range Co-4.54%Sn alloy. The maximum undercooling attains 208 K at molten state, and the dendritic growth velocity is quite sluggish in highly undercooled liquid Co-4.54%Sn alloy because it has a broad solidification range of 375 K (0.21 TL); the maximum value is only 0.95 m/s at the undercooling of 175 K, which then decreases with undercooling. The microstructure refines visibly and the volume fraction of the interdendritic βCo3Sn2 phase clearly decreases with undercooling. The microhardness and electrical resistivity increase with undercooling owing to the enhancement of solute content of the primary αCo phase and refinement of the microstructure where the increased crystal boundary hinders the electronic transmission. Meanwhile, the saturation magnetization also reduces with undercooling due to the crystal particle and boundary increasing significantly, and the dendritic growth velocity and solute content increase in the primary αCo phase under rapid solidification.

T. Tretyakova et al. [24] study the influence of the rigidity of the loading system on the kinetics of the initiation and propagation of the Portevin-Le Chatelier (PLC) strain bands due to the jerky flow in the Al-Mg alloy. To estimate the influence of the loading system, the original loading attachment, which allows for a reduction in the stiffness in a given range, was used. Registration of displacement and strain fields on the specimen surface was carried out via the Vic-3D non-contacting deformation measurement system based on the Digital Image Correlation (DIC) technique. The mechanical uniaxial tension tests were carried out using samples of Al-Mg alloy at the biaxial servo-hydraulic testing system Instron 8850. As a result of tensile tests, deformation diagrams were obtained for Al-Mg alloy samples tested at different values of stiffness of the loading system: 120 MN/m (nominal value), 50 MN/m, 18 MN/m, and 5 MN/m. All diagrams show discontinuous plastic deformations (the Portevin-Le Chatelier effect). It is noted that a decrease in the rigidity of the loading system leads to a change in the type of jerky flow. At constant parameters of the loading rate, temperature, and chemical composition of the material, the PLC effects of types A, B, and C are recorded in tests.

T. Hu et al. [25] describe the diffusion of TM and the nucleation and growth of particles in Al alloys based on first principles.

M. Gaggiotti et al. [26] present a review on the effect of ultrafast heating (UFH) treatment on carbon steels, non-oriented grain (NGO) electrical steels, and ferritic or austenitic stainless steels. The study highlights the effect of ultrarapid annealing on microstructure and textural evolution in relation to microstructural constituents, recrystallization temperatures, and its effect on mechanical properties. A strong influence of the UFH process was reported on grain size, promoting a refinement in terms of both prior austenite and ferrite grain size. Such an effect is more evident in medium–low carbon and NGO steels than that in ferritic/austenitic stainless steels. A comparison between conventional and ultrafast annealing on stainless steels shows a slight effect on the microstructure. On the other hand, an evident increase in uniform elongation was reported due to UFH. Textural evolution analysis shows the effect of UFH on the occurrence of the Goss component (which promotes magnetic properties), and the opposite with the recrystallization g-fiber. The recovery step during annealing plays an important role in determining textural features; the areas of higher energy content are the most suitable for the nucleation of the Goss component. As expected, the slow annealing process promoted equiaxed grains, whereas rapid heating promoted microstructures with elongated grains as a result of the cold deformation. It is worth mentioning, and may be useful for readers, that, after this review, the same authors published experimental investigations related to the effect of ultra-fast heating on stainless steels [27,28].

Acknowledgments: As Guest Editors, we would like to especially thank Sunny He, Topic Specialist, for her support and her active role in the publication. We are also grateful to the entire staff of MDPI for the valuable collaboration. Last but not least, we express our gratitude to all the contributing authors and reviewers: without your excellent work it would not have been possible to accomplish this topic that we hope will be a both an interesting read and of lasting importance as a piece of reference literature.

Conflicts of Interest: The authors declare no conflict of interest.

References

1. Mancini, S.; Langellotto, L.; Di Nunzio, P.E.; Zitelli, C.; Di Schino, A. Defect reduction and quality optimization by modeling plastic deformation and metallurgical evolution in ferritic stainless steels. *Metals* **2020**, *10*, 186. [CrossRef]
2. Di Schino, A.; Gaggiotti, M.; Testani, C. Heat treatment effect on microstructure evolution in a 7% Cr steel for forging. *Metals* **2020**, *10*, 808. [CrossRef]
3. Di Schino, A.; Testani, C. Corrosion behavior and mechanical properties of AISI 316 stainless steel clad Q235 plate. *Metals* **2020**, *10*, 552. [CrossRef]
4. Stornelli, G.; Gaggiotti, M.; Mancini, S.; Napoli, G.; Rocchi, C.; Tirasso, C.; Di Schino, A. Recrystallization and grain growth of AISI 904L super-austenitic stainless steel: A multivariate regression approach. *Metals* **2022**, *12*, 200. [CrossRef]
5. Stornelli, G.; Di Schino, A.; Mancini, S.; Montanari, R.; Testani, C.; Varone, A. Grain refinement and improved mechanical properties of Eurofer97 by thermos-mechanical properties. *Appl. Sci.* **2021**, *11*, 10598. [CrossRef]
6. Zhu, X.; Qian, L.; Yao, Q.; Huang, G.; Xu, F.; Yao, Z. Automatic Detection of Sorbite Content in High Carbon Steel Wire Rod. *Metals* **2023**, *13*, 900. [CrossRef]
7. Li, Z.; Yan, Q.; Xu, S.; Zhou, Y.; Liu, S.; Xu, G. In Situ Observation of the Grain Growth Behavior and Martensitic Transformation of Supercooled Austenite in NM500 Wear-Resistant Steel at Different Quenching Temperatures. *Materials* **2023**, *16*, 3840. [CrossRef]
8. Wu, Z.; Zuo, L.; Zhang, H.; He, Y.; Liu, C.; Yu, H.; Wang, Y.; Feng, W. Effect of Liquid-Solid Volume Ratio and Surface Treatment on Microstructure and Properties of Cu/Al Bimetallic Composite. *Crystals* **2023**, *14*, 794. [CrossRef]
9. Li, L.; Pan, X.; Liu, B.; Li, P.; Liu, Z. Strength and Toughness of Hot-Rolled TA15 Aviation Titanium Alloy after Heat Treatment. *Aerospace* **2023**, *10*, 436. [CrossRef]
10. Kusmanov, S.; Mukhacheva, T.; Tambovskiy, I.; Naumov, A.; Blov, R.; Sokova, E.; Kusanova, I. Increasing Hardness and Wear Resistance of Austenitic stainless-steel surface by anodic plasma electrolytic treatment. *Metals* **2023**, *13*, 872. [CrossRef]
11. Sun, Y.; Zhang, F.; Ren, J.; Song, G. Effect of Li Content on the Microstructure and Mechanical Properties of as-Homogenized Mg-Li-Al-Zn-Zr Alloys. *Alloys* **2023**, *2*, 89. [CrossRef]
12. Ferro, C.G.; Varetti, S.; Maggiore, P. Experimental Evaluation of Fatigue Strength of AlSi10Mg Lattice Structures Fabricated by AM. *Aerospace* **2023**, *10*, 400. [CrossRef]
13. Guo, Y.; Zhu, M.; Wang, Z.; Sun, Q.; Wang, Z.; Li, Z. Design of Glass Fiber-Doped High-Resistivity Hot-Pressed Permanent Magnets for Reducing Eddy Current Loss. *Metals* **2023**, *13*, 808. [CrossRef]
14. Stornelli, G.; Tselikova, A.; Mirabile Gattia, D.; Mortello, M.; Schmidt, R.; Sgambetterra, M.; Testani, C.; Zucca, G.; Di Schino, A. Influence of Vanadium Micro-Alloying on the Microstructure of Structural High Strength Steels Welded Joints. *Materials* **2023**, *16*, 2897. [CrossRef]
15. Najafi, S.; Sheikani, A.; Sabbaghian, M.; Nagy, P.; Kejete, K.; Gubicza, J. Modification of the Tensile Performance of an Extruded ZK60 Magnesium Alloy with the Addition of Rare Earth Elements. *Materials* **2023**, *16*, 2828. [CrossRef]
16. Xia, C.; Ni, C.; Pang, Y.; Jia, Y.; Deng, S.; Zheng, W. Study of Coarse Particle Types, Structure and Crystallographic Orientation Relationships with Matrix in Cu-Cr-Zr-Ni-Si Alloy. *Crystals* **2023**, *13*, 518. [CrossRef]
17. Liu, C.; Peng, J.; Xu, Z.; Shen, Q.; Wang, C. Combined Effect of Substrate Temperature and Sputtering Power on Phase Evolution and Mechanical Properties of Ta Hard Coatings. *Metals* **2023**, *13*, 583. [CrossRef]
18. Cohen, M.; Ashkenazi, D.; Gitler, H.; YTal, O. Archaeometallurgical Analysis of the Provincial Silver Coinage of Judah: More on the Chaîne Opératoire of the Minting Process. *Materials* **2023**, *16*, 2200. [CrossRef]
19. Khajesarvi, A.; Banadkouki, S.; Sajjadi, S.; Somani, M. Abnormal Trend of Ferrite Hardening in a Medium-Si Ferrite-Martensite Dual Phase Steel. *Metals* **2023**, *13*, 542. [CrossRef]
20. Liu, Q.; Liu, X.; Chen, R.; Kong, Z.; Xiang, T. Experimental Study on Anchoring Performance of Short-Lapped-Rebar Splices with Pre-Set Holes and Spiral Hoops. *Metals* **2023**, *13*, 530. [CrossRef]
21. Canumalla, R.; Jararaman, T. Decision Science Driven Selection of High-Temperature Conventional Ti Alloys for Aeroengines. *Aerospace* **2023**, *10*, 211. [CrossRef]
22. Zhao, Q.; Dong, R.; Lu, Y.; Yang, Y.; Wang, Y.; Yang, X. Effect of Trace Rare-Earth Element Ce on the Microstructure and Properties of Cold-Rolled Medium Manganese Steel. *Metals* **2023**, *13*, 116. [CrossRef]
23. Wang, W.; Li, W.; Wang, A. Dendritic solidification and physical properties of Co-4.54Sn alloy with broad mushy zone. *Metals* **2023**, *13*, 1046. [CrossRef]
24. Tretyakova, T.; Tretyakov, M. Spatial-time inhomogeneity due the Portevin-Le Chatelier effect depending on stiffness. *Metals* **2023**, *13*, 1054. [CrossRef]
25. Hu, T.; Ruan, Z.; Fan, T.; Wang, K.; He, K.; Wu, Y. First-Principles Investigation of the Diffusion of TM and the Nucleation and Growth of L12 Al3TM Particles in Al Alloys. *Crystals* **2023**, *13*, 1032. [CrossRef]
26. Gaggiotti, M.; Albini, L.; Di Nunzio, P.E.; Di Schino, A.; Stornelli, G.; Tiracorrendo, G. Ultrafast Heating Heat Treatment Effect on the Microstructure and Properties of Steels. *Metals* **2022**, *23*, 1313. [CrossRef]

27. Stornelli, G.; Albini, L.; Tiracorrendo, G.; Rodriguez Vargas, B.R.; Di Schino, A. Effect of ultra-fast heating on AISI 441 ferritic stainless steel. *Acta Metall. Slovaca* **2023**, *29*, 22–25. [CrossRef]
28. Rodriguez Vargas, B.R.; Albini, L.; Tiracorrendo, G.; Massi, R.; Stornelli, G.; Di Schino, A. Effect of ultra-fast heating on AISI 304 austenitic stainless steel. *Acta Metall. Slovaca* **2023**, *29*, 104–107. [CrossRef]

 *crystals*

Article

First-Principles Investigation of the Diffusion of TM and the Nucleation and Growth of L1₂ Al₃TM Particles in Al Alloys

Te Hu [1,2], Zixiong Ruan [3], Touwen Fan [1,3,*], Kai Wang [4,*], Kuanfang He [4] and Yuanzhi Wu [3]

[1] School of Material Science and Hydrogen Energy Engineering, Foshan University, Foshan 528001, China; hute98@hotmail.com

[2] Guangdong Key Laboratory for Hydrogen Energy Technologies, Foshan 528000, China

[3] School of Science and Research Institute of Automobile Parts Technology, Hunan Institute of Technology, Hengyang 421002, China; ruan_625627@163.com (Z.R.); 2013001767@hnit.edu.cn (Y.W.)

[4] School of Mechatronic Engineering and Automation, Foshan University, Foshan 528001, China; hkf791113@163.com

* Correspondence: fantouwen@hotmail.com (T.F.); hfutwk927@fosu.edu.cn (K.W.)

Citation: Hu, T.; Ruan, Z.; Fan, T.; Wang, K.; He, K.; Wu, Y. First-Principles Investigation of the Diffusion of TM and the Nucleation and Growth of L1₂ Al₃TM Particles in Al Alloys. *Crystals* **2023**, *13*, 1032. https://doi.org/10.3390/cryst13071032

Academic Editors: Andrea Di Schino and Claudio Testani

Received: 5 June 2023
Revised: 26 June 2023
Accepted: 27 June 2023
Published: 29 June 2023

Abstract: The key parameters of growth and nucleation of Al₃TM particles (TM = Sc-Zn, Y-Cd and Hf-Hg) have been calculated using the combination of the first principles calculations with the quasi-harmonic approximation (QHA). Herein, the diffusion rate D_s of TM elements in Al is calculated using the diffusion activation energy Q, and the results show that the D_s of all impurity atoms increases logarithmically with the increase in temperature. With the increase in atomic number of TM, the D_s of 3–5d TM elements decreases linearly from Sc, Y and Hf to Mn, Ru and Ir, and then increases to Zn, Ag and Au, respectively. The interface energy $\gamma_{\alpha/\beta}$, strain energy ΔE_{cs}, chemical formation energy variation ΔG_V and surface energy E_{sur}^{ave} were further computed from the based interface and slab models, respectively. It was found that, with the increase in the atomic number of TM, the interface energies $\gamma_{\alpha/\beta}$ of Al/Al₃TM (TM = (Sc-Zn, Y-Cd)) decreased from Sc and Y to Mn and Tc and then increased to Zn and Cd, respectively (except for the (001) plane of Al/Al₃(Fe-Co), the (111) plane of Al/Al₃Pd and the (110) and (111) planes of Al/Al₃Cd). The strain energies ΔE_{cs} of Al/Al₃TM (TM = (Sc-Zn)) increased at first, and then decreased for all cycles. The chemical formation energy ΔG_V of all Al₃TM changed slightly in the temperature range of 0~1000 K, except that the ΔG_V of Al₃Sc, Al₃Cu, Al₃(Y-Zr), Al₃Cd, Al₃Hf and Al₃Hg increased nonlinearly. With the increase in atomic number at both 300 and 600 K, the ΔG_V of 3–5d TM elements increased from Sc, Y and Hf to Mn, Tc and Re at first, and then decreased to Co, Rh and Ir, respectively, and slightly changes at the end. With the increase in atomic number of TM, the variation trends of the surface energies of Al₃TM intermetallic compounds present similar changes for all cycles, and the (111) surface always has the lowest values.

Keywords: DFT framework; nucleation and growth; diffusion behavior; L1₂ Al₃TM

1. Introduction

Al-based alloys have been widely applied in the electronics, aerospace and automotive industries due to their low density, high specific strength and welding strength [1]. Adding transition elements (TMs) can significantly improve the mechanical and thermodynamic properties of Al alloys [2–6]. For example, the existence of Sc (0.3%) in the Al matrix increases the ultimate rupture strength of annealed Al sheets from 55 to 240 MPa [7], and L1₂-Al₃Zr in the Al matrix is used as a grain refiner to improve the coarsening resistance and creep properties [8,9]. However, the high cost of Sc and Zr limits their applications in commercial Al alloys. Specifically, intermetallic compounds with TMs are suitable candidates for high temperature applications, as the crystal types in the Al matrix may be L1₂, D0₂₂, D0₂₃ or D0₁₉ structures [8,10–12], of which the L1₂ phase is an important intermetallic compound and has been widely studied [13–16]. Moreover, the TMs can be used to substitute the expensive Sc and Zr elements in $L1_2$-Al_3Sc and Al_3Zr.

The previous research proved that fixing the dislocations and grain boundaries can effectively refine the deformed and recrystallized grains, depending on the dispersed distribution of $L1_2$ Al_3TM particles during rising heat [17,18]. The diffusion rate of TM solute atoms in an Al matrix and the interfacial properties of Al_3TM/Al are important parameters for the investigation of nucleation, the growth of $L1_2$ Al_3TM phases [19–21], and the low-index bonds of particles to matrix [22,23]. However, the experimental exploration of appropriative substitution TMs is difficult because of the complex environment and the expensive cost [15,23–28]. Fortunately, in recent years, with the development of modern computer technologies, theoretical identification (e.g., first-principles (FP) calculations based on density functional theory (DFT) [15]) in the complicated systems (e.g., metals and ceramics) has become the most powerful method to accomplish this [29–32].

The stability and nucleation behavior of $L1_2$-Al_3Sc and Al_3Li binary phases have first been investigated using the framework of density functional theory (DFT) calculation by Mao et al. [15]. Their results showed that the $L1_2$-Al_3Sc and Al_3Li structures have lower formation energies than those of the corresponding $D0_{23}$, $D0_{19}$ and $D0_{22}$ structures. Furthermore, they found that the interface and strain energies of Al_3Sc are much higher than those of Al_3Li for all (001), (110) and (111) interfaces. Zhang et al. [33] have comprehensively studied the solubility of RE (RE = Y, Dy, Ho, Er, Tm and Lu) in Al based on the free energy difference between $L1_2$ bulk and Al solid solution matrix in the DFT theoretical framework. Their results indicated that the solubility of all rare earth (RE) (RE = Y, Dy, Ho, Er, Tm and Lu) elements increases with the increase in temperature (~1000 K). They also believed that Dy and Y elements can become better candidates for Sc due to the better stability of Al_3Dy and Al_3Y compounds and their almost identical solubility compared to the higher-cost Sc element. Sun et al. [34] have calculated low-index (001), (110) and (111) surface energies of $L1_2$-Al_3Sc particles adopting slab model with 15 Å vacuum region. Their results show that when the surface energies of non-stoichiometric (001) and (110) surfaces of Al_3Sc are calculated, their values should be considered as different under different Al chemical potentials, and in a wide range of Al chemical potentials, the surface energies of the (111) surface with AlSc-terminated have lower values, indicating that they are more stable than other surfaces.

However, up to date, the diffusion rates D_s of TMs in Al, the surface properties of $L1_2$-Al_3TM and the interface of Al_3TM/Al-matrix have not been systematically investigated. Specifically, the nucleation and growth of $L1_2$-Al_3TM (TM = Sc-Zn, Y-Ag, Hf-Au) particles at finite temperatures have not been obtained, and their relationship to the atomic number of TM hasn't been described in detail due to the large computational cost required. In the present work, by combining the first-principles calculations with the quasi-harmonic approximation (QHA), the relationship between the particles' nucleation/growth and atomic number/temperature are discussed. First, the diffusion rates D_s of TMs as a functional of atomic number and temperature have been researched. Then, the relationship between the driving force and the hindrance of particle nucleation and the atomic number of TM is explained based on the interface model. Finally, the effects of the surface stability of different intermetallic compounds with the change in atomic number based on the slab model are obtained.

2. Computed Methods

All calculations in this work were performed in Vienna ab initio Simulation Package (VASP) [34] with the 5.4.4 version, which adopts the framework of density functional theory (DFT) [35] calculations to solve the Kohn–Sham equation and obtain the total energy from different models. In the calculated processing of VASP, to relax all models to their most stable ground state, the electron–core interaction was described by the projector augmented wave (PAW) [36] method. The optimal choice of exchange–correlation functional was considered using the generalized gradient approximation (GGA) with the Perdew–Burke–

Ernzerh (PBE) version [31]. A $10 \times 10 \times 10$ k-point sampling grid with the Gamma-centered Monkhorst–Pack method [37] in the first Brillouin zone was selected via strict convergence testing (see Figure 1) for bulk properties calculation. A cut off energy of the plane-wave basis of 500 eV was chosen for the whole calculated process. The energy and force tolerance were set to 10^{-7} eV and 0.01 eV/Å, respectively, by using conjugate gradient (CG) minimization and Broyden–Fletcher–Goldfarb–Shannon (BFGS) schemes [38].

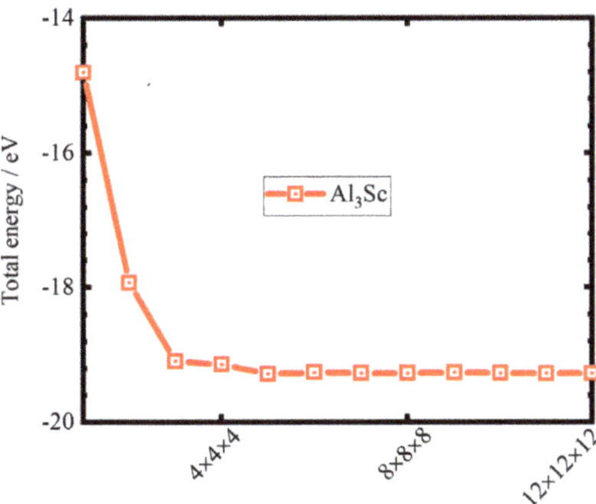

Figure 1. The total energy of Al₃Sc as a function of k-point sampling grids.

Here, based on the slab model, we investigated three low-index surfaces, containing (100), (110) and (111) surfaces of Al and $L1_2$-Al$_3$TM, which adopted 14, 14 and 16 layers, respectively [24,39]. All interfaces of (100), (110) and (111) surfaces of Al/$L1_2$–Al$_3$TM are calculated by using the 18 layers interface model. A $10 \times 10 \times 1$ k–mesh grid for both cases was tested to be suitable for this work. To simulate diffusion behavior, we constructed a $3 \times 3 \times 3$ supercell with $4 \times 4 \times 4$ k–mesh grids to obtain the diffusion barriers of the solute diffusion of TMs in the Al matrix based on the climbing-image nudged elastic band (CI-NEB) [40] method. Meanwhile, a spring force constant of 5 eV/Å was considered to keep all the images separated, and these CI-NEB iterations were continued until the forces on each atom were less than 0.05 eV/Å.

3. Conclusion Description

3.1. Diffusion

In the process of heating up, some atoms will detach from their original equilibrium positions and then diffuse to a new site while obtaining enough energy. Thus, the diffusion behavior is a common phenomenon in the field of material science and engineering. According to the Lifshitz and Slyozov and Wagner methods [41,42], the growth of particles is affected by the diffusion behavior of solute atoms, and the faster diffusion in the Al matrix is beneficial for the grain growth. In the current work, to investigate diffusion behavior, we first show a vacancy-substitution model, as depicted and visualized in Figure 2a by VESTA codes. The vacancy-substitution model can be divided into two types: the self-diffusion of the violet Al atom and the impurity diffusion of the TM pink ball [24,39]. The black arrow represents the diffusion path for the TM atom. To further investigate the diffusion behavior, the diffusion coefficient as a function of jump frequency I is expressed, which satisfies the Arrhenius equation as follows [40,43–45]:

$$D(T) = \frac{\lambda a^2}{2Z} I \tag{1}$$

where λ ($\lambda = 2$), Z ($Z = 1$) and a are the number of directions for atomic transitions, the dimension of diffusion and the corresponding atomic distance of diffusion, respectively.

Here, the jump frequency for both diffusions in solid-state was established using the classical transition state theory (TST) [46,47]:

$$I = v exp^{\left(-\frac{Q}{\kappa T}\right)} \tag{2}$$

where v, Q, T and κ are the effective frequencies associated with the vibration of the transition atom, the diffusion activation energy, the special temperature and the Boltzmann constant, respectively.

According to Winter–Zener theory (WZT), the v can be approximately expressed as [48]:

$$v = \left(\frac{2E_{Diff}}{ma^2}\right)^{\frac{1}{2}} \tag{3}$$

where m represents the atomic mass of transition atoms. Herein, two types of diffusion activation energies Q corresponding to self D_0 and impurity D_s diffusion coefficients are gained using first principles calculations. The Q for self-diffusion contains two separate energies: vacancy formation energy E_{vac} and the migration energy of Al atom E_m in Al matrix. For impurity diffusion activation energy, the activation energy Q consists of three parts: the substitutional solution energy E_s of a TM atom replacing a Al atom, vacancy formation energy E_f in the presence of TM in $Al_{107}TM$ supercell, and the migration energy of diffusion E_m [49]:

$$E_s = E_{Al_{107}TM} - 107E_{Al} - E_{TM} \tag{4}$$

$$E_f = E_{(Al_{106}TM:Vac)} - E_{Al_{107}TM} + E_{Al} \tag{5}$$

$$E_b = E_s + E_f \tag{6}$$

where E_{TM} and E_{Al} are the energies of single TM and Al atoms in the stable bulk, respectively, and E_b is the binding energy of a TM atom substituting a vacancy in Al matrix.

To further investigate the physical mechanism of behaviors, the electron localized function (ELF) has been drawn using the VESTA code [50]. The ELF is defined as:

$$ELF = \frac{1}{1 + \left(\frac{D_r}{Dh_r}\right)^2} \tag{7}$$

where D_r and Dh_r are the true electron gas density and the pre-assumed uniform electron gas density, respectively.

The E_m, E_b (E_{vac}), Q and D_s (D_0) for TM and Al at 300 K with available experimental and theoretical values are summarized in Table 1 [51–54]. It can be seen that errors between the present and previous values in literatures for E_m, E_b (E_{vac}) and E_{Diff} are within 20%, and the current value of E_{Diff} of Sc element is only ~2% larger than that of the experiment value. To visually illustrate the regularity of the variations of activation energy Q as a function of the atomic number of TM, it is further plotted in Figure 2b. The result shows that the Q increases at first and then decreases as the atomic number increases (Sc-Zn, Y-Ag, Hf-Au) in the Al matrix (except for Cr of 2.23 eV), indicating that there is a correlation between the valence electron configuration of impurity elements and the activation energy Q. Additionally, the TM elements in the fourth cycle generally have lower diffusion activation energies Q_s, ranging from 0.35 to 2.60 eV. For Mn-Co, Tc-Rh and Re-Ir, they have larger Q_s in the Al matrix, which are 2.45~2.60, 3.82~3.94 and 3.95~4.26 eV, respectively, indicating that their diffusion abilities are relatively weak in the Al matrix. Meanwhile, for Cu-Zn, Ag and Au, the activation energy Q is very low, or even negative for a Cd of -0.12 eV and an Hg of -0.30 eV, as shown in Table 1, which shows they are easier to move in the Al matrix. In the undoped-Al system, self-diffusion activation energies Q_0 is lower compared to all Q_s in the doped system, except for the Q_s of Cu, Zn, Y and Ag, indicating that the diffusion of most TM atoms is more difficult than self-diffusion.

The variation in activation energy Q with the temperature increasing can be calculated from the above results by combining them with the quasi-harmonic vibration (QHA) [55]; by doing this, the change in diffusion rate D with the temperature can be obtained via Equations (1)–(3), and the results are summarized in Table 1 and Figure 2c,d. It should be noted that only the self-diffusion rate D_0 as a function of Q is presented in the inlet of Figure 2c, owing to the fact that that all activation energies Q of TM elements are nearly the same. The self-diffusion rate D_0 of 3.55×10^{-28} m$^2 \cdot$s^{-1} for Al in this work is in general agreement with the experimental extension values from $1.76 \times 10^{-27} \sim 4.42 \times 10^{-12}$ m$^2 \cdot$s^{-1} in the range of 300~1000 K and $1.47 \times 10^{-14} \sim 1.36 \times 10^{-12}$ in the range of 739~917 K in literature [56,57], seen from Table 1 and Figure 2c. Meanwhile, the theoretical predicted D_0 of 3.55×10^{-28} m$^2 \cdot$s^{-1} of Al is lower than that of the experiment at 300 K. The reason for this may be that it is difficult to accurately determine the D_0 due to the influence of crystal structure defects, dislocations and grain boundaries in experiments. The D_s of all impurity atoms except for Cd and Hg increases logarithmically with the increase in temperature. A negative Q for Cd and Hg cases makes it impossible to theoretically calculate values according to Equations (1) and (2). Reasonably, the D indicate the inverse pattern to Q; higher barriers mean slower passage. Additionally, the larger the value at 300 K, the lower the increasing rate. This trend result is consistent with the variation trend of D_s for Mg, Si and Cu with temperature calculated by Mantina et al. [44]. Figure 2d further shows the diffusion rate D_s at 300 K as a function of the atomic number of TM, and it can be seen that the diffusion rate D_s first decreases linearly from 2.05×10^{-37}, 6.47×10^{-24} and 2.79×10^{-44} m$^2 \cdot$s^{-1} for Sc, Y and Hf to 2.43×10^{-50}, 6.77×10^{-73} and 1.60×10^{-78} m$^2 \cdot$s^{-1} for Mn, Ru and Ir and then increases with the increase in atomic number to 3.09×10^{-13}, 9.17×10^{-17} and 2.93×10^{-29} m$^2 \cdot$s^{-1} for Zn, Ag and Au, respectively (except for Cr of 3.36×10^{-44} m$^2 \cdot$s^{-1}).

From the above results, it can be seen that higher peaks occur for half- or near half-full d shells for all cycles considered. The reason for this may be that half- or near half-full d shells of the TM element in the Al matrix are more stable and more energy is required to force them to move from the stable site to the vacancy. Although the atomic diffusion barrier changes similarly with the increase in atomic number in the same period, TM with 3d shells present a faster diffusion behavior. To explore the underlying potential, the ELF of Sc and Ru doping systems on the (010) plane are presented in Figure 2e,f. The value of ELF, which is selected as 0 to 1, demonstrates the probability of finding an electron in the neighborhood space. To be specific, when it equals 0, it reflects a strongly delocalized electron area; when it equals 1, it corresponds to a strongly localized electron area. It can be seen that, when Sc and Ru are the first nearest neighbors of the vacancy, different values of ELF are exhibited. The Ru would make the surrounding electrons appear more likely than Sc, resulting in Ru being difficult to diffuse to the vacancy.

Table 1. The calculated diffusion barrier E_m (eV), vacancy–solute binding energy E_b (eV), diffusion activation energy Q (eV) and diffusion rate D_s (m$^2 \cdot$s^{-1}) for TM atoms in Al matrix at 300 K. It should be noted that for pure Al, E_b and D_s are in fact E_{vac} and D_0, respectively. Note: A negative activation energy Q can't meet calculating D_s according to Equations (1)–(3).

Element	E_m	E_b	Q	D_s
Al	0.68 0.55–0.70 [54] 0.57 [58]	0.63 0.60–0.80 [54] 0.63 [58]	1.31 1.15–1.50 [54] 1.20 [58] 1.31 [56]	3.55×10^{-28} 1.76×10^{-27} [56]
Sc	0.85	0.97	1.82 1.79 [55]	2.05×10^{-37}
Ti	1.43	0.84	2.27	6.04×10^{-45}
V	1.90	0.42	2.32	1.09×10^{-45}
Cr	2.14	0.09	2.23	3.36×10^{-44}

Table 1. *Cont.*

Element	E_m	E_b	Q	D_s
Mn	2.11	0.49	2.60	2.43×10^{-50}
Fe	1.90	0.63	2.53	3.10×10^{-49}
Co	1.55	0.90	2.45	7.79×10^{-48}
Ni	1.06	0.97	2.03	6.44×10^{-41}
Cu	0.57	0.22	0.79	2.94×10^{-20}
Zn	0.40	−0.04	0.35	3.09×10^{-13}
Y	0.36	0.63	0.99	6.47×10^{-24}
Zr	1.19	0.98	2.17	2.16×10^{-43}
Nb	1.88	0.75	2.63	4.94×10^{-51}
Mo	2.46	0.75	3.22	7.40×10^{-61}
Tc	2.54	1.27	3.82	7.25×10^{-71}
Ru	2.25	1.69	3.94	6.77×10^{-73}
Rh	1.68	2.16	3.84	2.81×10^{-71}
Pd	0.98	1.71	2.68	5.51×10^{-52}
Ag	0.51	0.06	0.56	9.17×10^{-17}
Cd	0.35	−0.46	−0.12	-
Hf	1.41	0.80	2.21	2.79×10^{-44}
Ta	2.10	0.41	2.51	2.90×10^{-49}
W	2.85	0.17	3.02	9.43×10^{-58}
Re	3.09	0.86	3.95	2.93×10^{-73}
Os	2.77	1.36	4.14	1.96×10^{-76}
Ir	2.15	2.11	4.26	1.60×10^{-78}
Pt	1.27	2.15	3.42	1.77×10^{-64}
Au	0.53	0.78	1.31	2.93×10^{-29}
Hg	0.21	−0.50	−0.30	-

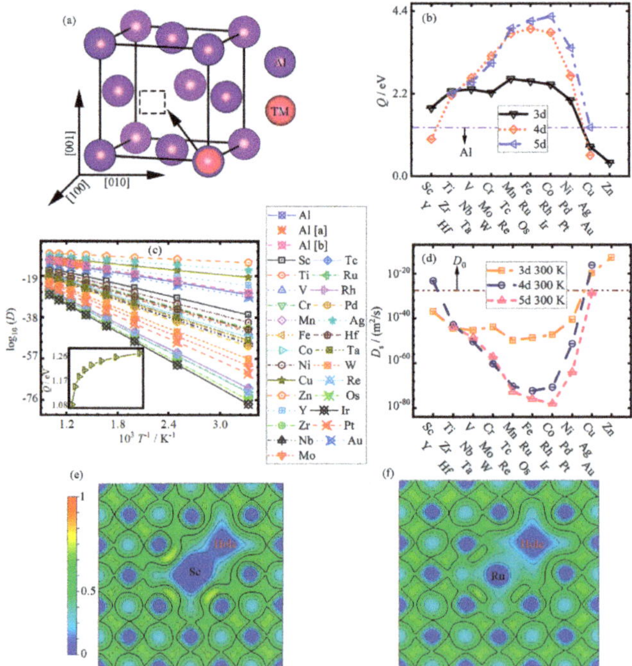

Figure 2. (**a**) The diffusion model. (**b**) The calculated diffusion barrier of a vacancy E_m, vacancy solute binding energy E_b (vacancy formation energy E_{vac} for self-diffusion in Al matrix) and diffusion activation energy E_{Diff} with the change in atomic number. (**c**) The diffusion rate D and E_{Diff} as a function of temperature. (**d**) The impurity diffusion rate D_s as a function of the atomic number of TM. (**a**,**b**) represent the experimental values from Murphy et al. [57] and Volin et al. [56], respectively. (**e**,**f**) The ELFs on the (010) planes of Sc and Ru doping systems, respectively.

3.2. Nucleation

According to the classical nucleation method (CNT) [44,56,57], the total energy of the nucleation process of second phases can be expressed as follows: $\Delta G_{tot} = \frac{4}{3}\pi R^3(\Delta G_V + \Delta E_{CS}) + 4\pi R^2 \gamma_{\alpha/\beta}$. Here, a positive strain energy contribution would be a hindrance when Al_3TM grains gradually form, while the difference in free energy in bulk between the matrix and particles and the interfacial free energy would promote particle nucleation.

Here, to calculate interface energy $\gamma_{\alpha/\beta}$, we adopt a total energy of interface model that subtracts the total energy of the phases on either side of the interface in a two-phase system [23]:

$$\gamma_{\alpha/\beta} = \frac{E_{\alpha/\beta} - (E_\alpha + E_\beta)}{2A} \tag{8}$$

where A is the area of the interface, $E_{\alpha/\beta}$ is the total internal energy of the relaxed α/β system containing an interface and E_α and E_β are the total internal energies of phases α and β from the strains of all directions, respectively.

The chemical formation energy difference ΔG_V of $L1_2$-Al_3TM precipitates can be expressed in dilute solid solution based thermodynamics, $Al_nTM \rightarrow Al_3TM + Al_{n-3}$. It can be shown as [15]

$$\Delta G_V = \Delta G_{Al_3TM} + (n-3)\Delta G_{Al} - \Delta G_{Al_nTM} \tag{9}$$

where n (n = 31) and ΔG are the number of atoms and Gibbs free energy, respectively. To investigate the dependence of ΔG_V on temperature, the non-equilibrium free energy ΔG_V is derived as the following equation [15,59]:

$$G(V, P, T) = \min[F(V, T)] + PV \tag{10}$$

where $F(V; T)$ is the free energy computed by the sum of electronic internal energy and phonon Helmholtz free energy $F(V, T) = U_{el} + F_{vib}$. P is the circumstance pressure.

Due to lattice mismatch, both the harmonic and non-harmonic contributions were observed to calculate the strain energy ΔE_{CS} of the $L1_2$ precipitation phases [15]:

$$\Delta E_{CS}\left(x, \hat{G}\right) = \min_{a_s}\left(x\Delta E_\alpha^{eqi}\left(a_s, \hat{G}\right) + (1-x)\Delta E_\beta^{eqi}\left(a_s, \hat{G}\right)\right) \tag{11}$$

where a_s is the constrained superlattice parameter, $\hat{G}$ is the direction and x is the mole fraction of phase α. ΔE_α^{eqi} and ΔE_β^{eqi} are the epitaxial deformation energies of phases α and β, respectively.

Figure 3a shows the interface model for calculating the interface properties in this work. The Al matrixes are highlighted in dashed rectangles, and different layer numbers are used for the calculation convenience. Comparing the present results with references [15,23] listed in Table 2, there are larger errors compared by Mao and Li et al. [15,23], and these errors are further discussed. The main reasons are as follows:

1. Li et al. [23] adopted the vacuum slab model for the calculation, resulting in the values of interface energies being affected by different terminal surfaces, and the interface energy of Al/Al_3Ti of 61.85 mJ·m^{-2} calculated by the vacuum model is in a good agreement with that of Li et al. according to $\gamma_{\alpha/\beta} = \frac{E_{\alpha/\beta}^* - (E_{slab,\alpha} + E_{slab,\beta})}{S} + E_{sur}^\alpha + E_{sur}^\beta$, where $E_{\alpha/\beta}^*$ is the total energy of the vacuum slab model system, E_{slab} denotes the total energy of the fully relaxed surface slabs and E_{sur}^α and E_{sur}^β represent the surface energies of the α and β surface slabs, respectively. Meanwhile, the strain energy caused by lattice mismatch in the vacuum slab model was not taken into account in the above equation.

2. Mao et al. [15] had investigated interface properties in a periodic supercell and, considering the strain energy of interface model, they calculated interface properties with less accuracy, performed on a 0.13 (1/Å) spacing Monkhorst–Pack k-point mesh and an energy cutoff of 300 eV.

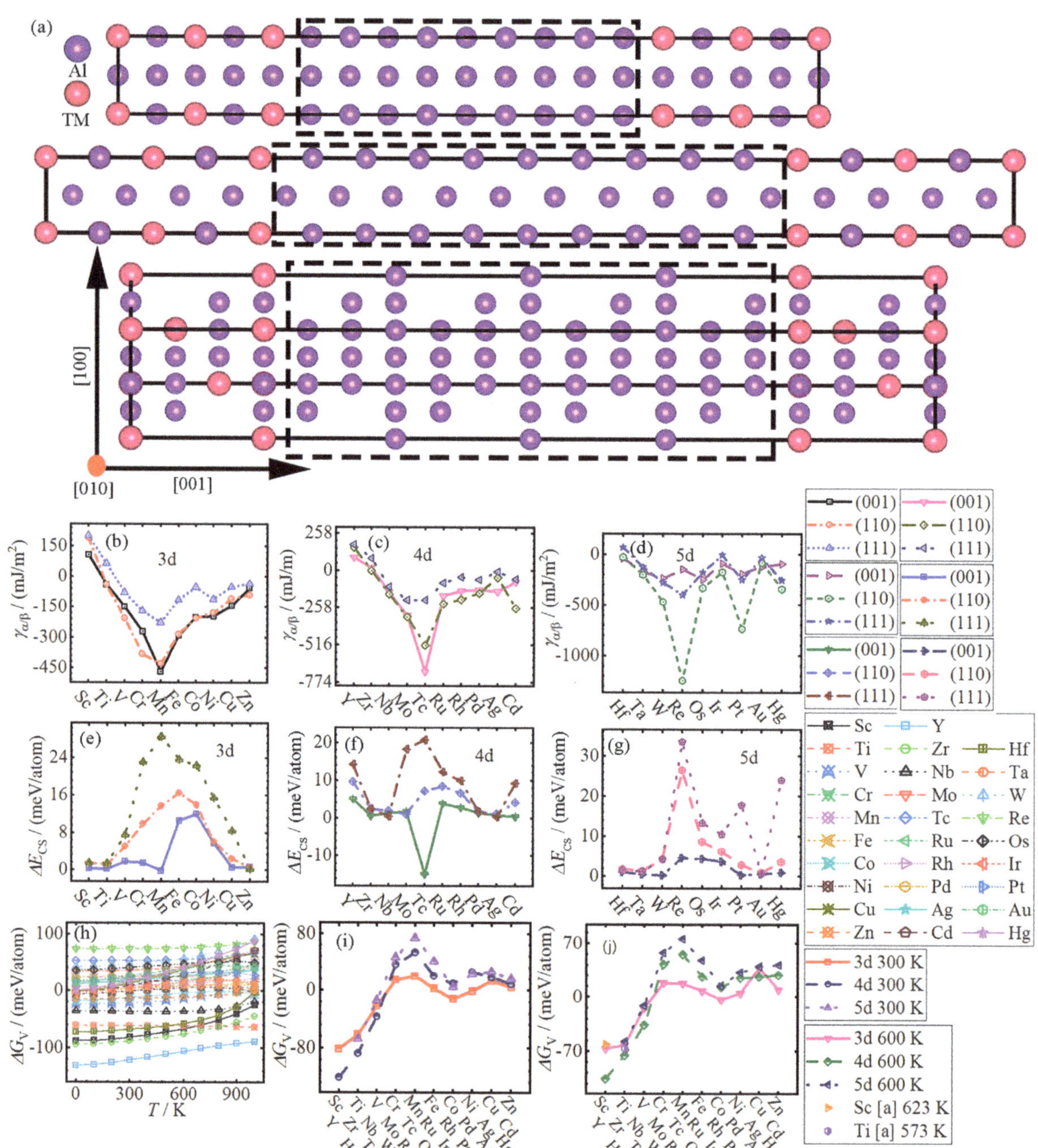

Figure 3. (**a**) The interface models. (**b–d**) The calculated interface energy $\gamma_{\alpha/\beta}$ and (**e–g**) strain energy ΔE_{cs} with the change in atomic number. (**h**) The chemical formation energy ΔG_V with the change in temperature. (**i,j**) The chemical formation energy ΔG_V as a function of atomic number under different constant temperature conditions. (**a**) represent the calculated result of Li et al. [23].

Table 2. The calculated interface energy $\gamma_{\alpha/\beta}$ (mJ·m^{-2}) and strain energy ΔE_{cs} (meV·atom^{-1}) in Al/Al$_3$TM interface systems. (Note: A * symbol represents the calculated result from the vacuum slab model).

Dir.	(001)		(110)		(111)	
Systems	$\gamma_{\alpha/\beta}$	ΔE_{CS}	$\gamma_{\alpha/\beta}$	ΔE_{CS}	$\gamma_{\alpha/\beta}$	ΔE_{CS}
Al/Al$_3$Sc	108.65 108.00 [15] 165.00 [23] 176.00 [23]	0.32 0.60 [23] - -	194.41 159.00 [15] 178.00 [15] 193.00 [23]	1.50	204.81 191.00 [23] 189.00 [15] 203.00 [15]	1.52
Al/Al$_3$Ti	−38.48 61.85 * 52.00 [23]	0.20 0.30 [15]	−38.90 61.00	1.12	66.67 79.00 [15]	1.36
Al/Al$_3$V	−147.83	1.77	−203.77	5.10	−75.76	7.56
Al/Al$_3$Cr	−270.04	1.52	−379.95	9.92	−167.43	23.16
Al/Al$_3$Mn	−468.86	−0.14	−429.65	13.80	−225.66	28.62
Al/Al$_3$Fe	−291.22	10.64	−283.37	16.58	−113.54	23.69
Al/Al$_3$Co	−200.04	12.06	−205.31	14.01	−49.49	22.43
Al/Al$_3$Ni	−195.59	5.83	−176.89	6.20	−109.50	15.46
Al/Al$_3$Cu	−143.59	0.59	−108.27	2.40	−48.53	8.43
Al/Al$_3$Zn	−53.81	0.64	−88.33	0.67	−33.48	0.24
Al/Al$_3$Y	93.37	5.13	159.60	9.72	181.29	14.30
Al/Al$_3$Zr	20.15	0.65	1.39	2.43	86.32	2.48
Al/Al$_3$Nb	−143.96	1.48	−160.59	2.07	−109.57	0.56
Al/Al$_3$Mo	−309.65	1.87	−319.59	1.20	−201.04	18.33
Al/Al$_3$Tc	−699.48	−14.84	−516.46	7.24	−201.10	20.79
Al/Al$_3$Ru	−173.70	3.92	−228.40	8.52	−82.71	12.26
Al/Al$_3$Rh	−138.07	2.92	−197.66	6.71	−42.72	9.98
Al/Al$_3$Pd	−132.75	1.53	−153.88	1.94	−60.52	1.79
Al/Al$_3$Ag	−142.47	0.86	−46.85	1.31	−5.67	0.40
Al/Al$_3$Cd	−75.68	0.51	−261.59	4.33	−55.69	9.33
Al/Al$_3$Hf	−37.53	1.14	−25.31	1.79	69.41	1.59
Al/Al$_3$Ta	−169.68	0.50	−198.59	1.36	−124.38	1.26
Al/Al$_3$W	−232.18	0.29	−467.97	4.35	−276.16	4.24
Al/Al$_3$Re	−146.35	4.59	−1242.00	26.49	−396.57	33.54
Al/Al$_3$Os	−243.80	4.34	−328.80	8.72	−174.06	13.36
Al/Al$_3$Ir	−87.86	3.71	−173.40	6.26	−3.44	10.60
Al/Al$_3$Pt	−190.37	0.38	−734.31	2.90	−246.39	17.81
Al/Al$_3$Au	−118.52	0.58	−80.77	1.02	−33.30	0.67
Al/Al$_3$Hg	−93.95	0.94	−341.48	3.64	−251.40	24.04

The calculated $\gamma_{\alpha/\beta}$ with the increase in atomic number is further depicted in Figure 3b–d. According to the CNT, the theoretical nucleation radius R* cannot be calculated by a negative $\gamma_{\alpha/\beta}$, and the $\gamma_{\alpha/\beta}$ of all Al/Al$_3$TM are less than 0 mJ·m^{-2}, except for the (111) of Al/Al$_3$Sc, Al/Al$_3$Ti, Al/Al$_3$(Y-Zr) and Al$_3$Hf systems.

It can be seen from Figure 3b,c that the $\gamma_{\alpha/\beta}$ of Al/Al$_3$TM (TM = (Sc-Zn, Y-Cd)) decreases from Sc and Y to Mn and Tc, and then increases to Zn and Cd, respectively, except for the (001) of Al/Al$_3$(Fe-Co), the (111) of Al/Al$_3$Pd and the (110) and (111) of Al/Al$_3$Cd. These trends of $\gamma_{\alpha/\beta}$ for Al/Al$_3$TM (TM = (H$_f$-Hg)) in the (110) and (111) systems present two Al/Al$_3$Re and Al/Al$_3$Pt compound troughs in Figure 3d, and they show the same change with the increase in atomic number. For the (001) system, the $\gamma_{\alpha/\beta}$ of Al/Al$_3$TM (TM = (H$_f$-Hg) is larger than −250 mJ·m^{-2}, and the Al/Al$_3$TM with 3d^{6}4s^2 has the lowest $\gamma_{\alpha/\beta}$. Figure 3e shows the variation of strain energy ΔE_{cs} of Al/Al$_3$TM (TM = (Sc-Zn) with the increase in atomic number. It can be seen that the ΔE_{cs} increases from 0.32~1.52 meV·atom^{-1} for Sc to 12.06 meV·atom^{-1} for Co on the (001) system, to 16.58 meV·atom^{-1} for Fe on the (110) system, and to 28.62 meV·atom^{-1} for Mn on the (111) system, respectively, and then they all decrease to 0.24 ~ 0.67 meV·atom^{-1} for Zn (except for Al/Al$_3$Mn, of the order of −0.14 meV·atom^{-1}). For the (110) and (111) systems of Al/Al$_3$TM (TM = (Y-Cd, Hf-Hg)), as seen in Figure 3f,g, respectively, the largest values of ΔE_{cs} for the (110) and (111) interface systems are all located at Al/Al$_3$Re, being 26.49 and 33.54 meV·atom^{-1}, respectively, while the (001) interface system of Al/Al$_3$Tc has the lowest value of ΔE_{cs}, being -14.84 meV·atom^{-1}.

The trends of ΔG_V as a function of temperature for all Al$_3$TM compounds have been calculated according to Equations (10) and (11), and results are shown as Figure 3h. The results show that the ΔG_V of all Al$_3$TM change slightly in the temperature range of 0~1000 K, except that the ΔG_V of Al$_3$Sc, Al$_3$Cu, Al$_3$(Y-Zr), Al$_3$Cd, Al$_3$Hf and Al$_3$Hg increase non-linearly from -89.69, -1.44, -130.51, -93.86, -1.65, -72.35, and 0.65 meV·atom^{-1} to -24.38, 66.09, -88.46, -44.47, 71.60, -2.05 and ~88.51 meV·atom^{-1}, respectively. Furthermore, the obtained ΔG_V as a function of the atomic number of TM is shown in Figure 3i,j, and the calculated values of -66.46 and -61.54 meV·atom^{-1} for Al$_3$Sc and Al$_3$Ti, respectively, at 600 K agree well with the value of -61.14 meV·atom^{-1} at 350 °C (623 K) for Al$_3$Sc and -66.15 meV·atom^{-1} at 300 (573 K) for Al$_3$Ti calculated by Li et al. [15]. From Figure 3i, one can see that the ΔG_V at 300 K increases from -80.96 meV·atom^{-1} for Sc, -120.46 meV·atom^{-1} for Y and -66.82 meV·atom^{-1} for Hf to 20.37 meV·atom^{-1} for Mn, 53.89 meV·atom^{-1} for Tc and 74.50 meV·atom^{-1} for Re, and then decreases slightly to -11.72 meV·atom^{-1} for Co, 9.62 meV·atom^{-1} for Rh and 4.89 meV·atom^{-1} for Ir, respectively. As a final step, they change slightly. At 600 K, the variation trends of ΔG_V for 3–5d TMs are the same as those at 300 K.

3.3. Surface Energy

In the framework of Peierls theory, a lower surface energy of bulk materials in comparison to an unstable stacking fault will cause metals to crack from material failure [42,60]. Thus, it is necessary to analyze surface energy for all Al$_3$TM particles and the Al matrix. The surface energy of Al is given by the following formula [61,62]:

$$E_{sur} = \frac{E_{Al}^{slab} - N\mu_{Al}^{bulk}}{2A} \tag{12}$$

where E_{Al}^{slab} and N are the total energy and the number of Al atoms in the slab model, respectively. μ_{Al}^{bulk} represents the chemical potential of a single atom in bulk Al.

For stoichiometric surfaces (111) of the Al$_3$TM slab, the calculated formula is given as follows:

$$3\mu_{Al}^{slab} + \mu_{TM}^{slab} = \mu_{Al_3TM}^{bulk} \tag{13}$$

$$E_{sur} = \frac{E_{Al_3TM}^{slab} - N\mu_{Al_3TM}^{bulk}}{2A} \tag{14}$$

where μ_{Al}^{slab}, μ_{TM}^{slab} and $\mu_{Al_3TM}^{bulk}$ are the chemical potential of Al, AlTM-terminated and Al$_3$TM bulk, respectively. N and A are the number of Al$_3$TM cells and the surface area, respectively.

To further discuss the non-stoichiometric (001) and (110) surfaces of the Al3TM (3NTM $\neq$ NAl), we used the following the equation [24,63]:

$$E_{sur} = \frac{E_{Al_3TM}^{slab} - N\mu_{Al_3TM}^{bulk} + n\mu_{Al}^{slab}}{2A} \tag{15}$$

where n is the number of the rest ($n < 0$) and missing ($n > 0$) Al atoms.

To obtain μ_{Al}^{slab} in the systems, we first need to avoid Al and Sc bulk phases. Therefore, μ_{Al}^{slab} and μ_{Sc}^{slab} are limited, as follows:

$$\mu_{Al}^{slab} - \mu_{Al}^{bulk} < 0 \tag{16}$$

$$\mu_{TM}^{slab} - \mu_{TM}^{bulk} < 0 \tag{17}$$

Further, the thermodynamic stability of AlTM compounds should meet the equation given by:

$$3\mu_{Al}^{bulk} + \mu_{TM}^{bulk} + \Delta H_f = \mu_{Al_3TM}^{bulk} \tag{18}$$

Combined with Equations (12) and (15)–(17), two limit values of μ_{Al}^{slab} are respectively given by:

$$\mu_{Al}^{a} = \mu_{Al}^{slab} = \mu_{Al}^{bulk} \tag{19}$$

$$\mu_{Al}^{b} = \mu_{Al}^{slab} = \mu_{Al}^{bulk} + \frac{1}{3}\Delta H_f \tag{20}$$

The calculated surface energies of AlTM from different Al chemical potentials, μ_{Al}^{a} and μ_{Al}^{b}, are summarized in Table 3 as references. To solve the dependence of surface energy on Al chemical potential, the average surface energy of non-stoichiometric surfaces is obtained by two identical index surfaces of different termination [24,63]:

$$E_{sur}^{ave} = \frac{1}{4A}\left[E_{slab}^{Al} + E_{slab}^{AlTM} - \left(N_{slab}^{Al} + N_{slab}^{AlTM}\right) \times \mu_{Al_3TM}^{bulk}\right] \tag{21}$$

where E_{slab}^{Al}, N_{slab}^{Al} and E_{slab}^{AlTM}, N_{slab}^{AlTM} are the relaxed energy and total number of *TM* atoms in Al and AlTM-terminated surfaces, respectively.

Table 3. The calculated surface energy E_{sur} (J·m^{-2}) of the (001) and (110) surfaces from different Al chemical potential μ_{Al}^{a} and μ_{Al}^{b} in Al$_3$TM.

| Systems | (001) | | | | (110) | | | |
| | Al-Ter. | | AlTM-Ter. | | Al-Ter. | | AlTM-Ter. | |
	μ_{Al}^{a}	μ_{Al}^{b}	μ_{Al}^{a}	μ_{Al}^{b}	μ_{Al}^{a}	μ_{Al}^{b}	μ_{Al}^{a}	μ_{Al}^{b}
Al$_3$Sc	1.10	1.69	1.42	0.84	1.19	1.61	1.63	1.22
Al$_3$Ti	1.04	1.53	1.70	1.21	0.99	1.34	1.84	1.49
Al$_3$V	1.03	1.24	1.61	1.40	0.86	1.01	1.77	1.63
Al$_3$Cr	0.97	0.92	1.54	1.59	0.63	0.60	1.50	1.53
Al$_3$Mn	0.93	0.99	1.55	1.49	0.42	0.47	1.19	1.14
Al$_3$Fe	1.06	1.23	1.57	1.39	0.78	0.90	1.51	1.39
Al$_3$Co	0.97	1.29	1.32	0.99	0.77	0.99	1.32	1.09
Al$_3$Ni	0.78	1.10	1.03	0.71	0.73	0.96	0.90	0.68
Al$_3$Cu	0.83	0.89	0.99	0.92	0.87	0.92	0.85	0.80
Al$_3$Zn	0.84	0.82	0.78	0.80	0.93	0.91	0.75	0.77
Al$_3$Y	1.13	1.63	0.96	0.46	1.12	1.47	1.34	0.98
Al$_3$Zr	0.88	1.46	1.36	0.77	0.88	1.29	1.59	1.18
Al$_3$Nb	0.81	1.16	1.32	0.97	0.70	0.95	1.61	1.37
Al$_3$Mo	0.36	0.51	0.88	0.73	0.26	0.36	1.17	1.08
Al$_3$Tc	1.02	1.29	1.62	1.35	0.00	0.18	0.77	0.59
Al$_3$Ru	1.17	1.67	1.62	1.11	0.80	1.15	1.53	1.18
Al$_3$Rh	0.85	1.54	1.01	0.31	0.66	1.15	1.04	0.55
Al$_3$Pd	0.58	1.05	0.67	0.20	0.44	0.80	0.47	0.12
Al$_3$Ag	0.68	0.63	0.53	0.58	0.72	0.68	0.50	0.53
Al$_3$Cd	0.81	0.63	0.23	0.41	0.72	0.60	0.40	0.53
Al$_3$Hf	0.96	1.45	1.63	1.14	0.94	1.29	1.78	1.43
Al$_3$Ta	0.91	1.10	1.66	1.46	0.75	0.89	1.82	1.68
Al$_3$W	0.64	0.54	1.45	1.54	0.57	0.50	1.50	1.56
Al$_3$Re	0.90	0.95	1.76	1.70	0.00	0.04	0.93	0.89
Al$_3$Os	1.08	1.40	1.81	1.49	0.59	0.82	1.52	1.29
Al$_3$Ir	1.07	1.75	1.38	0.69	0.77	1.26	1.39	0.91
Al$_3$Pt	0.66	1.30	0.71	0.07	0.49	0.97	0.55	0.08
Al$_3$Au	0.55	0.70	0.35	0.20	0.46	0.57	0.31	0.20
Al$_3$Hg	0.67	0.45	−0.15	0.07	0.35	0.20	0.19	0.34

Figure 4a shows the slab model for calculating E_{sur}, and the detail calculated results of E_{sur}^{ave} of Al$_3$TM (RE = Sc-Zn, Y-Cd and Hf-Hg) are depicted in Table 4. It can be seen that for pure Al, Al$_3$(Sc-V) and Al$_3$(Y-Nb), the calculated values of this work are in good agreement with references [24,64,65]. The E_{sur}^{ave} of low index surfaces of cubic Al follows in the sequence of (110) > (100) > (111), which follows the general law of surface energies for face-centered cubic metals [66]. Figure 4c–e illustrates the change in E_{sur}^{ave} with the atomic number of TM elements, and it can be found that the variation tendency of E_{sur}^{ave} of Al$_3$TM intermetallic compounds presents similar characteristics for different cycles. For example, the E_{sur}^{ave} of Al$_3$TM for 3d elements firstly decreases from Sc to Mn, and then increases to Fe, and then decreases to Ni, and finally changes slightly in the (001) surface. The variation ranges of E_{sur}^{ave} for the (001), (110) and (111) surfaces of Al$_3$TM are 0.25~1.44, 0.26~1.45 and -0.32~1.18 J·m^{-2}, respectively, and the (111) surface has the lowest surface energy for all elements. As can be seen from Figure 4e,f, we have calculated the values of ELF by using the Equation (7) on the (111) plane of Al$_3$Sc and Al$_3$Mo. It can be seen that the (111) plane of Al$_3$Sc has more strongly localized electron areas than the (111) plane of Al$_3$Mo, indicating that strongly localized electrons make the surface energy lower. Clearly, if the E_{sur}^{ave} of Al$_3$TM is larger than that of Al, they would increase the toughness of Al alloys such as Al$_3$Sc. However, the complete toughness is not only determined by the above results but also by assessing the information of generalized stacking fault energy (GSFE) for particles in the Al matrix. The work needed for this is underway and will be published elsewhere.

Table 4. The calculated surface energy E_{sur} (J·m^{-2}) in Al or Al$_3$TM.

Systems	(001)	(110)	(111)
Al	0.79; 0.93 [23]	0.87; 0.98 [23]	0.68; 0.73 [39] 0.81 [23]
Al$_3$Sc	1.26; 1.32 [24]	1.41; 1.45 [24]	1.18; 1.22 [24]; 1.17 [39]
Al$_3$Ti	1.37	1.42	0.92; 0.93 [39]
Al$_3$V	1.32	1.32	0.72; 0.65 [39]
Al$_3$Cr	1.25	1.07	0.33
Al$_3$Mn	1.24	0.81	0.33
Al$_3$Fe	1.31	1.15	0.56
Al$_3$Co	1.14	1.04	0.67
Al$_3$Ni	0.91	0.82	0.57
Al$_3$Cu	0.91	0.86	0.69
Al$_3$Zn	0.81	0.84	0.73
Al$_3$Y	1.05	1.23	1.06; 1.11 [39]
Al$_3$Zr	1.12	1.23	0.80; 0.94 [39]
Al$_3$Nb	1.06	1.16	0.54; 0.59 [39]
Al$_3$Mo	0.62	0.73	-0.32
Al$_3$Tc	1.32	0.39	-0.22
Al$_3$Ru	1.39	1.17	0.44
Al$_3$Rh	0.93	0.85	0.46
Al$_3$Pd	0.62	0.46	0.21
Al$_3$Ag	0.61	0.61	0.46
Al$_3$Cd	0.52	0.56	0.47
Al$_3$Hf	1.30	1.36	0.87
Al$_3$Ta	1.28	1.29	0.66
Al$_3$W	1.04	1.05	0.13
Al$_3$Re	1.32	0.46	-0.32
Al$_3$Os	1.45	1.06	0.41
Al$_3$Ir	1.22	1.08	0.73
Al$_3$Pt	0.69	0.52	0.30
Al$_3$Au	0.45	0.38	0.35
Al$_3$Hg	0.26	0.27	0.27

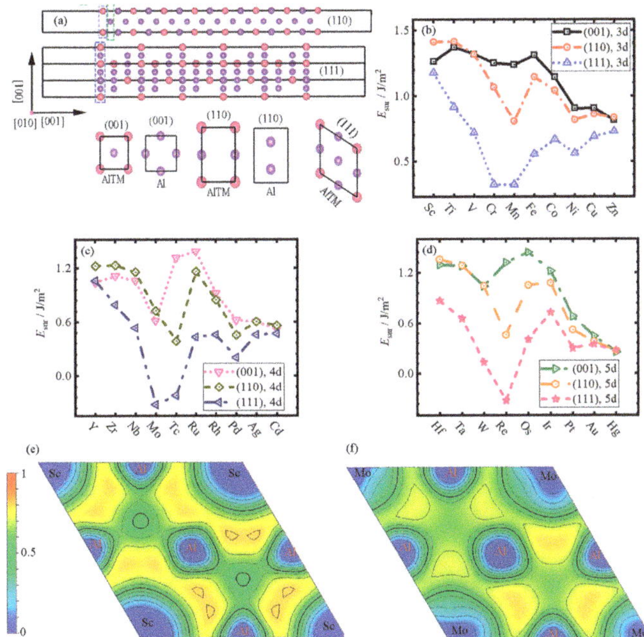

Figure 4. (**a**) The slab model. (**b–d**) The calculated average surface energy E_{sur}^{ave} with the change in atomic number. (**e**,**f**) The ELF on the (111) plane of Al$_3$Sc and Al$_3$Mo systems, respectively.

4. Conclusions

In this work, we have calculated the diffusion rates of TM elements in Al and the key parameters of nucleation and growth of second phase particles Al$_3$TM (TM = Sc-Zn, Y-Cd and Hf-Hg) using the first principles combing with quasi-harmonic approximation in the theoretical framework of density functional theory. Firstly, as can be seen from the discussed results, the trends of the Q, chemical formation and surface energies with the change in atomic number in the same cycle are similar. The reason may be related to the valence electron configures (VECs) of TM elements because TM in the same cycle has the same VECs. Meanwhile, the calculation interface and strain energy of Al$_3$TM/Al composed of the same cycle of TM elements show a lack of similarity. The reason for this may be that the interface and strain energy are co-determined by the matrix and second phases. Here, the main conclusions are as follows:

1. In the vacancy–substitution model, the diffusion activation energy Q first increases, and then decreases with the increase in atomic number (Sc-Zn, Y-Ag and Hf-Au) in the Al matrix, except for Cr; the TM elements in the fourth cycle generally have lower Qs.

2. Mn-Co, Tc-Rh and Re-Ir elements have larger activation energies Q_s in the Al matrix, while Cu-Zn, Ag and Au have lower activation energies Q_s; even Cd and Hg elements have negative activation energies. In the undoped-Al system, the self-diffusion activation energy Q_0 is lower compared to all Q_s in the doped system, except for the Q_s of Cu, Zn, Y and Ag.

3. The diffusion rate D_s of all impurity atoms increases logarithmically with the increase in temperature. With the increase in atomic number, the diffusion rate D_s first decreases linearly from Sc, Y and Hf to Mn, Ru and Ir, and then increases to Zn, Ag and Au for 3–5d TM elements, respectively.

4. With the increase in atomic number, the interface energy $\gamma_{\alpha/\beta}$ of Al/Al$_3$TM (TM = (Sc-Zn, Y-Cd)) decreases from Sc and Y to Mn and Tc, and then increases to Zn and

Cd, respectively, except for (001) in Al/Al$_3$(Fe-Co), (111) in Al/Al$_3$Pd and (110) and (111) in Al/Al$_3$Cd. Meanwhile, the strain energy ΔE_{cs} increases from Sc to Co in the (001) system, to Fe in the (110) system, and to Mn in the (111) system, respectively, and then they all decreases to Zn, except for Al/Al$_3$Mn. The largest values of ΔE_{cs} for (110) and (111) interface systems are all located at Al/Al$_3$Re, while the (001) interface system of Al/Al$_3$Tc has the lowest value.

5. The variation in chemical formation energy ΔG_V of all Al$_3$TM changes slightly in the temperature range of 0~1000 K, except that the ΔG_V of Al$_3$Sc, Al$_3$Cu, Al$_3$(Y-Zr), Al$_3$Cd, Al$_3$Hf and Al$_3$Hg increase nonlinearly. With the increase in atomic number at 300 K, the ΔG_V increases from Sc, Y and Hf to Mn, Tc and Re at first, and then decreases to Co, Rh and Ir, respectively, and finally, it slightly changes. The variation trends of the ΔG_V for 3–5d TMs are the same as those at 300 K.

6. With the increase in atomic number, the trend of E_{sur}^{ave} of Al$_3$TM intermetallic compounds presents a similar change in different cycles and the (111) surface always has the lowest surface energy in all surfaces of Al$_3$TM particles.

Author Contributions: Methodology, T.F.; Investigation, Z.R. and T.F.; Writing—original draft, Z.R.; Writing—review & editing, T.F.; Visualization, T.H., Z.R., K.W., K.H. and Y.W.; Project administration, T.H., K.W. and Y.W. All authors have read and agreed to the published version of the manuscript.

Funding: This work is supported by the Foshan Technology Project (1920001000409), the Scientific Research Project of Hunan Institute of Technology (HQ21016, 21A0564, HP21047, 21B0796), the Natural Science Foundation of China (52171115). The APC was funded by Foshan Technology Project (1920001000409).

Data Availability Statement: In this article, the key parameters of growth and nucleation of Al$_3$TM particles based on the corresponding models (see Figures 2 and 3) have been discussed in our work. Figure 4 describe calculated surface energy of Al$_3$TM and Al. Tables 1–4 present important data as a reference. All data can be found in the manuscript.

Conflicts of Interest: The authors declare that they have no known competing financial interest or personal relationships that have influenced the work reported in this paper.

References

1. Schlapbach, L.; Züttel, A. Hydrogen-storage materials for mobile applications. *Nature* **2001**, *414*, 353–358. [CrossRef] [PubMed]
2. Wen, K.; Xiong, B.-Q.; Fan, Y.-Q.; Zhang, Y.-A.; Li, Z.-H.; Li, X.-W.; Wang, F.; Liu, H.-W. Transformation and dissolution of second phases during solution treatment of an Al-Zn-Mg-Cu alloy containing high zinc. *Rare Met.* **2018**, *37*, 376–380. [CrossRef]
3. Seidman, D.N.; Marquis, E.A.; Dunand, D.C. Precipitation strengthening at ambient and elevated temperatures of heat-treatable Al(Sc) alloys. *Acta Mater.* **2002**, *50*, 4021–4035. [CrossRef]
4. Drits, M.E.; Kadaner, E.S.; Turkina, N.I.; Fedotov, S.G. The Mechanical Properties of Aluminum-Lithium Alloy. *Transl. Splav. Tsvetn. Met.* **1972**, *14*.
5. Yang, Y.; Licavoli, J.J.; Hackney, S.A.; Sanders, P.G. Coarsening behavior of precipitate Al$_3$(Sc, Zr) in supersaturated Al-Sc-Zr alloy via melt spinning and extrusion. *J. Mater. Sci.* **2021**, *56*, 11114–11136. [CrossRef]
6. Yan, K.; Chen, Z.; Lu, W.J.; Zhao, Y.; Le, W.; Naseem, S. Nucleation and growth of Al$_3$Sc precipitates during isothermal aging of Al-0.55wt% Sc alloy. *Mater. Charact.* **2021**, *179*, 111331. [CrossRef]
7. Clemens, H.; Kestler, H. Processing and Applications of Intermetallic -TiAl-Based Alloys. *Adv. Eng. Mater.* **2010**, *2*, 551–570. [CrossRef]
8. Ug, Ş.; Arıkan, N.; Soyalp, F.; Ug, G. Phonon and elastic properties of AlSc and MgSc from first-principles calculations. *Comput. Mater. Sci.* **2010**, *48*, 866–870.
9. RMichi, A.; Toinin, J.P.; Farkoosh, A.R.; Seidman, D.N.; Dunand, D.C. Effects of Zn and Cr additions on precipitation and creep behavior of a dilute Al–Zr–Er–Si alloy. *Acta Mater.* **2019**, *181*, 249–261.
10. Zedalis, M.S.; Fine, M.E. Precipitation and ostwald ripening in dilute Al Base-Zr-V alloys. *Metall. Trans. A* **1986**, *17*, 2187–2198. [CrossRef]
11. Parameswaran, V.R.; Weertman, J.R.; Fine, M.E. Coarsening behavior of L1$_2$ phase in an Al-Zr-Ti alloy. *Scr. Metall.* **1989**, *23*, 147–150. [CrossRef]
12. Chen, Z.; Zhang, P.; Chen, D.; Wu, Y.; Wang, M.; Ma, N.; Wang, H. First-principles investigation of thermodynamic, elastic and electronic properties of Al$_3$V and Al$_3$Nb intermetallics under pressures. *J. Appl. Phys.* **2015**, *117*, 085904. [CrossRef]
13. Li, R.-Y.; Duan, Y.-H. Electronic structures and thermodynamic properties of HfAl$_3$ in L1$_2$, D0$_{22}$ and D0$_{23}$ structures. *Trans. Nonferrous Met. Soc. China* **2016**, *26*, 2404–2412. [CrossRef]

14. Czerwinski, F. Thermal Stability of Aluminum Alloys. *Materials* **2020**, *13*, 3441. [CrossRef] [PubMed]
15. Mao, Z.; Chen, W.; Seidman, D.N.; Wolverton, C. First-principles study of the nucleation and stability of ordered precipitates in ternary Al–Sc–Li alloys. *Acta Mater.* **2011**, *59*, 3012–3023. [CrossRef]
16. Zhang, X.; Huang, Y.; Liu, Y.; Ren, X. A comprehensive DFT study on the thermodynamic and mechanical properties of $L1_2$-Al_3Ti/Al interface. *Vacuum* **2021**, *183*, 109858. [CrossRef]
17. Wang, Y.; Wang, J.; Zhang, C.; Huang, H. Mechanical properties of defective $L1_2$-Al_3X (X= Sc, Lu) phase: A first-principles study. *J. Rare Earths* **2021**, *39*, 217–224. [CrossRef]
18. Dorin, T.; Babaniaris, S.; Jiang, L.; Cassel, A.; Robson, J.D. Stability and stoichiometry of $L1_2$ Al_3(Sc,Zr) dispersoids in Al-(Si)-Sc-Zr alloys. *Acta Mater.* **2021**, *216*, 117117. [CrossRef]
19. Liu, T.; Ma, T.; Li, Y.; Ren, Y.; Liu, W. Stable mechanical and thermodynamic properties of Al-RE intermetallics: A First-principles study. *J. Rare Earths* **2022**, *40*, 345–352. [CrossRef]
20. Nakai, M.; Eto, T. New aspect of development of high strength aluminum alloys for aerospace applications. *Mater. Sci. Eng. A* **2000**, *285*, 62–68. [CrossRef]
21. Saha, S.; Todorova, T.Z.; Zwanziger, J.W. Temperature dependent lattice misfit and coherency of Al_3X (X=Sc, Zr, Ti and Nb) particles in an Al matrix. *Acta Mater.* **2015**, *89*, 109–115. [CrossRef]
22. Shi, T.T.; Wang, J.N.; Wang, Y.P.; Wang, H.C.; Tang, B.Y. Atomic diffusion mediated by vacancy defects in pure and transition element (TM)-doped (TM=Ti, Y, Zr or Hf) $L1_2$ Al_3Sc. *Mater. Des.* **2016**, *108*, 529–537. [CrossRef]
23. Li, S.-S.; Li, L.; Han, J.; Wang, C.-T.; Xiao, Y.-Q.; Jian, X.-D.; Qian, P.; Su, Y.-J. First-Principles study on the nucleation of precipitates in ternary Al alloys doped with Sc, Li, Zr, and Ti elements. *Appl. Surf. Sci.* **2020**, *526*, 146455. [CrossRef]
24. Sun, S.P.; Li, X.P.; Wang, H.J.; Jiang, H.F.; Lei, W.N.; Jiang, Y.; Yi, D.Q. First-principles investigations on the electronic properties and stabilities of low-index surfaces of $L1_2$–Al_3Sc intermetallic. *Appl. Surf. Sci.* **2014**, *288*, 609–618. [CrossRef]
25. Hood, G.M. The diffusion of iron in aluminium. *Philos. Mag.* **1970**, *21*, 305–328. [CrossRef]
26. Mantl, S.; Petry, W.; Schroeder, K.; Vogl, G. Diffusion of iron in aluminum studied by Mössbauer spectroscopy. *Phys. Rev. B* **1983**, *27*, 5313–5331. [CrossRef]
27. Yan, K.; Chen, Z.W.; Zhao, Y.N.; Ren, C.C.; Aldeen, A.W. Morphological characteristics of Al_3Sc particles and crystallographic orientation relationships of Al_3Sc/Al interface in cast Al-Sc alloy. *J. Alloys Compd.* **2020**, *861*, 158491. [CrossRef]
28. Mandal, P.K.; Kumar, R.J.F.; Varkey, J.M. Effect of artificial ageing treatment and precipitation on mechanical properties and fracture mechanism of friction stir processed $MgZn_2$ and Al_3Sc phases in aluminium alloy. *Mater. Today Proc.* **2021**, *46*, 4982–4987. [CrossRef]
29. Alexander, W.B.; Slifkin, L.M. Diffusion of Solutes in Aluminum and Dilute Aluminum Alloys. *Phys. Rev. B* **1970**, *1*, 3274–3282. [CrossRef]
30. Zhao, X.; Chen, H.; Wilson, N.; Liu, Q.; Nie, J.-F. Direct observation and impact of co-segregated atoms in magnesium having multiple alloying elements. *Nat. Commun.* **2019**, *10*, 3243. [CrossRef]
31. Maruhn, J.A.; Reinhard, P.G.; Suraud, E. *Density Functional Theory*; Springer: Berlin/Heidelberg, Germany, 2010.
32. Finnis, M.W. The theory of metal-ceramic interfaces. *J. Phys. Condens. Matter* **1996**, *8*, 5811–5836. [CrossRef]
33. Zhou, W.F.; Ren, X.D.; Ren, Y.P.; Yuan, S.Q.; Ren, N.F.; Yang, X.Q.; Adu-Gyamfi, S. Initial dislocation density effect on strain hardening in FCC aluminium alloy under laser shock peening. *Philos. Mag.* **2017**, *97*, 917–929. [CrossRef]
34. Zhao, S.J.; Stocks, G.M.; Zhang, Y.W. Stacking fault energies of face-centered cubic concentrated solid solution alloys. *Acta Mater.* **2017**, *134*, 334–345. [CrossRef]
35. Mardirossian, N.; Head-Gordon, M. Thirty years of density functional theory in computational chemistry: An overview and extensive assessment of 200 density functionals. *Mol. Phys.* **2017**, *115*, 2315–2372. [CrossRef]
36. Kresse, G.; Joubert, D. From ultrasoft pseudopotentials to the projector augmented-wave method. *Phys. Rev. B* **1999**, *59*, 1758–1775. [CrossRef]
37. Hafner, J. Ab-initio simulations of materials using VASP: Density-functional theory and beyond. *J. Comput. Chem.* **2008**, *29*, 2044–2078. [CrossRef]
38. TFischer, H.; Almlof, J. General methods for geometry and wave function optimization. *J. Phys. Chem.* **1992**, *96*, 9768–9774. [CrossRef]
39. Liu, Y.; Wen, J.C.; Zhang, X.Y.; Huang, Y.C. A comparative study on heterogeneous nucleation and mechanical properties of the fcc-Al/$L1_2$-Al_3M (M = Sc, Ti, V, Y, Zr, Nb) interface from first-principles calculations. *Phys. Chem. Chem. Phys.* **2021**, *23*, 4718–4727. [CrossRef]
40. Henkelman, G.; Uberuaga, B.P.; Jónsson, H. A climbing image nudged elastic band method for finding saddle points and minimum energy paths. *J. Chem. Phys.* **2000**, *113*, 9901–9904. [CrossRef]
41. Lifshitz, I.M.; Slyozov, V.V. The kinetics of precipitation from supersaturated solid solutions. *J. Phys. Chem. Solids* **1961**, *19*, 35–50. [CrossRef]
42. Wagner, C. Zeitschrift für Elektrochemie. *Berichte der Bunsengesellschaft für physikalische Chemie.* **1961**, *65*, 7–8.
43. Mantina, M.; Wang, Y.; Arroyave, R. First-Principles Calculation of Self-Diffusion Coefficients. *Phys. Rev. Lett.* **2008**, *100*, 215901. [CrossRef]
44. Mantina, M.; Wang, Y.; Chen, L.Q.; Liu, Z.K.; Wolverton, C. First principles impurity diffusion coefficients. *Acta Mater.* **2009**, *57*, 4102–4108. [CrossRef]

45. Evangelakis, G.; Papanicolaou, N. Adatom self-diffusion processes on (001) copper surface by molecular dynamics. *Surf. Sci.* **1996**, *347*, 376–386. [CrossRef]
46. Gomer, R. Diffusion of adsorbates on metal surfaces. *Rep. Prog. Phys.* **1990**, *53*, 917. [CrossRef]
47. Bennett, C.H. Exact defect calculations in model substances. *Diffus. Solids Recent Dev.* **1975**, *1*, 73–113.
48. Vineyard, G.H. Frequency factors and isotope effects in solid state rate processes. *J. Phys. Chem. Solids* **1957**, *3*, 121–127. [CrossRef]
49. Yang, B.; Wang, L.-G.; Yi, Y.; Wang, E.-Z.; Peng, L.-X. First-principles calculations of the diffusion behaviors of C, N and O atoms in V metal. *Acta Phys. Sin.* **2015**, *64*, 026602. [CrossRef]
50. Chen, L.; Li, Y.; Xiao, B.; Gao, Y.; Zhao, S. Chemical bonding, thermodynamic stability and mechanical strength of $Ni_3Ti/\alpha\text{-}Al_2O_3$ interfaces by first-principles study. *Scr. Mater.* **2021**, *190*, 57–62. [CrossRef]
51. Wert, C. Diffusion coefficient of C in α-iron. *Phys. Rev.* **1950**, *79*, 601. [CrossRef]
52. Wert, C.; Zener, C. Interstitial atomic diffusion coefficients. *Phys. Rev.* **1949**, *76*, 1169. [CrossRef]
53. Liu, P.; Wang, S.; Li, D.; Li, Y.; Chen, X.-Q. Fast and Huge Anisotropic Diffusion of Cu (Ag) and Its Resistance on the Sn Self-diffusivity in Solid β–Sn. *J. Mater. Sci. Technol.* **2016**, *32*, 121–128. [CrossRef]
54. Ehrhart, P. *Atomic Defects in Metals*; Springer: Berlin/Heidelberg, Germany, 1991.
55. Fujikawa, S.I. Impurity Diffusion of Scandium in Aluminium. *Defect Diffus. Forum* **1997**, *143*, 115–120. [CrossRef]
56. Volin, T.E.; Balluffi, R.W. Annealing kinetics of voids and the Self-diffusion coefficient in aluminum. *Phys. Status Solidi* **2010**, *25*, 163–173. [CrossRef]
57. Murphy, J.B. Interdiffusion in dilute aluminium-copper solid solutions. *Acta Metall.* **1961**, *9*, 563–569. [CrossRef]
58. Feng, Y.; Liu, M.; Shi, Y.; Ma, H.; Li, D.; Li, Y.; Lu, L.; Chen, X.Q. High-throughput modeling of atomic diffusion migration energy barrier of fcc metals. *Prog. Nat. Sci. Mater. Int.* **2019**, *29*, 341–348. [CrossRef]
59. Pareige, C.; Soisson, F.; Martin, G.; Blavette, D. Ordering and phase separation in Ni–Cr–Al: Monte Carlo simulations vs. three-dimensional atom probe. *Acta Mater.* **1999**, *47*, 1889–1899. [CrossRef]
60. Marder, M. Correlations and Ostwald ripening. *Phys. Rev. A* **1987**, *36*, 858. [CrossRef]
61. Hirth, J.P.; Lothe, J. *Theory of Dislocations*, 2nd ed.; Wiley: New York, NY, USA, 1982.
62. Hutchinson, J.W. Singular behavior at the end of a tensile crack in a hardening material. *J. Mech. Phys. Solids* **1968**, *16*, 13–31. [CrossRef]
63. Rice, J.R.; Rosengren, G.F. Plane strain deformation near a crack tip in a power-law hardening material. *J. Mech. Phys. Solids* **1968**, *16*, 1–12. [CrossRef]
64. Zhang, X.; Ren, X.; Li, H.; Zhao, Y.; Huang, Y.; Liu, Y.; Xiao, Z. Interfacial properties and fracture behavior of the $L1_2\text{-}Al_3Sc \mid \mid Al$ interface: Insights from a first-principles study. *Appl. Surf. Sci.* **2020**, *515*, 146017. [CrossRef]
65. Wang, Y.X.; Arai, M.; Sasaki, T.; Wang, C.L. First-principles study of the (001) surface of cubic $CaTiO_3$. *Phys. Rev. B* **2006**, *73*, 035411. [CrossRef]
66. Vitos, L.; Ruban, A.V.; Skriver, H.L.; Kollár, J. The surface energy of metals. *Surf. Sci.* **1998**, *411*, 186–202. [CrossRef]

 *metals* 

Article

Spatial-Time Inhomogeneity Due to the Portevin-Le Chatelier Effect Depending on Stiffness

Tatyana Tretyakova * and Mikhail Tretyakov

Center of Experimental Mechanics, Perm National Research Polytechnic University, 614990 Perm, Russia;
cem_tretyakov@mail.ru
* Correspondence: cem.tretyakova@gmail.com; Tel.: +7-902-641-5560

Abstract: This work is devoted to the study of the influence of the rigidity of the loading system on the kinetics of the initiation and propagation of the Portevin-Le Chatelier (PLC) strain bands due to the jerky flow in Al-Mg alloy. To estimate the influence of the loading system, the original loading attachment, which allows reducing the stiffness in a given range, was used. Registration of displacement and strain fields on the specimen surface was carried out with the Vic-3D non-contacting deformation measurement system based on the Digital Image Correlation (DIC) technique. The mechanical uniaxial tension tests were carried out using samples of Al-Mg alloy at the biaxial servo-hydraulic testing system Instron 8850. As a result of tensile tests, deformation diagrams were obtained for Al-Mg alloy samples tested at different values of stiffness of the loading system: 120 MN/m (nominal value), 50 MN/m, 18 MN/m, and 5 MN/m. All diagrams show discontinuous plastic deformations (the Portevin-Le Chatelier effect). It is noted that a decrease in the rigidity of the loading system leads to a change in the type of jerky flow. At constant parameters of the loading rate, temperature, and chemical composition of the material, the PLC effects of types A, B, and C are recorded in tests.

Keywords: jerky flow; Portevin-Le Chatelier effect; strain band; plasticity; stiffness; loading system; digital image correlation; aluminum-magnesium alloy

Citation: Tretyakova, T.; Tretyakov, M. Spatial-Time Inhomogeneity Due to the Portevin-Le Chatelier Effect Depending on Stiffness. *Metals* **2023**, *13*, 1054. https://doi.org/10.3390/met13061054

Academic Editor: Andrea Di Schino

Received: 24 April 2023
Revised: 24 May 2023
Accepted: 25 May 2023
Published: 31 May 2023

1. Introduction

By understanding the physical, mechanical, and strength properties of materials, we can ensure the high reliability of structures and prevent industrial accidents caused by mechanical failures. It is necessary to take the influence of real external impacts and the material response into account. Inelastic behavior and failure of critical structural elements are related to elastoplastic deformation of the material, accompanied by a change in its structure, physical and mechanical properties during loading. Some structural materials, for example, Al-Mg, Al-Cu, Ni-Cr, and Ni-Fe-Cr alloys, are characterized by the manifestation of the effects of discontinuous yield in certain ranges of temperatures and loading rates [1–3].

Under uniaxial tension, the effect of the jerky flow manifests itself in the formation of teeth on the deformation curve in the case of kinematic loading (the Portevin-Le Chatelier effect) or in the form of steps in the case of force loading (the Savart-Masson effect). The PLC effect reduces the surface quality of the material, causes non-uniform development of deformations, and causes a significant decrease in strength and ductility. In this case, the formation of strain bands and macroscopic localization of plastic flows lead to a difference in thickness of structural elements, stress concentration, defects, and, therefore, the initiation of macroscopic failure processes [4–8]. It is a promising approach to obtaining and systematizing experimental data on the jerky flow under various types of mechanical effects. The study analyzes the kinetics of strain fields during the initiation and development of macroscopic localization and instability of plastic flow based on the example of Al-Mg alloys [6,8].

This work considers the failure as a result of the stability loss of inelastic deformation processes [9–12]. The theoretical and experimental study of the basic laws of post-critical deformation makes it possible to predict conditions of destruction of deformable bodies and analyze possibilities for controlling destruction. At the postcritical stage of deformation, the formation of macroscopic failure conditions occurs, which are not uniquely associated with the stress-strain state. The diagram breaks off at the highest point only at zero stiffness of the loading system in the case of "soft" (force) loading. Under 'hard' (kinematic) loading, a complete diagram characterizes the relationship between the load and displacement and shows a decrease in the force applied to the sample to zero. The loading system [13–15] plays a key role in the transition from the stage of equilibrium accumulation of damage to the nonequilibrium, avalanche-like stage of destruction. As the effects of the intermittent flow are characterized by instability of the ongoing deformation processes [1,6,10], we expect a significant dependence of this phenomenon on parameters of the loading system's stiffness.

Special attention should be paid to the experimental study and theoretical description of the spatio-temporal inhomogeneity of inelastic deformation, the development of defects, and destruction, taking into account the influence of the type of complex stress-strain state of the material. It is well known, that in most cases, the material in a structure works under complex stress states and complex thermomechanical influences [16–22]. Mechanical tests are associated with significant methodological difficulties and technical limitations of the schemes and loading modes, as well as the methods of recording and interpreting the results obtained [22–27]. Of interest is the development of existing techniques and the creation of new scientifically grounded approaches to the experimental study of inelastic behavior and fracture of advanced structural metals and alloys based on original techniques that allow creating the required characteristics of the loading system in the working part of the sample. Thus, the aim of the work is to study how the stiffness of the loading system affects the manifestation of the jerky flow effect in an Al-Mg alloy using an original device of variable stiffness and the digital image correlation technique.

2. Materials and Methods

A structural Al-Mg alloy was chosen as the material for the study. The chemical composition, according to GOST 4784-97, is shown in Table 1. Specimens were made from a rod with a diameter of 20 mm as delivered (without additional heat treatment) on a lathe.

Table 1. Chemical composition of Al-Mg alloy.

Element	Al	Mg	Mn	Fe	Si	Zn	Ti	Cu	Be
Composition, %.wt.	92.55	6.12	0.84	0.27	0.17	0.005	0.039	0.001	0.005

In tensile tests with the attachment having a variable stiffness, solid cylindrical specimens were used, a scheme of which is shown in Figure 1. On one of the gripping parts, a thread is cut to fix it to the tooling rod. This sample geometry meets the requirements of GOST 1497-84 "Metals. Tensile Test Methods ".

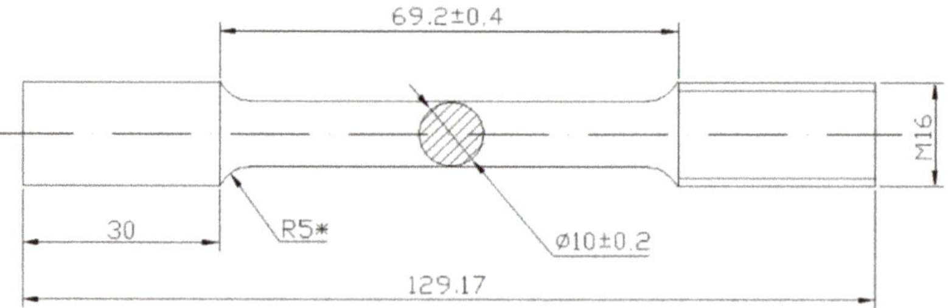

Figure 1. Scheme of a solid cylindrical sample for testing with fixtures with variable stiffness.

Mechanical uniaxial tensile tests are carried out at the Instron 8850 biaxial servo-hydraulic testing system, which allows tensile-compression tests with a maximum load of 100 kN and torsion with a maximum torque of 1000 Nm. Kinematic loading is realized with a traverse speed of 2.4 mm/min, which corresponds to a material strain rate of $6.67 \cdot 10^{-4} \, \text{s}^{-1}$. The stiffness of the loading system is changed by using specialized equipment with a variable stiffness.

The equipment shown in Figure 2 consists of the following elements: body—1, consisting of two parts connected by means of a threaded connection, a package of plate springs—2; a guide rod—3; compensation rings—4; hardened washers—5; fastening nuts for fixing the package of springs—6; a rod to install the equipment in the grips of the testing machine—7; and a threaded hole for fastening in the equipment of the test specimen—8. Elements 3 and 5 are hardened to prevent damage and bite to the springs when the package is loaded. There is a patent (RU 153985 U1) for the rigging with a variable stiffness used in the work for tensile tests with various loading parameters.

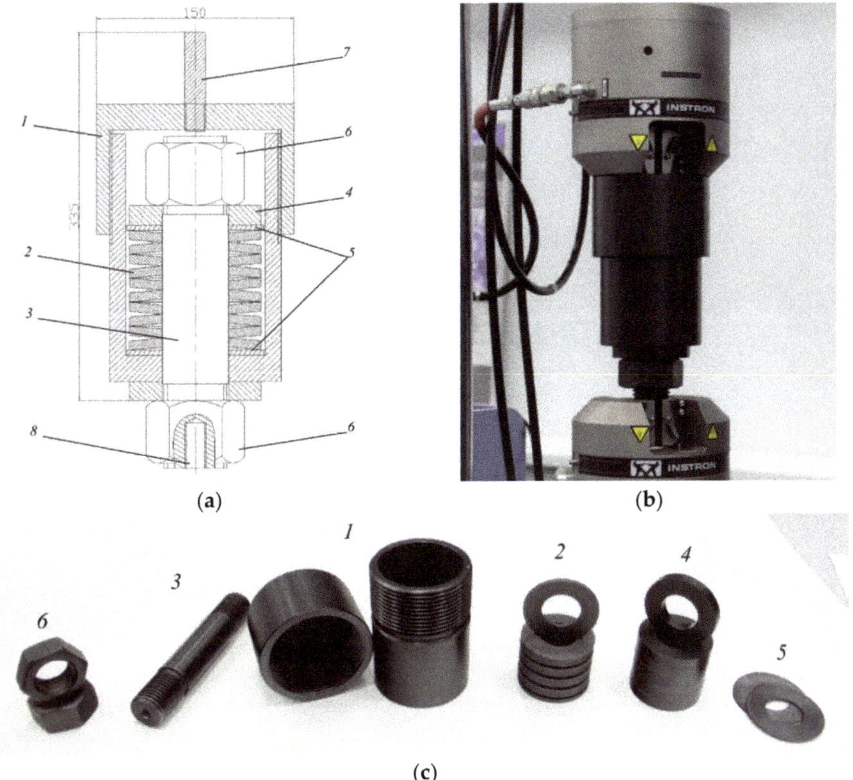

Figure 2. A scheme containing the main elements (**a**), a general view (**b**) and elements of attachment with a variable stiffness (**c**).

By changing the stacking order and the number of plate springs (see Table 2) we provided the required stiffness of the loading system in the range of 5 MN/m to 120 MN/m.

In total, four values of the rigidity of the loading system are implemented during the work: 120 MN/m (nominal value), 50 MN/m, 18 MN/m and 5 MN/m (Figure 3). The maximum stiffness of the tooling coincides with the stiffness of the Instron 8850 testing machine when using hydraulic grips.

Table 2. Setting up of plate springs inside the attachment with a variable stiffness.

Set	Number of Plate Springs	Setting-Up of Plate Springs	Stiffness, MN/m
1	-	-	120
2	10	< <	50
3	2	< >	18
4	10	< >	5

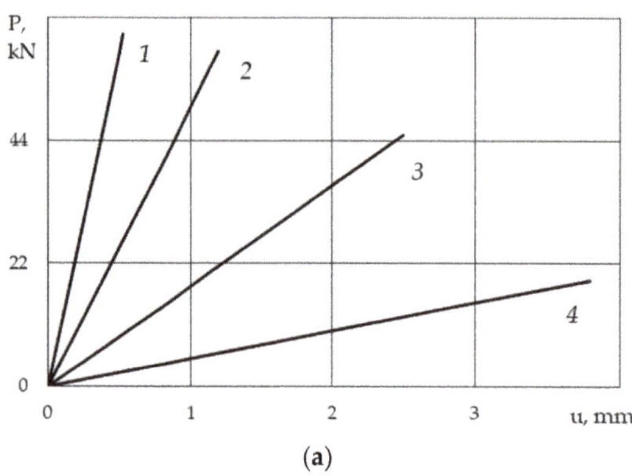

(a)

(b)

Figure 3. Stiffness characteristics for various setting-ups of plate springs corresponding to the values: 1—120 MN/m, 2—50 MN/m, 3—18 MN/m, 4—5 MN/m (**a**), testing equipment set-up (**b**).

Plate springs are installed according to the required scheme. Further, the tooling is installed in the grips of the testing machine in series with the sample. The sample is fixed to the tooling rod by means of a threaded connection. Then a constant velocity displacement is applied to the sample, the change in loading is recorded, and the evolution of the deformation field on the surface of the working part of the sample is recorded. The test system controller allows synchronization with the Vic-3D non-contact video system based on the digital image correlation (DIC) method.

3. Results

As a result, the 'load-elongation' curves were obtained for samples of the Al-Mg alloy at different values of stiffness of the loading system: 120 MN/m (Figure 4), 50 MN/m (Figure 5), 18 MN/m (Figure 6), and 5 MN/m (Figure 7). To demonstrate the jerky type flow, enlarged fragments of these diagrams are additionally shown. The curves are plotted according to the built-in sensor of the testing system.

For the values of the loading system's stiffness of 120 MN/m (Figure 4b) and 50 MN/m (Figure 5b), the loading diagrams show single load drops (or 'teeth'), which are maintained by the initiation and propagation of a single PLC band. Up to a certain load level, uniform deformation of the sample is observed, which corresponds to a smooth section of elastic deformation and the initial stage of hardening. At the stage of supercritical deformation, the diagram is characterized by the presence of a large number of closely spaced teeth (Figures 4b and 5b).

A decrease in the level of the loading system's stiffness in relation to the sample leads to a change in the type of jerky flow of the Al-Mg alloy and an increase in the frequency and amplitude of the load jumps. Figure 6 shows the loading curve for a sample installed in a special gripping device, which realizes a stiffness of 18 MN/m. At a minimum stiffness level of 5 MN/m, we noted a decrease in the frequency of jumps on the diagram but an increase in their amplitude (Figure 7). An increase in the load leads to an increase in the amplitude of the load disruptions (Figure 7b).

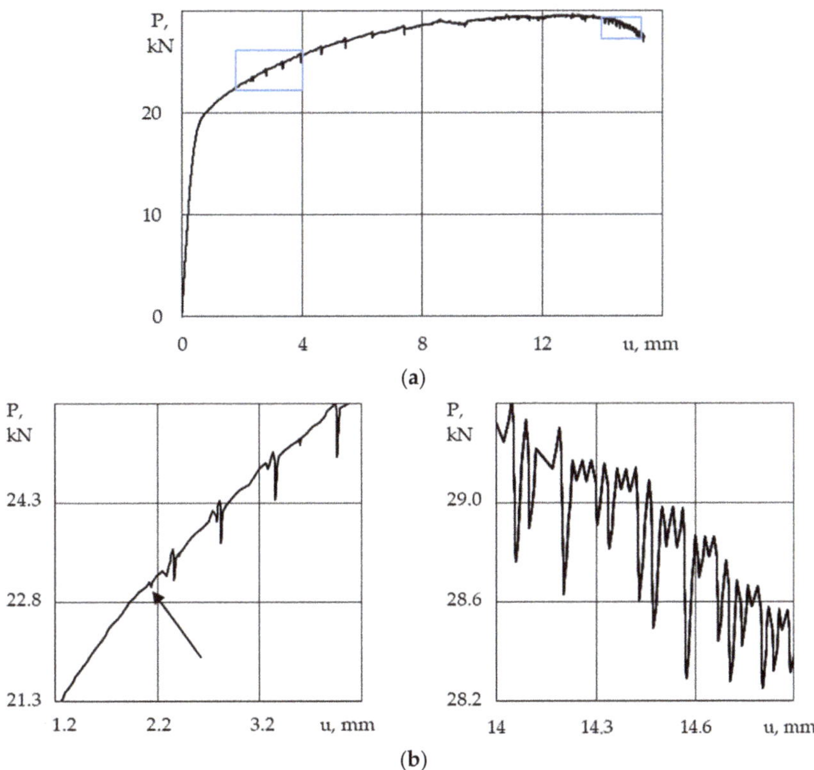

(a)

(b)

Figure 4. The load-displacement curve for the loading system stiffness of 120 MN/m (**a**) and enlarged sections (blue boxes) of the curve illustrating the discontinuity of the plastic flow (**b**).

To determine the level of critical deformation (ε_{cr}), at which the effect of the jerky flow begins to appear, deformation diagrams were constructed by using an additional software module of the video system called the 'virtual' extensometer (Figure 8a). The 'virtual' extensometer allows for the non-contact determination of the deformation in the gauge length of the specimen based on the displacement fields' analysis. Thus, the 'stress-strain' curves for each group of the samples were obtained. As an example, Figure 8b shows a deformation diagram for a specimen tested at the maximum level of the loading system (120 MN/m).

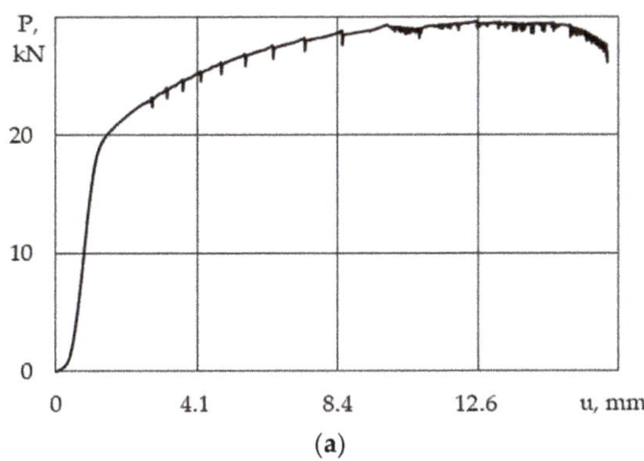

(a)

Figure 5. *Cont.*

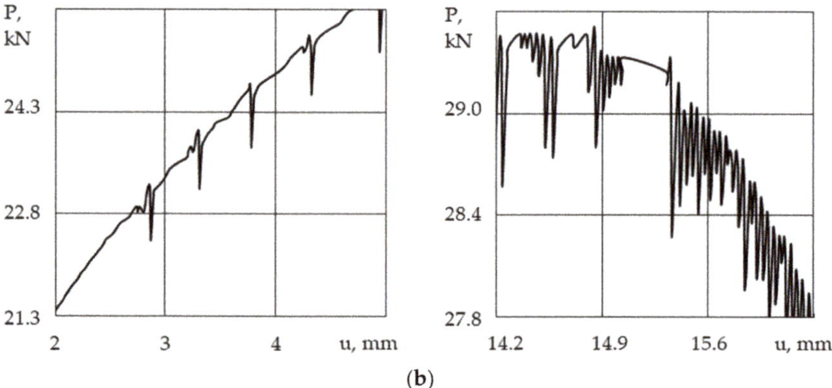

(b)

Figure 5. The load-displacement curve for the loading system's stiffness of 50 MN/m (**a**) and enlarged sections of the curve illustrating the discontinuity of the plastic flow (**b**).

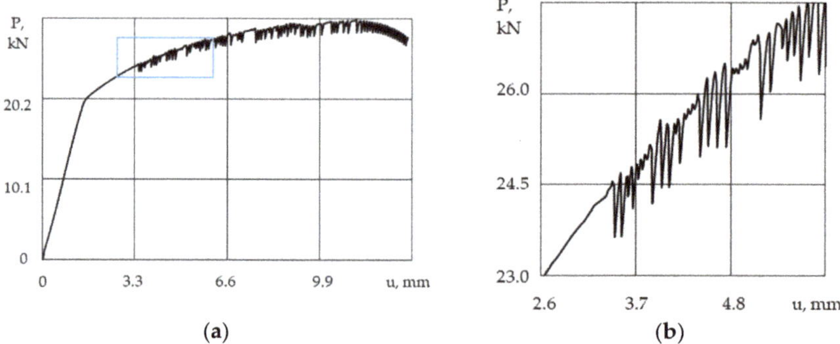

(a)

(b)

Figure 6. The load-displacement curve for the loading system's stiffness of 18 MN/m (**a**) and the enlarged section (blue box) of the curve illustrating the discontinuity of the plastic flow (**b**).

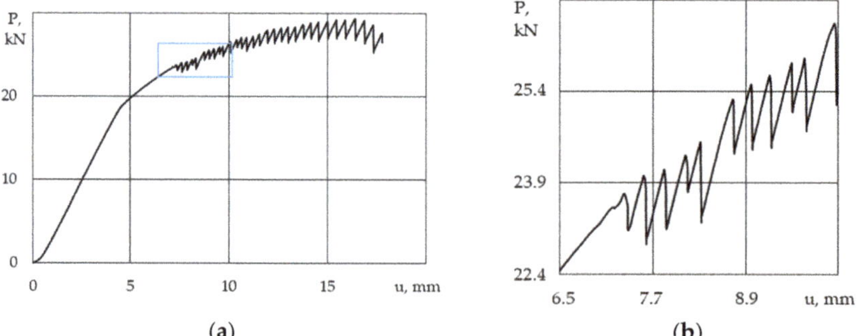

(a)

(b)

Figure 7. The load-displacement curve for the loading system's stiffness of 5 MN/m (**a**) and the enlarged section (blue box) of the curve illustrating the discontinuity of the plastic flow (**b**).

This figure shows the moment of the onset of unstable plastic deformation at a result of the initiation of the effect of the jerky flow of the material. The level of critical deformation of the onset of the PLC effect is 2.36%. The ε_{cr} values for different levels of stiffness of the loading system are shown in Table 3.

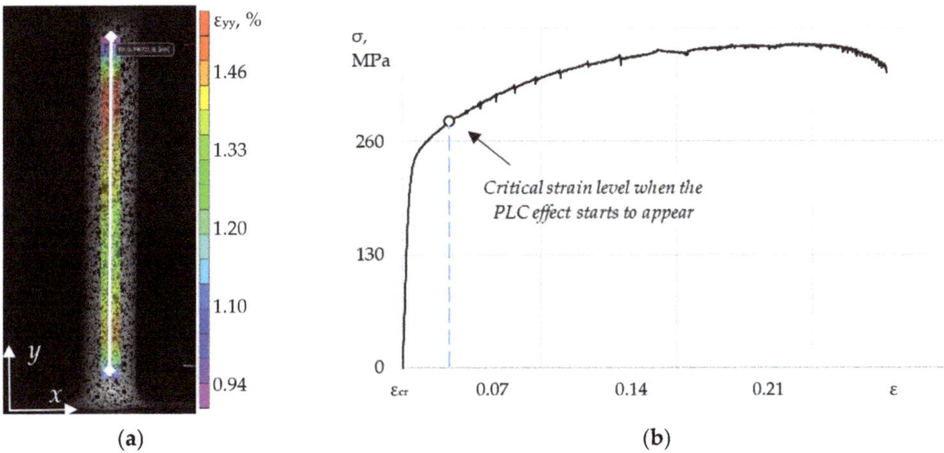

Figure 8. The location of the 'virtual extensometer' of the *Vic-3D* measurement system (**a**); the stress-strain curve in the case of the loading system's stiffness of 120 MN/m (**b**).

Table 3. Influence of the loading system's stiffness on the critical strain value at the onset of the PLC effect.

Stiffness of the Loading System, MN/m	120	50	18	5
Critical strain of the onset of the PLC effect, %	2.36	2.51	2.85	3.18

Spatial-Time Inhomogeneity Depending on the Stiffness of Loading System

The analysis of the evolution of inhomogeneous deformation fields and local rates of longitudinal deformation is carried out to study the kinetics of the development of the macroscopic localization of the plastic flow depending on the loading system's stiffness. At the moment of the breakdown of the load on the diagram, a sharp localization of the plastic flow occurs in the sample, and the front of the PLC strip is formed. Depending on the type of jerky flow (type A, B, or C), the deformation band begins to either move uniformly along the length of the sample (in the case of the discontinuous flow according to type A). A random occurrence of PLC bands is observed in the sample, causing frequent drops in the load of small amplitude (type B) or drops in the load of large amplitude (type C).

To visualize the formation and propagation of PLC bands, the authors proposed the presentation of the results in the form of a series of deformation profiles (ε_{yy}) plotted along the length of the sample at equal time intervals. With the help of this representation, it is possible to characterize the basic laws and kinetics of the spatial-time inhomogeneous fields. As an example, Figure 9 shows a series of ε_{yy} profiles for a specimen tested at the loading system stiffness of 120 MN/m. The time interval between the deformation profiles is 2 s. The origin of coordinates for the abscissa axis (Oy) is fixed in the center of the gauge length of the specimen. Under uniaxial tension of the aluminum-magnesium alloy, the effect of quasi-periodic homogenization of plastic deformation is observed, which consists of alternating stages of localization of the plastic flow when PLC bands appear and propagate, as well as stages of macroscopic leveling of longitudinal deformations on the surface of the cylindrical sample.

For the values of the loading system's stiffness of 120 MN/m and 50 MN/m, the jerky flow of type A is found. Figure 10 shows the development of longitudinal deformation fields at a stiffness of 120 MN/m, illustrating the formation and development of a single PLC band (marked with a white arrow).

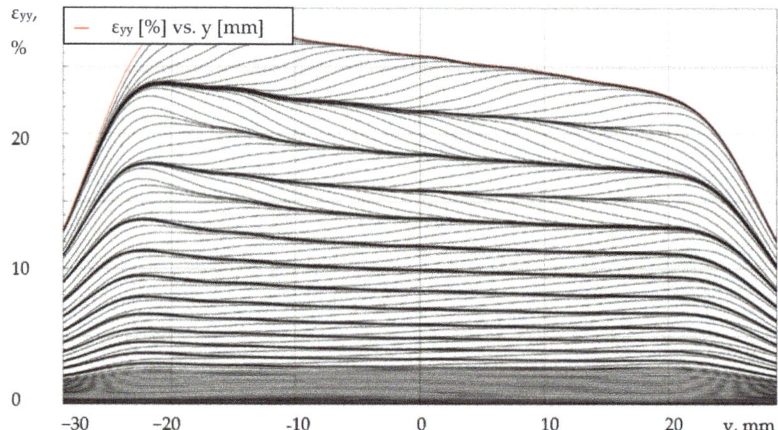

Figure 9. Series of plots of longitudinal deformations plotted at equal time intervals along the specimen length in the case of a stiffness of 120 MN/m.

We consider a series of longitudinal strain profiles (Figure 11) for a single strain band shown in Figure 9. Numbers 1–6 (Figure 11) denote the profiles of longitudinal deformations, for which Figure 10 shows the corresponding strain fields. The propagation of a single strain band proceeds uniformly from one capture to the opposite one with a constant speed of the band's front movement.

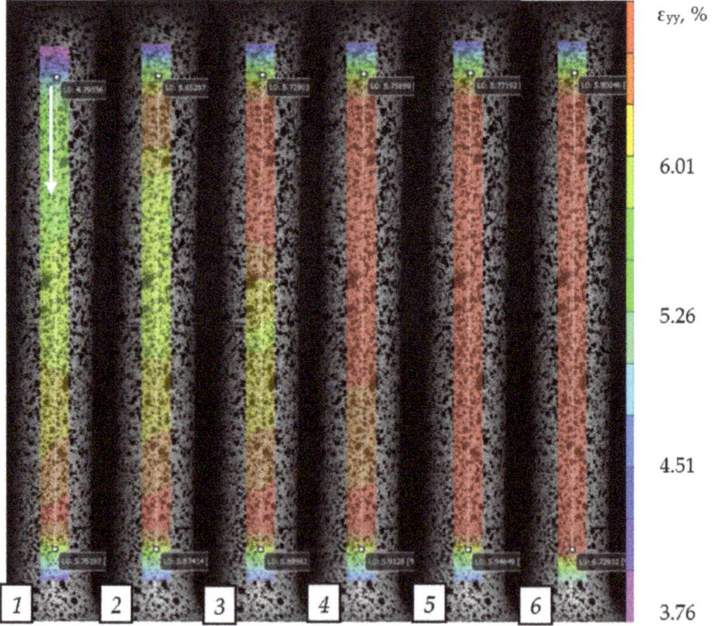

Figure 10. Evolution of the longitudinal strain fields on the surface of an Al-Mg alloy specimen in the case of a stiffness of 120 MN/m.

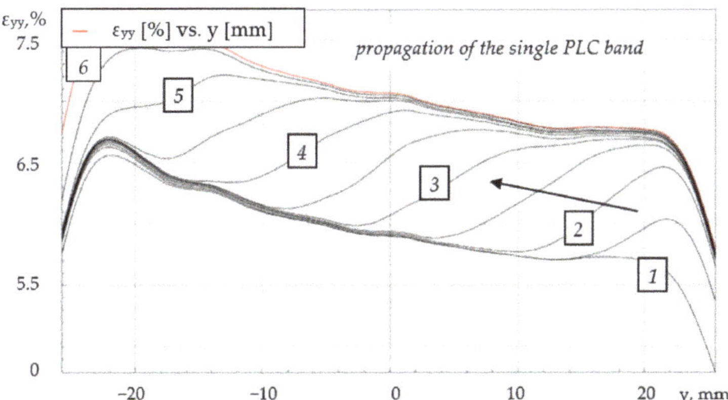

Figure 11. Series of plots of longitudinal deformations plotted at equal time intervals along the specimen length during the single PLC band propagation.

The deformation of the material is interrupted after the crossing of the front of a single PLC band. The process of active plastic deformation is concentrated in a small area, in the area of the strain band. To illustrate this process, data on the evolution of local rates of longitudinal strain ($d\varepsilon_{yy}$) are given (Figure 12).

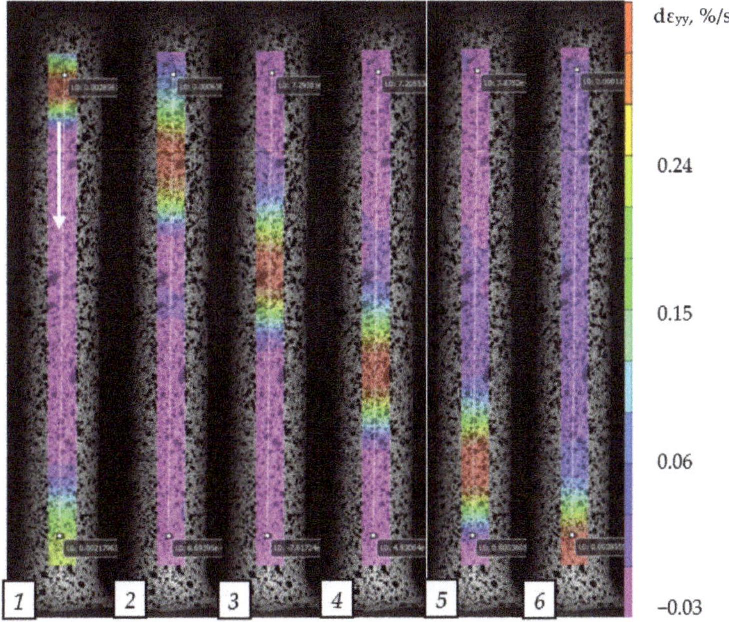

Figure 12. Evolution of local rates of the longitudinal strain fields on the surface of an Al-Mg alloy specimen in the case of a stiffness of 120 MN/m, numbers 1-6 corresponded to the plots of longitudinal deformations in the Figure 11.

In Figure 12, the white arrow indicates the direction of movement of the PLC band. Based on the obtained data, a series of profiles of local rates of longitudinal strain with equal time intervals was plotted as well (Figure 13).

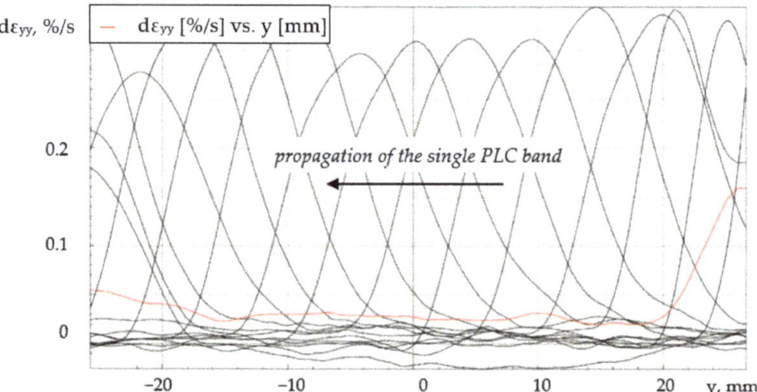

Figure 13. Series of plots of the local rate of longitudinal strain plotted at equal time intervals along the specimen length in the case of 120 MN/m.

Figure 14 shows the change in the maximum and minimum local rates of longitudinal deformation depending on the frame number. The values of the rates are given in the interval of frames during which the initiation and propagation of the PLC band occurred (Figure 12). At the moment when a single band occurs, the deformation rate of the material in the region of the band is 0.66%/s, while in the rest of the sample, elastic unloading is recorded (negative values of the strain rates). The formed strain band flows at a constant rate of deformation of the order of 0.32%/s.

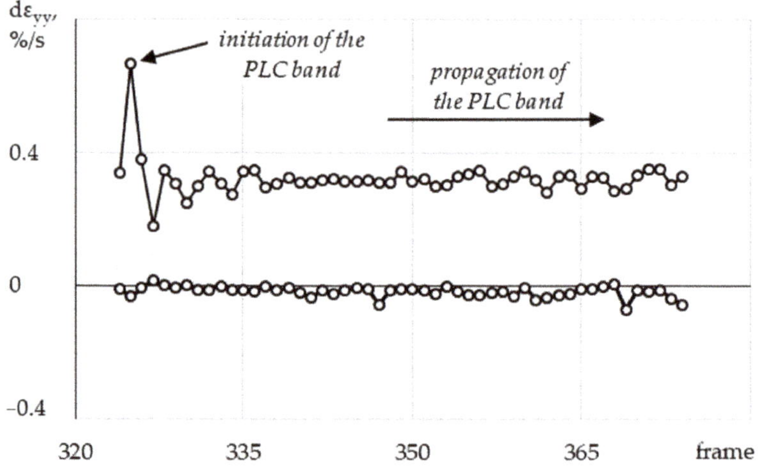

Figure 14. Frame dependence of maximum and minimum values of the local rate of longitudinal strain during the initiation and propagation of the single PLC band in the case of 120 MN/m.

A similar analysis of the spatio-temporal inhomogeneity and influence of the loading system's stiffness on the kinetics of PLC bands was carried out for the remaining samples from the test program presented in the work. When the stiffness is 50 MN/m, a series of longitudinal strain profiles (Figure 15) and a graph of maximum/minimum local rates of longitudinal strain (Figure 16) are similar to the results obtained when the stiffness is 120 MN/m. In the course of uniaxial tension, the manifestation of discontinuous flow type A is observed. At the moment when the single strain band occurs, the strain rate of the material in the region of the band front is slightly higher and amounts to about 0.74%/s. The advance of the formed band proceeds at a constant strain rate of 0.36%/s.

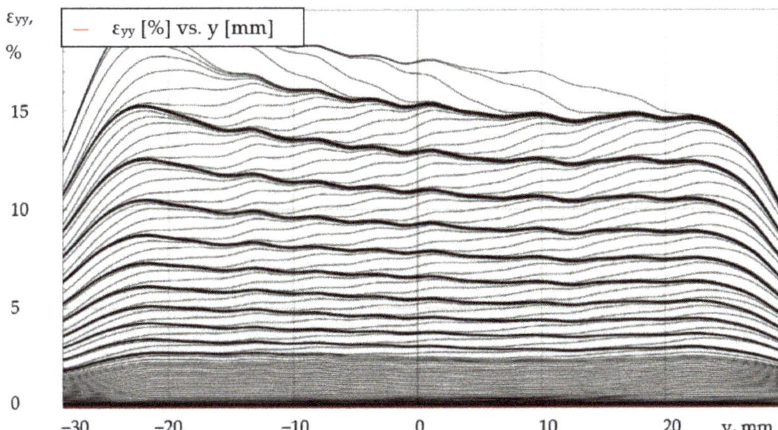

Figure 15. Series of plots of longitudinal deformations at equal time intervals along the specimen length in the case of stiffness 50 MN/m.

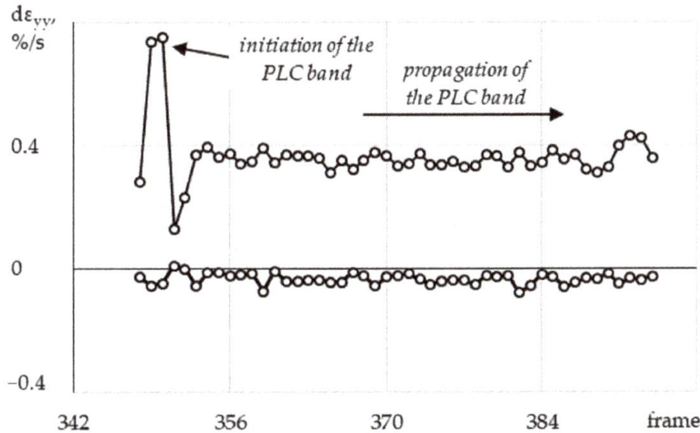

Figure 16. Frame dependence of the maximum and minimum values of the local rate of longitudinal strain during the initiation and propagation of the single PLC band in the case of 50 MN/m.

The reduction of the loading system's stiffness to 18 MN/m significantly affects the jerky flow. The frequency of load drops and their amplitude are higher. The effect of quasi-periodic homogenization of the plastic flow becomes less pronounced (Figure 17). The discontinuous flow type is mixed (A + B).

At times on the sample surface, the propagation of the single bands of the localized plastic flow can be recorded (Figure 18). Nevertheless, unlike the previous values of stiffness (120 MN/m and 50 MN/m), the movement of the strip along the length of the sample takes place at different speeds. It is of interest that the profiles designated by numbers 1–6 in Figure 19 bring inhomogeneous fields of local rates of longitudinal deformation (Figure 20) in order to estimate at what rate the material is deformed at the front of the band.

The deformation rate at the front of the strain band increases and takes values in the range from 0.39%/s to 1.33%/s (Figure 20). The deformation rate of the material when the PLC band appears exceeds the applied rate in the test (0.67%/s) due to an increase in the compliance of the loading system. Consequently, elastic unloading of the peripheral regions of the sample is observed at a rate of −0.14%/s.

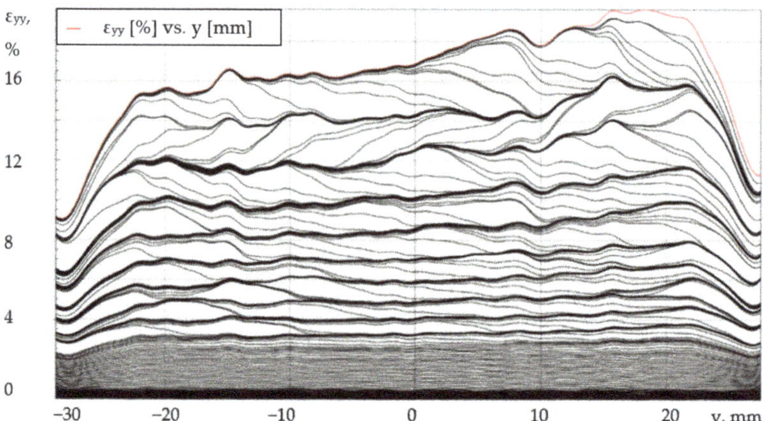

Figure 17. Series of plots of longitudinal deformations at equal time intervals along the specimen length in the case of a stiffness of 18 MN/m.

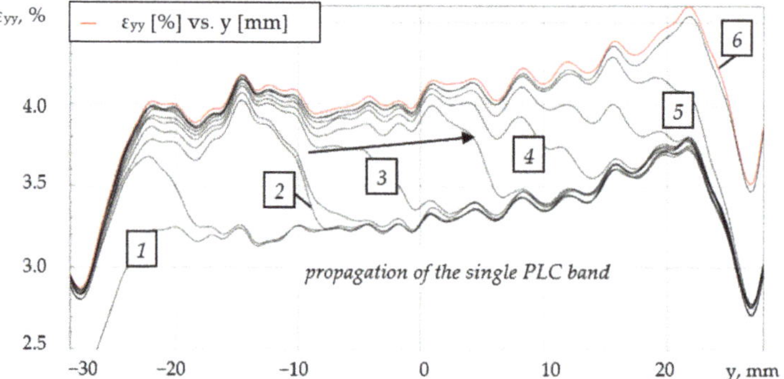

Figure 18. Series of plots of longitudinal deformations at equal time intervals along the specimen length during the single PLC band propagation.

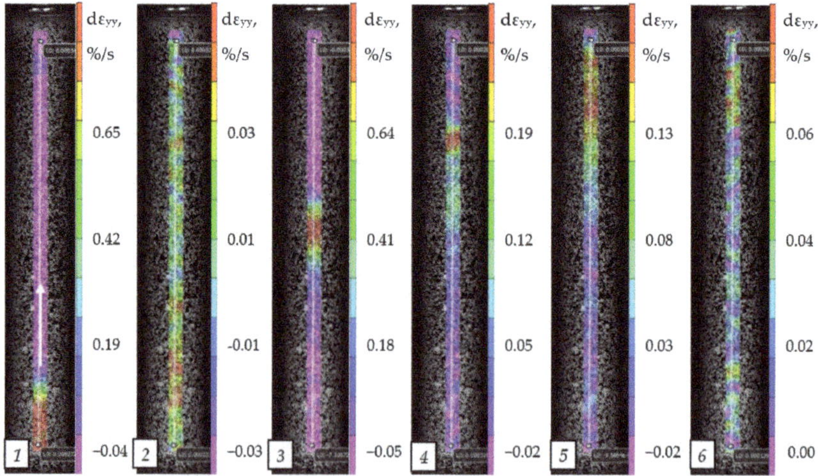

Figure 19. Evolution of local rates of the longitudinal strain fields on the surface of Al-Mg alloy specimen in case of stiffness 50 MN/m, numbers 1-6 corresponded to the plots of longitudinal deformations in the Figure 18.

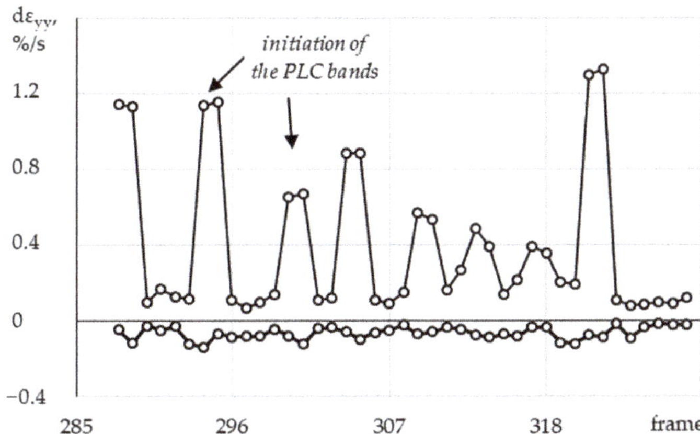

Figure 20. Frame dependence of maximum and minimum values of the local rate of the longitudinal strain during the initiation and propagation of the single PLC band in the case of 18 MN/m.

At the minimum level of stiffness of the loading system, implemented in this work, R_{LS} = 5 MN/m, load drops of a large amplitude are observed in the diagram, which corresponds to type C discontinuous flow. Figure 21 shows a series of longitudinal deformation profiles illustrating the kinetics of PLC bands. It should be noted that due to the high compliance of the loading system at the moment when the next PLC strip appears, a significant drop in the load is observed (Figure 7). Strain bands are formed randomly on the sample surface.

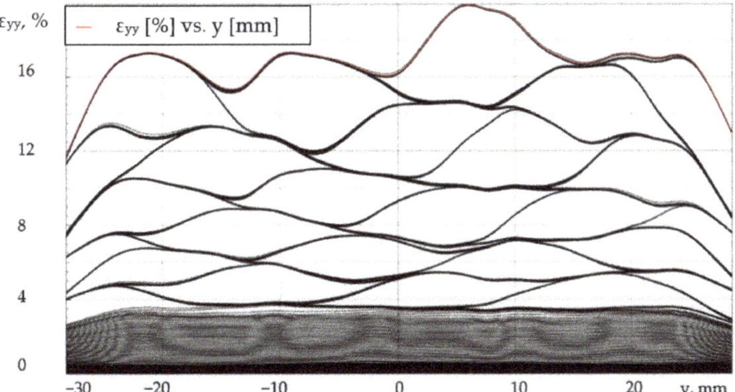

Figure 21. Series of plots of longitudinal deformations plotted at equal time intervals along the specimen length in the case of a stiffness of 5 MN/m.

For a more detailed analysis of the kinetics of the discontinuous flow at stiffness of 5 MN/n, a series of longitudinal deformation profiles was derived for a part of the loading (Figure 22). In this case, PLC stripes appear in the sample, but do not advance along the length. On the graph 'local rate of longitudinal deformation—frame' (Figure 23), one can distinguish a rather high intensity of the plastic flow at the front of the strip at the moment of its formation; values of the order of 3%/s are recorded.

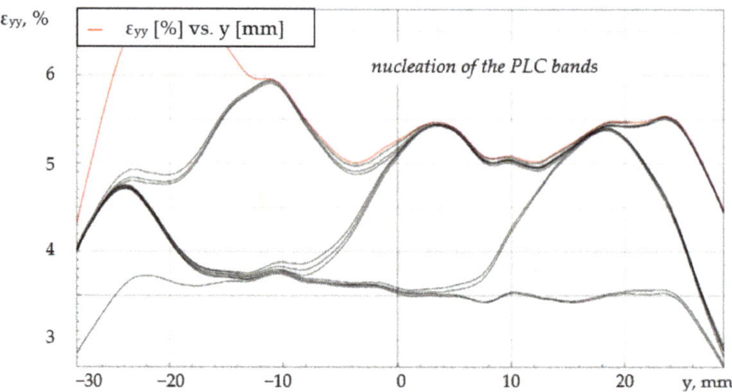

Figure 22. Series of plots of longitudinal deformations plotted at equal time intervals along the specimen length during the several PLC bands propagation.

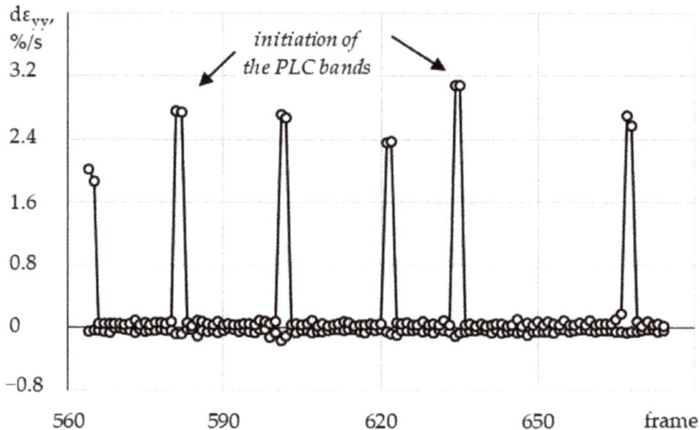

Figure 23. Frame dependence of maximum and minimum values of the local rate of longitudinal strain during the initiation and propagation of the single PLC band in the case of 5 MN/m.

4. Discussion

To estimate the degree of inhomogeneity of the process of macro-localization of the plastic flow under the Portevin-Le Chatelier effect, the coefficient of inhomogeneity of plastic deformation of the material (k_{PLC}) during loading is considered. The coefficient is equal to the ratio of the maximum value of longitudinal deformation (ε_{yy} (max)) to the average value of deformations (ε_{yy} (mean)) for each frame recorded by the video system (1):

$$k_{PLC} = \varepsilon_{yy}(\text{max}) / \varepsilon_{yy}(\text{mean}), \tag{1}$$

Based on the analysis of the evolution of inhomogeneous fields of longitudinal deformations, the time dependence of the k_{PLC} coefficient for the stage of material hardening, which is characterized by the presence of the effect of intermittent flow of the material, was constructed (Figure 24). For the stage of tooth formation and yield area, as well as for the stage of material softening, the change in k_{PLC} was not considered.

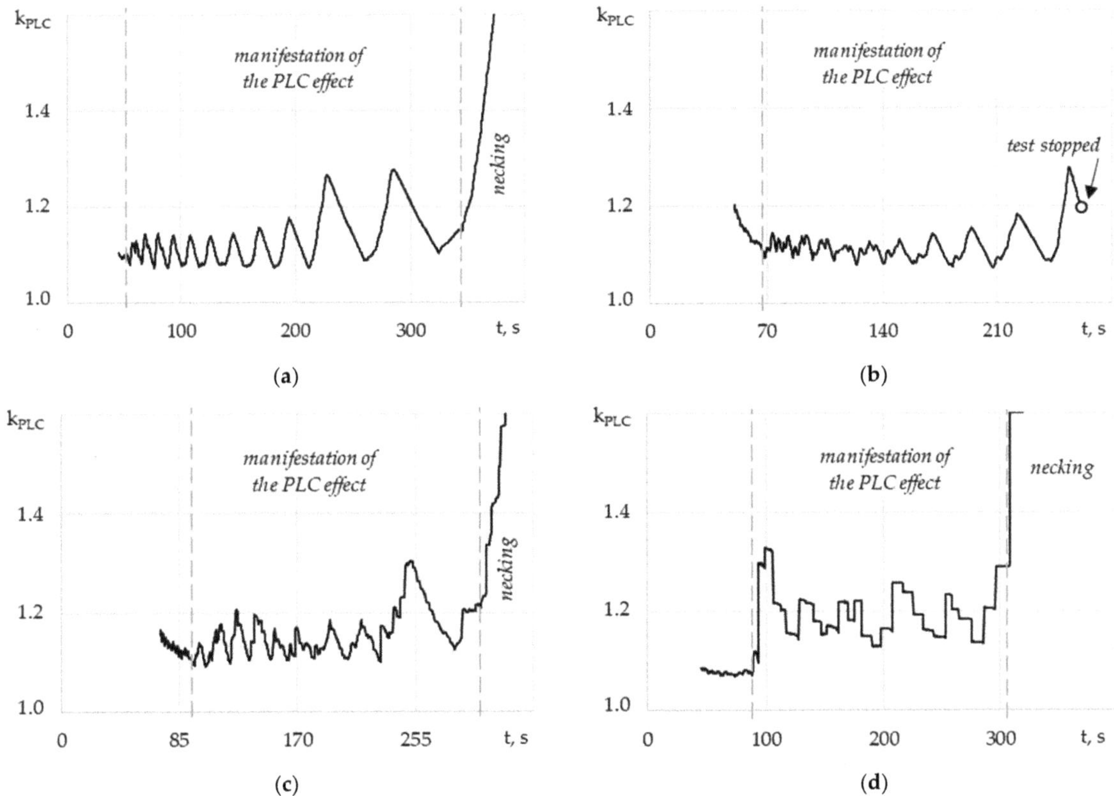

Figure 24. Time dependence of the coefficient of inhomogeneity of plastic deformation due to the Portevin-Le Chatelier effect in the case of the loading system's stiffness: 120 MN/m (**a**), 50 MN/m (**b**), 18 MN/m (**c**), and 5 MN/m (**d**), blue dotted line corresponds to beginning and ending of the manifestation of the PLC effect.

5. Conclusions

Thus, this work shows the high efficiency of using the specialized equipment to reduce the stiffness of the loading system in order to experimentally study the spatial-time inhomogeneity of the plastic flow of the aluminum-magnesium alloy. Mechanical tests for uniaxial tension of standard cylindrical specimens in the range with the stiffness of the loading system equal to 120 MN/m, 50 MN/m, 18 MN/m, and 5 MN/m have been implemented. We carried out the analysis of the kinetics of the occurrence and propagation of the deformation bands of the localized plastic flow under the Portevin-Le Chatelier effect using the digital image correlation method. Experimental data have been obtained that illustrate how the properties of the loading system affect the unstable plastic flow.

It is shown that despite the constant chemical composition of the material, external loading factors (temperature, stretching rate), the type of discontinuous yield depends on the value of the loading chain compliance. At higher values of the loading system's stiffness (120 MN/m and 50 MN/m), the initiation of single PLC bands and their uniform distribution along the length of the sample (type A) are recorded on the material surface. With an increase in the compliance of the loading system, the intermittent fluidity passes to types B and C. On the loading curves, the amplitude of the load disruptions increases, which is accompanied by plastic deformation of the material in the area of the front of the PLC bands and the unloading of elastic deformations in the rest of the sample's volume (outside the front of the strip). Reducing the stiffness of the loading system leads to multiple increases in the maximum values of the local rate of longitudinal strain during the initiation and propagation of the single PLC band. At high loading stiffness (120 MN/m), the maximum values of the local rate of longitudinal strain during the initiation of the single PLC band were 0.69%/s, at a loading stiffness of 50 MN/m–0.76%/s, at a loading stiffness of 18 MN/m–1.33%/s, and at a minimum loading stiffness (5 MN/m)–3.11%/s. In this case, the minimum values of the local rate of longitudinal strain during the propagation of the single PLC band decreased from 0.39%/s to 0.1%/s. These results are consistent with the experimental stretching diagrams obtained. Thus, at the maximum loading stiffness (120 MN/m), the diagrams show extended sections between single teeth, which correspond to the stage of propagation of the single PLC band. At the minimum loading stiffness (5 MN/m), the diagrams show successive teeth that correspond to the initiation of the single PLC band. This kind of research is relevant and requires further comprehensive theoretical and experimental studies and the implementation of a wider range of properties of the loading system.

The practical significance of the work is determined by the fact that the manifestation of the effects of intermittent flow in Al-Mg alloys leads to a significant decrease in the strength and plasticity of material in structures. The processes of strip formation and spontaneous macroscopic localization of plastic flow lead to the appearance of different thicknesses of structural elements, a decrease in the surface quality of parts, the appearance of concentrators, defects, and, as a result, to subsequent destruction or violation of the structural strength of parts. The study of the influence of the stiffness of the loading system on the of the Portevin-Le Chatelier effects has not been previously considered. The presented work shows that the stiffness of the loading system, as well as other parameters (deformation rate, temperature, chemical composition), has a direct effect on the kinetics of initiation and propagation of PLC deformation bands and the type of intermittent fluidity. Therefore, the loading stiffness must be considered along with other parameters when designing structures and planning experimental research programs that use materials that exhibit of the Portevin-Le Chatelier effects.

Author Contributions: Conceptualization, T.T.; methodology, M.T. and T.T.; software, T.T.; validation, T.T. and M.T.; formal analysis, T.T.; investigation, T.T. and M.T.; resources, T.T.; data curation, T.T. and M.T.; writing—original draft preparation, T.T.; writing—review and editing, T.T. and M.T.; visualization, T.T.; supervision, T.T.; project administration, T.T.; funding acquisition, T.T. All authors have read and agreed to the published version of the manuscript.

Funding: This research was funded by the Russian Science Foundation (No. 20-79-10235).

Data Availability Statement: Not applicable.

Acknowledgments: The work was carried out by using the capabilities of a large unique scientific facility (UNU) 'A set of test and diagnostic equipment for studying the properties of structural and functional materials under complex thermomechanical influences', which is part of the Center of Experimental Mechanics, PNRPU (Perm, Russia).

Conflicts of Interest: The authors declare no conflict of interest.

References

1. Tretyakova, T.V.; Wildemann, V.E. Plastic Strain Localization and Its Stages in Al–Mg Alloys. *Phys. Mesomech.* **2018**, *21*, 314–319. [CrossRef]
2. Bell, J.F. *Eksperimental'nye Osnovy Mekhaniki Deformiruemykh Tverdykh tel. Ch.2. Konechnye Deformatsii [The Experimental Foundations of Solid Mechanics. Part 2. Finite Deformation]*; Nauka: Moscow, Russia, 1984; 432p.
3. Klueh, R.L. Discontinuous creep in short-range order alloys. *Mater. Sci. Eng.* **1982**, *54*, 65–80. [CrossRef]
4. Yilmaz, A. The Portevin-Le Chatelier effect: A review of experimental findings. *Sci. Technol. Adv. Mater* **2011**, *12*, 063001. [CrossRef] [PubMed]
5. Aguirre, F.; Kyriakides, S.; Yun, H.D. Bending of steel tubes with Lüders bands. *Int. J. Plast.* **2004**, *20*, 1199–1225. [CrossRef]
6. Tretyakova, T.V.; Wildemann, V.E. *Spatial-Time Inhomogeneity of the Processes of Inelastic Deformation of Metals*; Fizmatlit: Moscow, Russia, 2016; p. 120.

7. Trusov, P.V.; Chechulina, E.A. Serrated yielding: Physical mechanisms, experimental dates, macro-phenomenological models. *PNRPU Mech. Bull.* **2014**, *3*, 186–232.

8. Tretyakova, T.V.; Wildemann, V.E. Experimental study of the influence of strain-stress state on the jerky flow in metals and alloys. *Procedia Struct. Integr.* **2019**, *17*, 906–913. [CrossRef]

9. Vildeman, V.E.; Sokolkin, Y.V.; Tashkinov, A.A. *Mechanics of Inelastic Deformation and Fracture of Composite Materials*; Nauka: Moscow, Russia, 1997; p. 288.

10. Wildemann, V.E.; Lomakin, E.V.; Tretyakov, M.P.; Tretyakova, T.V.; Lobanov, D.S. *Experimental Studies of Postcritical Deformation and Fracture of Structural Materials*; PNRPU: Perm, Russia, 2018; 156p.

11. Tretyakov, M.P.; Wildemann, V.E.; Lomakin, E.V. Failure of materials on the postcritical deformation stage at different types of the stress-strain state. *Procedia Struct. Integr.* **2016**, *2*, 3721–3726. [CrossRef]

12. Tretyakov, M.P.; Tretyakova, T.V.; Wildemann, V.E. Regularities of mechanical behavior of steel 40Cr during the postcritical deformation of specimens in condition of necking effect at tension. *Frat. Ed Integrità Strutt.* **2018**, *43*, 145–153. [CrossRef]

13. Vildeman, V.E.; Tretyakov, M.P. Analysis of the effect of loading system rigidity on postcritical material strain. *J. Mach. Manuf. Reliab.* **2013**, *42*, 219–226. [CrossRef]

14. Wildemann, V.E.; Lomakin, E.V.; Tretyakov, M.P. Effect of vibration stabilization of the process of postcritical deformation. *Dokl. Phys.* **2016**, *61*, 147–151. [CrossRef]

15. Bazant, Z.P.; Di Luizo, G. Nonlocal microplane model with strain-softening yield limits. *Int. J. Solids Struct.* **2004**, *41*, 7209–7240. [CrossRef]

16. Mansouri, L.Z.; Coër, J.; Thuillier, S.; Laurent, H.; Manach, P.Y. Investigation of Portevin-Le Châtelier effect during Erichsen test. *Int. J. Mater. Form.* **2020**, *13*, 687–697. [CrossRef]

17. Rousselier, G.; Morgeneyer, T.F.; Ren, S.; Mazière, M.; Forest, S. Interaction of the Portevin–Le Chatelier phenomenon with ductile fracture of a thin aluminum CT specimen: Experiments and simulations. *Int. J. Fract.* **2017**, *206*, 95–122. [CrossRef]

18. Dahdouh, S.; Mehenni, M.; Ait-Amokhtar, H. Kinetics of formation and propagation of type A Portevin-Le Chatelier bands in the presence of a small circular hole. *J. Alloy. Compd.* **2021**, *885*, 160982. [CrossRef]

19. Choi, Y.; Ha, J.; Lee, M.G.; Korkolis, Y.P. Observation of Portevin-le Chatelier effect in aluminum alloy 7075-w under a heterogeneous stress field. *Scr. Mater.* **2021**, *205*, 114178. [CrossRef]

20. Ren, S.C.; Morgeneyer, T.F.; Mazière, M.; Forest, S.; Rousselier, G. Effect of Lüders and Portevin–Le Chatelier localization bands on plasticity and fracture of notched steel specimens studied by DIC and FE simulations. *Int. J. Plast.* **2021**, *136*, 102880. [CrossRef]

21. Ren, S.C.; Morgeneyer, T.F.; Mazière, M.; Forest, S.; Rousselier, G. Portevin-Le Chatelier effect triggered by complex loading paths in an Al–Cu aluminium alloy. *Philos. Mag.* **2019**, *99*, 659–678. [CrossRef]

22. Le Cam, J.B.; Robin, E.; Leotoing, L.; Guines, D. Calorific signature of PLC bands under biaxial loading conditions in Al-Mg alloys. *Residual Stress Infrared Imaging Hybrid Tech. Inverse Probl.* **2018**, *8*, 29–35. [CrossRef]

23. Skripnyak, V.V.; Skripnyak, V.A. Localization of Plastic Deformation in Ti-6Al-4V. *Alloy Met.* **2021**, *11*, 1745. [CrossRef]

24. Brünig, M.; Gerke, S.; Koirala, S. Biaxial Experiments and Numerical Analysis on Stress-State-Dependent Damage and Failure Behavior of the Anisotropic Aluminum Alloy EN AW-2017A. *Metals* **2021**, *11*, 1214. [CrossRef]

25. Chen, Y.; Ji, C.; Zhang, C.; Sun, S. The Application of DIC Technique to Evaluate Residual Tensile Strength of Aluminum Alloy Plates with Multi-Site Damage of Collinear and Non-Collinear Cracks. *Metals* **2019**, *9*, 118. [CrossRef]

26. Mäkinen, T.; Ovaska, M.; Laurson, L.; Alava, M.J. Portevin–Le Chatelier effect: Modeling the deformation bands and stress-strain curves. *Mater Theory* **2022**, *6*, 15. [CrossRef]

27. Xu, J.; Holmedal, B.; Hopperstad, O.S.; Maník, T.; Marthinsen, K. Dynamic strain ageing in an AlMg alloy at different strain rates and temperatures: Experiments and constitutive modelling. *Int. J. Plast.* **2022**, *151*, 103215. [CrossRef]

Communication

Dendritic Solidification and Physical Properties of Co-4.54%Sn Alloy with Broad Mushy Zone

Weili Wang *, Wenhui Li and Ao Wang

Department of Applied Physics, Northwestern Polytechnical University, Xi'an 710072, China
* Correspondence: wlwang@nwpu.edu.cn

Abstract: The high undercooling of binary Co-4.54%Sn alloy has a significant influence on its microstructural characteristics and physical properties. Here, we report that the dendritic growth and physical properties of broad-temperature-range Co-4.54%Sn alloy remarkably depends on the undercooling during the rapid solidification. The maximum undercooling attains 208 K at molten state, and the dendritic growth velocity is quite sluggish in highly undercooled liquid Co-4.54%Sn alloy because it has a broad solidification range of 375 K (0.21 T_L); the maximum value is only 0.95 m/s at the undercooling of 175 K, which then decreases with undercooling. The microstructure refines visibly and the volume fraction of the interdendritic βCo_3Sn_2 phase obviously decreases with undercooling. The microhardness and electrical resistivity increase with undercooling owing to the enhancement of solute content of the primary αCo phase and refinement of the microstructure where the increased crystal boundary hinders the electronic transmission. Meanwhile, the saturation magnetization also reduces with undercooling due to the crystal particle and boundary increasing significantly, and the dendritic growth velocity and solute content increase in the primary αCo phase under rapid solidification.

Keywords: rapid solidification; dendritic growth; microhardness; electrical resistivity; magnetic property

---

Citation: Wang, W.; Li, W.; Wang, A. Dendritic Solidification and Physical Properties of Co-4.54%Sn Alloy with Broad Mushy Zone. *Metals* **2023**, *13*, 1046. https://doi.org/10.3390/met13061046

Academic Editor: Maciej Motyka

Received: 17 February 2023
Revised: 17 April 2023
Accepted: 28 April 2023
Published: 30 May 2023

1. Introduction

The broad solidification temperature range is the temperature range of a solid–liquid two-phase mixture which is larger than 0.2 T_L, in which the alloy ingot is in a semisolid state for this liquid/solid range during the solidification; this range is called the broad mushy zone [1,2]. For a binary single-phase alloy, a solidification nucleates from the liquid to solid phase over a broad temperature interval, and then grows rapidly and forms a single-phase microstructure for a considerable time in this mushy zone [2]. Therefore, the phenomenon of solidification shrinkage and hot tearing is caused by thermal contraction, as the higher density of the solid driving the cracking process and hindered liquid flow prevents any void from being filled with liquid. Several defects, such as low fluidity, shrinkage porosity, heat cracking, segregation, and so on, take place inside this interval [1–5]. The rapid solidification of high undercooling is an effective method to avoid or eliminate these defects.

The rapid solidification of metallic alloys under high undercooling has the advantages of extended solubility, refined microstructures and the formation of a metastable phase [6–11]; thus, it can reduce defects, effectively improve microstructure and achieve excellent physical properties. Some binary single-phase alloys with a broad solidification temperature range show special phenomena during the rapid solidification of high undercooling; for example, dendritic growth velocity decreases when the undercooling exceeds a critical value, and the final microstructure is vermicular dendrite instead of equiaxed grains. Moreover, the undercooling and cooling rates have significant influence on the microstructure, microhardness and electrical resistivity properties during the rapid solidification [12,13].

Co-Sn alloys have been widely studied in recent years owing to their application properties; for example, the production of Co-Sn-based metallic glasses and anode materials in lithium ion batteries [14–16]. In addition, the investigation of Co-Sn alloys have also focused on the structural transformation of different Sn content or eutectic alloys [17], physical properties of the viscous flow, surface tension and enthalpy of mixing near the eutectic concentration [18,19]. However, there are scarce studies on the structural evolution, dendritic growth and physical properties of binary Co-Sn alloys with a broad solidification temperature range and rapid solidification at high undercooling.

In this work, we selected a special composition of Co-4.54%Sn alloy with a broad solidification temperature range of 375 K (0.21 T_L) to explore the microstructural evolution and dendritic growth characteristics under substantial conditions of undercooling. Moreover, the physical properties of the microhardness, electrical resistivity and magnetic characteristics are investigated at different undercoolings.

2. Experimental Procedure

The rapid solidification experiments were performed using the glass fluxing method with a high vacuum chamber. The master alloy of binary Co-4.54%Sn alloy composition was prepared using the arc-melting technique from component metals of high purity above 99.999%. An in situ alloying procedure was applied to prepare about 1 g Co-4.54%Sn alloy using radio frequency induction heating. The sample was contained in an 8 mm ID × 10 mm OD × 12 mm alumina crucible and placed into the experimental chamber, then it was evacuated to a 2×10^{-4} Pa vacuum with a turbo pump and subsequently backfilled with argon gas to 10^5 Pa. The sample was superheated to 250~350 K above its liquidus temperature and kept submerged in a pool of molten fluxing agent for 3~5 min. It was naturally cooled down and solidified by switching off the induction heating power. Each sample was subjected to the melting–solidifying cycle 3~5 times. Their heating and cooling curves were recorded using a Land NQO8/15C infrared pyrometer (Sensortherm Gmbh, Steinbach, Germany), while the dendrite growth velocity was measured with an infrared photodiode device (Mikrotron, Unterschleißheim, Germany). The phase constitutions of binary Co-4.54%Sn alloy were analyzed by Rigaku D/max 2500 X-ray diffractometer (XRD) (Rigaku, Tokyo, Japan), whereas their solidification structures and solute distribution profiles were investigated with FEI Sirion electron microscope (SEM) (FEI Sirion, the Netherlands) and INCA Energy 300 energy-dispersive spectrometer (EDS) (FEI Sirion, the Netherlands). The microhardness was measured using an HXD-2000TMC/LCD Vickers hardness tester (Kexin, Beijing, China). The magnetic characteristics was analyzed using a VSM instrument. The resistivity was tested using the four-point probe method.

3. Results and Discussion

3.1. Rapid Solidification Structures

The phase constitution of Co-4.54%Sn alloy was selected with the largest solid solubility point of Sn solute in Co solvent; as shown in the binary Co-Sn phase diagram of Figure 1a,b, the liquidus and solidus temperatures were 1760 and 1385 K, respectively, and the solidification temperature range attained 375 K (0.21 T_L). From the equilibrium phase diagram, it can be seen that the primary face-centered cubic αCo solid solute phase formed in the liquid phase when the temperature reduced to 1760 K; firstly, during the solidification, the solute content was about 0.56 wt.%Sn in the αCo phase, and then the βCo$_3$Sn$_2$ intermetallic phase was generated as the temperature lowered to about 1443 K. Subsequently, the close-packed hexagonal εCo phase had the potential to form once the temperature decreased to about 695 K. Furthermore, the X-ray diffraction patterns showed three phases of αCo, βCo$_3$Sn$_2$ and εCo displayed at the different undercoolings under the rapid solidification, which corresponds with the equilibrium phase diagram. The microstructures of binary Co-4.54%Sn alloy relate to the undercooling at rapid solidification, which are shown in Figure 2a–d. At the small undercooling of 11 K, the microstructure of the primary αCo phase was characterized by the coarse and well-developed dendrites, the

βCo_3Sn_2 phase distributed into the interdendrites, shown as the white phase and the εCo phase displayed the dark structures, as seen in Figure 2a. Subsequently, the microstructure was remarkably refined and the dendritic feature disappeared gradually; the volume fraction of βCo_3Sn_2 phase decreased at the same time, but the volume fraction of the εCo phase increased with the increase of undercooling. Ultimately, the microstructure of the primary αCo phase grew with the disorder at the largest undercooling of 189 K, and a small number of βCo_3Sn_2 particles and a large number of εCo phase distributed randomly into the matrix of the primary αCo phase. Clearly, the phase constitution has not changed at the rapidly solidified condition. It is worth noting that the eutectic-like microstructure exists in the whole undercooling of the binary Co-4.54%Sn alloy, as seen in Figure 3a,b, which grew around the βCo_3Sn_2 phase. Meanwhile, the eutectic-like structure appears coarsening and the volume fraction increased obviously with increase of undercooling and decrease of the βCo_3Sn_2 volume fraction, and formed eutectic cell characteristics at the largest undercooling of 189 K.

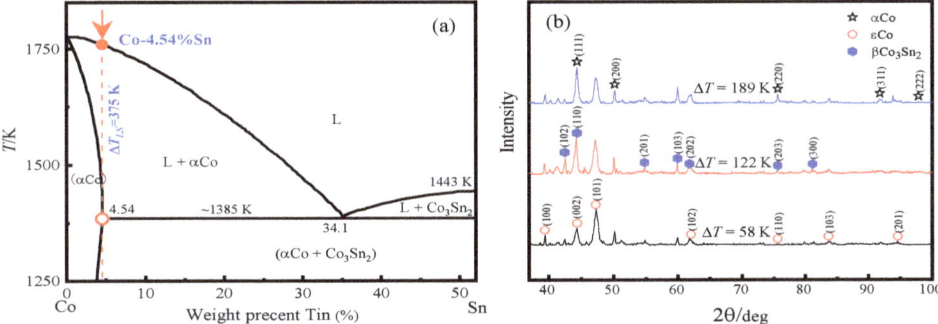

Figure 1. Selection of alloy composition and X-ray diffraction of rapidly solidified alloy droplets. (**a**) Co-4.54%Sn alloy location in phase diagram and (**b**) XRD pattern.

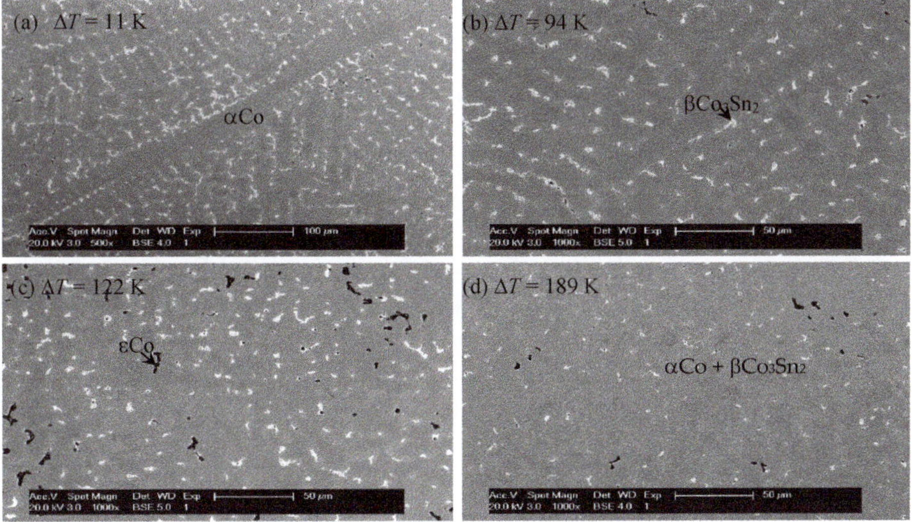

Figure 2. Microstructures at different undercoolings, (**a**) 11 K, (**b**) 94 K, (**c**) 122 K, and (**d**) 189 K.

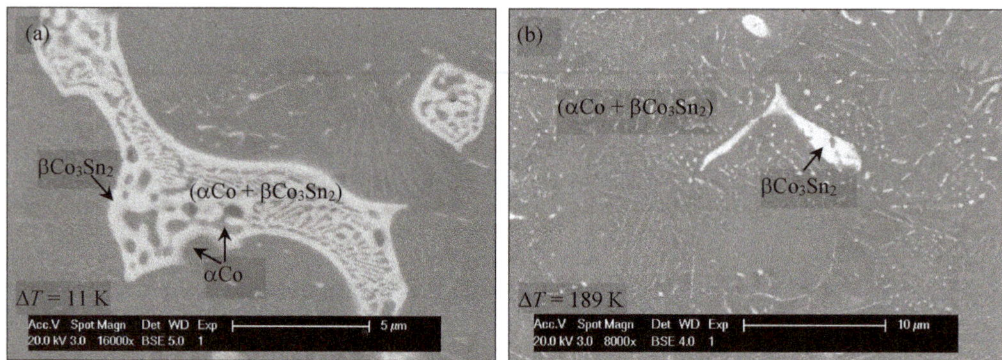

Figure 3. Eutectic structures within interdendritic spacings, (**a**) ΔT = 11 K, and (**b**) ΔT = 189 K.

3.2. Dendritic Growth Kinetics

The dendritic growth velocity is an important feature under the rapid solidification condition, which has an influence on the microstructure and physical properties of binary Co-4.54%Sn alloy. Figure 4a displays the measured and calculated growth velocity of the primary αCo phase, which presents two characteristics. At first, the dendritic growth velocity of primary αCo phase increased with the increase of undercooling, and a maximum velocity of 0.95 m/s was obtained at 175 K undercooling, which demonstrates a sluggish tendency for this alloy. However, the dendritic growth velocity slowed down once the undercooling exceeded 175 K and the dendrite decreased to 0.85 m/s growth velocity at the largest undercooling of 208 K. A Gaussian function was derived by fitting the actual dendrite growth versus undercooling to

$$V = 1.37 \times 10^3 \exp\left(-\frac{1.71 \times 10^{-21}}{k_B \cdot \Delta T}\right) \exp\left(-\frac{-1.14 \times 10^{-19}}{k_B \cdot T}\right) \tag{1}$$

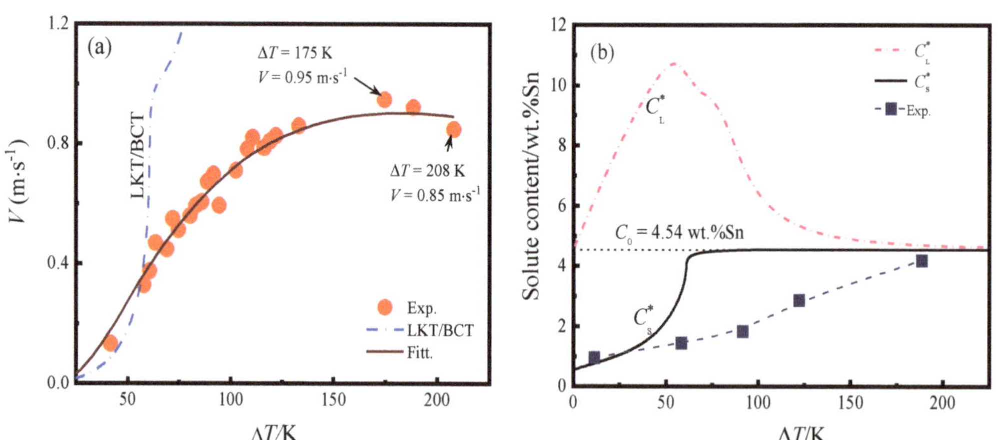

Figure 4. Measured and calculated dendritic growth velocity and solute content of rapidly solidified binary. Co-4.54%Sn alloy versus undercoolings, (**a**) dendritic growth velocity, and (**b**) solute content.

The LKT model for dendritic growth developed by Lipton, Kurz and Trivedi [20], which was further extended by Boettinger, Coriell and Trivedi (BCT) [21], has proven to be the most competent model for predicting dendritic growth characteristics with linear liquidus and solidus lines during rapid solidification. Cao et al. [22] modified the LKT/BCT dendritic growth theory to be applicable to predicting the kinetic characteristics of dendritic growth in the alloy systems which have extremely curved liquidus and solidus lines. This model is comprised by the following seven principal equations:

$$\Delta T = \Delta T_t + \Delta T_c + \Delta T_r + \Delta T_k \tag{2}$$

$$R = \frac{\Gamma}{\sigma^*}\left[\frac{\Delta H}{C_{PL}}P_t\xi_t - \frac{2m'_L C_0(1-k_V)P_c}{1-(1-k_v)I_v(P_c)\xi_c}\right]^{-1} \tag{3}$$

ΔT: bulk undercooling; ΔT_t: thermal undercooling; ΔT_c: solutal undercooling; ΔT_r: curvature undercooling; ΔT_k: kinetic undercooling; V: dendritic growth velocity; R: the dendrite tip radius; ΔH: heat of fusion; C_{PL}: specific heat of the liquid phase; P_t and P_c: thermal and solutal Peclet numbers; ξ_t and ξ_c: thermal and solutal stability functions; m'_L: actual liquidus slope; C_0: alloy composition; k_v: actual solute partition coefficient; $I_v(P_c)$: solutal Ivantsov function; Γ: Gibbs–Thomson coefficient; and σ^*: the stability constant equal to $1/(4\pi^2)$.

From the LKT/BCT dendritic growth model, it can be seen that the dendritic growth velocity increased remarkably when the undercooling was larger than 25 K, and was 1~3 orders of magnitude larger than the actual dendritic growth velocity at the undercooling of 58~208 K. Therefore, the actual dendritic growth velocity was quite sluggish under the rapid solidification. According to the theoretical calculation of the LKT/BCT model, the dendritic growth is mainly controlled by the solutal diffusion if the undercooling is smaller than 91 K, which corresponds with the slower velocity. Subsequently, the effect of thermal diffusion becomes an increasingly important factor and finally replaces solutal diffusion as the dominant controlling factor once the undercooling exceeds 91 K.

The solute content of primary αCo phase was detected using the EDS method, and the results show the increasing tendency of undercooling, as seen in Figure 4b. For example, the solute content C_S^* was 0.96 wt.%Sn when the undercooling ΔT was 11 K, which is the smallest level of undercooling acquired by the experiment. If the undercooling ΔT attained the largest undercooling 189 K of the experimental sample, the solute content also achieved 4.19 wt.%. According to the equilibrium diagram of Figure 1a,b, the solute content of primary αCo phase was only 0.56 wt.%Sn once the temperature decreased to 1760 K and the primary αCo phase began to grow from the liquid phase. Obviously, the solute trapping takes place during the rapid solidification. From Figure 2, it can be seen that the dendritic structure of the primary αCo phase refined remarkably with the increase of undercooling caused by the strong recalescence, the volume fraction of the βCo$_3$Sn$_2$ phase decreased apparently at the same time and a large number of Sn solutes diffused in the solvent interstitial of the primary αCo phase and led to the higher solute content. The variations of the liquid concentration C_L^* and the solid concentration C_S^* at the liquid/solid interface with undercooling were calculated using the LKT/BCT model, as illustrated in Figure 4b; the experimental value was close to the calculated C_S^* result at the largest undercooling of 189 K. Apparently, the segregationless solidification may occur once the undercooling exceeds 189 K for binary Co-4.54%Sn alloy at t rapid solidification.

3.3. Microhardness and Electrical Resistivity

The microhardness of the binary Co-4.54%Sn alloy is illustrated in Figure 5a. As the undercooling was small, ΔT = 11 K, the microhardness H_v was 228.6 HV, and then increased with undercooling. The maximum microhardness attained was 335.5 HV where the undercooling ΔT was 189 K, which is 1.47 times larger than the value of 11 K undercooling. The liner fitting function at different undercoolings is demonstrated as:

$$H_v = 221.13 + 0.65\Delta T \tag{4}$$

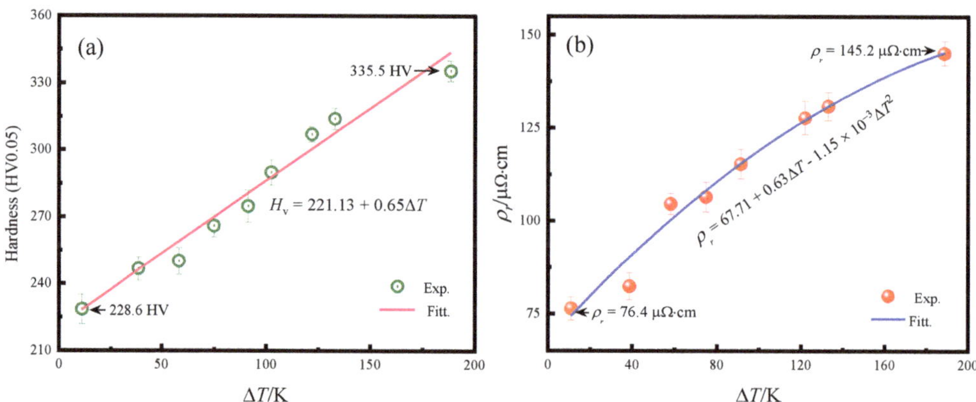

Figure 5. Microhardness and resistivity of Co-4.54%Sn alloy at different undercoolings, (**a**) microhardness and (**b**) resistivity.

Apparently, undercooling has a significant influence on microhardness. From the analysis above, it can be seen that large undercooling had more obvious recalescence, which led to the refined microstructure, and also enhanced the dendritic growth velocity and solute content. Therefore, the refined structure had increased grain boundary and larger solute content, which improved the microhardness under rapid solidification [23]. In addition, the microstructure also affected the hardness of Co-4.54%Sn alloy. Similarly, the undercooling also prompted the resistivity of binary Co-4.54%Sn alloy to rise, as displayed in Figure 5b. To describe the relationship between resistivity ρ_r and undercoolings ΔT, an expression is proposed:

$$\rho_r = 67.71 + 0.63\Delta T - 1.15 \times 10^{-3}\Delta T^2 \tag{5}$$

At the small undercooling of 11 K, the electrical resistivity ρ_r was only 76.4 μΩ·cm. The electrical resistivity grew rapidly with the increase of undercooling, and the largest electrical resistivity ρ_r of 145.2 μΩ·cm was obtained at the undercooling of 189 K, which is 1.9 times larger than that of 11 K. There are two reasons for this phenomenon. According to the classical nucleation energy definition, the nucleation energy of melt varies inversely with the square of the undercooling. Clearly, the nucleation energy decreases rapidly with the increase of undercooling, which enhances the number of crystal nuclei and refines the microstructure obviously. The growing crystal boundary hinders the electronic transmission, resulting in the electrical resistivity increasing. On the other hand, the reciprocal of relaxation time for lattice scattering caused by the impurity defect is proportional to the content of impurity concentration. Since the lattice distortion will cause strong electron scattering and reduce the mobility of free electrons, the more serious the lattice distortion of the total solute matrix in the constituent phase, the higher the resistivity of the alloy [24]. As shown in Figure 4b, the solute content of the primary (αCo) phase enhances with the increase of undercooling, which generates the lattice distortion influence significant on the electronic transmission, thus improving the electrical resistivity.

3.4. Magnetic Properties

The magnetic characteristics of the Co-4.54%Sn alloy were measured at different undercoolings, as demonstrated in Figure 6. The Co-4.54%Sn alloy shows typical soft magnetic features (Figure 6a). Figure 6b describes the process of domain growth and magnetization rotation under an external magnetic field. The magnetic domains parallel to the applied magnetic field will grow, and the remaining magnetic domains will shrink. With the increase of magnetic field H, B increases rapidly, and magnetic domain growth takes the way of magnetic domain wall movement. When the external magnetic field is increased, the magnetic rotation begins, and the slope of the curve of *B* relative to *H* becomes smaller. Finally, the magnetic moment direction of each magnetic domain remains horizontal with the direction of the external magnetic field, and the *B* value does not change.

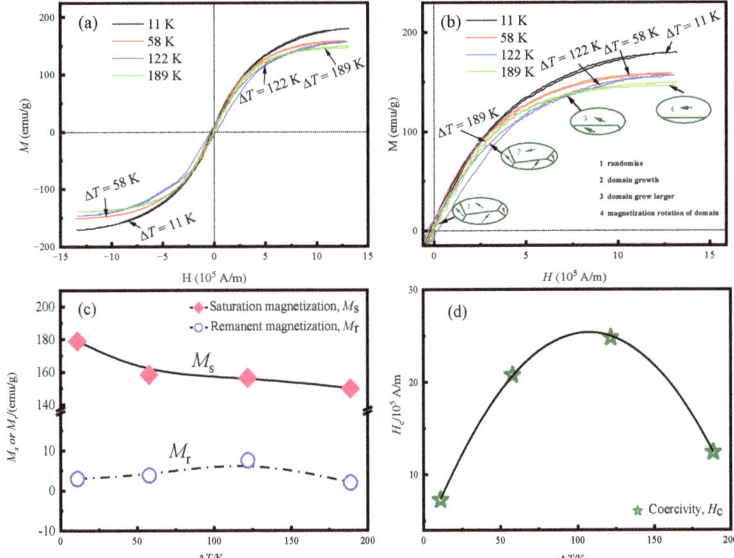

Figure 6. Magnetic characteristics of Co-4.54%Sn alloy solidified at different undercoolings. (**a**) Hysteresis loop, (**b**) magnetic domain growth and rotation process, (**c**) Ms and Mr and (**d**) coercivity Hc.

The saturation magnetization is generally related to the composition, the magnetic domain wall and the lattice constant of the material. The magnetic properties of the material are determined by the exchange interaction between the electrons in the material [25–27]. Its coercivity and saturation magnetization mainly depend on the material composition, crystal defect, internal stress and grain size [28]. Table 1 specifies the magnetic parameters of the Co-4.54%Sn alloys at different undercoolings.

Table 1. The magnetic property parameters of Co-4.54%Sn alloy under different undercoolings.

ΔT	M_s (emu/g)	M_r (emu/g)	H_c (kA/m)	M_r/M_s
11K	178.92	3.04	7.3	0.017
58K	158.40	3.91	20.81	0.025
122K	156.49	7.63	24.84	0.049
189K	149.94	2.01	12.42	0.013

With the increase of undercooling, the saturation magnetization of the alloy decreases from 178.92 emu/g to 149.94 emu/g, which is caused by the refinement of the microstructure, and the enhancement of dendritic growth velocity and solute content for the primary phase. Han et al. showed that alloy variants with larger average particle size usually have higher saturation magnetization, which is consistent with this result [29]. Compared with the undercooling of 11 K (3.04 emu/g, 7.3 kA/m), the undercooling of 189 K has lower remanent magnetization and higher coercivity, which are 2.01 emu/g and 12.42 kA/m, respectively. In addition, with the increase of undercooling, the squareness ratio showed an "increase first and then decrease" trend. The remanent magnetization and coercivity also exhibited a trend of increasing first, then decreasing with the increase of undercooling and solute content, as shown in Figure 6c,d. The reason is that the crystal particle and boundary are raised significantly due to the refinement of microstructure and enhancement of solute content with the increase of undercooling, and a large amount of solute is distributed into the primary phase, which leads to the decrease of saturation magnetization and the increase of the remanent magnetization and coercivity. When the undercooling ΔT was 189 K, the remanent magnetization and coercivity were decreased in that the volume fraction of eutectic cell was higher than the undercooling of 11 K. Furthermore, a more in-depth study of the magnetic mechanism is still needed.

4. Conclusions

In summary, the broad temperature range of binary Co-4.54%Sn alloy was rapidly solidified using the fluxing method and the maximum undercooling attained 208 K. XRD analyses show that three phases of αCo, βCo_3Sn_2 and εCo were displayed during the solidification, which corresponds with the equilibrium phase diagram. The microstructure of the primary αCo phase remarkably refines and the volume fraction of βCo_3Sn_2 phase of interdendritic structure decreases evidently with undercooling. The dendritic growth velocity presents an increasing tendency at the beginning stage and the maximum velocity reaches 0.95 m/s when the undercooling is lower than 175 K, and then it decreases once the undercooling exceeds 175 K. Otherwise, the microhardness and resistivity of Co-4.54%Sn alloy increase with undercooling owing to the refinement of the microstructure, and the increase of solute content in the primary αCo phase is caused by the high undercooling and rapid solidification. In addition, the saturation magnetization reduces with the increase of undercooling, which is also due to the crystal particle and boundary increasing significantly and solute content being enhanced in the primary αCo phase, which results in the undercooling and dendritic growth velocity under rapid solidification. Moreover, the remanent magnetization and coercivity enhance shows a "increase first and then decrease" trend.

Author Contributions: Methodology, A.W.; software, W.W. and W.L.; formal analysis, W.W. and W.L.; data curation, W.W.; writing—original draft preparation, W.W. and W.L.; writing—review and editing, W.W. and W.L. All authors have read and agreed to the published version of the manuscript.

Funding: Financial support was provided by the National Natural Science Foundation of China [grant numbers 51931005, 52171048, 51571163] and the Key Industry Innovation Chain Project of Shaanxi Province (2020ZDLGY12-02).

Data Availability Statement: Not applicable.

Conflicts of Interest: The authors declare no conflict of interest.

References

1. Sistaninia, M.; Terzi, S.; Phillion, A.; Drezet, J.-M.; Rappaz, M. 3-D granular modeling and in situ X-ray tomographic imaging: A comparative study of hot tearing formation and semi-solid deformation in Al–Cu alloys. *Acta Mater.* **2013**, *61*, 3831–3841. [CrossRef]
2. Eskin, D.G.; Suyitno; Katgerman, L. Mechanical properties in the semi-solid state and hot tearing of aluminium alloys. *Prog. Mater. Sci.* **2004**, *49*, 629–711. [CrossRef]
3. Kou, S. A criterion for cracking during solidification. *Acta Mater.* **2015**, *88*, 366–374. [CrossRef]
4. Bordreuil, C.; Niel, A. Modelling of hot cracking in welding with a cellular automaton combined with an intergranular fluid flow model. *Comput. Mater. Sci.* **2013**, *82*, 442–450. [CrossRef]
5. Rajani, H.Z.; Phillion, A. 3-D multi-scale modeling of deformation within the weld mushy zone. *Mater. Des.* **2016**, *94*, 536–545. [CrossRef]
6. Wang, W.; Shen, C.; Luo, B.; Wei, B. Sluggish dendrite growth in substantially undercooled liquid Fe–Sb alloy. *Philos. Mag. Lett.* **2009**, *89*, 409–418. [CrossRef]
7. Ramirez-Ledesma, A.L.; Lopez-Molina, E.; Lopez, H.F.; Juarez-Islas, J.A. Athermal ε-martensite transformation in a Co20Cr alloy: Effect of rapid solidification on plate nucleation. *Acta Mater.* **2016**, *111*, 138–147. [CrossRef]
8. Spinelli, J.E.; Silva, B.L.; Garcia, A. Microstructure, phases morphologies and hardness of a Bi–Ag eutectic alloy for high temperature soldering applications. *Mater. Des.* **2014**, *58*, 482–490. [CrossRef]
9. Wang, H.P.; Yao, W.J.; Wei, B. Remarkable solute trapping within rapidly growing dendrites. *Appl. Phys. Lett.* **2006**, *89*, 201905. [CrossRef]
10. Kundin, J.; Mushongera, L.; Emmerich, H. Phase-field modeling of microstructure formation during rapid solidification in Inconel 718 superalloy. *Acta Mater.* **2015**, *95*, 343–356. [CrossRef]

11. Lü, P.; Wang, H. Observation of the transition from primary dendrites to coupled growth induced by undercooling within Ni Zr hyperperitectic alloy. *Scr. Mater.* **2017**, *137*, 31–35. [CrossRef]
12. Wang, W.L.; Hu, L.; Yang, S.J.; Wang, A.; Wang, L.; Wei, B. Liquid Supercoolability and Synthesis Kinetics of Quinary Refractory High-entropy Alloy. *Sci. Rep.* **2016**, *16*, 6. [CrossRef]
13. Oloyede, O.; Cochrane, R.F.; Mullis, A.M. Effect of rapid solidification on the microstructure and microhardness of BS1452 grade 250 hypoeutectic grey cast iron. *J. Alloys Compd.* **2017**, *707*, 347–350. [CrossRef]
14. Yakymovych, A.; Shtablavyi, I.; Mudry, S. Structural studies of liquid Co–Sn alloys. *J. Alloys Compd.* **2014**, *610*, 438–442. [CrossRef] [PubMed]
15. Jiang, M.; Sato, J.; Ohnuma, I.; Kainuma, R.; Ishida, K. A thermodynamic assessment of the Co–Sn system. *Calphad* **2004**, *28*, 213–220. [CrossRef]
16. Fan, Q.; Chupas, P.J.; Whittingham, M.S. Characterization of Amorphous and Crystalline Tin–Cobalt Anodes. *Electrochem. Solid-State Lett.* **2007**, *10*, A274–A278. [CrossRef]
17. Qiu, X.X.; Li, J.S.; Wang, J.; Guo, T.; Kou, H.C. Eric Beaugnon, Effect of liquideliquid structure transition on the nucleation in undercooled Co-Sn eutectic alloy. *Mater. Chem. Phys.* **2016**, *170*, 261–265. [CrossRef]
18. Yakymovych, A.; Plevachuk, Y.; Mudry, S.; Brillo, J.; Kobatake, H.; Ipser, H. Viscosity of liquid Co–Sn alloys: Thermodynamic evaluation and experiment. *Phys. Chem. Liq.* **2013**, *52*, 562–570. [CrossRef]
19. Yakymovych, A.; Fürtauer, S.; Elmahfoudi, A.; Ipser, H.; Flandorfer, H. Enthalpy of mixing of liquid Co–Sn alloys. *J. Chem. Thermodyn.* **2014**, *74*, 269–285. [CrossRef]
20. Lipton, J.; Kurz, W.; Trivedi, R. Rapid dendrite growth in undercooled alloys. *Acta Metall.* **1987**, *35*, 957–964. [CrossRef]
21. Boetinger, W.J.; Coriell, S.R.; Trivedi, R. Rapid Solidification Processing: Principles and Technoligies. In Proceedings of the Third Conference on Rapid Solidification Processing, Gaithersburg, MD, USA, 6–8 December 1982; p. 13.
22. Cao, C.-D.; Lu, X.-Y.; Wei, B.-B. Solute Diffusion Controlled Dendritic Growth Under High Undercooling Conditions. *Chin. Phys. Lett.* **1998**, *15*, 840–842. [CrossRef]
23. Lu, P.; Wang, H.P. Effects of undercooling and cooling rate on peritectic phase crystallization within Ni-Zr alloy melt. *Metall. Mater. Trans. B-Proc. Metall. Mater. Proc. Sci.* **2018**, *49*, 499–508. [CrossRef]
24. Chou, H.P.; Chang, Y.-S.; Chen, S.-K.; Yeh, J.W. Microstructure, thermophysical and electrical properties in AlxCoCrFeNi ($0 \leq x \leq 2$) high-entropy alloys. *Mater. Sci. Eng. B* **2009**, *63*, 184–189. [CrossRef]
25. Hamzaoui, R.; Elkedim, O.; Gaffet, E. Friction mode and shock mode effect on magnetic propeties of mechanically alloyed Fe-based nanocrystalline materials. *J. Mater. Sci.* **2004**, *39*, 5139–5142. [CrossRef]
26. Billoni, O.V.; Bordone, E.E.; Urreta, S.E.; Fabietti, L.M.; Bertorello, H.R. Magnetic viscosity in a nanocrystalline two phase composite with enhanced remanence. *J. Magn. Magn. Mater.* **2000**, *208*, 1–12. [CrossRef]
27. Amils, X.; Garitaonandia, J.; Nogués, J.; Suriñach, S.; Plazaola, F.; Muñoz, J.; Baró, M. Micro- and macroscopic magnetic study of the disordering (ball milling) and posterior reordering (annealing) of Fe-40 at.% Al. *J. Non-Cryst. Solids* **2001**, *287*, 272–276. [CrossRef]
28. Herzer, G. Grain structure and magnetism of nanocry-stalline ferromagnets. *IEEE Trans. Magn.* **1989**, *25*, 3327–3329. [CrossRef]
29. Han, L.; Maccari, F.; Filho, I.R.S.; Peter, N.J.; Wei, Y.; Gault, B.; Gutfleisch, O.; Li, Z.; Raabe, D. A mechanically strong and ductile soft magnet with extremely low coercivity. *Nature* **2022**, *608*, 310–316. [CrossRef]

 *metals*

Article

Automatic Detection of Sorbite Content in High Carbon Steel Wire Rod

Xiaolin Zhu [1,2,*], Ling Qian [1], Qiang Yao [1], Guanxi Huang [1], Fan Xu [1], Xue Chen [1] and Zhengjun Yao [2,*]

[1]   Jiangsu Product Quality Testing & Inspection Institute, Nanjing 210007, China; yaoqiangjszj@163.com (Q.Y.)
[2]   College of Material Science and Technology, Nanjing University of Aeronautics and Astronautics, Nanjing 211106, China
*   Correspondence: zxlin@nuaa.edu.cn (X.Z.); yaozj@nuaa.edu.cn (Z.Y.)

Abstract: This paper presents a method for the automatic detection of sorbite content in high-carbon steel wire rods. A semantic segmentation model of sorbite based on DeepLabv3+ is established. The sorbite structure is segmented, and the prediction results are analyzed and counted based on the metallographic images of high-carbon steel wire rods marked manually. For the problem of sample imbalance, the loss function of Dice loss + focal loss is used, and the perturbation processing of training data is added. The results show that this method can realize the automatic statistics of sorbite content. The average pixel prediction accuracy is as high as 94.28%, and the average absolute error is only 4.17%. The composite application of the loss function and the enhancement of the data perturbation significantly improve the prediction accuracy and robust performance of the model. In this method, the detection of sorbite content in a single image only takes 10 s, which is 99% faster than that of 10 min using the manual cut-off method. On the premise of ensuring detection accuracy, the detection efficiency is significantly improved and the labor intensity is reduced.

Keywords: high carbon steel wire rod; sorbite content; deep learning; semantic segmentation

1. Introduction

High-carbon steel wire rods are the main raw materials for the production of spring steel, prestressed steel strands, steel cords, steel wire ropes, and other products. Their plasticity, strength, and uniformity of structure are crucial for subsequent processing. The delivery status of high carbon steel wire rods is generally hot-rolled, and its microstructure is a sorbite structure. Sorbite is a non-equilibrium pearlite-type structure and a dual-phase mixed structure of sorbite ferrite and cementite. Its difference from pearlite is that a slice of sorbite and cementite in sorbite are thinner and the spacing between slices is smaller. Sorbite and cementite slices can be distinguished by a magnification of more than 600 times under an optical microscope. Due to the small slice spacing and many interphase boundaries, the sorbite structure has enhanced resistance to plastic deformation under the action of an external force, excellent strength, and plasticity, and is the most ideal structure for high-carbon wire rod steel [1]. In addition to sorbite, there are also pearlite, martensite, net cementite, and other structures in the high carbon steel wire rods. The content of sorbite is one of the important indicators to evaluate the performance of high-carbon wire rods. Generally, the content is required to be not less than 80%. The higher, the better.

At present, the standard for sorbite structure testing of high-carbon steel wire rods is the YB/T 169-2014 Metallegrephic Tat Method of Sorbite in High Carbon Steel Wire Rod [2]. Three detection methods are specified in the standard. The first method is the metallographic manual method. According to the stereological quantitative metallography principle, the grid number point method or grid truncation method is adopted, i.e., in a grid with a determined size or area, the sorbite volume fraction is calculated by manually counting the proportion of the grid intersection or grid length occupied by the sorbite structure to the total number of grid intersections or lengths. The second method is standard

Citation: Zhu, X.; Qian, L.; Yao, Q.; Huang, G.; Xu, F.; Chen, X.; Yao, Z. Automatic Detection of Sorbite Content in High Carbon Steel Wire Rod. *Metals* **2023**, *13*, 990. https://doi.org/10.3390/met13050990

Academic Editors: Marcello Cabibbo and Andrea Di Schino

Received: 13 February 2023
Revised: 30 March 2023
Accepted: 24 April 2023
Published: 20 May 2023

sample detection with an image analyzer. The standard sample with a determined sorbite content is used for synchronous sample preparation, corrosion, and detection with the sample to be tested to determine the reference grayscale value Gs of the standard reference sample. The sample to be tested is grayed with the same grayscale value Gs, and the sorbite volume fraction is quantitatively measured by the grayscale value. The third method is comparison. The microstructure of the sample to be tested through a metallographic microscope is observed, it is then compared with the standard rating chart of sorbite content, the standard chart with the closest structure to the sample to be tested is determined, and the sorbite volume fraction or sorbite grade of the sample to be tested is estimated. The three detection methods have their own advantages and disadvantages. Methods 1 and 2 produce more accurate detection results but have many detection steps and low efficiency. Method 3 is simple, fast, and efficient, but the error rate is large (the sorbite content of 10% is a grade in the standard chromatogram) and is greatly affected by the inspector. The consistency of the results is low.

At present, the detection of sorbite content is controversial, there are three reasons for this. Firstly, there is a heavy reliance on the individual researcher's background and experience which both introduce significant bias and potentially error into the process of detecting sorbite content. Secondly, the detection of sorbite content is greatly affected by the level of sample preparation; it is difficult to obtain a perfect microstructure to clearly distinguish the structure details and to determine the size of the region of a certain type of microstructure, due to the poor sample prepared by the polishing process. Corrosion defects can be caused by a number of factors, such as inadequate etching or over etching. Inadequate etching would lead to an incomplete microstructure display, and sorbite is easily misjudged as pearlite. Over etching would lead to the microstructure image showing a sense of relief, forming microstructure artifacts. Thirdly, there is a pearlite-sorbite transitional microstructure (PS) in high-carbon wire rods [3], and its lamellar spacing is between the pearlite and sorbite in high resolution; whether this type of transitional structure is classified as pearlite or sorbite also directly affects the judgment of the results. Therefore, the accurate detection of the sorbite content of high-carbon wire rods has been a problem in the steel industry.

With the rapid development of electronic information technology and artificial intelligence technology, computer vision (CV)—a discipline that studies how to make computers "see" like humans—is widely used in various industries, such as text recognition, and medical diagnosis [4,5]. The computers are objective and fast, and they are supposed to play an important role in the field of detection. In the field of materials, the automatic metallographic analysis of metal materials based on computer vision and machine learning technologies has become the focus of researchers at home and abroad. Chowdhury [6] used pre-trained convolutional neural networks (deep learning algorithms) to extract microstructure features, and used support vector machine, voting, nearest neighbors, and random forest models for classification, and obtained a high classification accuracy. Azimi [7] employed pixel-wise segmentation based on full convolutional neural networks (FCNN), together with the deep learning method to classify and identify the microstructure of mild steel, achieving a 93.94% classification accuracy. Park [8] employed a pixel-wise segmentation method via U-NET architecture built upon FCNN, and employed several techniques ranging from data augmentation, the Amazon computing service, to semantic segmentation. The system achieved a maximum classification accuracy of 98.689% for the pearlite and ferrite phases. Therefore, the automatic detection of the sorbite content based on machine learning is feasible. This paper builds a semantic segmentation model based on the deep learning algorithm and uses the sorbite tissue image calibrated by technical experts as the training data set to segment and analyze the sorbite, thus realizing the automatic detection of the sorbite content of high-carbon steel wire rods.

2. Acquisition and Calibration of Data Set

2.1. Sample Preparation

A hundred metallographic samples were prepared by selecting rolls of 55SiCr, 65Mn, 72A, and SWRH82B hot rolled wire rods with a diameter of 6–12 mm, 25 in each. After cutting, grinding, and polishing, a 4% nitric acid alcohol solution was used for corrosion for 25–35 s (the higher the carbon content is, the longer the corrosion time is). When a clear microstructure was displayed, the ZEISS Axio Observer optical metallographic microscope (Manufactured by Carl Zeiss, Munich, Germany) was used for observation. Twenty-five metallographic photographs were taken randomly from the edge to the center of each metallographic specimen, totaling 2500.

2.2. Image Calibration

The metallographic images were distributed to experts to determine the sorbite region and sorbite content in each image using the grid truncation method. The sorbite area (dark zone) was then labeled using LabelMe software version 5.1.1. Figure 1 shows an example of the sorbite labeling in a single metallographic image. For this study, 2500 valid metallographic images and their labeling files were collected, with an image size of 2048 × 1536.

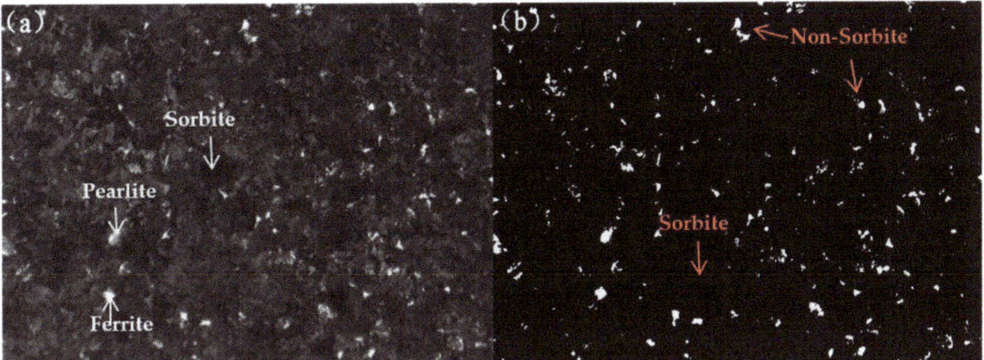

Figure 1. Example of sorbite content labeling. (**a**) Original image; (**b**) Labeling image.

3. Model and Method

3.1. Network Architecture

In this paper, the residual neural network ResNet [9] was selected as the backbone network for feature extraction, and DeepLabv3+ [10] and U-Net ++[11] were used as semantic segmentation models for the segmentation and statistics of sorbite content.

The network structure of ResNet is based on VGG-19, which is improved, and residual elements are added by a short circuit mechanism. In ResNet-34, it adopts 3 × 3 filters, and its design principle is that: first, in the same output feature map, the number of filters in each layer is the same; second, when the size of the feature map is reduced by half, the number of filters will double to ensure the time complexity of each layer. The network ends with a global average pooling layer and a 1000-dimensional fully connected layer with softmax. The rest of the residual neural network structure is deformed on this basis.

The U-Net++ network structure is shown in Figure 2, which consists of several parts: the convolution unit, down-sampling and up-sampling modules, and skip connection between convolution units. In the U-Net model structure, nodes X0.4 only form a skip connection with nodes X0.0, while in the U-Net++ model structure, nodes X0.4 connect the outputs of four convolution units X0.0, X0.1, X0.2, and X0.3 at the same layer, where each node Xi.j represents one convolution down-sampling or deconvolution up-sampling. The U-Net++ network has a nested structure and dense skip paths, which is conducive to aggregating features with different semantic scales on the decoder subnetwork and has achieved excellent performance levels in other fields [12,13]. In this paper, after grayscale processing, the sorbite was quite different from other tissues, which was intuitively applicable to this problem.

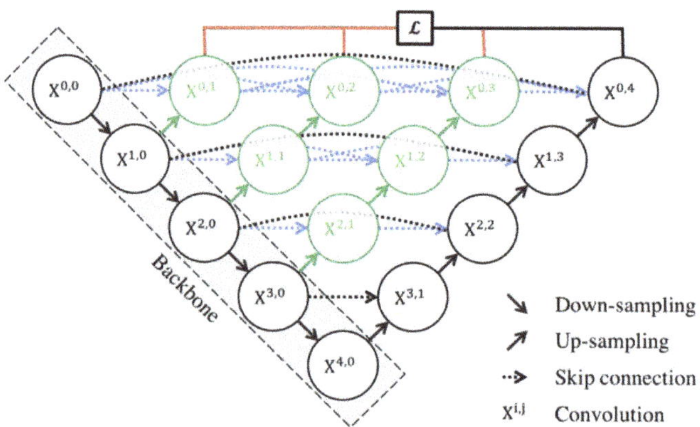

Figure 2. Schematic diagram of the U-Net++ network structure [11], with permission from Springer, 2023.

Enter X0.0 from X0.0 on the first layer of the model, and calculate according to the following formula in turn:

$$x^{i,j} = \begin{cases} \mu\left(x^{i-1,j}\right), & j = 0 \\ \mu\left(\left[\left[x^{i,k}\right]_{k=0}^{j-1}, T\left(x^{i+1,j-1}\right)\right]\right), & j > 0 \end{cases} \tag{1}$$

where μ denotes the convolution, [] denotes the feature cascade, T denotes the deconvolution up-sampling, and $x^{i,j}$ denotes the output of the node $X^{i,j}$, where i denotes the down-sampling layer along the encoder index and j denotes the convolution layer along the skip index dense block.

The skip connection can introduce high-resolution information in the image into the result of the up-sampling, thereby ensuring high segmentation accuracy. Taking the first layer of the model as an example, the skip connection results are shown in Figure 3.

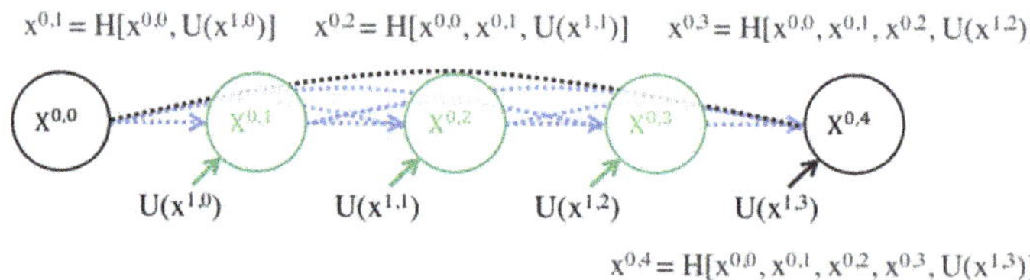

Figure 3. Schematic diagram of the U-Net++ skip connection results [11], with permission from Springer, 2023.

The deep supervision process was introduced into the U-Net++ network model. The shallow feature perception of the image can be increased by deconvolutional up-sampling of the results $x^{i,0}$ obtained by down-sampling at each level, and then adding the final up-sampling segmentation results $x^{0,j}$ corresponding to each level of the training loss calculation process [14]. The split loss function designed accordingly is shown in the following formula:

$$\text{Loss} = \frac{1}{J}\sum_{i=1}^{J} L_j\left(x^{0,j}\right) \tag{2}$$

where **Loss** is the total segmentation loss, $\mathbf{L_j}$ is the loss function used to calculate the segmentation loss of $\mathbf{x^{0,j}}$, and **J** is the number of all nodes except the down-sampling nodes in the first layer of the model.

Compared with the U-Net++ network, the DeepLabv3+ network greatly reduces the actual running graphic memory occupancy under the same setting and slightly improves the performance, which is also suitable for this problem. The structure of the DeepLabv3+ network is shown in Figure 4. The network with a codec structure was generated based on the DeepLabv3 network structure, in which DeepLabv3 was used as the encoder part to extract and fuse multi-scale features. A simple structure was added as the decoder to further merge the underlying features with the higher-layer features, improve the accuracy of the segmentation boundary, and obtain more details, forming a new network that merges atrous spatial pyramid pooling (ASPP) [15] and codec structures.

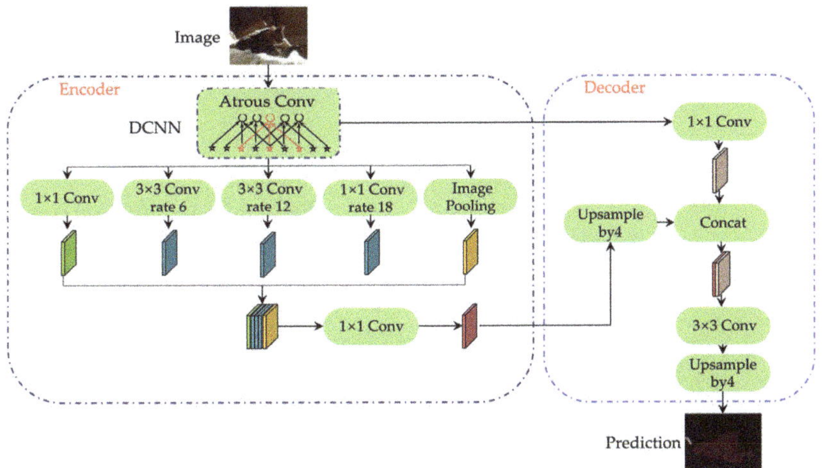

Figure 4. Schematic diagram of the DeepLabv3+ network structure [10], with permission from Springer, 2023.

Atrous convolution is the core of the DeepLab series model [10,16,17], which is a convolution method to increase the receptive field and is conducive to extracting multi-scale information. Atrous convolution adds voids based on ordinary convolution, in which a parameter dilation rate is added to control the size of the receptive field. As shown in the Figure 5, taking **3 × 3** convolution as an example, the grey lattice represents the **3 × 3** convolution kernel, and the receptive field after ordinary convolution is 3. When **rate = 2**, the receptive field after convolution by the atrous convolution module is 5, which is increased by 2 compared with the ordinary convolution receptive field. When **rate** $= 3$, the receptive field after convolution by the atrous convolution module is 7, which is increased by 4 compared with the ordinary convolution receptive field. Due to the mesh effect of atrous convolution, some information will be lost in the image after the atrous convolution operation.

For two-dimensional signals, in particular, for each position **i** on the output feature map **y** and the convolution filter **w**, an atrous convolution was applied on the input feature map **x**, with the following formula:

$$\mathbf{y[i]} = \sum_{\mathbf{k}} \mathbf{x[i + rate \cdot k]w[k]} \tag{3}$$

Atrous spatial pyramid pooling (ASPP) uses atrous convolution with different expansion rates to make up for the defects of atrous convolution, captures the context of multiple layers, fuses the obtained results, reduces the probability of information loss, and helps to improve the accuracy of convolutional neural networks.

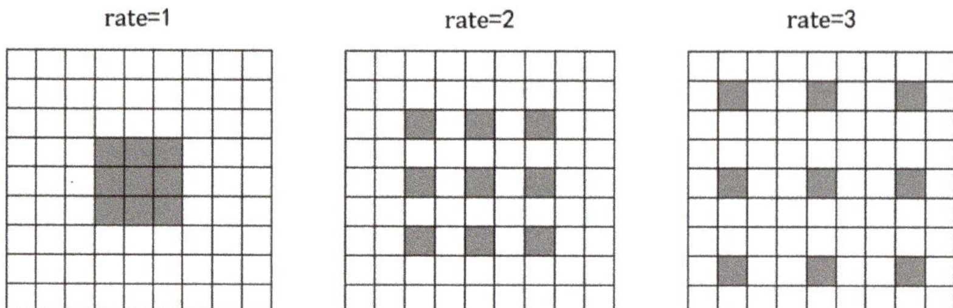

Figure 5. Schematic diagram of the atrous convolution receptive field.

3.2. Loss Function

The loss function was used to evaluate the difference between the prediction result of the model and the actual situation. Different loss functions are used in different models. In general, the better the loss function, the more accurate the model predictions. The main microstructure of high-carbon steel is sorbite, the content of which is generally more than 50% and is more than 70% in most samples. Therefore, the proportion of sorbite in the metallographic image is extremely unbalanced with the proportion of the background. In the sample shown in Figure 1, the sorbite content (dark part) was about 97%. For the data set of this paper, the statistics of sample proportions with different sorbite contents are shown in Table 1. Samples with a sorbite content higher than 80% accounted for 56.5% of the total. In general, unbalanced samples can cause training models to focus on predicting pixels as dominant types, while "disregarding" the minority types, which negatively affects the model's ability to generalize on test data. Therefore, it is necessary to use the appropriate loss function or its combination to deal with the imbalance of the sample. Table 1 shows the proportion of samples with different sorbite contents in the data set.

Table 1. Proportions of samples with different sorbite contents in the data set.

Sorbite Content/%	[0–50)	[50–60)	[60–70)	[70–80)	[80–90)	(≥90)
Sample Proportion/%	0	4.7	12.2	26.6	38.3	18.2

The detection of sorbite content is essentially a problem of a significant imbalance between positive and negative samples in a binary classification, and a large number of background pixels affect the segmentation accuracy of the model. Therefore, focal loss was selected as the semantic segmentation loss function in this paper. This function was originally proposed by He [18] to solve the model performance problems caused by the imbalance of data classes and the difference in classification difficulty in the image domain. Focal loss adds a parameter γ to the cross-entropy loss function and constructs an adjustment factor $(1 - p(x))^{\gamma}$ to solve the problem of the sample imbalance. The calculation formula of the loss function is as follows:

$$\mathbf{FL(p(x))} = -(1 - p(x))^{\gamma}\log(p(x)) \tag{4}$$

where the sample $p(x)$ with accurate classification tends to 1, the regulation factor tends to 0; the sample $1 - p(x)$ with inaccurate classification tends to 1, the regulation factor tends to 1. Compared with the cross-entropy loss function, focal loss does not change for inaccurately classified samples and decreases for accurately classified samples. Overall, it is equal to adding the weight of the sample with inaccurate classification to the loss function, $p(x)$. It also reflects the difficulty of classification. The greater the $p(x)$, the higher the confidence of classification, the more easily the representative sample is classified; the smaller the $p(x)$, the lower the confidence of classification, the lower the confidence of classification, and the more difficult it is to classify the representative sample. Therefore, focal loss is equivalent to increasing the weight of difficult samples in the loss function, so that the loss function tends to be difficult samples, which is helpful to improve the accuracy of difficult samples.

In addition, the problem of region size imbalance between the sample foreground and the background of the sorbite image can be handled by the Dice loss [19] function. Dice loss is a region-dependent loss, that is, the loss of the current pixel is not only related to the predicted value of the current pixel, but also related to the value of other pixel points. The specific loss function formula is:

$$\textbf{Dice Loss} = 1 - \frac{2|\textbf{X} \cap \textbf{Y}|}{|\textbf{X}| + |\textbf{Y}|} \tag{5}$$

where $\textbf{X}$ represents the target segmented image, $\textbf{Y}$ represents the predicted segmented image, and the intersection form of **Dice Loss** can be understood as a mask operation. Therefore, regardless of the size of the image, the calculated loss of the fixed-size positive sample area is the same, and the supervision contribution to the network does not change with the size of the image. **Dice Loss** training tends to tap into foreground areas and thus adapts to the smaller foreground situation in this paper. Training loss, however, is prone to instability, especially against small targets. In addition, gradient saturation occurs in extreme cases. Therefore, considering the sample situation of sorbite content, this paper combines **Dice Loss** with focal loss.

3.3. Data Augmentation

Due to different sample preparation levels, corrosion depths, shooting equipment, illumination, and other factors, the sorbite metallographic images collected were very different. It was not possible to exhaust all the possibilities and obtain images that were representative enough. Therefore, data enhancement was needed to expand the distribution of data sources and features, improve the size and quality of the training data sets, and solve the problem of limited data. At the same time, data enhancement can also cope with the problem of multi-distributed data scenarios. Data augmentation typically [20] includes flipping, cropping, rotation, zooming, color transformation, noise injection, etc. In this project, considering the characteristics of sorbite, simple flipping, rotation, cropping, zooming, and color transformation could not only enhance the data but may also destroy the label information of the sample to a certain extent, bringing negative effects to the model learning. For this paper, an enhancement method based on noise injection was designed. In the training, a certain amount of data perturbation was added to the initial sample image, and the original pixel labels were not changed incorrectly so that the model could focus more on learning the difference between the sorbite pixel values and background pixel values, rather than just learning the pixel values of the sorbite. In this way, the model was able to ignore irrelevant factors in the training process to a certain extent and make more accurate judgments. The specific perturbation expression is as follows, where X_{ij} represents the pixel value of the ith row and jth column of the sorbite sample (where $\textbf{R} = \textbf{rand}(-\textbf{a}, \textbf{a})$):

$$X_{ij} = \begin{cases} X_{ij} + \textbf{R}, & 0 < X_{ij} + \textbf{R} < X_{\text{thres}} \text{ and } X_{ij} < X_{\text{thres}} \\ X_{ij} + \textbf{R}, X_{\text{thres}} < X_{ij} + \textbf{R} < 255 \text{ and } X_{ij} > X_{\text{thres}} \end{cases} \tag{6}$$

In the practical application process, let $\textbf{a} = 10$, and the samples before and after adding the noise injection enhancement mode are shown in Figure 6.

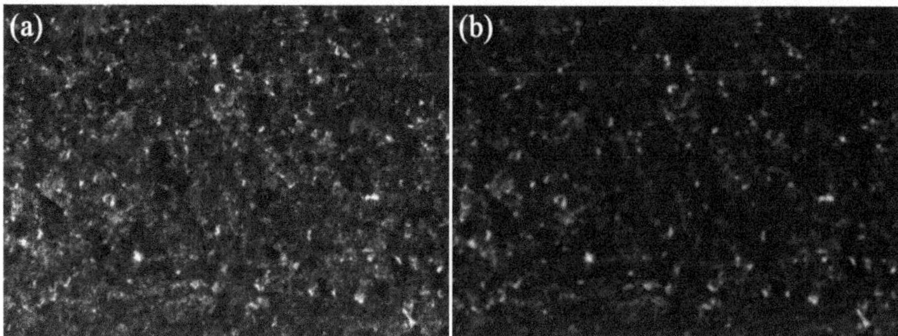

Figure 6. Comparison of the effects before and after sample enhancement. (**a**) Original image; (**b**) Image after noise injection.

3.4. Training and Evaluation Indexes of the Model

In this paper, preprocessed sorbite images were used as data sets, and each data set was randomly divided into training data sets and test data sets in a 9:1 ratio. Training then took place with each training set in turn, and testing was conducted on the test set of all data sets separately. All models were based on the PyTorch framework, the programming language was Python, and the graphics card was NVIDIA GeForce RTX 3090. The model learning rate used in this paper was 0.0001, the batch size was 4, and the number of iteration rounds was 100.

In this paper, the accuracy of semantic segmentation was evaluated by pixel accuracy (PA), intersection over union (IoU), and mean square error (MSE). For a binary task, the following four situations may occur. TP (true positive) indicates that a sample is predicted to be a positive class with a positive class true label. FN (false negative) indicates that a sample is predicted as a negative class with a positive class true label. FP (false positive) indicates that a sample is predicted as a positive class with a negative class true label. TN (true negative) indicates that a sample is predicted as a negative class with a negative class true label.

Assuming that there are a total of $\mathbf{n}$ classes in the test data set, $\mathbf{p_{ii}}$ indicating the number of classes predicted as class $\mathbf{i}$ in class $\mathbf{i}$ data, and $\mathbf{p_{ij}}$ indicating the number of classes predicted as class $\mathbf{j}$ in class $\mathbf{i}$ data, $\mathbf{PA}$ is defined as the ratio of the number of correctly classified pixels to the total number of pixels, and the formula is as follows:

$$\mathbf{PA} = \frac{\sum_{i=1}^{n} \mathbf{P_{ii}}}{\sum_{i=1}^{n} \sum_{j=1}^{n} \mathbf{P_{ij}}} \tag{7}$$

That is, pixel accuracy (**PA**) represents the percentage of the pixel value predicted correctly to the total pixel value, and the calculation formula is as follows:

$$\mathbf{PA} = \frac{\mathbf{TP} + \mathbf{TN}}{\mathbf{TP} + \mathbf{FP} + \mathbf{FN} + \mathbf{TN}} \tag{8}$$

Mean pixel accuracy (mPA) is averaged by summing the pixel accuracy of each category.

IoU is the most commonly used semantic segmentation evaluation standard, which is the ratio of the intersection of real and predicted labels to their union. IoU can better evaluate the performance of semantic segmentation methods. The calculation formula of IoU is as follows:

$$\mathbf{IoU} = \frac{\mathbf{TP}}{\mathbf{TP} + \mathbf{FP} + \mathbf{FN}} \tag{9}$$

The mean intersection over union ratio (MIoU) was averaged over the sum of the intersection over union for each category.

Mean absolute error (MAE) is a measure reflecting the degree of difference between the estimator and the estimated quantity, which was used to characterize the difference between the predicted proportional result and the true proportional result of the model. The MAE reflects the average distance that the predicted value deviates from the true value, that is,

$$\mathbf{MAE} = \sum_{i=1}^{n} \frac{1}{n} |\mathbf{f(x_i)} - \mathbf{y_i})| \tag{10}$$

4. Results and Discussion

Researchers have conducted some research and exploration in improving the accuracy of the determination of sorbite content. Wuhan Iron & Steel Group [21] used SEM and OM to observe the microstructure in the same region under different magnifications, and the analysis results indicated that the so-called "pearlite" structure was a kind of etching morphology of the sorbite structure with different orientations on the cross-section of the metallographic sample surface. Zhejiang University of Technology [22] found that the etching condition has a significant influence on the display and identification of sorbite structure, and some of the bright white areas under the optical microscope may be local flat areas of sorbite structure, while some of the "pearlite" may also be local depressions resulting in a "magnified lamellar structure", so it is unreliable to rely on light region and dark regions to determine the sorbite content. Liuzhou Steel Group [23] used digital image storage technology for the accurate detection of sorbite content. All these studies were carried out based on the traditional grayscale image, pearloid slice spacing, etc. The problem of fluctuations in the detection process and the need for manual intervention are still not well resolved. Therefore, the automatic detection of the sorbite content based on artificial intelligence is a focus of current research. Luo [1] established a library of high-carbon wire rod sorbite material for neural network learning, and successfully realized the intelligent identification of sorbite by using artificial intelligence and deep learning technology.

In this paper, DeepLabv3+ and U-Net++ semantic segmentation models were used, respectively. ResNet34 was used as the backbone network for training and verification, and the prediction of each pixel of the output sample image was judged to be correct or not. The evaluation results are shown in Table 2. It can be seen that the semantic segmentation model based on DeepLabv3+ improved the mPA by 0.7% and reduced the MAE by 10.7% compared with U-Net++ and obtained more accurate prediction results. This semantic segmentation model was also more refined compared to the classification model in Luo's article. The visualization effect is shown in Figure 7. The white part in the figure is the part of the correct prediction by the model, and the red part is the part of the incorrect prediction by the model. The segmentation results show that the model had a good segmentation effect, and an MIoU of 74.89% indicates that the predicted results of the model were more accurate than the manually annotated boundary.

Subsequently, the DeepLabv3+ semantic segmentation recognition model was used to test the unlabeled sorbite metallographic images. For each sorbite metallographic image tested, the sorbite image and the length of a single pixel point of that image were input, then the recognition model output the sorbite content of the image. The output results were compared with the manually calibrated test results, and some of the test results are shown in Table 3. The test recognition deviations were almost all less than 5%, and the recognition results were very good.

Table 2. Comparison of the test results of different frames.

Seg_Model	Backbone	mPA	MIoU	MAE	Epoch Time(s)
DeepLabv3+	ResNet34	94.28%	74.89%	4.17	9
U-Net++	ResNet34	93.58%	73.24%	4.68	9

Figure 7. Schematic diagram of the sorbite sample segmentation effect.

Table 3. Part of the test results of unlabeled metallographic images.

No.	Prediction	Truth	MAE
1	81.18	85.69	4.51
2	85.79	85.43	0.36
3	87.11	86.62	0.49
4	90.71	88.99	1.71
5	91.58	89.32	2.26
6	92.35	90.29	2.06
7	92.08	90.36	1.72
8	90.05	89.63	0.42
9	90.11	87.98	2.13
10	90.67	88.70	1.96
11	88.31	86.57	1.73
12	87.65	85.89	1.75
13	85.06	89.95	4.88
14	91.27	91.27	0.00
15	90.34	87.14	3.20

At the same time, Luo [1] did not consider the imbalance of positive and negative distributions and foreground and background imbalance in sorbite samples, while this paper used loss function and its combination to deal with the imbalance problem, and the prediction results of different loss functions were also tested and analyzed. The three sets of results were based on DeepLabv3+, and the Backbone used was ResNet-34. The results are shown in Table 4. The model of the focal loss function alone predicts a mPA of 94.18% and a MAE of 4.46%, which demonstrates a good segmentation effect. The results of using the dice loss function alone were poor. After the combination of the two, the prediction results were optimized, and the mPA and MAE were improved to 94.28% and 4.17%, respectively. This fully illustrates that the focal loss + Dice loss combined loss function selected in this paper is correct and reasonable considering the imbalance of positive and negative distributions and the foreground and background imbalance in sorbite samples analyzed in 3.2. In this paper, the detection of sorbite content in a single image only took 10 s, which was 99.9% faster than that of 10 min using the manual cut-off method. On the premise of ensuring detection accuracy, the detection efficiency was significantly improved and the labor intensity was reduced.

Table 4. Training results of the different loss schemes.

Loss	mPA	MIoU	MAE	Epoch Time(s)
Focal Loss + Dice Loss	94.28%	74.89%	4.17	9
Focal Loss	94.18%	73.18%	4.46	9
Dice Loss	71.01%	52.76%	21.89	9

Lou [1] standardized the sample preparation process without considering the diversity of data sources and feature distributions due to different sample preparation levels, corrosion depths, shooting equipment, illumination, and other factors. In this paper, a data augmentation was performed to expand the distribution of data sources and features, and the actual effects of the data augmentation were verified. The training and verification results of the data set before and after the perturbation added to the sample image were compared and analyzed. The data set was divided into three sub-datasets according to the different sample batches, and then the three sub-datasets were divided into training sets and test sets at a ratio of 9:1. The training set and test set of the first sub-dataset were written as Training set 1 and Test set 1, respectively, and so on for the others. Then training according to the respective training set followed by the testing of all test sets was conducted. In this project, the model prediction results of the original data and the model prediction results after adding perturbation were statistically analyzed, and the evaluation indexes were characterized by MAE, i.e., the absolute error between the sorbite content output by the model compared to the sample and the real sorbite content. The results are presented in Tables 5 and 6. It can be seen that the results of the training and validation of the raw data had better prediction results for the autologous test dataset. However, the difference between the prediction results of the three test sets was reduced after the perturbation was added, indicating that the generalization ability and robustness of the model were improved after the perturbation was added.

Table 5. Prediction results of the original model.

Training Set\Test Set	Test Set 1	Test Set 2	Test Set 3
Training set 1	3.36	3.84	4.18
Training set 2	7.40	5.77	8.09
Training set 3	8.56	9.19	4.43

Table 6. Model prediction results after adding perturbation.

Training Set\Test Set	Test Set 1	Test Set 2	Test Set 3
Training set 1	3.47	3.79	3.51
Training set 2	6.65	5.50	7.19
Training set 3	7.82	8.02	4.12

The original sample images collected in this project were all 2048×1536, with a high resolution and large size. In general, large-size images contain more data information and reflect more features of the predicted object. However, using large-size images requires higher computing power, and if the characteristic size of the measured object is much smaller than the image size, using large-size images as the training set cannot achieve better results. In this project, considering the characteristic size of sorbite, we used a sliding window to divide each original image into sixteen 512×384 images, and then trained and tested the model. This operation can not only reduce the requirements for computing power and improve the efficiency of model training and prediction, but also evaluate the uniformity of the distribution of sorbite content in the whole image through the calculation of each small image. Specifically, after the model predicted the sorbite content of the 16 local areas cut from each image, the variance was calculated, and the standard deviation was used to evaluate the sample uniformity. The formula can be expressed as:

$$\textbf{Uniformity} = \sqrt{\sum_{i=1}^{4\times4} (\textbf{R}_i - \mu)^2/4 \times 4} \tag{11}$$

where $\textbf{R}_i$ represents the ferrite content of the $\textbf{i}$ th local region. Uniformity is the evaluation index of sample uniformity. The lower the value, the better the uniformity. Figure 8 shows that the experimental effect of uniformity evaluation can characterize the uniformity of the microstructure of the sample, so that the macroscopic performance of the original material can be inferred.

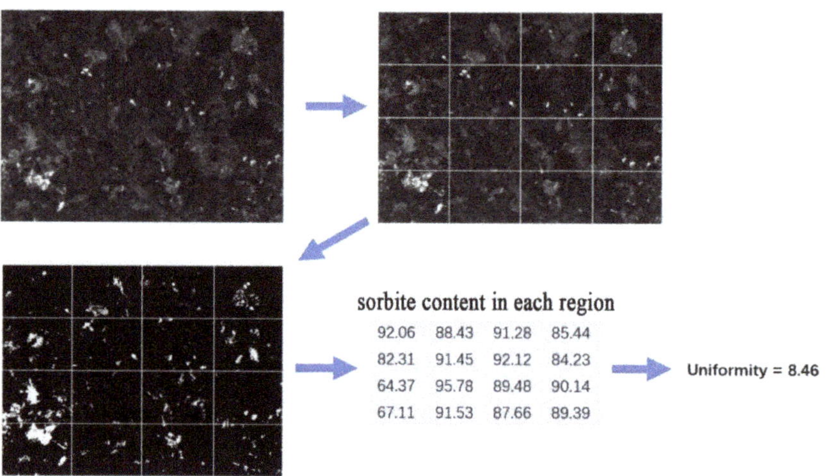

Figure 8. Effect of sample uniformity analysis.

5. Conclusions

In this paper, DeepLabv3+ was selected as the semantic segmentation model, and ResNet-34 was used as the backbone network to establish an intelligent detection model of sorbite content based on deep learning. The metallographic images of high-carbon steel wire rods were manually labeled and cut as data sets. To solve the multi-distribution problem of the source and characteristics of the samples, this paper used the Dice loss and focal loss functions to design data perturbation processing to enhance the accuracy of the prediction results and the robustness of the model. Meanwhile, the uniformity of the samples was evaluated by separately predicting and analyzing the sorbite content in the slit region. The results show that the proposed method can realize the automatic statistics of sorbite content. The average pixel prediction accuracy was as high as 94.28%, and the average absolute error was only 4.17%. The composite application of the loss function and the enhancement of the data perturbation significantly improved the prediction accuracy and robust performance of the model. In this method, the detection of sorbite content in a single image only took 10 s, which was 99% faster than that of 10 min using the manual cut-off method. On the premise of ensuring detection accuracy, the detection efficiency was significantly improved and the labor intensity was reduced.

Author Contributions: Conceptualization, Z.Y. and X.Z.; methodology, Z.Y. and X.Z.; software, X.Z.; validation, G.H., F.X., and Q.Y.; investigation, X.Z.; resources, G.H., X.C., and Q.Y.; data curation, X.Z.; writing—original draft preparation, X.Z.; writing—review and editing, G.H., L.Q., X.C., and F.X.; supervision, Q.Y. and L.Q.; project administration, Z.Y.; funding acquisition, Z.Y. and X.Z.; All authors have read and agreed to the published version of the manuscript.

Funding: This project was supported by the Science and Technology Program of Jiangsu Provincial Administration for Market Regulation (grant no. KJ204115, KJ21125122, KJ2022062).

Data Availability Statement: The data presented in this study are available on request from the corresponding author.

Acknowledgments: Thanks to Baodong Feng, Xuebin Xu, Bo Liu, Linning Qian, and Jun Wan for the help with data calibration.

Conflicts of Interest: The authors declare no conflict of interest.

References

1. Luo, X.Z.; Xiao, M.D.; Zhang, Z.Y.; Li, F.Q.; Zhu, X.R. Recognition discussion about sorbite in high-carbon wire rod steel based on artificial intelligence. *Phys. Exam. Test.* **2021**, *39*, 34–37. [CrossRef]
2. *YB/T 169-2014*; Metallographic Test Method of Sorbite in High Carbon Steel Wire Rod. National Steel Standardization Technical Committee: Beijing, China, 2014.
3. Wang, K.J.; Liu, H.Y. Categories and Morphological Features of Metallographic Microstructure in High Carbon Stelmor Wire Rods. *Phys. Exam. Test.* **2013**, *31*, 1–5. [CrossRef]
4. Ren, Z.G.; Ren, G.Q.; Wu, D.H. Deep Learning Based Feature Selection Algorithm for Small Targets Based on mRMR. *Micromachines* **2022**, *13*, 1765. [CrossRef] [PubMed]
5. Vaiyapuri, T.; Srinivasan, S.; Sikkandarn, M.Y.; Balaji, T.S.; Kadry, S.; Meqdad, M.N.; Nam, Y.Y. Intelligent Deep Learning Based Multi-Retinal Disease Diagnosis and Classification Framework. *Comput. Mater. Contin.* **2022**, *73*, 5543–5557. [CrossRef]
6. Chowdhury, A.; Kautz, E.; Yener, B.; Lewis, D. Image driven machine learning methods for microstructure recognition. *Comput. Mater. Sci.* **2016**, *123*, 176–187. [CrossRef]
7. Azimi, S.M.; Britz, D.; Engstler, M.; Fritz, M.; Mücklich, F. Advanced steel microstructural classification by deep learning methods. *Sci. Rep.* **2018**, *8*, 2128. [CrossRef] [PubMed]
8. Park, H.; Öztürk, A. Machine Learning Approach on Steel Microstructure Classification. In Proceedings of the Europe-Korea Conference on Science and Technology, Vienna, Austria, 15–18 July 2021; pp. 13–23. [CrossRef]
9. He, K.; Zhang, X.; Ren, S.; Sun, J. Deep Residual Learning for Image Recognition. In Proceedings of the IEEE Conference on Computer Vision and Pattern Recognition, Las Vegas, NV, USA, 27–30 June 2016; pp. 770–778. [CrossRef]
10. Chen, L.C.; Zhu, Y.; Papandreou, G.; Schroff, F.; Adam, H. Encoder-Decoder with Atrous Separable Convolution for Semantic Image Segmentation. In Proceedings of the ECCV 2018, Munich, Germany, 8–14 September 2018; Springer: Cham, Switzerland, 2018; pp. 833–851. [CrossRef]
11. Zhou, Z.W.; Rahman, S.M.M.; Tajbakhsh, N.; Liang, J.M. UNet++: A Nested U-Net Architecture for Medical Image Segmentation. In Proceedings of the DLMIA ML-CDS, Granada, Spain, 20 September 2018; Springer: Cham, Switzerland, 2018. [CrossRef]
12. Huo, Y.; Li, X.; Tu, B. Image Measurement of Crystal Size Growth during Cooling Crystallization Using High-Speed Imaging and a U-Net Network. *Crystals* **2022**, *12*, 1690. [CrossRef]
13. Li, Z.Z.; Yin, H.; Zuo, J.K.; Sun, Y.F. Ship Detection Model Based on UNet++ Network and Multiple Side-Output Fusion Strategy. *Comput. Eng.* **2022**, *48*, 276–283.
14. Lv, B.L.; Li, Z.F.; Xi, Z.H.; Yao, Y.M.; Ji, J.J. Component Analysis of Coal Rock Microscopic Image Based on UNet++. *Comput. Digit. Eng.* **2022**, *50*, 389–393, 404.
15. Liu, R.; Tao, F.; Liu, X.; Na, J.; Leng, H.; Wu, J.; Zhou, T. RAANet: A Residual ASPP with Attention Framework for Semantic Segmentation of High-Resolution Remote Sensing Images. *Remote Sens.* **2022**, *14*, 3109. [CrossRef]
16. Chen, L.C.; Papandreou, G.; Kokkinos, I.; Murphy, K.; Yuille, A.L. DeepLab: Semantic Image Segmentation with Deep Convolutional Nets, Atrous Convolution, and Fully Connected CRFs. *IEEE Trans. Pattern Anal. Mach. Intell.* **2018**, *40*, 834–848. [PubMed]
17. Chen, L.C.; Papandreou, G.; Schroff, F.; Adam, H. Rethinking Atrous Convolution for Semantic Image Segmentation. *arXiv* **2017**, arXiv:1706.05587.
18. Lin, T.-Y.; Goyal, P.; Girshick, R.; He, K.; Dollar, P. Focal Loss for Dense Object Detection. In Proceedings of the 2017 IEEE International Conference on Computer Vision (ICCV), Venice, Italy, 22–29 October 2017; pp. 2999–3007.
19. Li, X.; Sun, X.; Meng, Y.; Liang, J.; Wu, F.; Li, J. Dice Loss for Data-Imbalanced NLP Tasks. In Proceedings of the Meeting of the Association for Computational Linguistics, Online, 5–10 July 2020; Association for Computational Linguistics: Toronto, ON, Canada, 2020.
20. Shorten, C.; Khoshgoftaar, T.M. A survey on image data augmentation for deep learning. *J. Big Data* **2019**, *6*, 60.

21. Sun, Y.Q.; Chen, X.Y.; Zhang, P.; Zhou, Y. Analysis and discussion on the sorbite volume fraction measurement method of 72A steel wire rod. *Phys. Exam. Test.* **2015**, *33*, 25–28. [CrossRef]

22. Cai, W.; Zheng, J.W.; Qiao, L.; Qiang, L.Q. Indentification on Sorbite Structure in 82A High Carbon Steel Wire. *Spec. Steel* **2010**, *31*, 59–62. [CrossRef]

23. Zhao, X.P.; Liu, Z.P.; Xiao, J.; Wu, H.L. Application of Digital Image Storage Technology in the Determination of Sorbite Content in High Carbon Wire Rod. *Liugang Sci. Technol.* **2017**, *4*, 42–45.

 *materials*

Article

In Situ Observation of the Grain Growth Behavior and Martensitic Transformation of Supercooled Austenite in NM500 Wear-Resistant Steel at Different Quenching Temperatures

Zhongbo Li [1,2], Qing Yuan [1,*], Shaopu Xu [2], Yang Zhou [2], Sheng Liu [1] and Guang Xu [1]

[1] State Key Laboratory of Refractories and Metallurgy, Key Laboratory for Ferrous Metallurgy and Resources Utilization of Ministry of Education, Wuhan University of Science and Technology, Wuhan 430081, China; lzb7460377@163.com (Z.L.); liusheng@wust.edu.cn (S.L.); xuguang@wust.edu.cn (G.X.)

[2] Nanyang Hanye Special Steel Co., Ltd., Nanyang 474500, China; lcb-9999@163.com (S.X.); zy18203778277@126.com (Y.Z.)

* Correspondence: yuanqing@wust.edu.cn; Tel.: +86-15994235997

Abstract: In situ observations of the austenite grain growth and martensite transformations in developed NM500 wear-resistant steel were conducted via confocal laser scanning high-temperature microscopy. The results indicated that the size of the austenite grains increased with the quenching temperature (37.41 μm at 860 °C → 119.46 μm at 1160 °C) and austenite grains coarsened at ~3 min at a higher quenching temperature of 1160 °C. Furthermore, a large amount of finely dispersed $(Fe, Cr, Mn)_3C$ particles redissolved and broke apart at 1160 °C, resulting in many large and visible carbonitrides. The transformation kinetics of martensite were accelerated at a higher quenching temperature (13 s at 860 °C → 2.25 s at 1160 °C). In addition, selective prenucleation dominated, which divided untransformed austenite into several regions and resulted in larger-sized fresh martensite. Martensite can not only nucleate at the parent austenite grain boundaries, but also nucleate in the preformed lath martensite and twins. Moreover, the martensitic laths presented as parallel laths (0~2°) based on the preformed laths or were distributed in triangles, parallelograms, or hexagons with angles of 60° or 120°.

Keywords: in situ observation; austenite; martensite; twins; quenching temperature

Citation: Li, Z.; Yuan, Q.; Xu, S.; Zhou, Y.; Liu, S.; Xu, G. In Situ Observation of the Grain Growth Behavior and Martensitic Transformation of Supercooled Austenite in NM500 Wear-Resistant Steel at Different Quenching Temperatures. *Materials* **2023**, *16*, 3840. https://doi.org/10.3390/ma16103840

Academic Editors: Andrea Di Schino and Claudio Testani

Received: 24 April 2023
Revised: 12 May 2023
Accepted: 17 May 2023
Published: 19 May 2023

1. Introduction

Wear, fracture and corrosion are usually the main failure modes during the service of metal materials. Wear will not directly cause the failure of metal parts, but equipment parts are difficult to repair due to wear. Besides, frequent replacement significantly reduces the working efficiency and service life of equipment, thus leading to a large amount of material and energy loss [1–4]. At present, among the metal wear-resistant materials, austenitic high manganese (Mn) steel, high chromium (Cr) cast iron and low alloy wear-resistant steel are the most widely used. Among them, austenitic high manganese steel has a surface austenitic structure, quickly producing work hardening by a phase transition under strong extrusion or impact. The core of austenitic high manganese steel still retains good toughness and plasticity because of the austenitic structure [5–7]. However, the wear resistance of austenitic high manganese steel is relatively low under low or medium stress conditions, which severely limits their application scope in the wear-resistant materials field [8]. High chromium cast iron, as a second-generation wear-resistant material, is currently recognized as the best wear-resistant material [9–11]. A lot of non-network carbide M_7C_3 with a hardness of 1600 HV is precipitated in high chromium cast iron and the toughness and plasticity is better than that of white cast iron [12]. Therefore, mechanical equipment made of high chromium cast iron can meet the needs of long-term wear resistance in complex environments. However, due to the large number of valuable elements such as Cr and nickel (Ni), the production of high chromium cast iron is complex, which also increases the production cost and limits its wide application in industrial production [13]. In view of the many problems regarding the service and production of the above two steels, low-alloy wear-resistant steel has gradually become the research topic of the new generation of wear-resistant metal materials [14–18].

Various wear-resistant steel products are made by Chinese iron and steel manufacturers, among which the production technology of the products below NM400 is relatively mature. Tempered martensite is obtained by off-line re-austenitizing after rolling to improve the strength hardness, and then the toughness is improved by subsequent tempering. Martensite has ultra-high strength and hardness among the different microstructures in steel, and it is usually selected as an important microstructure in the production of ultra-high strength steel. Therefore, among all the kinds of low-alloy wear-resistant steels, martensitic wear-resistant steel is promising. The wear resistance of martensite mainly relies on its high hardness, but the wear resistance under high impact is erratic due to its poor toughness. Therefore, many studies on martensitic wear-resistant steel have focused on improving its toughness and plasticity [19–21]. It has been revealed that martensitic laths and blocks are the organizational units affecting strength and hardness, while martensitic packets affect the plasticity and toughness [22,23]. In addition, Liang et al. found that crack propagation could be effectively inhibited by using smaller sizes and angles of the martensitic packet [24]. The main structure control unit affecting fracture, the size of martensitic block, was identified by Inoue et al. through a study on the cleavage fracture of tempered martensitic steel [25]. Moreover, some scholars improved the fracture toughness and elongation of martensitic steel by optimizing the composition and heat treatment process, so that about 30% residual austenite was obtained at room temperature [26].

In addition, the wear resistance of martensitic steel had been studied extensively. Liang et al. reported that low-alloy martensitic wear-resistant steel exhibited better wear resistance under moderate impact wear, and its comprehensive mechanical property was more than twice that of austenitic high manganese steel [27]. Cao et al. prepared Ti-Cr-B (boron) microalloyed high-strength wear-resistant steel with tempered martensite, in which a high dislocation density and tempered carbide precipitation hardened the matrix [28]. In the work of Ma et al., they found that the solid solution carbon content in the martensitic structure was a direct factor affecting the wear resistance and subsurface hardness [29].

Research on high-grade low-alloy wear-resistant steel is insufficient. A high-grade NM500 wear-resistant steel is presented in the present study. Martensite transformation greatly influenced the microstructure and properties of the low-alloy wear-resistant steel, and the quenching temperature and the subsequent cooling also played a decisive role in these properties. However, few studies about NM500 wear-resistant steel have studied the relationship between grain growth and martensite transformation. Moreover, a dynamic investigation into austenite grain growth and martensite transformation in NM500 wear-resistant steel has not been conducted. Therefore, the phase transformation behavior of NM500 wear-resistant steel in a continuous cooling process was analyzed by confocal laser scanning high-temperature microscopy (CSLM). The novelty of the present work is summarized in two aspects: (1) the grain growth behavior of high grade NM500 wear-resistant steel at two quenching temperatures was first recorded and compared and (2) the martensite transformations in different austenite grains were dynamically analyzed. The results of the present study will provide a reference to understand austenite grain growth and martensite transformation at different quenching temperatures.

2. Experimental Procedures

Figure 1 shows the VL2000DX-SVF17SP confocal laser scanning high-temperature microscope and the corresponding quenching process. This CSLM equipment consists of a flow control device, a console, a display, high temperature microscope, etc., which can observe and capture all kinds of physical and metallurgical phenomena in real time. It studies the dynamic process at a high resolution in real time through high-speed laser scanning imaging. A higher automation degree was achieved through digital image information storage and processing technology. The light source of CSLM was a blue laser, whose wavelength and resolution were about 410 nm and 0.25 μm, respectively. Various phases emerged under the effect of thermal etching, rather than chemical corrosion.

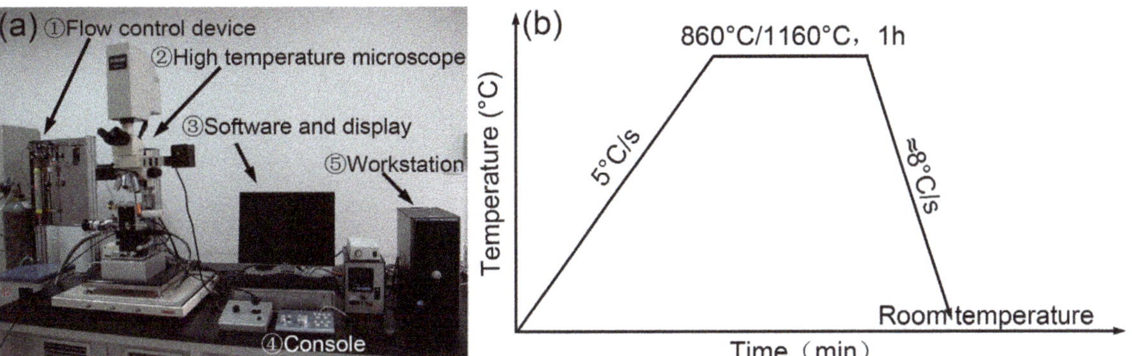

Figure 1. (**a**) The confocal laser scanning high-temperature microscope and (**b**) in situ observation process.

The experimental steel was a developed high-grade NM500 steel with the chemical composition Fe-0.23C-0.20Si-1.49Mn-1.15Cr-0.25Ni-0.37(Nb+V+Ti+Mo)-0.022Cu-0.00174B-0.01P-0.002S (wt.%). Figure 2 illustrates the structure of the confocal laser scanning high-temperature microscope. A small cylindrical sample with dimensions of Φ 6 mm × 5 mm was finish machined, and the two faces of the sample were polished until a mirror surface was obtained. Subsequently, the sample was placed into the Al_2O_3 crucible for in situ observation. Before the experiment, the sample chamber was vacuumed to 6×10^{-3} Pa, and then argon gas was introduced to prevent sample oxidation. Microstructure evolution was recorded throughout the whole process with a recording frequency of 5 photos/s. Figure 1b demonstrates the heating process with two different quenching temperatures. Firstly, the samples were reheated to 860 °C and 1160 °C, respectively, at a rate of 5 °C/s and then held for 1 h. Subsequently, the maximum cooling rate was applied to cool the specimens to room temperature after thermal holding to simulate the quenching process. The heating rate of 5 °C/s was determined by an empirical value considering its small size. Two different quenching temperatures of 860 °C and 1160 °C were chosen according to the minimum and maximum tempering parameters in industrial production. The holding time of 1h was utilized also based on the requirements of industrial production. It should be noted that the average cooling rate during quenching was estimated to be only 8 °C/s due to the low cooling ability at a low temperature range. However, the cooling rate before martensite transformation could reach 15 °C/s, totally ensuring the martensite transformation. In addition, to facilitate the analysis of grain size evolution, the eyepiece was manually changed during the in situ observation experiment. An appropriate magnification was selected due to the different size of parent austenite grains (PAGs) at 860 °C and 1160 °C. Meanwhile, the precipitates quenched at 1160 °C were identified by transmission electron microscopy (TEM) using a thin film specimen.

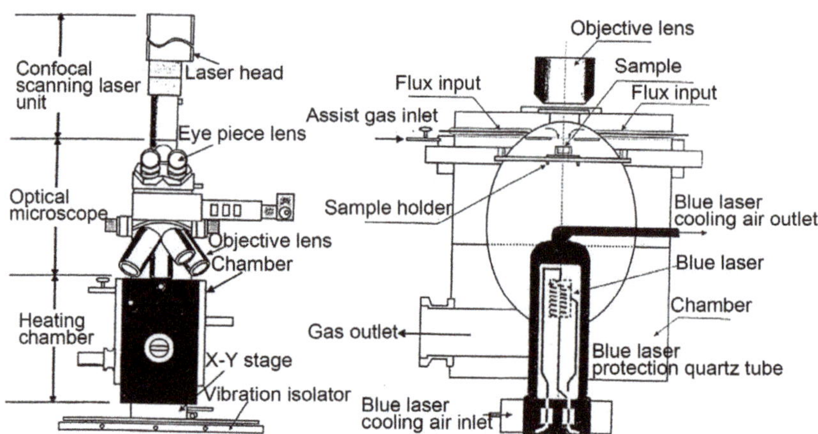

Figure 2. The structure of a confocal laser scanning high-temperature microscope.

3. Results and Discussion

3.1. Austenite Nucleation

Figure 3 shows the morphologic changes from room temperature to the preset temperature of 860 °C. Some particles were present on the surface of the sample, and these dark particles were (Fe, Cr, Mn)$_3$C precipitates (Figure 3a) determined by the following results. Due to the increased amount of Cr and Mn, (Fe, Cr, Mn)$_3$C precipitates dominated in the as-received steel treated by hot rolling and subsequent air cooling. When the temperature was increased to 548.7 °C, some corrugated folds appeared on the sample surface. These folds correspond to the grain boundaries of the initial ferrite (pearlite), which gradually emerged under the effect of thermal etching (Figure 3b). As the temperature increased to 701.6 °C, another corrugated fold gradually covered the grain boundaries of the existing ferrite (Figure 3c). Additionally, this corrugated fold became more and more clear and gradually formed the grain boundaries of polygonal grains as the temperature rose to 827.7 °C (Figure 3d) and 862.9 °C (Figure 3e). It was inferred that the corrugated fold at 701.6 °C was an austenitic grain boundary, that is, the A$_{c1}$ temperature (the beginning temperature at which the pearlite transforms to austenite during the heating process) was about 701.6 °C when the steel was reheated at 5 °C/s. The measured A$_{c1}$ temperature of this experimental steel was about 658 °C using a thermal simulated test, a little lower than that obtained via in situ observations. The measured A$_{c1}$ temperature was obtained with a very slow heating rate (about 0.1 °C/s), and the A$_{c1}$ temperature increased with the increase in the heating rate (5 °C/s). Moreover, austenization process of the sample finished more quickly at a higher heating rate. The austenization process completed at 862.9 °C (Figure 3e), but the grain boundary morphology of the initial microstructure remained. Meanwhile, the visible (Fe, Cr, Mn)$_3$C precipitates became clearer and their size increased.

Austenite transformation is related to the nucleation rate and the growth rate, and can be expressed as Equations (1) and (2) [30,31]:

$$N = f_N \exp\left(-Q_N/K\Delta T\right) \tag{1}$$

$$G = f_G \exp\left(-Q_G/K\Delta T\right) \tag{2}$$

where N is the nucleation rate, G is the growth rate, Q_N and Q_G are the nucleation and growth activation energies, respectively, f_N and f_G are the impact factors between structure and nucleation with growth, respectively, and ΔT is the superheat. This equation reveals that the superheat increases with the increase in the heating rate, which increases the nucleation and growth rates of the austenitic transformation. Therefore, the rate of austenitic transformation increased significantly, the time required from initial austenization to complete austenization was greatly reduced, and thus the required phase transition interval was correspondingly reduced. In addition, the transformation of steel during continuous heating is equivalent to the accumulation of countless isothermal transformations. The relationship between the isothermal incubation period and the transition temperature can be established using Scheil's superposition principle [32]:

$$\sum_{i=1}^{i=n} \frac{\Delta t}{A_i} = 1 \tag{3}$$

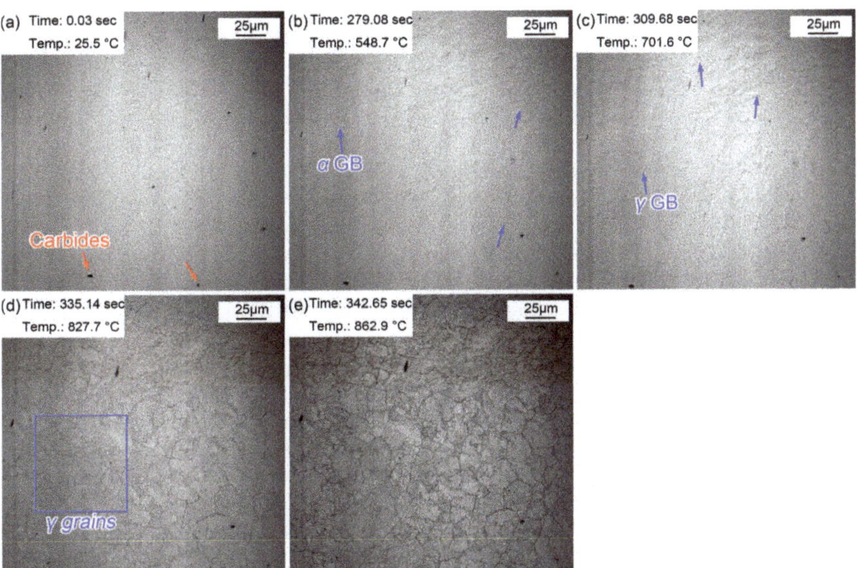

Figure 3. Morphology variations from room temperature to 860 °C. (**a**) 25.5 °C, before heating; (**b**) 548.7 °C, initial grain boundaries appeared; (**c**) 701.6 °C, austenization began; (**d**) 827.7 °C, obvious austenite grains; (**e**) 862.9 °C, the preset quenching temperature.

Differential Equation (4) is obtained when Δt is small enough.

$$\int_{t=0}^{t=t_n} \frac{dt}{A(T)} = 1 \tag{4}$$

where Δt and dt are the transformation time at temperature T and A_i and $A(T)$ are the corresponding incubation periods. The relationship between the incubation period and the transition temperature is shown in Equation (5):

$$\int_{T_1}^{T_s} \frac{dt}{A(T)} = \int_{T_1}^{T_s} \frac{1}{A(T)} \cdot \frac{1}{\frac{dT}{dt}} \cdot dT = \int_{T_1}^{T_s} \frac{1}{A(T)} \cdot \frac{1}{v} \cdot dT = 1 \tag{5}$$

The relationships between the transformation rate, C, transformation beginning and ending temperatures, T_s and T_f, and the heating rate, v, are interpreted by Equations (6) and (7), where the transformation volume is f.

$$c = \frac{df}{dt}, v = \frac{dT}{dt} \tag{6}$$

$$\int_{t=0}^{t=t_n} \frac{df}{dt} dt = \int_{T_s}^{T_f} \frac{df}{dt} \cdot \frac{dT}{v} = \int_{T_s}^{T_f} \frac{C}{v} \cdot dT = 1 \tag{7}$$

where T_1 is the equilibrium temperature and t_n is the time to the transformation ending temperatures T_f. This equation proves that with the increase in the heating rate, both the initial temperature and the end temperature of the phase transition increase. In addition, the dissolution and diffusion of carbonitrides is inevitable during the austenization process of experimental steel, and atoms migrate between phases through the diffusion mechanism. With the increase in the heating rate, the diffusion of carbon and alloying elements at the equilibrium temperature decreases, thus increasing the austenitic transition temperature. In the process of continuous heating, with the increase in temperature, the diffusion coefficient and diffusion rate of atoms increase greatly, so the driving force of the austenite phase transformation is enhanced.

The morphological variations from room temperature to 1160 °C are displayed in Figure 4. Some (Fe, Cr, Mn)$_3$C particles appeared on the sample surface (Figure 4a) and corrugated folds appeared at 550.4 °C (Figure 4b). Another corrugated fold gradually covered the grain boundaries of the existing ferrite structure at 704.7 °C (Figure 4c). The corrugated fold became more and more clear and gradually formed the grain boundaries of polygonal grains at 828.4 °C (Figure 4d). The A_{c1} temperature was basically the same as that in specimen reheated to 860 °C. This is because the heating processes of the two samples were the same before heating to 860 °C. However, when the sample was reheated to 1160 °C (Figure 4e), more large-sized grains appeared and the grain boundaries became sharper and clearer. In addition, the number of clearly visible (Fe, Cr, Mn)$_3$C particles increased significantly, and they gradually became more coarse. This can be explained by the gradual dissolution of some invisible finely dispersed (Fe, Cr, Mn)$_3$C particles at 1160 °C. The austenite grain boundary mobility increased, quickly resulting in the coarsening of austenite grains. The increased amount of visible (Fe, Cr, Mn)$_3$C particles was attributed to the ripening of more micro/nano (Fe, Cr, Mn)$_3$C particles at high temperatures. Compared with the sample quenched at 860 °C, the austenite grains quenched at 1160 °C clearly coarsened. Figure 5 shows the (Fe, Cr, Mn)$_3$C particles in the specimen quenched at 1160 °C by TEM and the related energy spectrum. (Fe, Cr, Mn)$_3$C particles was a compound of cementite (Fe$_3$C) with other alloy elements. A few microalloy elements such as Ti, V, and Mo were captured due to their increased solvation at higher temperatures. In addition, apparent quenching dislocations were observed in the lath martensite.

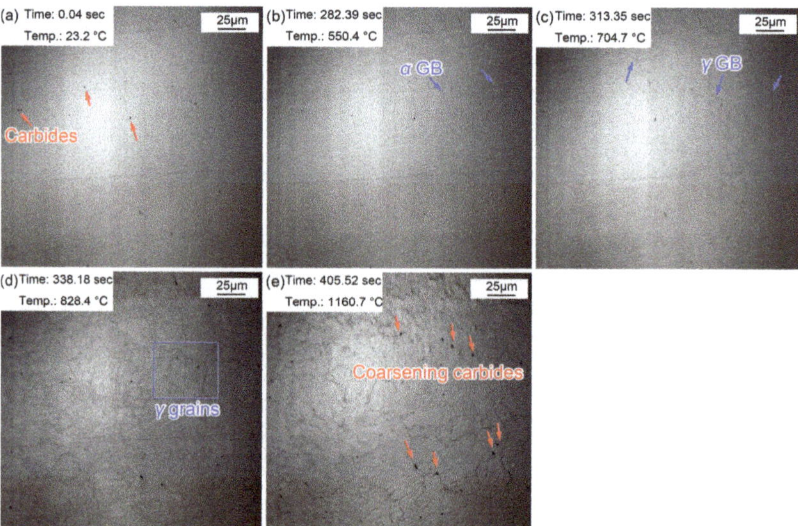

Figure 4. Morphology variations from room temperature to 1160 °C. (**a**) 23.2 °C, before heating; (**b**) 550.4 °C, initial grain boundaries appeared; (**c**) 704.7 °C, austenization began; (**d**) 828.4 °C, obvious austenite grains; (**e**) 1160.7 °C, the preset quenching temperature.

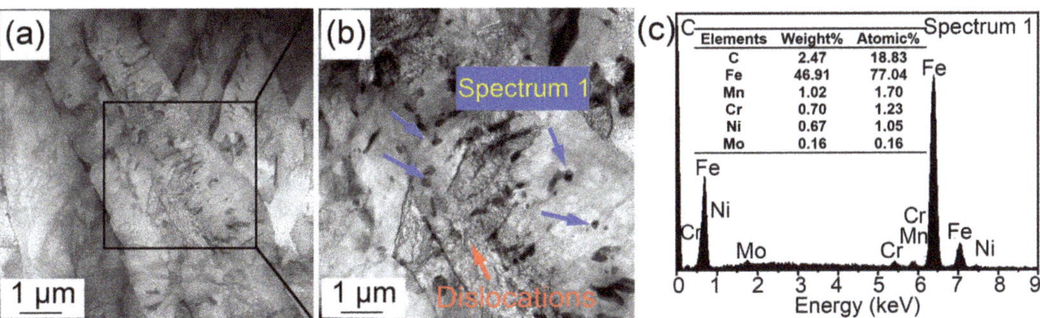

Figure 5. (**a**,**b**) TEM images and (**c**) energy spectrum showing (Fe, Cr, Mn)$_3$C particles at 1160 °C.

3.2. Austenite Growing

Figure 6 shows the morphologic changes from 1~10 min with a time interval of 1 min when the quenching temperature was 860 °C. Compared with the morphology after just reaching the preset temperature, the grain boundaries of austenite grains were clearer after 1 min (Figure 6a). This is because the grain boundary grooves are more easily exposed after longer thermal etching. In addition, the austenite grain boundaries were narrow and straight with a grain boundary angle of 120°. Some local small grains gradually merged into large ones, as shown in the rectangle in Figure 6f,j. In addition, the austenite grain boundaries expanded and migrated to form large grains, as shown by the pink arrows in Figure 6d,j. The gradual merging of small grains and the migration of some grain boundaries indicated a unconspicuous growth process and trend.

During the thermal holding process, a fog-like substance shown by the oval in Figure 6c appeared. The fog-like substance gradually turned black, and then subsequently disappeared. This dark mist is the vapor of alloying elements, which tends to steam outward from the steel matrix when reheated. A similar phenomenon was also reported in the research of Lan et al. [33]. Manganese (Mn) volatilization was determined by a simultaneous thermal analysis, and they clarified that Mn tended to migrate to the substrate surface and volatilize when the temperature was high enough. In addition to the clearly visible precipitates at the beginning of reheating, many fine (Fe, Cr, Mn)$_3$C particles also appeared in the austenite grains during thermal holding, as shown in Figure 6d. These fine (Fe, Cr, Mn)$_3$C particles gradually appeared as some of the unprecipitated (Fe, Cr, Mn)$_3$C particles matured and emerged. The (Fe, Cr, Mn)$_3$C particles matured during thermal holding, as shown in Figure 6j.

The morphologic variation in a time interval of 10 min from 20~60 min at the quenching temperature of 860 °C is exhibited in Figure 7. The coarsening of (Fe, Cr, Mn)$_3$C particles was more obvious, and there were more areas where small grains merged into large grains. In addition, twins could be observed in austenite grains (Figure 7a). The existing fog-like steam gradually volatilized and disappeared during thermal holding, while it appeared in other areas. This may be explained by the uneven distribution of some alloying elements such as Mn. In addition, austenite grain coarsening was obvious during thermal holding, in which the proportion of small grains decreased gradually. Moreover, and the intramural twins were more clearly visible (Figure 7e). The intracrystalline twins can be considered as annealing twins [34]. There were more alloying elements in the experimental steel, which significantly reduced the stacking fault energy. Compared with ordinary carbon steel, intracrystalline twins are more likely to occur in alloyed steels. The appearance of twins segregated and refined the grains, thus increasing the resistance of dislocation movement and strengthening the steel.

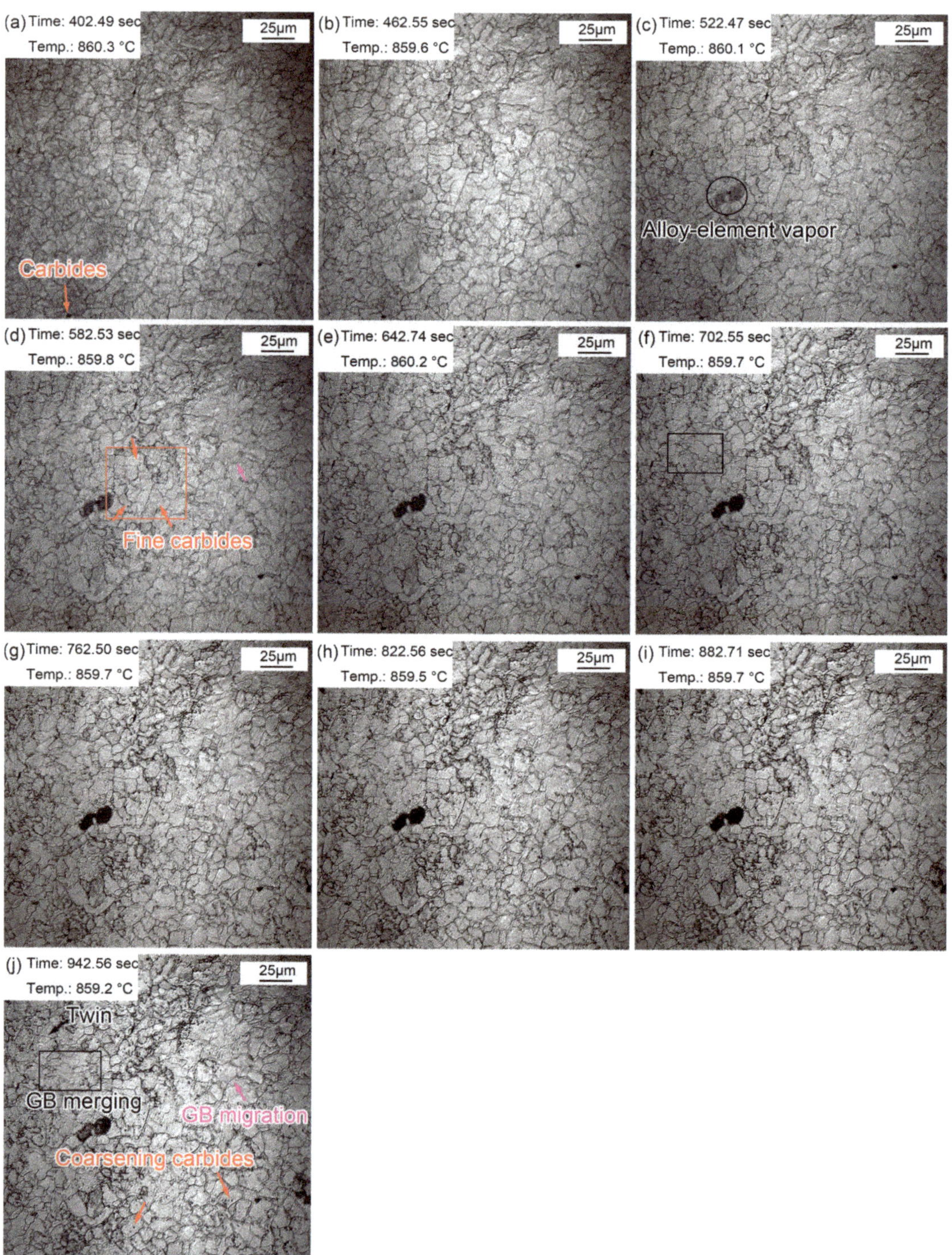

Figure 6. Morphologies during thermal holding at 860 °C from 1~10 min. (**a**) 1 min; (**b**) 2 min; (**c**) 3 min; (**d**) 4 min; (**e**) 5 min; (**f**) 6 min; (**g**) 7 min; (**h**) 8 min; (**i**) 9 min; (**j**) 10 min.

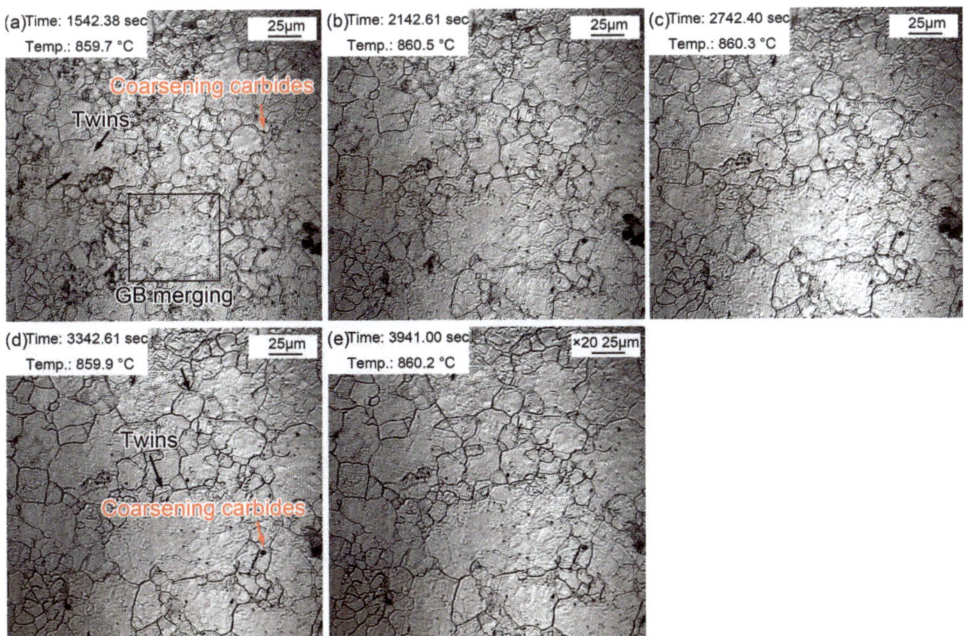

Figure 7. Morphologies during thermal holding at 860 °C from 20~60 min. (**a**) 20 min; (**b**) 30 min; (**c**) 40 min; (**d**) 50 min; (**e**) 60 min.

Figure 8 presents the morphologic evolution from 1~10 min at the quenching temperature of 1160 °C. The austenite grains were much clearer after holding at 1160 °C for 1 min (Figure 8a) as a result of continuous thermal etching. The migration of grain boundaries was obvious during holding, as shown in Figure 8a,b (blue arrow 1) and Figure 8b,c (blue arrow 2). Part of the original grain boundaries gradually faded away during grain boundary migration, and the old ones were gradually filled in. In addition, except for the outward expansion of grain boundaries, small grains were partitioned by surrounding large grains, as shown in the oval in Figure 8d. Alloy element steam, as shown in Figure 7c, also appeared in the specimen reheated to 1160 °C. In addition, many annealing twins traversing or occupying the whole grain were captured in the austenite grains. (Fe, Cr, Mn)$_3$C particles ripened during the holding process, as shown in Figure 8j. The gradual appearance of fine (Fe, Cr, Mn)$_3$C particles was attributed to (Fe, Cr, Mn)$_3$C ripening at 1160 °C, which was captured by limited magnification. Compared with the grain morphology at 860 °C for 20~60 min, the grain size was obviously coarsened at 1160 °C, and there were many dense fine (Fe, Cr, Mn)$_3$C particles inside the grains.

The coarsening of austenite grains at 1160 °C is related to the redissolution of (Fe, Cr, Mn)$_3$C particles. Firstly, the atomic size of Cr/Mn/Ti is very different to Fe, which causes a certain solute atomic dragging effect. Reconcentration of a large number of solute atoms such as Cr, Mn, vanadium (V), and titanium (Ti) at the grain boundaries or subgrain boundaries could prevent the migration of grain boundaries and thus inhibit recrystallization. In addition, (Fe, Cr, Mn)$_3$C particles were preferentially precipitated at the grain boundaries and dislocation lines, pinning the austenite grain boundaries and hindering the growth of austenite grains. Grain boundary migration caused austenite grain growth. The surface energy increased when the grain boundaries contacted the (Fe, Cr, Mn)$_3$C particles. Only when the thermal activation energy was greater than the increased surface energy were the (Fe, Cr, Mn)$_3$C particles cut or bypassed by the grain boundary. Therefore, the (Fe, Cr, Mn)$_3$C particles significantly slowed down the formation of austenite and prevented the growth of grains. Similar observations were made in the work of G.

Khalaj et al., where they established a model to predict the austenite grain size in Nb/Ti microalloyed steel [35]. Since the solute concentration around small particles was greater than those around large particles, the solute atoms spread from small particles to large particles, resulting in the redissolution of small particles and the growth of larger particles. Therefore, the fine precipitates gradually redissolved and continuously formed large size carbonitride particles when the holding time was long enough at 1160 °C.

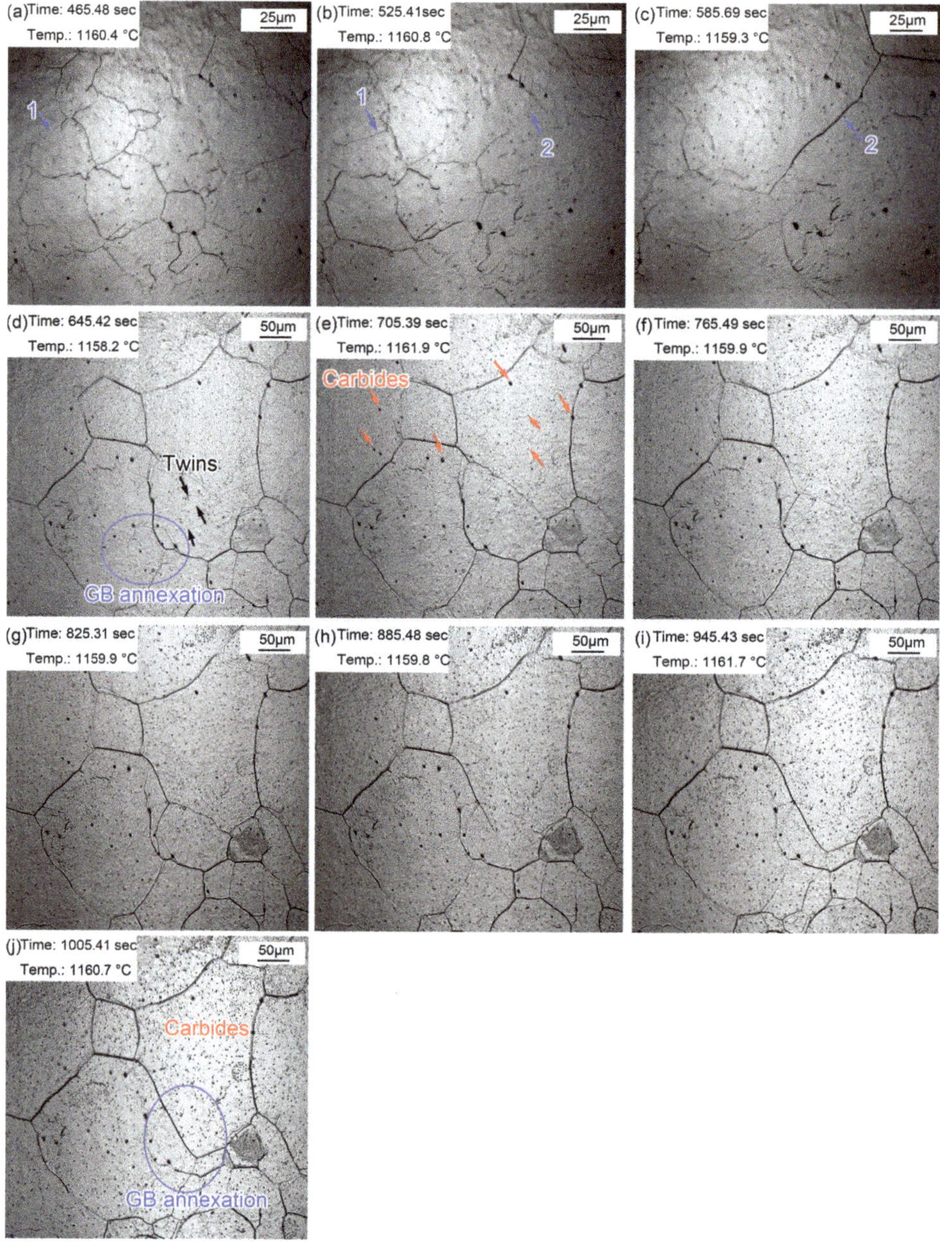

Figure 8. Morphologies during thermal holding at 1160 °C from 1~10 min. (**a**) 1 min; (**b**) 2 min; (**c**) 3 min; (**d**) 4 min; (**e**) 5 min; (**f**) 6 min; (**g**) 7 min; (**h**) 8 min; (**i**) 9 min; (**j**) 10 min.

Figure 9 displays the morphologic changes with a time interval of 10 min from 20 to 60 min at 1160 °C. Dense, small $(Fe, Cr, Mn)_3C$ particles formed in the austenite grains. This signifies that the coarsening of $(Fe, Cr, Mn)_3C$ particles was more obvious compared to that during the holding time of 10 min. In addition, the grain boundaries of small-size austenite were gradually absorbed by the surrounding large-size austenite. In addition, apparent twins were observed in austenite grains (Figure 9a). Furthermore, austenite grain coarsening still occurred during holding from 20 to 60 min, in which the proportion of small grains further decreased. The intra twins were more clearly visible (Figure 9e) because of the larger austenite grains.

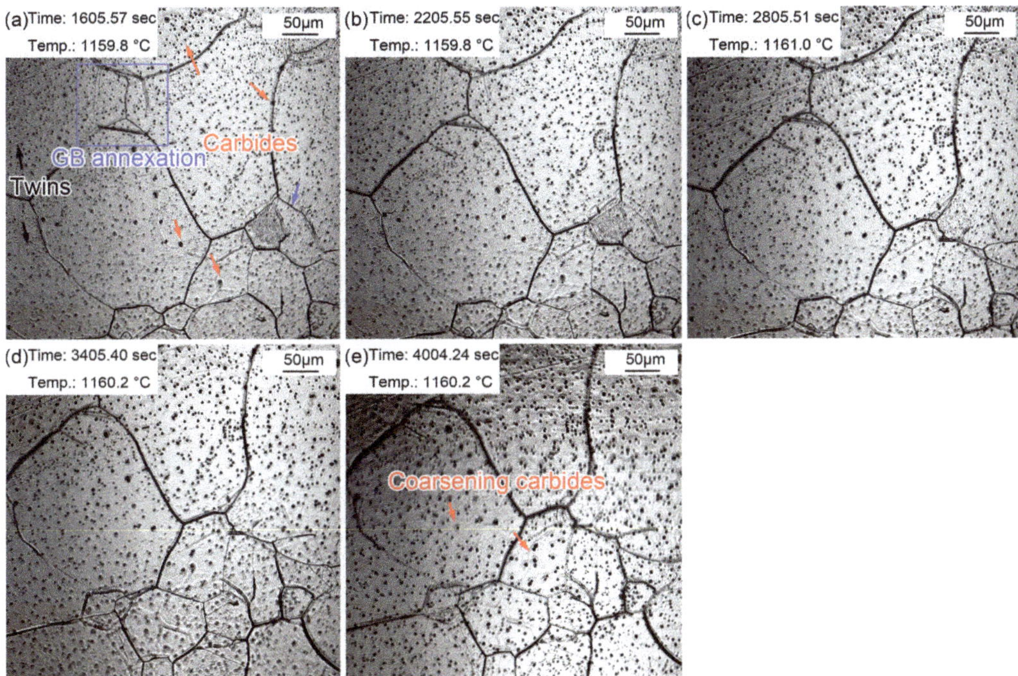

Figure 9. Morphologies during thermal holding at 1160 °C from 20~60 min. (**a**) 20 min; (**b**) 30 min; (**c**) 40 min; (**d**) 50 min; (**e**) 60 min.

3.3. Grain Size

Figure 10 summarizes the grain size and growth rate of austenite at different quenching temperatures. Enough grains were present to ensure an improvement in the accuracy of the statistical process. Grains less than half the average size grain were not counted, and grains larger than half the average size grain were considered. There was little difference in the austenite grain size during the reheating stage before the preset quenching temperatures. However, the austenite grain size at 860 °C was always smaller than that at 1160 °C. Figure 10b shows the growth trend of austenite grains with time at 860 °C. The growth process was relatively slow, and the obvious coarsening of austenite grains was complete after holding for about 30 min. The growth process of austenite grains was very rapid and intense, and the coarsening of austenite grains was completed within 10 min at 1160 °C. Austenite grain growth rate curves at different quenching temperatures were obtained through the first derivative (Figure 10d–f). The austenite grain growth rate was significantly faster at 1160 °C. In addition, when the quenching temperature was 1160 °C, the maximum austenitic growth rate appeared at the holding time of ~3 min, whereas it occurred at ~30 min at 860 °C.

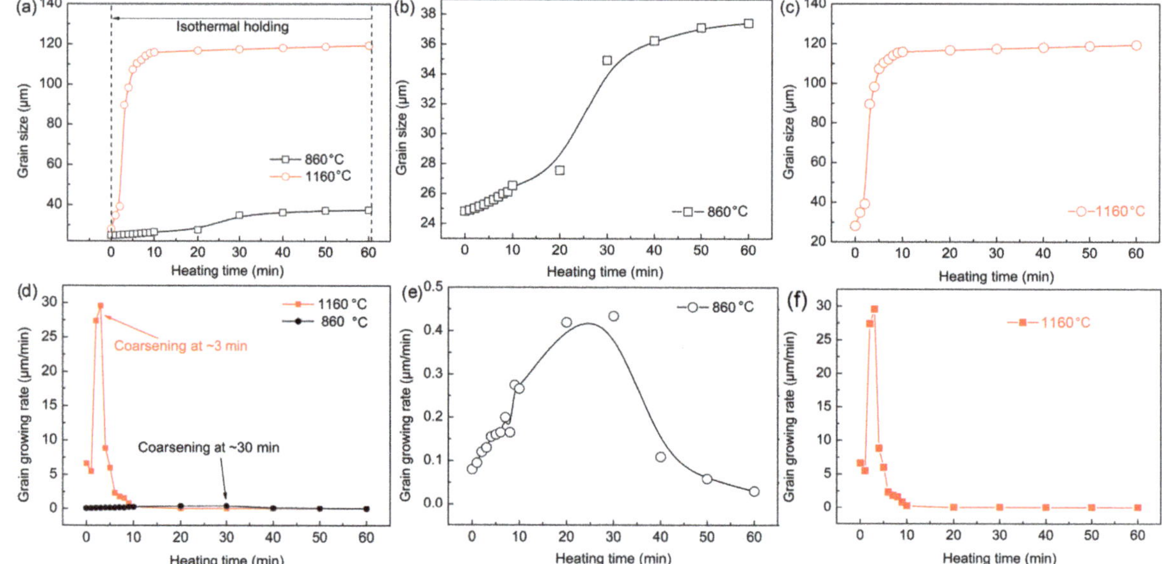

Figure 10. (**a–c**) Grain size and (**d–f**) growth rate of austenite from formation to thermal holding at different quenching temperatures.

It can be concluded that the austenite grains coarsened in a short time and the coarsening rate was higher at 1160 °C. This is because the austenite grain boundary migration ability was increased at a higher temperature. The atomic diffusion process was more rapid, and part of the grain boundary faded and disappeared more easily. In addition, many small dispersed (Fe, Cr, Mn)$_3$C particles redissolved and ripped, which led to a significant decrease in the migration ability of austenite grain boundaries. Furthermore, the growth rate began to decrease gradually when austenite grains coarsened extensively. This was because the energy for grain growth can no longer be provided as the heating temperature was unchanged, and the redissolution and breaking of the (Fe, Cr, Mn)$_3$C particles was basically resolved. Consequently, the pinning effect of (Fe, Cr, Mn)$_3$C particles on austenite grain boundaries was stabilized, so the austenite coarsening gradually weakened.

3.4. Martensite Transformation

The above-mentioned austenite grain growth rules indicated that the austenite grain size greatly varied at different quenching temperatures. It has been pointed out that the austenite grain size affected the martensitic transformation temperature and the phase transformation behavior of supercooled austenite during cooling. Figure 11 shows the martensitic transformation during cooling in the sample quenched at 860 °C. Lath martensite appeared, as shown by the blue arrow in Figure 11a, when the temperature decreased to 369.2 °C. This martensite was primary martensite, also called fresh martensite (FM). The martensitic phase transition point (M$_s$) of the sample was about 369.2 °C, while the M$_s$ temperature of this steel was determined to be 340 °C via a thermal simulation experiment. The effective M$_s$ was obtained via a thermal simulation experiment through the overall volume expansion effect of the martensitic transformation, while in situ observation determined the M$_s$ just according to the temperature at which the martensite appeared in one certain grain. Generally speaking, the M$_s$ determined by in situ observations is higher than that reflected in thermal simulation experiments. This is because the martensitic transformation does not start at the same time in all grains, although the nucleation and growth of martensite explosively proceeded following this. In addition, martensite nucleated from the grain boundary and grew in between grains until stopping at the grain boundary. More and more lath martensite explosively appeared as the temperature decreased, and most lath martensite traversed the entire grain. Furthermore, some lath martensite was found to nucleate and grow from the twins (Figure 11b). Since the formed martensite stimulated the nucleation of the surrounding untransformed austenite, the austenite nucleated and grew in parallel after this trigger. Therefore, the lath martensite grew in a parallel manner in some austenite grains. The lath martensite appeared simultaneously with an angle of 60° at 267.5 °C. In addition, some lath martensite simultaneously formed parallel to each other. More FM was observed as the temperature continued decreasing accompanied by secondary martensite (SM). SM refers to martensite with slightly thin laths formed around FM, which appeared at a certain angle with FM (Figure 11c). More and more surface reliefs due to martensitic transformations gradually appeared at the PAG boundaries (Figure 11d). Most martensite stopped growing when they encountered grain boundaries, and some martensite met each other, which also stopped the growth of lath martensite (Figure 11e). The nucleation and growth of martensite were very weak when the temperature approached room temperature. Most martensite transformations finished within 13 s, and the rate of martensitic transformations gradually slowed down. However, the distortion caused by martensitic transformations prevented martensitic transformations in the surrounding austenite. Small parts of the regions were retained as residual austenite, in which the sharing of elements such as carbon in ferrite to residual austenite was mainly completed (Figure 11f). The growth rate of lath martensite was relatively fast, but the growth rate of longitudinal lath martensite was faster than that of lateral lath martensite. Although the martensitic transformations explosively proceeded, the martensitic transformations were not simultaneous. Martensitic transformations selectively started in PAGs, but this selective process was very short. Nevertheless, the temperature of the sample may remain unchanged or even slightly increase during the cooling process since martensitic transformations release more latent heat of transformation. Therefore, isothermal martensite formation was inevitable. This latent heat caused by martensitic transformations was also one of the reasons for the selective initiation of martensitic transformations. The supercooling degree became smaller at a constant or slightly increased temperature; thus, martensitic transformations were inhibited. In addition, the selective initiation of martensitic transformations was also related to the distortion caused by martensitic transformations. Martensitic transformations in untransformed austenite were strongly inhibited by the surrounding martensitic transformations.

The martensitic transformation of the sample quenched at 1160 °C is displayed in Figure 12. The martensitic phase transition point, M_s, was about 310.0 °C (Figure 11a), which was lower than that in the sample quenched at 860 °C (369.2 °C). The M_s temperature should be higher in a larger austenite. However, the results of in situ observations of martensitic transformations were extraordinary. The possible reason for this is that the martensitic transformations observed by the in situ method were local to the sample surface, with a limited view field. An unobserved view field may have shown the martensitic transformations at a relatively higher temperature. In addition, it is difficult to unify the different starting temperatures of martensitic transformations due to the uneven composition caused by the evaporation of alloying elements. It was accidentally observed that lath martensite grew through grain boundaries in Figure 12c. These newly formed grain boundaries were relatively straight. In addition, some lath martensite nucleated and grew from the twins. The reason why the twins acted as martensitic nuclei was that martensitic transformations require structural and energy fluctuations. As a kind of crystal defect, twins provide a large defect energy which meets the structural and energy fluctuation requirements. Increasingly more FM and SM gradually appeared with the decrease in temperature, in which the SM appeared at a certain angle to FM. Most martensite transformations finished within 2.25 s, and the rate of martensite transformations slowed down. However, the distortion caused by martensite transformations inhibited the martensite transformations in surrounding austenite (Figure 12j). Martensite growth was a nondiffusion interfacial cooperative pushing process, and the martensite specific volume was larger than that of austenite. Therefore, elastic deformation was caused, accompanied by volume expansion, during the martensite phase transition. Additionally, then a large distortion energy formed, which hindered further martensite transformations.

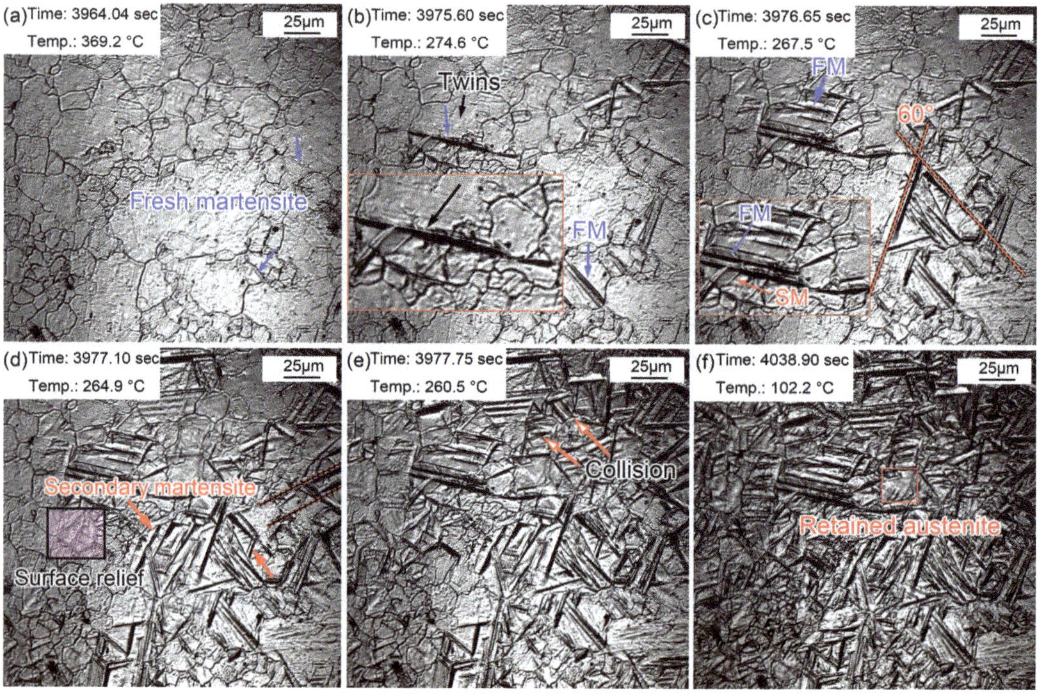

Figure 11. Martensite transformations of supercooled austenite quenched at 860 °C. (**a**) 369.2 °C, martensite appeared; (**b**) 274.6 °C, martensite increased; (**c**) 267.5 °C, martensite nucleated and grew at the twins; (**d**) 264.9 °C, SM and surface relief; (**e**) 260.5 °C, martensitic lath collisions; (**f**) 102.2 °C, martensitic transformation stopped and residual austenite formed.

Martensitic transformations in the same sample did not appear at first in large-sized grains but appeared in the PAGs in a seemingly chaotic manner. This may be explained by the differences in the size, composition, and defect density in PAGs, which led to the selectivity of martensite nuclei. In addition, the grains were coarser due to a higher quenching temperature; thus, the driving force of martensitic transformations was greater. Additionally, the migration rate of phase interfaces increased accordingly, so the martensitic transformations were faster.

Figure 13 shows the martensitic transformation of supercooled austenite at different quenching temperatures, which is a summary based on in situ observations. Martensite nuclei did not occur simultaneously during the quenching process of supercooled austenite, but selectively proceeded and increased in batches in some areas. This nucleation pattern divided untransformed austenite into multiple regions. In different regions, the size of firstly formed martensite (fresh martensite) was large and the size of subsequent martensite (secondary martensite) was small. This is because the shape of martensite depends on the stress field between the nucleated lath martensite and other martensitic nuclei. The parent austenite presented obvious different grain sizes at different quenching temperatures. In addition, the size and volume fraction of the coarse (Fe, Cr, Mn)$_3$C particles in the matrix increased with the quenching temperature.

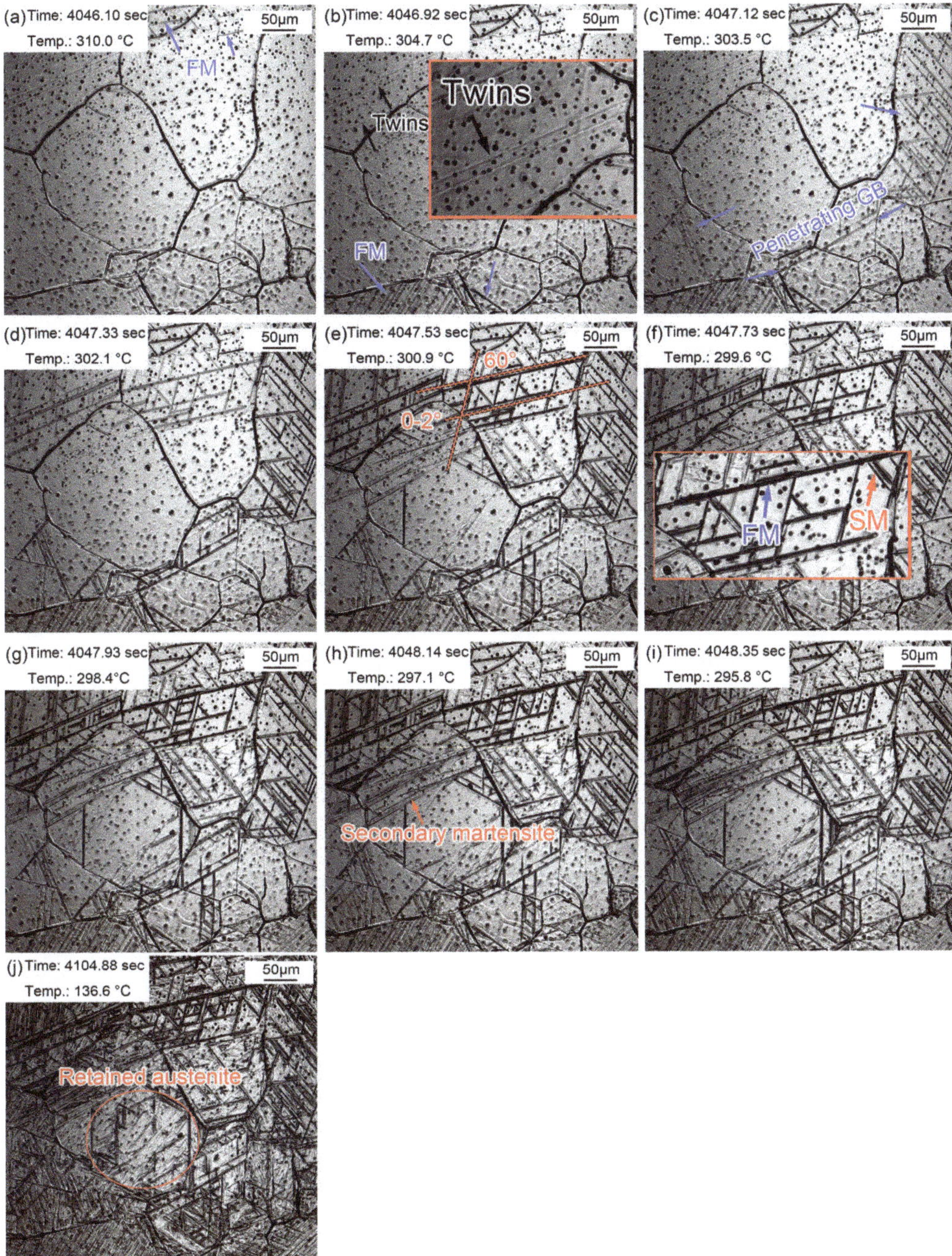

Figure 12. Martensite transformation of supercooled austenite quenched at 1160 °C. (**a**) 310.0 °C, martensite appeared; (**b**) 304.7 °C, martensite increased; (**c**) 303.5 °C, martensite traversed grain boundaries; (**d**) 302.1 °C, martensite increased and appeared at 60° angles; (**e**) 300.9 °C, martensitic packet; (**f**) 299.6 °C, SM increased; (**g–i**) 298.4 °C, explosive martensite; (**j**) retained austenite.

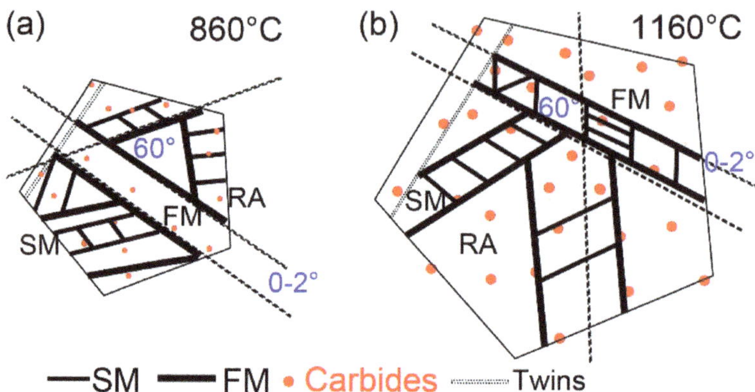

Figure 13. Schematic diagram of martensitic transformations of supercooled austenite at (**a**) 860 °C and (**b**) 1160 °C.

Martensitic nucleation and growth in different parent austenite grains did not affect each other in the early stage of martensitic transformations, during which less martensite formed. The martensitic transformations gradually increased as the temperature decreased, and the martensitic laths restricted each other. In general, there were three types of martensitic nucleation. Firstly, martensite nucleated along the PAG boundaries and grew in between the grains until stopping when it collided with other lath martensite or grain boundaries. In addition, martensite nucleated at annealing twins, which had lattice defects and provided better structural and energy fluctuations. Moreover, martensite nucleated at the preformed lath martensite and grew in the austenitic grains at about 60° or 120° to form new lath martensite. The lath packet exhibited two types: parallel laths (0~2°) based on the preformed laths and martensitic laths at 60° or 120° in the other direction stimulated by the preformed laths, finally forming triangle, parallelogram, or hexagon morphologies. The formation of SM laths also strongly inhibited the martensitic transformations of the surrounding untransformed austenite and promoted the formation of residual austenite.

4. Conclusions

1. The austenite grains in NM500 steel at a quenching temperature of 860 °C (37.41 μm) were smaller than those at a quenching temperature of 1160 °C (119.46 μm). Austenite grains coarsened at ~3 min and ~30 min, respectively, at quenching temperatures of 1160 and 860 °C. In addition, a large amount fine dispersed (Fe, Cr, Mn)$_3$C particles redissolved and broke apart at 1160 °C, resulting in many large, visible carbonitrides.
2. The nucleation of martensite did not proceed simultaneously during the quenching process. Selective prenucleation dominated, which divided untransformed austenite into several regions and resulted in a larger size fresh martensite compared to secondary martensite.
3. Martensite can not only nucleate at parent austenite grain boundaries, but it can also nucleate in the preformed lath martensite and twins. The larger the parent austenite grain size, the smaller the constraints of martensite growth, resulting in longer fresh martensite and secondary martensite. In addition, the martensite transformation (2.25 s) was shorter at a higher quenching temperature of 1160 °C than that (13 s) at 860 °C. In addition, martensitic lath could traverse the unstable parent austenitic grain boundaries.
4. The martensitic lath was present in parallel laths (0~2°) based on preformed laths or distributed in triangles, parallelograms, or hexagons with an angle of 60° or 120°.

Author Contributions: Methodology, Z.L., Q.Y., Y.Z. and S.L.; Formal analysis, S.X. and S.L.; Investigation, Z.L., Q.Y., Y.Z. and G.X.; Resources, S.X.; Data curation, Y.Z.; Writing—original draft, Z.L.; Writing—review & editing, Q.Y., S.X. and G.X.; Visualization, G.X.; Supervision, Z.L.; Funding acquisition, Q.Y. and S.L. All authors have read and agreed to the published version of the manuscript.

Funding: National Natural Science Foundation of China (no. 52004193), China Postdoctoral Science Foundation (no. 2022M710596), and the Joint Foundation of Nature Science Foundation of Hubei Province (2022CFD078).

Institutional Review Board Statement: Not applicable.

Informed Consent Statement: Not applicable.

Data Availability Statement: The raw/processed data required to reproduce these findings cannot be shared at this time as the data also form part of an ongoing study.

Acknowledgments: The authors gratefully acknowledge the financial support from National Natural Science Foundation of China (no. 52004193), China Postdoctoral Science Foundation (No. 2022M710596), and the Joint Foundation of Nature Science Foundation of Hubei Province (2022CFD078).

Conflicts of Interest: The authors declare no conflict of interest.

Glossary

Words/Equations	Abbreviation/Parameter Meaning
Confocal laser scanning high-temperature microscope	CSLM
Parent austenite grains	PAGs
Fresh martensite	FM
Secondary martensite	SM
$N = f_N \exp\left(-Q_N/K\Delta T\right)\ \&\ G = f_G \exp\left(-Q_G/K\Delta T\right)$	N: the nucleation rate; G: the growth rate; Q_N and Q_G: the nucleation and growth activation energies; f_N and f_G: the impact factors between structure and nucleation with growth; ΔT: superheat.
$\sum_{i=1}^{i=n} \frac{\Delta t}{A_i} = 1\ \ \&\ \ \int_{t=0}^{t=t_n} \frac{dt}{A(T)} = 1$	Δt and dt: the transformation time at temperature T; A_i and $A(T)$: the corresponding incubation periods. C: transformation rate;
$\int_{T_1}^{T_s} \frac{dt}{A(T)} = \int_{T_1}^{T_s} \frac{1}{A(T)} \cdot \frac{1}{\frac{dT}{dt}} \cdot dT = \int_{T_1}^{T_s} \frac{1}{A(T)} \cdot \frac{1}{v} \cdot dT = 1\ \&$ $c = \frac{df}{dt}, v = \frac{dT}{dt}\ \&$ $\int_{t=0}^{t=t_n} \frac{df}{dt} dt = \int_{T_s}^{T_f} \frac{df}{dt} \cdot \frac{dT}{v} = \int_{T_s}^{T_f} \frac{C}{v} \cdot dT = 1$	T_s and T_f: the temperatures at the beginning and ending of the transformation; v: heating rate; T_1: equilibrium temperature; t_n: the time to the transformation ending temperature T_f.

References

1. Bourithis, L.; Papadimitriou, G.D. The effect of microstructure and wear conditions on the wear resistance of steel metal matrix composites fabricated with PTA alloying technique. *Wear* **2009**, *266*, 1155–1164. [CrossRef]
2. Lee, K.; Nam, D.H.; Lee, S.; Kim, C.P. Hardness and wear resistance of steel-based surface composites fabricated with Fe-based metamorphic alloy powders by high-energy electron beam irradiation. *Mater. Sci. Eng. A* **2006**, *428*, 124–134. [CrossRef]
3. Wei, M.X.; Wang, S.Q.; Lan, W.A.; Chen, K.M. Effect of microstructures on elevated temperature wear resistance of a hot working die steel. *J. Iron Steel Res. Int.* **2011**, *18*, 47–53. [CrossRef]
4. Wang, W.; Song, R.; Peng, S.; Pei, Z. Multiphase steel with improved impact abrasive wear resistance in comparison with conventional Hadfield steel. *Mater. Des.* **2016**, *105*, 96–105. [CrossRef]
5. Du, X.; Ding, H.; Wang, K.; Wu, K. Influence of impact energy on impact corrosion abrasion of high manganese steel. *J. Wuhan Univ. Technol. Mater. Sci.* **2007**, *22*, 412–416. [CrossRef]
6. Chen, J.; Wang, J.J.; Zhang, H.; Zhang, W.G.; Liu, C.M. Evolution of deformation twins with strain rate in a medium-manganese wear-resistant steel Fe–8Mn–1C–1.2Cr–0.2V. *J. Iron Steel Res. Int.* **2019**, *26*, 983–990. [CrossRef]
7. Si, H.; Xiong, R.; Song, F.; Wen, Y.; Peng, H. Wear resistance of austenitic steel Fe-17Mn-6Si-0.3C with high Silicon and high Manganese. *Acta Metall. Sin. (Engl. Lett.)* **2014**, *27*, 352–358. [CrossRef]
8. Liu, Y.; Tang, Y.; Huang, G. Improvement of mechanical properties of low alloy steel with high strength and wear resistance. *J. Iron Steel Res.* **1999**, *3*, 79–81. (In Chinese)
9. Zumelzu, E.; Goyos, I.; Cabezas, C.; Opitz, O.; Parada, A. Wear and corrosion behaviour of high Cr cast iron alloys. *J. Mater. Process. Technol.* **2002**, *128*, 250–255. [CrossRef]

10. Zhang, A.F.; Xing, J.D.; Fang, L.; Su, J.Y. Inter-phase corrosion of chromium white cast irons in dynamic state. *Wear* **2004**, *257*, 198–204. [CrossRef]
11. Tian, H.H.; Addie, G.R.; Visintainer, R.J. Erosion–corrosion performance of high-Cr cast iron alloys in flowing liquid–solid slurries. *Wear* **2009**, *267*, 2039–2047. [CrossRef]
12. Chen, H.; Hu, J. Study on Carbides of High Chromium Cast Iron. *J. Shanghai Inst. Technol. Nat. Sci. Ed.* **2003**, *3*, 57–62. (In Chinese)
13. Cui, X.; Wang, N.; Gong, P.; Yang, H.; Bai, P. Effect of C, Cr contents and heat treatment on the microstructure and mechanical properties of high chromium cast iron. *J. Inn. Mong. Univ. Technol.* **2016**, *4*, 277–282. (In Chinese)
14. Tkowska, B.; Dziurka, R.; Baa, P. Analysis of phase transformations of supercooled austenite in low-alloy steel with boron additives. *Met. Sci. Heat Treat.* **2016**, *57*, 525–530. [CrossRef]
15. Liu, Q.; Wen, H.; Zhang, H.; Gu, J.; Li, C.; Lavernia, E.J. Effect of multistage heat treatment on microstructure and mechanical properties of high-strength low-alloy steel. *Metall. Mater. Trans. A* **2016**, *47*, 1960–1974. [CrossRef]
16. Xu, X.; Ederveen, F.H.; van der Zwaag, S.; Xu, W. Correlating the abrasion resistance of low alloy steels to the standard mechanical properties: A statistical analysis over a larger data set. *Wear* **2016**, *368*, 92–100. [CrossRef]
17. Sundström, A.A.; José, R.; Olsson, M. Wear behaviour of some low alloyed steels under combined impact-abrasion contact conditions. *Wear* **2001**, *250*, 744–754. [CrossRef]
18. Fu, H.; Xiao, Q.; Fu, H. Heat treatment of multi-element low alloy wear-resistant steel. *Mater. Sci. Eng. A* **2005**, *396*, 206–212. [CrossRef]
19. Kaijalainen, A.J.; Suikkanen, P.P.; Limnell, T.J.; Karjalainen, L.P.; Kömi, J.I.; Porter, D.A. Effect of austenite grain structure on the strength and toughness of direct-quenched martensite. *J. Alloys Compd.* **2013**, *577*, 642–648. [CrossRef]
20. Zhang, C.Y.; Wang, Q.F.; Kong, J.L.; Xie, G.Z.; Wang, M.Z.; Zhang, F.C. Effect of martensite morphology on impact toughness of ultra-high strength 25CrMo48V steel seamless tube quenched at different temperatures. *J. Iron Steel Res. Int.* **2013**, *20*, 62–67. [CrossRef]
21. Liang, Y.; Long, S.; Xu, P.; Lu, Y.; Jiang, Y.; Liang, Y.; Yang, M. The important role of martensite laths to fracture toughness for the ductile fracture controlled by the strain in EA4T axle steel. *Mater. Sci. Eng. A* **2017**, *695*, 154–164. [CrossRef]
22. Fazaeli, A.; Ekrami, A.; Kokabi, A.H. Microstructure and mechanical properties of dual phase steels, with different martensite morphology, produced during TLP bonding of a low C-Mn steel. *Met. Mater. Int.* **2016**, *22*, 856–862. [CrossRef]
23. Li, S.; Zhu, G.; Kang, Y. Effect of substructure on mechanical properties and fracture behavior of lath martensite in 0.1C-1.1Si-1.7Mn steel. *J. Alloys Compd.* **2016**, *675*, 104–115. [CrossRef]
24. Liang, Y.; Lei, M.; Zhong, S.; Jiang, S. The relationship between fracture toughness and notch toughness, tensile ductilities in lath martensite steel. *Acta Metall. Sin.* **1998**, *9*, 950–958.
25. Inoue, T.; Matsuda, S.; Okamura, Y.; Aoki, K. The fracture of a low carbon tempered martensite. *Trans. JIM* **1970**, *11*, 36–43. [CrossRef]
26. Xie, Z.; Shang, C.; Zhou, W.; Wu, B. Effect of retained austenite on ductility and toughness of a low alloyed multi-phase steel. *Acta Metall. Sin.* **2016**, *2*, 224–232.
27. Liang, G.; Xu, Z.; Li, J.; Jiang, Q. The latest development of anti-wear steel. *Spec. Steel* **2002**, *23*, 1–7.
28. Cao, Y.; Wang, Z.; Jiang, Z.; Zhang, T.; Wang, C.; Wu, D. Development and research of wear-resistant steel with high strength and toughness. *Steel Roll.* **2011**, *28*, 3–7.
29. Li, X.L.; Wang, C.H.; Lu, L.U. Microstructure and impact wear resistance of TiN reinforced high manganese steel matrix. *J. Iron Steel Res. Int.* **2012**, *19*, 60–65.
30. Liu, N.; Liu, Z.D.; He, X.K.; Yang, Z.Q. Austenite transformation of SA508Gr.4N steel for nuclear reactor pressure vessels during heating process. *Heat Treat. Met.* **2017**, *42*, 88–92.
31. Yuan, Q.; Ren, J.; Mo, J.; Zhang, Z.; Tang, E.; Xu, G.; Xue, Z. Effects of rapid heating on the phase transformation and grain refinement of a low-carbon mciroalloyed steel. *J. Mater. Res. Technol.* **2023**, *23*, 3756–3771. [CrossRef]
32. Ågren, J.; Vassilev, G.P. Computer simulations of cementite dissolution in austenite. *Mater. Sci. Eng.* **1984**, *64*, 95–103. [CrossRef]
33. Lan, F.J.; Zhuang, C.L.; Li, C.R.; Chen, J.B.; Yang, G.K.; Yao, H.J. Study on manganese volatilization behavior of Fe–Mn–C–Al twinning-induced plasticity steel. *High Temp. Mater. Process.* **2021**, *40*, 461–470. [CrossRef]
34. Khalaj, G.; Yoozbashizadeh, H.; Khodabandeh, A.; Tamizifar, M. Austenite grain growth modelling in weld heat affected zone of Nb/Ti microalloyed linepipe steel. *Mater. Sci. Technol.* **2014**, *30*, 424–433. [CrossRef]
35. Rizi, M.S.; Minouei, H.; Lee, B.J.; Pouraliakbar, H.; Toroghinejad, M.R.; Hong, S.I. Hierarchically activated deformation mechanisms to form ultra-fine grain microstructure in carbon containing FeMnCoCr twinning induced plasticity high entropy alloy. *Mater. Sci. Eng. A* **2021**, *824*, 141803. [CrossRef]

Article

Effect of Liquid-Solid Volume Ratio and Surface Treatment on Microstructure and Properties of Cu/Al Bimetallic Composite

Zhiyuan Wu [1], Lijie Zuo [1,*], Hongliang Zhang [2,*], Yiqiang He [1], Chengwen Liu [1], Hongmiao Yu [2], Yuze Wang [1] and Wen Feng [1]

[1] Jiangsu Institute of Marine Resources Development, Jiangsu Ocean University, Lianyungang 222005, China
[2] Zhongtian Technology Submarine Cables Co., Ltd., Nantong 226010, China
* Correspondence: zuoliji@126.com (L.Z.); zhanghl@chinaztt.com (H.Z.)

Abstract: Due to exceptional conductivity, lightweight nature, corrosion resistance, and various other advantages, Cu/Al bimetallic composites are extensively utilized in the fields of communication, new energy, electronics, and other industries. To solve the problem of poor metallurgical bonding of Cu/Al bimetallic composites caused by high-temperature oxidation of Cu, different coating thicknesses of Ni layer on Cu rods were used to fabricate the Cu/Al bimetallic composite by gravity casting. The effect of liquid–solid volume ratio and coating thickness on microstructure and properties of a Cu/Al bimetallic composite were investigated in this study. The results indicated that the transition zone width increased from 242.3 μm to 286.3 μm and shear strength increased from 17.8 MPa to 30.3 MPa with a liquid–solid volume ratio varying from 8.86 to 50. The thickness of the transition zone and shear strength increased with the coating thickness of the Ni layer varying from 1.5 μm to 3.8 μm, due to the Ni layer effectively preventing oxidation on the surface of the Cu rod and promoting the metallurgical bonding of the Cu/Al interface. The presence of a residual Ni layer in the casted material hinders the diffusion process of the Cu and Al atom. Therefore, the thickness of the transition zone and shear strength exhibited a decreasing trend as the coating thickness of the Ni layer increased from 3.8 μm to 5.9 μm. Shear fracture observation revealed that the initiation and propagation of shear cracks occurred within the transition zone of the Cu/Al bimetallic composite.

Keywords: Cu/Al bimetallic composite; liquid–solid volume ratio; coating thickness

Citation: Wu, Z.; Zuo, L.; Zhang, H.; He, Y.; Liu, C.; Yu, H.; Wang, Y.; Feng, W. Effect of Liquid-Solid Volume Ratio and Surface Treatment on Microstructure and Properties of Cu/Al Bimetallic Composite. *Crystals* **2023**, *13*, 794. https://doi.org/10.3390/cryst13050794

Academic Editors: Andrea Di Schino and Claudio Testani

Received: 27 March 2023
Revised: 7 May 2023
Accepted: 8 May 2023
Published: 9 May 2023

1. Introduction

Due to a combination of the high electrical and thermal conductivity of Cu [1] with the light weight of Al [2–6], Cu/Al bimetallic composites [7–14] were widely used in the electric power transportation industry. Cu/Al composite materials can replace copper materials in generators and aluminum materials in external power grids, as well as the contact surfaces between the two, thereby reducing the use of copper resources and reducing the accident rate in power generation and supply. Although the compound casting method has exhibited great superiority in fabricating irregular shapes of Cu/Al bimetallic composites, it still has some drawbacks in the aspects of the rapid growth of intermetallic compounds and the oxidation of the solid Cu substrate. The hard and brittle transition zone was formed at the interface of the Cu/Al bimetallic composite, which reduced the mechanical properties of the material. Therefore, optimization of the transition zone has become a hot topic in the research and development of Cu/Al bimetallic composite.

It was widely reported that the transition zone played an important role in the microstructure and mechanical properties of Cu/Al bimetallic composites [15]. There have been extensive studies on the formation mechanism of the transition zone. For example, Wang et al. [16] used synchrotron X-ray technology to study the interfacial diffusion behavior and microstructure evolution of Cu/Al bimetals. Cu and Al first diffuse to each other and then form α-Al dendrites and intermetallic compounds (IMCs) between the matrix.

The liquid–solid ratio was one of the key factors for the formation of the transition zone in the preparation of bimetallic composite by gravity casting. For example, it has been found that the bonding quality of AZ91 and AZ31 alloys was better when the liquid–solid ratio was larger [17]. Other studies have shown the preparation of high chromium cast iron and medium carbon steel bimetals by gravity casting. With the increase of liquid–solid product ratio, the diffusion activity of elements increased, leading to the increase of the interfacial transition zone width and shear strength [18]. It has been reported that the transition zone consisted of the intermetallic compounds and the remelting zone, and the cooling rate influenced the thickness of intermetallic compounds, the microstructure of the remelting zone, and the morphology of the remelting zone/Al interface [19]. Tavasoli et al. [20] reported the effect of pouring temperature on the transition zone, and the results showed that an increase in Al melting temperature resulted in a gradual increase in the thickness of intermetallic compounds and interfaces. Chen et al. [21] found that pouring temperature, cooling mode of Cu plate surface, and starting time of forced cooling after pouring had no effect on the microstructure species of the transition zone.

In addition, during the preparation of the Cu/Al bimetallic composite by gravity casting, an oxidation reaction occurred on the surface of Cu during preheating [22], and the formation of an oxide film reduced the metallurgical bonding property of the interface between Cu and Al. Therefore, it was particularly important to cover the surface of the copper with a protective film to prevent oxidation [23]. A suitable protective film can not only prevent the surface oxidation of Cu, but also promote the metallurgical bonding between Cu and Al. Boucherit et al. [24] achieved friction stir welding of Cu/Al using a zinc interlayer and found that Zn can significantly reduce the formation of intermetallic compounds such as Al_2Cu and Al_4Cu_9, thereby improving the shear lap tensile strength of the joint. Ye et al. [25] adopted a new Zn-Al-Si filler metal to braze Cu/Al and found that Si could inhibit the growth of intermetallic compounds, thus significantly improving the corrosion resistance of Cu/Al bimetallic materials. Breedis et al. [26] found that adding a certain amount of Ni to copper alloys can inhibit the growth rate of Cu/Al intermetallic compounds, effectively reduce the content of intermetallic compounds, improve the microstructure of copper alloys, and effectively improve its properties. It was found that by depositing Ni-P coating on a copper substrate by electroless plating, the intermediate coating acted as a protective film, which could reduce the rate of intermetallic compound generation [27]. If Ni was used as the intermediate layer during the pouring process of Al and Cu, it can be seen from the Cu-Ni binary phase diagram that due to the infinite solubility of Cu and Ni, intermetallic compounds will not be generated and copper matrix oxidation can be prevented. Liu et al. [28] reported that a uniform Ni protective layer on the electroplating of the copper matrix can prevent the surface oxidation of the copper matrix, and the Ni layer dissolved during the interfacial reaction during the pouring process, promoting the metallurgical combination of copper and aluminum.

At present, the research on interface processing of Cu/Al composite materials mainly focuses on coating materials and coating methods, while there are few studies on the effect of coating thickness on the microstructure and properties of Cu/Al composite materials. In this study, the fabrication of Cu/Al bimetallic composite was achieved by gravity casting, and the effect of liquid–solid ratio and coating thickness on the microstructure and properties of Cu/Al bimetallic composites was discussed. The appropriate thickness of the transition zone will significantly improve the mechanical properties of the Cu/Al bimetallic composites.

2. Materials and Methods

The Cu/Al bimetallic composite was fabricated by pure copper rods and aluminum rods. In order to prevent oxidation of the copper while being kept at elevated temperature, a layer of nickel was plated on the surface of the copper rods before casting. The electroplating solution consisted of 840 g Ni_2SO_4, 100 g $NiCl_2 \cdot 6H_2O$, and 3 L deionized water. The nickel-plating voltage was 4V, and the nickel-plating time was 10, 25, and 40 min, respectively.

The electroplating device was shown in Figure 1. Both the casting mold and nickel-plated copper rod were kept at 500 °C with the resistance furnace at the beginning of the test. The pure aluminum rod was melted and refined in a steel crucible at approximately 740 °C. The melt was left to stand at 720 °C for about 10 min to ensure the equilibrium temperature after the refining slag was skimmed. Then, the aluminum melt was cast into the steel mold equipped with the nickel-plated copper rod, and after waiting until it had completely cooled and solidified, it was removed from the mold.

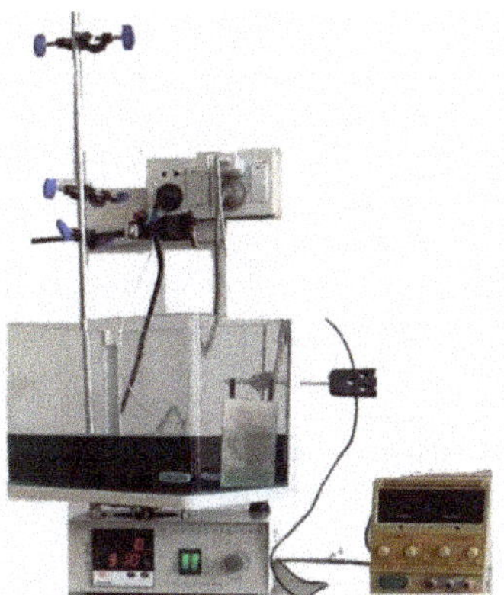

Figure 1. Electroplating Operation Console.

The metallographic specimen was first polished on different grit sandpaper, then with a combination of mechanical polishing and hand polishing on a velvet polish cloth with a solution of 0.5 μm aqueous magnesium oxide. The microstructure of Cu/Al bimetallic composite was observed by AXIO-type metallographic microscopy (OM). The Ultima IV X-ray diffraction (XRD) was used to analyze to the types of intermetallic compounds in the transition zone of materials, with a voltage of 35 kV and a scanning speed of 10°/min; diffraction angle range was $10° \leq \theta \leq 45°$. Jade 6.0 software was then used to perform phase calibration on the acquired dates. The sampling locations of hardness and shear samples were shown in Figure 2. The hardness (HV) of the Cu/Al bimetallic composite were tested on the HXD-1000TM digital microhardness tester with a load of 200 g and loading time of 15 s. The cylindrical shear specimen with diameter of 16 mm and height of 6 mm was fabricated by the electric spark machine. The shear specimen was performed on HXD-1000TM electronic universal material testing machine with the speed of 1.0mm/s. Then, the FEI Scios-2 HiVac scanning electron microscopy (SEM) was used to analyze the shear surface of the Cu/Al bimetallic composite.

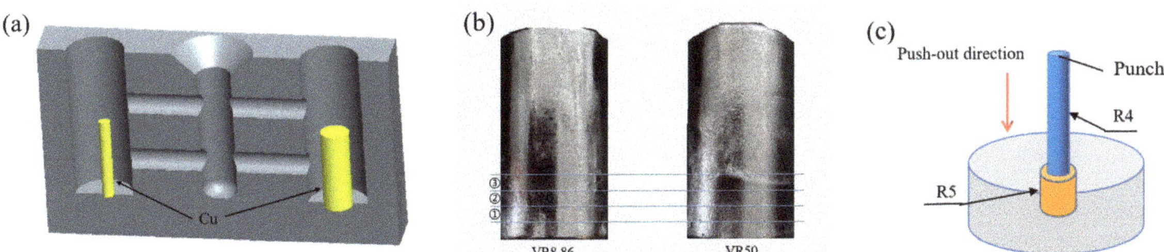

Figure 2. Casting mold drawing and shear material: (**a**) casting mold diagram; (**b**) Cu/Al bimetallic composite; (**c**) schematic diagram of shear strength test.

3. Results and Discussion

3.1. Effect of Liquid–Solid Volume Ratio on the Microstructure and Properties of Cu/Al Bimetallic Composite

3.1.1. Effect of Liquid–Solid Volume Ratio on the Microstructure of Cu/Al Bimetallic Composite

The Ni electroplating time for this part was 25 min. The metallographic structure of the Cu/Al bimetallic composite prepared at different liquid–solid volume ratio was shown in Figure 3. It was found that a distinct transition zone was formed at the junction of Cu and Al when casting at 720 °C. The thickness of the transition zone increases with the increase of the liquid–solid volume ratio, as the solidification time increases with the increase of the high-temperature liquid volume ratio, which means that Al and Cu atoms have a longer diffusion time. Figure 4 displayed the statistical results of dissociation thickness of the Cu and the thickness of the transition zone. With increasing the liquid–solid volume ratio of Cu and Al from 8.86 to 50, the thickness of the transition zone increased from 242.3 μm to 286.3 μm and the dissolved thickness of Cu increased from 74.3 μm to 90.3 μm. The ratio between the thickness of the transition zone and the dissolved thickness of Cu was 3.26 and 3.17 with increasing the liquid–solid volume ratio. Then, the ratio of Cu to Al in the transition zone was constant with increasing the liquid–solid volume ratio. Therefore, it is reasonable to infer that the phase composition of the transition zone did not change with the liquid–solid volume ratio, which was consistent with the literature [28].

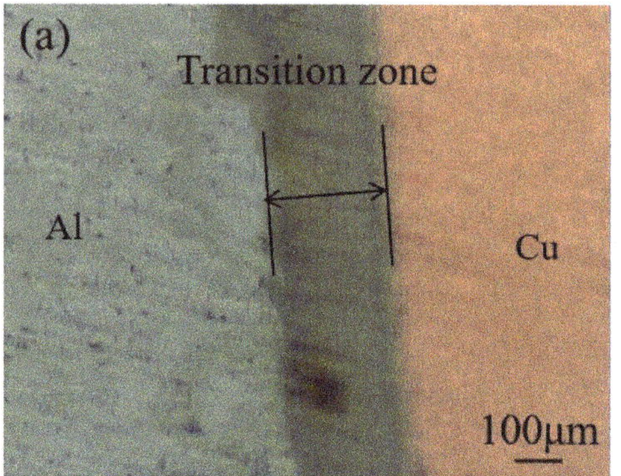

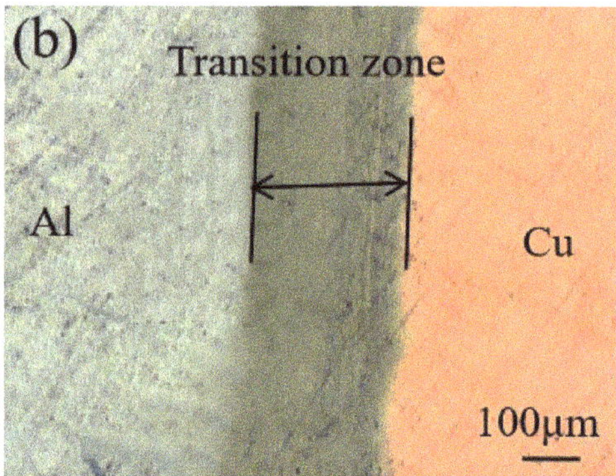

Figure 3. Metallographic structure of Cu/Al bimetallic composites prepared with different liquid–solid volume ratios: (**a**) VR8.86; (**b**) VR50.

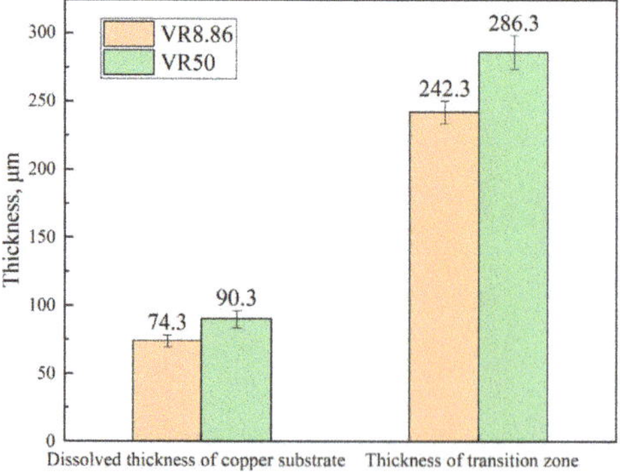

Figure 4. Thickness of the copper matrix and transition zone in composite fabricated with different liquid–solid volume ratio.

3.1.2. Effect of Liquid–Solid Volume Ratio on the Properties of Cu/Al Bimetallic Composite

(1)   Shear strength

The shear strength of the Cu/Al bimetallic composite was described in Figure 5. The shear strength of the Cu/Al bimetallic composite increased with increasing the liquid–solid volume ratio. As shown in Figure 5, the shear strength was 17.8 MPa and 30.3 MPa for VR8.86 and VR50, respectively. With the increasing volume ratio from 8.86 to 50, the shear strength of the Cu/Al bimetallic composite increased 70%. As discussed above, the thickness of the transition zone was increased with the liquid–solid volume ratio. In addition, the microhardness of the phase (Al_2Cu, $AlCu$, and Al_4Cu_9) in the transition zone was higher than the pure Cu. This may account for the fact that the shear strength increased with the thickness of the transition zone. To further confirm this phenomenon, metallographic and SEM observation of the shear fracture of the Cu/Al bimetallic composite were performed.

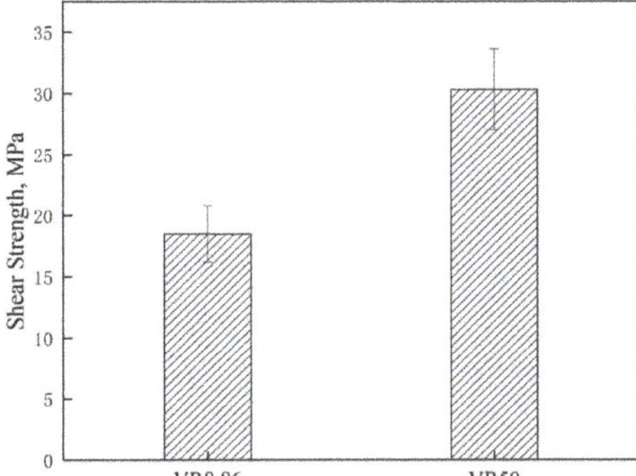

Figure 5. Shear strength of Cu/Al bimetallic composite fabricated with different liquid–solid volume ratio.

Figure 6 depicted the metallographic observation of the shear fracture of Cu/Al bimetallic composite fabricated with different liquid–solid volume ratio. It was shown that the yellow was Cu and some intermetallic compounds remained at the edge of Cu after the shear test. In addition, the content of the remaining intermetallic compounds remained at the edge of Cu increased with the thickness of the transition zone. This phenomenon indicated that the shear fracture was directly related to the transition zone.

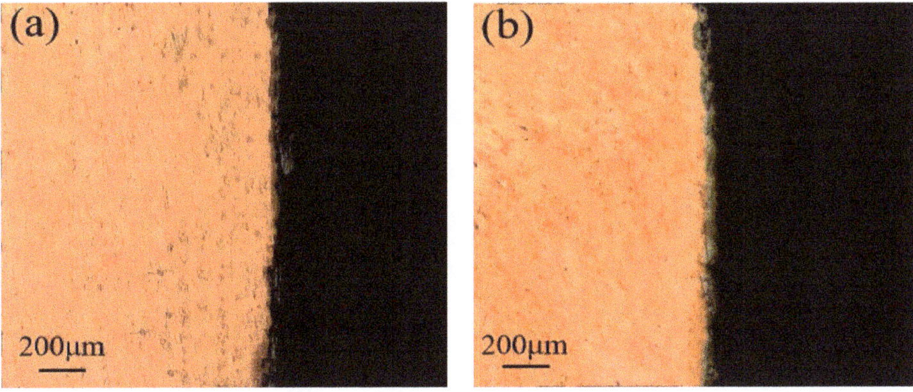

Figure 6. Metallographic observation of shear fracture of Cu/Al bimetallic composite fabricated by different liquid–solid volume ratio: (**a**) VR8.85; (**b**) VR50.

As displayed in Figure 7, XRD results of shear fracture surface of Cu/Al bimetallic composite indicated that the Al_2Cu, Al_4Cu_9, and AlCu phase remained at the surface of the Cu rod after the shear test. The SEM photograph of the shear fracture of the Cu/Al bimetallic composite fabricated by different liquid–solid ratio was shown in Figure 8. Combined with energy spectrum analysis and XRD results, the phase calibration of the shear fracture was illustrated in Figure 8. In summary, it was concluded that the initiation and propagation of shear cracks occurred in the transition zone of the Cu/Al bimetallic composite.

(2) Microhardness

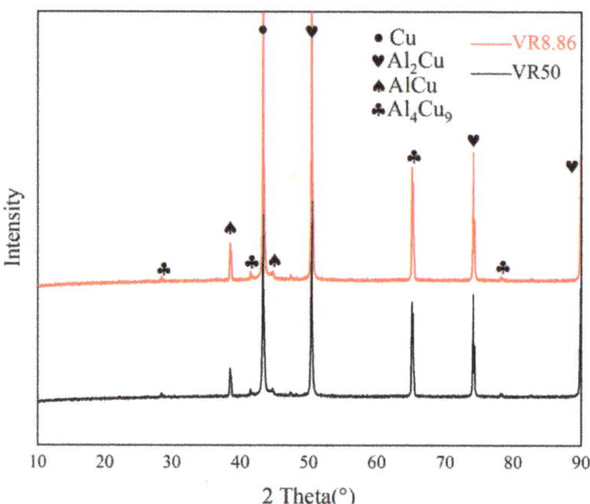

Figure 7. XRD patterns on shear fracture of Cu/Al bimetallic composite fabricated with different liquid–solid volume ratio.

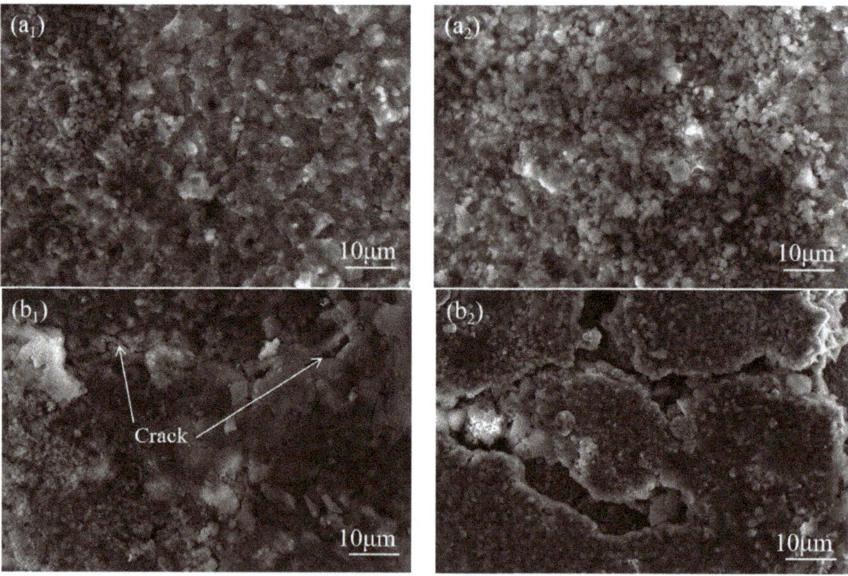

Figure 8. SEM photograph of shear fracture of Cu/Al bimetallic composite fabricated with different liquid–solid volume ratio: (**a₁,a₂**) VR8.86; (**b₁,b₂**) VR50.

The distribution of microhardness indentations and statistical results for the interfaces of Cu/Al bimetallic composite fabricated with different liquid–solid volume ratio were depicted in Figure 9, from left to right: Al, transition zone, Cu. It was reported [29] that the hardness of Al$_2$Cu was 400–500 HV, and the microhardness of [α(Al) + Al$_2$Cu] eutectic structure was about 150–200 HV. The microhardness of the transition zone of Cu/Al bimetallic composite was 140–180 HV, which was higher than Cu and Al. In addition, with the increase of liquid–solid volume ratio, the thickness of the intermediate transition layer increased, the content of the mesophase was higher, and the microhardness was increased.

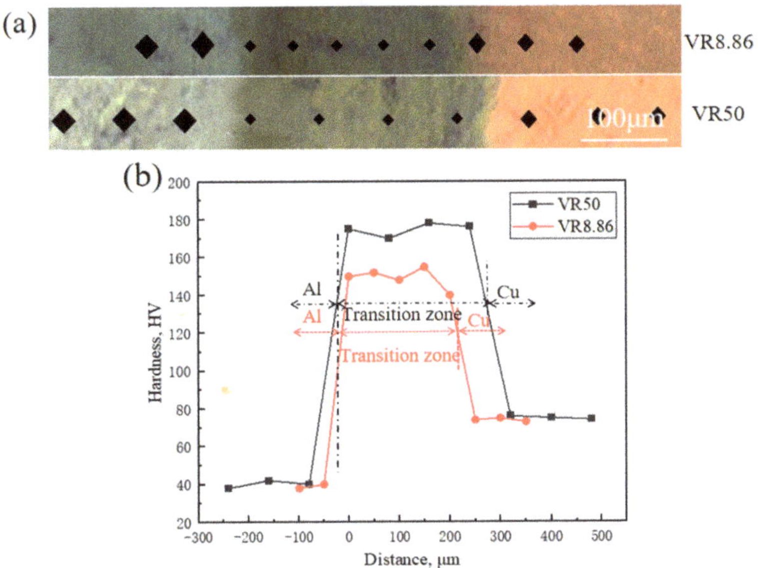

Figure 9. Microhardness of composite fabricated by different liquid–solid volume ratio: (**a**) indentation metallographic observation; (**b**) hardness distribution.

3.2. Effect of Coating Thickness on the Microstructure and Properties of Cu/Al Bimetallic Composite

3.2.1. Effect of Coating Thickness on the Microstructure of Cu/Al Bimetallic Composite

The coating thickness of the Ni layer on the Cu rod was 1.5 μm, 3.8 μm, and 5.9 μm with prolonging of the electroplating time from 10 min to 40 min, respectively. Then, the Al melt was cast into the mold equipped with an Ni-plated Cu rod, and the liquid–solid volume ratio was 50. The interfacial metallographic structure of Cu/Al bimetallic composite with different electroplating time was displayed in Figure 10. The transition zone width increased from 246 μm to 286 μm with the coating thickness varying from 1.5 μm to 3.8 μm. It was found that the thickness of the Ni plating was very uneven according to the metallographic observation of the Cu rod with Ni plating time of 10 min. Even some parts of the Ni layer had fallen off while kept at 500 °C. Then, the anti-oxidation effect of the Ni layer was sharply reduced, and part of the surface of the Cu rod was oxidized before casting the Al melt. This accounted for the smaller transition zone thickness for the electroplating time of 10 min.

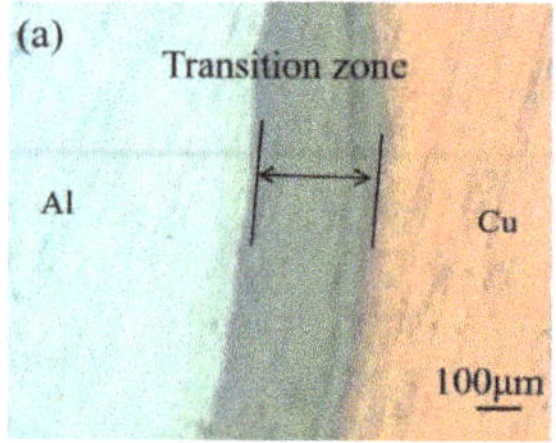
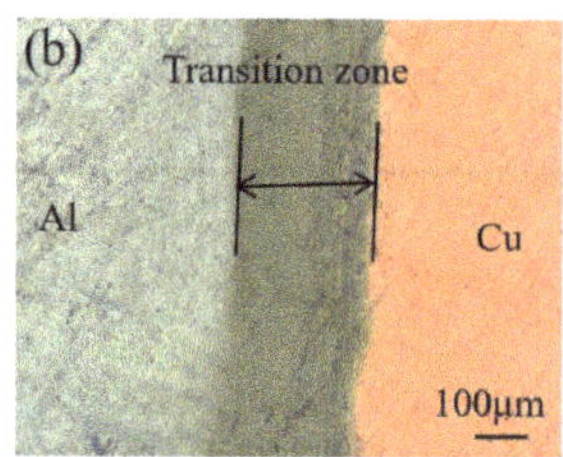
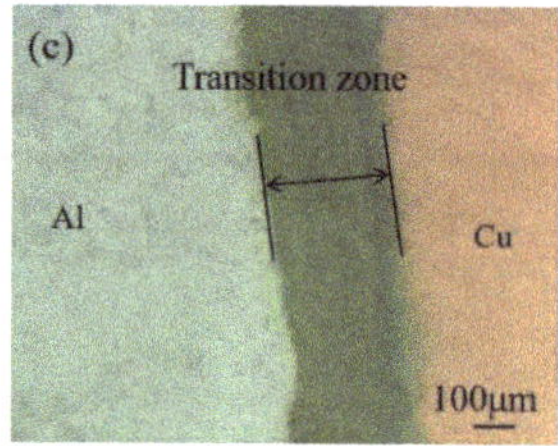

Figure 10. Interface microstructure of Cu/Al bimetallic composite with different electroplating time: (**a**) 10 min, (**b**) 25 min, and (**c**) 40 min.

The coating thickness and the width of the transition zone were summarized in Table 1. As shown in Table 1, the transition zone width first increased and then decreased with increasing the coating thickness of Ni from 1.5 μm to 5.9 μm. Exactly, the transition zone width decreased from 286 μm to 268 μm with the coating thickness varying from 3.8 μm to 5.9 μm. SEM image and chemical elements mapping of Al, Ni, and Cu of the Cu/Al bimetallic composite fabricated with 3.8 μm coating thickness was shown in Figure 11 Combined with Figure 10, it was inferred that the Ni layer on the Cu rod was dissolved by pouring in the high temperature Al liquid at the beginning of the casting. Subsequently, part of the Cu rod in contact with the Ni layer began to dissolve. Then, the Cu atoms diffused through the liquid Ni layer to the liquid Al, and the Al atoms diffused through the liquid Ni layer toward to the Cu side. At the same time, the Ni atoms diffused to both sides, which promoted the metallurgical combination of the Cu and Al. As the temperature dropped, different intermetallic compounds began to precipitate and transition zones gradually formed. As displayed in Figure 11, the 3.8 μm Ni layer had almost diffused out at elevated temperature. Therefore, the 3.8 μm Ni layer only partially limited the diffusion of the Cu and Al atoms, resulting in a peak thickness of the transition zone.

Table 1. Coating thickness and transition zone width of fabricated materials after different electroplating time.

Cu/Al	10 min	25 min	40 min
$d_{clad.}$ (μm)	1.5 ± 0.15	3.8 ± 0.2	5.9 ± 0.3
$d_{tran.}$ (μm)	246 ± 5.4	286.3 ± 6.8	268 ± 5.5

Figure 11. SEM image and chemical elements mapping of Al, Ni, and Cu of the Cu/Al bimetallic composite fabricated with 3.8 μm coating thickness.

The SEM image and chemical elements mapping of Al, Ni, and Cu of the Cu/Al bimetallic composite fabricated with 5.9 μm coating thickness was demonstrated in Figure 12. It was shown that there were still Ni layers distributed at the transition zone of the Cu/Al bimetallic composite with increasing the thickness of the Ni layer to 5.9 μm in Figure 12. The Ni layer was effectively limited the diffusion of the Cu and Al atoms at elevated temperature, which reduced the transition zone width to 268 μm.

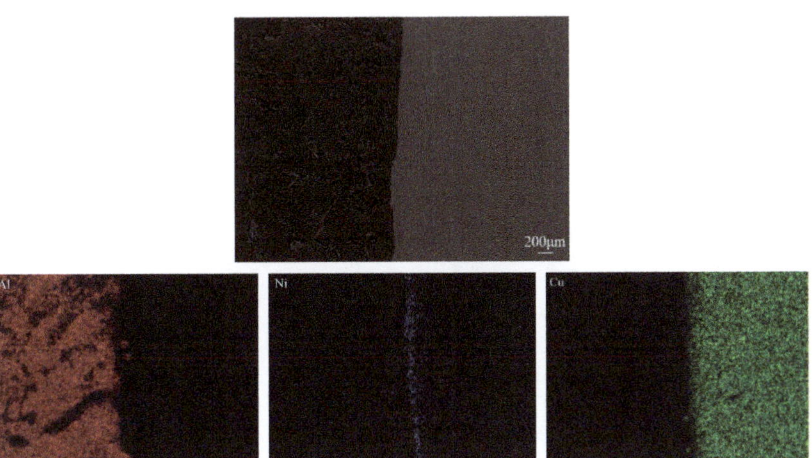

Figure 12. SEM image and chemical elements mapping of Al, Ni, and Cu of the Cu/Al bimetallic composite fabricated with 5.9 μm coating thickness.

The content of Ni in the transition zone increased with increasing the coating thickness of electroplating Ni. When the thickness of the Ni layer was 1.5 μm, the Ni layer had been diffused in the transition zone during the casting process. Therefore, the transition zone of the Cu/Al bimetallic composite with Ni layer thickness of 3.8 μm and 5.9 μm was characterized. Table 2 summarized the interface Al, Cu, and Ni elements content of Cu/Al bimetallic composite in Figures 11 and 12. The results also indicated that with increasing the thickness of the Ni layer, the diffusion of the Cu and Al was limited.

Table 2. Interface element content of Cu/Al bimetallic composite (wt, %).

Element	Time	25 min	40 min
Cu		49.84	44.73
Al		48.55	53.92
Ni		0.87	1.22

3.2.2. Effect of Coating Thickness on the Properties of Cu/Al Bimetallic Composite

(1) Shear Strength

The shear strength in transition zone of the Cu/Al bimetallic composite fabricated with different coating thickness was depicted in Figure 13. With increasing the coating thickness from 1.5 μm to 3.8 μm, the shear strength in transition zone increased from 18.6 MPa to 30.3 MPa, an increase of 62.9%. The shear strength decreased from 30.3 MPa to 26.7 MPa with the coating thickness varying from 3.8 μm to 5.9 μm. As discussed above, when the coating thickness was 1.5 μm, the Ni layer cannot effectively prevent the oxidation of the Cu rod, which results in poor metallurgical bonding of the Cu/Al interface. When the coating thickness was 3.8 μm, the Ni layer was effective in preventing the oxidation of the copper rod and promoting the mutual diffusion of Cu and Al atoms. Furthermore, due to the partial diffusion of the Ni layer, the binary phase structure in the transition zone changed into the ternary phase structure, which resulted in a great improvement in the metallurgical bonding property of the Cu/Al interface. When the coating thickness was 5.9 μm, the Ni layer was not only effective in preventing the oxidation of the Cu bar, but also greatly limiting the mutual diffusion of Cu and Al, which had an adverse effect on the metallurgical bonding of Cu/Al interface.

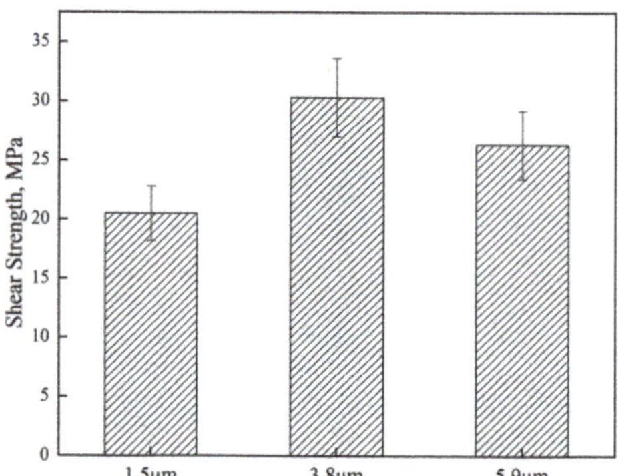

Figure 13. Shear strength of Cu/Al bimetallic composite fabricated with different coating thickness.

As illustrated in Figure 14, the phase composition of the transition zone changed with increasing the coating thickness of the Ni layer. For example, the diffraction of AlCu phase decreased with increasing the coating thickness of the Ni layer, indicating that the proportion of AlCu phase in the transition zone decreased. The diffraction of Al_3Ni_2 phase was found with the coating thickness of the Ni layer reached to 5.9 μm. Above all, the XRD results were consistent with the analysis of the influence of the Ni layer on the diffusion of Cu and Al.

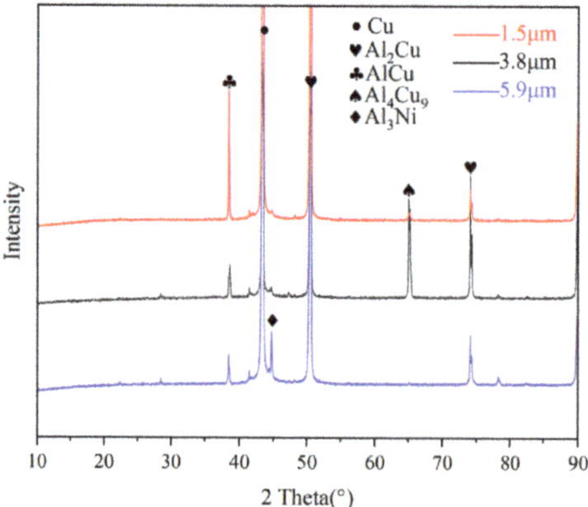

Figure 14. XRD results of shear fracture surface of Cu/Al bimetallic composite fabricated with different coating thickness.

Figure 15 depicted metallographic observations of shear fracture in the transition zone (Cu side) of Cu/Al bimetallic composite fabricated with different coating thickness. As shown in Figure 14, the content of residual intermetallic compounds at the Cu side increased with increasing the coating thickness of the Ni layer.

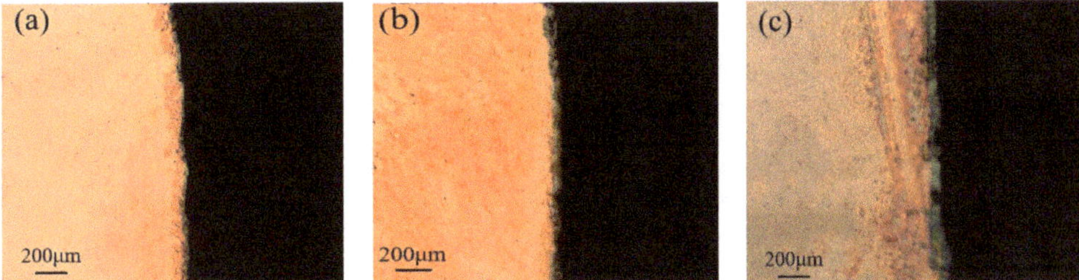

Figure 15. Shear fracture morphology of Cu/Al bimetallic composite fabricated with different coating thickness: (**a**) 1.5 μm; (**b**) 3.8 μm; (**c**) 5.9 μm.

The SEM morphology of the shear surface of the Cu/Al bimetallic composite fabricated with different coating thickness was displayed in Figure 16. Foam-like fluffy holes and cracks were found on the shear fracture surface in Figure 16a, which was consistent with poor metallurgical bonding due to surface oxidation of the Cu rod. As shown in Figure 16b, the foam-like fluffy holes disappeared and the length of cracks become shorter, indicating better shear strength than the former. As for Figure 16c, the Ni layer of 5.9 μm limited the diffusion of Al atom to the Cu side, resulting in a relatively low content of intermetallic compounds in the Cu side of the transition zone. Therefore, the morphology resembling a slip band appeared on the dissociation surface.

(2)   Microhardness

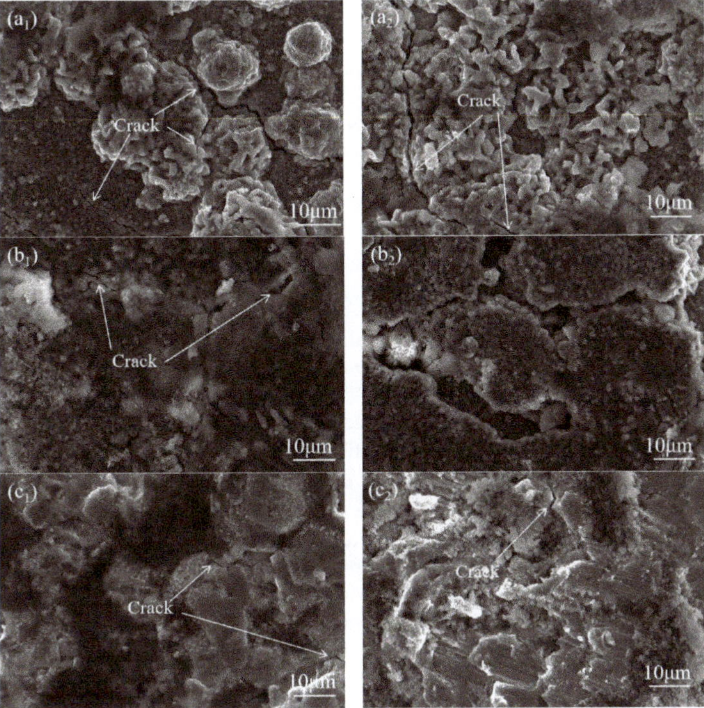

Figure 16. SEM morphology of shear surface of Cu/Al bimetallic composite fabricated with different coating thickness: (**a₁,a₂**) 1.5 μm; (**b₁,b₂**) 3.8 μm; (**c₁,c₂**) 5.9 μm.

Observations and statistical results of the interfacial microhardness of Cu/Al bimetallic composite fabricated with different coating thickness were described in Figure 17. The results showed that the micro-hardness of the Cu/Al bimetallic composite demonstrated a slight decreasing trend after a small increase, while the hardness of the interfacial compounds did not increase significantly. Ni elements were dissolved and diffused to the matrix under different electroplating times and different coating thickness, and the hardness of the electroplated Ni layer is 160–180 HV, which was close to the microhardness of the transition zone without the coating Ni layer. Therefore, it can be inferred that the Ni layer thickness has little effect on the microhardness of Cu/Al bimetallic composites.

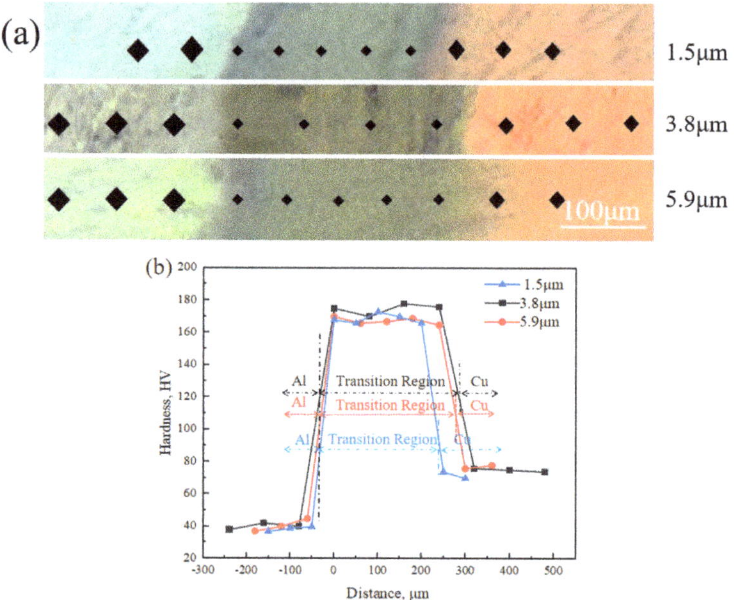

Figure 17. Microhardness of Cu/Al bimetallic composite with different coating thickness: (**a**) Indentation metallographic observation; (**b**) Hardness distribution.

4. Conclusions

The Cu/Al bimetallic composite was successfully fabricated by gravity casting. The effect of surface treatment and liquid–solid volume ratio on the microstructure and properties of the Cu/Al bimetallic composite were studied. The thickness of the transition zone was directly related to the mechanical properties of the Cu/Al bimetallic composite. Therefore, the quantitative regulation method of the thickness of the transition zone may become a new research hotspot in Cu/Al bimetallic composite fabrication. The following results and conclusions can be drawn:

(1) The thickness of transition zone, shear strength, and microhardness of transition zone increased with increasing the liquid–solid volume ratio of Cu/Al bimetallic composite fabricated by gravity casting.

(2) The thickness of transition zone and shear strength increased with the coating thickness of the Ni layer varied from 1.5 μm to 3.8 μm, due to the Ni layer effectively preventing oxidation on the surface of the Cu rod and promoting the metallurgical bonding of Cu/Al interface.

(3) The thickness of the transition zone and shear strength decreased with increasing the coating thickness of Ni layer from 3.8 μm to 5.9 μm, due to the thick Ni layer limiting the diffusion of Cu and Al atoms.

(4) The initiation and propagation of shear cracks occurred in the transition zone of Cu/Al bimetallic composite.

Author Contributions: Conceptualization, H.Y., Y.W. and W.F.; Methodology, H.Z. and Y.H.; Writing—original draft, Z.W.; Writing—review & editing, L.Z.; Project administration, C.L. All authors have read and agreed to the published version of the manuscript.

Funding: This work was funded by the Natural Science Foundation of the Jiangsu Higher Education Institutions of China (grant number 20KJB430015), the Open Fund of Jiangsu Institute of Marine Resources Development (grant number JSIMR202208), the Natural Science Foundation of Jiangsu Province, (grant number BK20201467), and the Scientific Research Funding Project of "333 High-level Talents Training Project" of Jiangsu Province (grant number BRA2020260).

Institutional Review Board Statement: Not applicable.

Informed Consent Statement: Not applicable.

Data Availability Statement: Not applicable.

Conflicts of Interest: The authors declare no conflict of interest.

References

1. Chandra, K.; Mahanti, A.; Singh, A.P.; Joshi, N.S.; Kain, V. Metallurgical Investigation on Embrittlement of Copper Cable of an Electric Motor. *J. Fail. Anal. Prev.* **2019**, *19*, 598–603.
2. Zuo, L.; Ye, B.; Feng, J.; Kong, X.; Jiang, H.; Ding, W. Microstructure, tensile properties and creep behavior of Al-12Si-3.5Cu-2Ni-0.8Mg alloy produced by different casting technologies. *J. Mater. Sci. Technol.* **2018**, *34*, 1222–1228. [CrossRef]
3. Zuo, L.; Ye, B.; Feng, J.; Xu, X.; Kong, X.; Jiang, H. Effect of δ-Al$_3$CuNi phase and thermal exposure on microstructure and mechanical properties of Al-Si-Cu-Ni alloys. *J. Alloys Compd.* **2019**, *791*, 1015–1024. [CrossRef]
4. Zhang, J.Y.; Zuo, L.J.; Jian, F.; Bing, Y.E.; Kong, X.Y.; Jiang, H.Y.; Ding, W.J. Effect of thermal exposure on microstructure and mechanical properties of Al-Si–Cu–Ni–Mg alloy produced by different casting technologies. *Trans. Nonferr. Met. Soc. China* **2020**, *30*, 1717–1730. [CrossRef]
5. Zuo, L.; Ye, B.; Feng, J.; Zhang, H.; Kong, X.; Jiang, H. Effect of ε-Al$_3$Ni phase on mechanical properties of Al–Si–Cu–Mg–Ni alloys at elevated temperature. *Mater. Sci. Eng. A* **2020**, *772*, 138794. [CrossRef]
6. Vishnu, P.; Mohan, R.R.; Sangeethaa, E.K.; Raghuraman, S.; Venkatraman, R. A review on processing of aluminium and its alloys through Equal Channel Angular Pressing die. *Mater. Today Proc.* **2020**, *21*, 212–222. [CrossRef]
7. Ghosh, A.; Ghosh, M.; Gudimetla, K.; Kalsar, R.; Kestens, L.A.I.; Kondaveeti, C.S.; Singh, P.B.; Ravisankar, B. Development of ultrafine grained Al–Zn–Mg–Cu alloy by equal channel angular pressing: Microstructure, texture and mechanical properties. *Arch. Civ. Mech. Eng.* **2020**, *20*, 7.
8. Wang, T.; Li, S.; Ren, Z.; Han, J.; Huang, Q. A novel approach for preparing Cu/Al laminated composite based on corrugated roll. *Mater. Lett.* **2018**, *234*, 79–82. [CrossRef]
9. Ye, Z.; Lv, Z.; Lu, M.; Jiang, H.; Li, S.; Li, Y.; Zhao, J.; He, J.; Zhang, L. Microstructure and properties of solid-liquid composite cast aluminum/copper bimetallic composite. *Foundry Technol.* **2022**, *43*, 176–179.
10. Liu, G.; Wang, Q.; Jiang, H. New progress in research of copper/aluminum bimetallic composite. *Mater. Guide* **2020**, *34*, 7115–7122.
11. Wang, J.; Lei, Y.; Liu, X.-H.; Xie, G.-L.; Jiang, Y.-Q.; Zhang, S. Microstructure and properties of horizontal continuous casting composite formed Cu/Al layered composite. *J. Eng. Sci.* **2020**, *42*, 216–224.
12. Wang, B.; Huang, S.; Zhou, M.; Lei, Y.; Sun, S. 6061 aluminum alloy powder cold pressing forming constitutive model. *Forg. Technol.* **2018**, *43*, 56–61.
13. Eivani, A.R.; Mirzakoochakshirazi, H.R.; Jafarian, H.R. Investigation of joint interface and cracking mechanism of thick cladding of copper on aluminum by equal channel angular pressing (ECAP). *J. Mater. Res. Technol.* **2020**, *9*, 3394–3405. [CrossRef]
14. Vini, M.H.; Daneshmand, S.; Forooghi, M. Roll bonding properties of Al/Cu bimetallic laminates fabricated by the roll bonding technique. *Technologies* **2017**, *5*, 32. [CrossRef]
15. Kouters, M.H.M.; Gubbels, G.H.M.; Dos Santos Ferreira, O. Characterization of intermetallic compounds in Cu-Al ball bonds: Mechanical properties, interface delamination and thermal conductivity. *Microelectron. Reliab.* **2013**, *53*, 1068–1075. [CrossRef]
16. Wang, T.; Cao, F.; Zhou, P.; Kang, H.; Chen, Z.; Fu, Y.; Xiao, T.; Huang, W.; Yuan, Q. Study on diffusion behavior and microstructural evolution of Al/Cu bimetal interface by synchrotron X-ray radiography. *J. Alloys Compd.* **2014**, *616*, 550–555. [CrossRef]
17. Zhao, K.N.; Liu, J.C.; Nie, X.Y.; Li, Y.; Li, H.X.; Du, Q.; Zhuang, L.Z.; Zhang, J.S. Interface formation in magnesium–magnesium bimetal composite fabricated by insert molding method. *Mater. Des.* **2016**, *91*, 122–131. [CrossRef]
18. Xiong, B.; Cai, C.; Lu, B. Effect of volume ratio of liquid to solid on the interfacial microstructure and mechanical properties of high chromium cast iron and medium carbon steel bimetal. *J. Alloys Compd.* **2011**, *509*, 6700–6704. [CrossRef]

19. Liu, G.; Wang, Q.; Zhang, L.; Ye, B.; Jiang, H.; Ding, W. Effect of Cooling Rate on the Microstructure and Mechanical Properties of Cu/Al Bimetal Fabricated by Compound Casting. *Metall. Mater. Trans.* **2018**, *49*, 661–672. [CrossRef]
20. Tavassoli, S.; Abbasi, M.; Tahavvori, R. Controlling of IMCs layers formation sequence, bond strength and electrical resistance in Al-Cu bimetal compound casting process. *Mater. Des.* **2016**, *108*, 343–353. [CrossRef]
21. Chen, S.; Chang, G.; Yue, X.; Li, Q. Solidification process and microstructure of transition layer of Cu–Al composite cast prepared by method of pouring molten aluminum. *Trans. Nonferr. Met. Soc. China* **2016**, *26*, 2247–2256. [CrossRef]
22. Li, H.; Chen, W.; Dong, L.; Shi, Y.; Liu, J.; Fu, Y.Q. Interfacial bonding mechanism and annealing effect on Cu-Al joint produced by solid-liquid compound casting. *J. Mater. Process. Technol.* **2018**, *252*, 795–803. [CrossRef]
23. Zhao, J.L.; Jie, J.C.; Fei, C.H.E.N.; Hang, C.H.E.N.; Li, T.J.; Cao, Z.Q. Effect of immersion Ni plating on interface microstructure and mechanical properties of Al/Cu bimetal. *Trans. Nonferr. Met. Soc. China* **2014**, *24*, 1659–1665. [CrossRef]
24. Yan, S.; Shi, Y. Influence of Ni interlayer on microstructure and mechanical properties of laser welded joint of Al/Cu bimetal. *J. Manuf. Process.* **2020**, *59*, 343–354. [CrossRef]
25. Ye, Z.; Huang, J.; Yang, H.; Liu, T.; Yang, J.; Chen, S. Effect of Si addition on corrosion behaviors of Cu/Al dissimilar joint brazed with novel Zn-Al-xSi filler metals. *J. Mater. Res. Technol.* **2019**, *8*, 5171–5179. [CrossRef]
26. Breedis, J.F.; Fister, J.C. Copper Alloys for Suppressing Growth of Cu-Al Intermetallic Compounds. U.S. Patent 4498121, 5 February 1985.
27. Yuan, H.U.; Chen, Y.Q.; Li, L.I.; Hu, H.D.; Zhu, Z.A. Microstructure and properties of Al/Cu bimetal in liquid-solid compound casting process. *Trans. Nonferr. Met. Soc. China* **2016**, *26*, 1555–1563.
28. Liu, G.; Wang, Q.; Zhang, L.; Ye, B.; Jiang, H.; Ding, W. Effects of melt-to-solid volume ratio and pouring temperature on microstructures and mechanical properties of Cu/Al bimetals in compound casting process. *Metall. Mater. Trans. A* **2019**, *50*, 401–414. [CrossRef]
29. Zhang, Y.; Yamane, T.; Hirao, K.; Minamino, Y. Microstructures and Vickers hardness of rapidly solidified Al-Cu alloys near the Al-Al$_2$Cu equilibrium eutectic composition. *J. Mater. Sci.* **1991**, *26*, 5799–5805. [CrossRef]

Article

Strength and Toughness of Hot-Rolled TA15 Aviation Titanium Alloy after Heat Treatment

Liangliang Li [1,2,*], Xin Pan [1], Biao Liu [1], Bin Liu [3], Pengfei Li [3] and Zhifeng Liu [2]

[1] Innovation Research Institute, Shenyang Aircraft Corporation, Shenyang 110850, China
[2] Key Laboratory of CNC Equipment Reliability, Ministry of Education, School of Mechanical and Aerospace Engineering, Jilin University, Changchun 130012, China
[3] School of Mechanical Engineering, Jiangsu University, Zhenjiang 212013, China
[*] Correspondence: liangleejob@163.com

Abstract: To investigate the impact of various heat treatments on the strength and toughness of TA15 aviation titanium alloys, five different heat treatment methods were employed in the temperature range of 810–995 °C. The microstructure of the alloy was examined using a scanning electron microscope (SEM) and X-ray diffraction (XRD), and its mechanical properties were analyzed through tensile, hardness, impact, and bending tests. The findings indicate that increasing the annealing temperature results in an increase in the phase boundary and secondary α phase, while the volume fraction of the primary α phase decreases, leading to a rise in hardness and a decrease in elongation. The tensile strength of heat-treated samples at 810 °C was notably improved, displaying high ductility at this annealing temperature. Heat treatment (810 °C/2 h/WQ) produced the highest tensile properties (ultimate tensile strength, yield strength, and elongation of 987 MPa, 886 MPa, and 17.78%, respectively). Higher heat treatment temperatures were found to enhance hardness but decrease the tensile properties, bending strength, and impact toughness. The triple heat treatment (810 °C/1 h/AC + 810 °C/1 h/AC + 810 °C/1 h/AC) resulted in the highest hardness of 601.3 MPa. These results demonstrate that various heat treatments have a substantial impact on the strength and toughness of forged TA15 titanium alloys.

Keywords: TA15 titanium alloy; tensile properties; impact toughness; triple heat treatment

Citation: Li, L.; Pan, X.; Liu, B.; Liu, B.; Li, P.; Liu, Z. Strength and Toughness of Hot-Rolled TA15 Aviation Titanium Alloy after Heat Treatment. *Aerospace* **2023**, *10*, 436. https://doi.org/10.3390/aerospace10050436

Academic Editor: Andrea Di Schino

Received: 21 March 2023
Revised: 30 April 2023
Accepted: 6 May 2023
Published: 8 May 2023

1. Introduction

The TA15 titanium alloy (Ti-6.5Al-2Zr-1Mo-1V) is classified as a medium strength titanium alloy with a high aluminum equivalent α type, making it highly desirable for its exceptional specific strength, creep resistance, ductility, fracture toughness, and fatigue properties. Its exceptional qualities make it a preferred material in the aviation field [1–3]. However, the alloy's mechanical properties are highly sensitive to its microstructure characteristics, such as phase composition, morphology, distribution, and grain size [4–6]. Given the high temperature sensitivity of titanium alloys, their microstructure and mechanical properties rely heavily on thermal mechanical processing and heat treatment [7].

In recent years, numerous researchers have explored the microstructure and mechanical properties of forged TA15 titanium alloys [8,9]. Sun et al. [10] investigated the tri-modal microstructure evolution in near-β and two-phase field heat treatments of forged TA15 alloys and found that a microstructure consisting of approximately 15% equiaxed α_p and a significant amount of thick and long lamellar α_s exhibited excellent comprehensive mechanical properties. Sun et al. [11] further studied the evolution of the lamellar α phase during two-phase field heat treatment in TA15 alloys and found that the volume fraction of the primary lamellar α decreased, the average thickness increased, and the average length variation trend was complex. Li et al. [12] studied mesoscale deformation mechanisms in relation to slip and grain boundary sliding in TA15 alloys during tensile deformation and found that grain boundary slip occurred at the equiaxed α (α_p)/β and lamellar α (α_l)/β

boundaries during deformation. Sun et al. [13] conducted a study on the microstructure and mechanical properties of TA15 alloys after thermomechanical processing and successfully fabricated equiaxed and fine-grained TA15 alloys with a mean grain size of 2 μm. Lei et al. [14] examined the microscopic damage development and crack propagation characteristics of TA15 titanium alloys with a trimodal structure consisting of equiaxed α (α_p), lamellar α (α_l), and β transformed matrix α (β_t) for the first time. Xu et al. [15] discovered that heat treatment could improve the mechanical and fatigue properties of TA15 alloy joints produced through friction stir welding. In their study, Wu et al. [16] discovered that using a dual heat treatment approach, which involves near-β heat treatment followed by (α + β) heat treatment, is an effective way to achieve a tri-modal microstructure in TA15 titanium alloys. This particular microstructure was found to have exceptional overall performance. Additionally, Zhao et al. [17] investigated the recrystallization behavior of TA15 titanium alloy sheets with numerous predeformed substructures during annealing and hot drawing at 800 °C. Meanwhile, Wu et al. [18] focused on the microstructure and mechanical properties of heat-treated and thermomechanically processed TA15 titanium alloy composites. They discovered that composite microstructures consisting of two or three phases exhibited excellent mechanical properties, while fully lamellar microstructures had the highest fracture toughness.

Despite the numerous studies conducted on the microstructure and mechanical properties of wrought TA15 titanium alloys, there is still a lack of research focusing on the effects of multiple heat treatments at high temperatures [19]. In a study conducted by Zhao et al. [20], the impact of vacuum annealing on the microstructure and mechanical properties of TA15 titanium alloy sheets was investigated. It was discovered that the recovery, recrystallization, and phase transformation of TA15 sheets annealed in different modes significantly affected their microstructure and mechanical properties. Furthermore, the strength of TA15 titanium alloy sheets was observed to improve, and the plastic toughness increased as the annealing temperature increased, although the strength decreased.

In addition, many scholars have studied dynamic recrystallization at high temperatures [21–23]. Wang et al. [24] investigated the thermal deformation mechanism of laser-welded TA15 joints through high-temperature tensile tests ranging from 800 °C to 900 °C, and found that at 900 °C and a strain rate of 0.001 s^{-1}, the maximum elongation was 292%. Vo et al. [25] simulated the flow stress of TA15 titanium alloys in compression tests at temperatures of 975, 1000, 1025, 1060, and 1100 °C and strain rates of 0.1 and 1 s^{-1}. Yang et al. [26] studied the hot tensile properties of TA15 by conducting tensile tests at temperatures ranging from 750 °C to 850 °C at strain rates of 0.001, 0.01, and 0.1 s^{-1}, and the metallographic results showed that the primary α phase transformed into equiaxed crystals, while the secondary α phase and the lamellar β phase bent and fractured. Hao et al. [27] conducted tensile tests on TA15 titanium alloys over a wide temperature range from −60 °C to 900 °C and at a strain rate of 0.001 s^{-1}. They used electron backscatter diffraction (EBSD) to analyze the evolution of microstructure and deformation mechanisms at different temperatures. At relatively low temperatures (60 °C, 23 °C, and 400 °C), the dislocation slip mechanism dominated the deformation, even when twinning was detected. However, the spheroidization and dynamic recrystallization (DRX) of α lamellae dominated the deformation mechanism at high temperatures (600 °C and 800 °C), leading to flow softening. Zhao et al. [17,28] found significant recrystallization behavior in the TA15 alloy during high temperature tensile tests at 800 °C. Li et al. [29] studied the effect of strain and temperature on hot deformation using an exponential Zener-Hollomon equation, and tested the effectiveness of the proposed constitutive equation in isothermal compression tests at temperatures ranging from 800 °C to 1000 °C, with strain rates of 10, 1, 0.1, 0.01, and 0.001 s^{-1} and a strain of 0.9. Feng et al. [30,31] studied the tensile properties and fracture mechanisms of TA15 titanium alloys at temperatures ranging from 400 °C to 700 °C, and attempted to improve its high-temperature plastic deformation ability by adding TiBw. Zhao et al. [32] analyzed the deformation non-uniformity and slip mode of TA15 titanium alloy sheets during hot tensile tests at 750 °C along the rolling direction.

Li et al. [33] revealed the flow softening and plastic damage mechanisms in the uniaxial thermal tensile tests of titanium alloys at temperatures ranging from 910 °C to 970 °C and strain rates of 0.01 to 0.1 s^{-1}. Zhao et al. [34] studied the evolution of texture in the rolling direction of TA15 sheets through experiments and crystal plasticity simulations in a tensile test at 750 °C. Liu et al. [35] investigated the correlation between the α value and strain in TA15 titanium alloys deformed at 750 °C.

The research presented above highlights the need to further investigate the impact of different heat treatments on the strength and toughness of forged TA15 titanium alloy sheets. Consequently, several heat treatment experiments were conducted to systematically explore the effect of heat treatment parameters on the microstructure of the forged TA15 titanium alloy sheets. Furthermore, microhardness, room temperature tensile, impact, and bending tests were conducted on TA15 titanium alloy sheets subjected to different heat treatments. The resulting data was analyzed to assess the impact of different heat treatments on the strength and toughness of the forged TA15 titanium alloy sheets.

2. Experimental Procedures

2.1. Materials and Heat Treatment

Table 1 presents the chemical composition of TA15 titanium alloys, while the rectangular samples used in this study were 100 mm × 77 mm × 5 mm in size. The heat treatment furnace type is an SA2-9-12TP 1200 °C box atmosphere furnace. The furnace adopts a double-shell structure and an intelligent programmable temperature control system. The heating element is a resistance wire thermocouple with a temperature range of 0–1200 °C, and the temperature control accuracy is ±1 °C. Table 2 outlines the various heat treatment methods utilized, including two multiple heat treatment methods (TriAC and TriWQ) to examine the effect of high temperatures on the microstructure and mechanical properties of the TA15 titanium alloy. Additionally, other samples (810AC, 810WQ, and 940WQ) were utilized to evaluate the impact of cooling rates on microstructure and mechanical properties. The samples were polished using silicon carbide paper (300, 600, 1000, 1500, and 2000) and etched using Kohler reagent (H$_2$O:HNO$_3$:HCl:HF is 190:5:3:1). The microstructure was analyzed using an S-7800N Scanning Electron Microscope (SEM), and phase identification was conducted using X-ray diffraction (XRD).

Table 1. Chemical composition (wt%) of TA15 titanium alloy used in the study.

Alloy Element	Al	Zr	Mo	V	Fe	C	Ti
Mass fraction	6.3	1.9	1.32	1.68	0.04	0.01	Bal

Table 2. Heat treatment conditions for different samples.

Sample Name	Heat Treatment Condition
None	No heat treatment
810AC	810 °C for 2 h—Air cooling(AC)
810WQ	810 °C for 2 h—water quench (WQ)
940WQ	940 °C for 1 h—water quench (WQ)
TriAC	Step1: 965 °C for 1 h—Air cooling(AC)
	Step2: 950 °C for 1 h—Air cooling(AC)
	Step3: 810 °C for 1 h—Air cooling(AC)
TriWQ	Step1: 995 °C for 1 h—water quench (WQ)
	Step2: 940 °C for 1 h—water quench (WQ)
	Step3: 810 °C for 1 h—water quench (WQ)

2.2. Mechanical Properties Test

To determine the microhardness of the TA15 titanium alloy, an indentation test was conducted, measuring the microhardness of each sample at 100 g load and 15 s residence time. Nine indentation tests were performed on each sample at 0.5 mm intervals, and the average value was taken as the microhardness of the material. Due to limitations in the testing equipment, the extensometer was not used in the experiment. Figure 1 displays the dimensions of the sample used for the tensile, bending, and impact toughness tests. The room temperature tensile and bending tests were carried out using a DDL100 electronic universal testing machine, and the calculation formula for bending properties was applied. Impact properties were evaluated using an NI150 metal pendulum impact testing machine (ASTM, E23-2018).

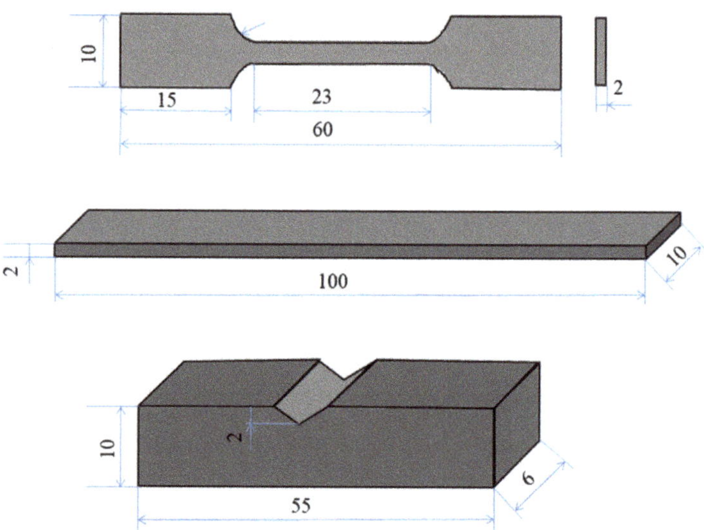

Figure 1. Size of samples with different mechanical properties.

3. Results and Discussion

3.1. Microstructure Observation

Figure 2 displays the X-ray diffraction patterns of TA15 titanium alloys subject to various heat treatments. All samples exhibit α phase, while some also exhibit β phase. However, after TriAC or TriWQ treatment, only α phase is detected. Figure 3 illustrates the microstructures of TA15 titanium alloys under different heat treatment methods. Combined with XRD diffraction results and SEM images, it is evident that the initial sample comprises primary equiaxed α phase and β grain microstructures (Figure 3a). However, the microstructures present various morphologies after different heat treatments. Figure 3b–d demonstrate that the microstructures comprise secondary α phase and transformed β phase after 810AC, 810WQ, and 940WQ treatments, respectively. Figure 3e,f indicate that the main microstructure after TriAC and TriWQ treatments is acicular α phase, with the α phase being slenderer after TriWQ treatment than after TriAC treatment.

To better illustrate the variation in elemental content between the α and β phases, we conducted an EDS scanning analysis of the initial sample's microstructure, as presented in Figure 4. The TA15 alloy consists predominantly of Ti, Al, and V. The EDS mapping results indicate that the α phase is primarily composed of Ti and Al elements, whereas the β phase contains more Mo and O. Additionally, V is found to be evenly distributed across the entire surface.

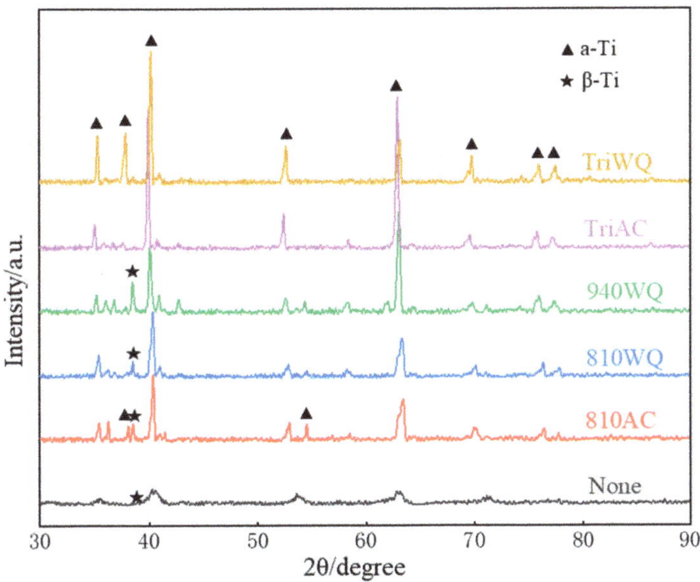

Figure 2. X-ray diffraction patterns of the TA15 Titanium Alloy under different heat treatment conditions.

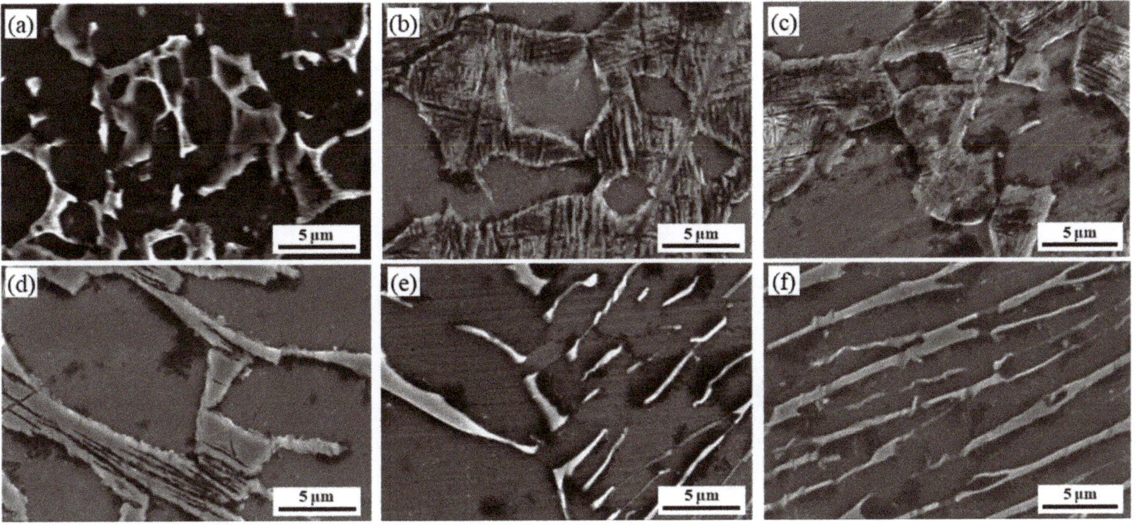

Figure 3. SEM images of the TA15 Titanium Alloy under different heat treatment methods: (**a**) None, (**b**) 810AC, (**c**) 810WQ, (**d**) 940WQ, (**e**) TriAC, (**f**) TriWQ.

3.2. Microhardness

The average microhardness of TA15 alloys with various heat treatments is presented in Figure 5. It is evident that the microhardness of the untreated sample is the lowest, measuring at 310.5 HV. All heat treatment methods result in an improvement of microhardness in TA15 titanium alloys. However, there is no significant increase in microhardness observed with 810AC and 810WQ, which measured at 365.9 HV and 338.2 HV, respectively. The 940WQ sample exhibited a more extensive decomposition of metastable β phase than the 810WQ sample, resulting in an increased microhardness [19]. The TriAC sample, with the largest volume fraction of α phase, has the highest microhardness at about 601.3 HV.

Furthermore, the TriAC and TriWQ samples underwent triple heat treatment processes, which fully decomposed the β phase and led to even higher microhardness [36].

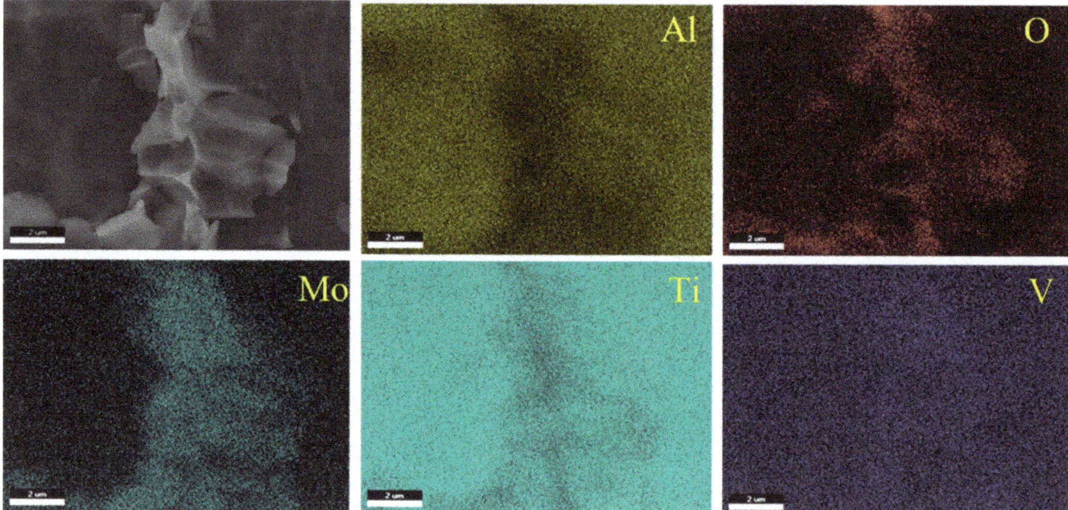

Figure 4. EDS mapping and elemental content analysis of TA15 microstructure.

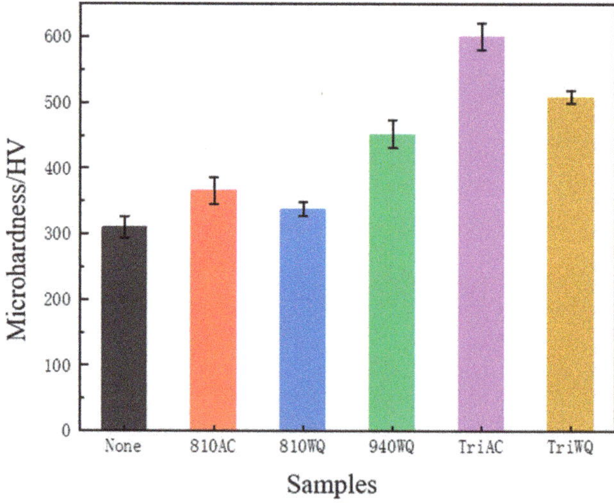

Figure 5. Effect of different heat treatment methods on microhardness of the TA15 titanium alloy.

3.3. Impact Toughness

Figure 6 illustrates a comparison of impact toughness among different heat treatment methods. The impact toughness of the initial sample measures at 32.19 J/cm^2, while the 810AC sample displays a slightly higher value at about 34.69 J/cm^2. In general, the impact toughness of the untreated sample and low-temperature treated samples are higher than those of the other samples. Conversely, the impact toughness of the 940WQ, TriAC, and TriWQ samples significantly decreased, measuring at 13.13, 18.75, and 5.31 J/cm^2, respectively. It can be inferred that the presence of transformed α phase and acicular α phase will greatly reduce the impact properties of the TA15 titanium alloy.

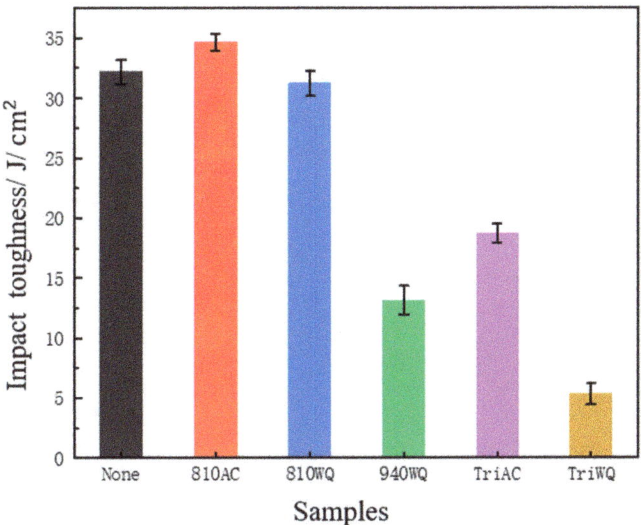

Figure 6. Influence of different heat treatment methods on impact resistance of the TA15 titanium alloy.

3.4. Bending Properties

Figure 7 and Table 3 show that the overall bending properties of TA15 titanium alloys after heat treatment are reduced significantly. The data in Table 3 are calculated by the formula in Li et al. [37]:

$$m_E = \frac{L_S^3}{48I} \cdot \left(\frac{\Delta F}{\Delta f} \right) \tag{1}$$

$$R_{bb} = \frac{F_{bb} L_S}{4W} \tag{2}$$

where $I = bh^3/12$, b is the width of the sample, h is the thickness of the sample, L_S is the distance between the two support points of the sample on the bending test device, and L_S =100 mm, F_{bb} is the maximum bending stress, and W is the section coefficient of the sample ($W = bh^2/6$).

Based on the analysis presented in Figure 7 and Table 3, it can be observed that the bending strengths of the 810AC and 810WQ samples are approximately 4920.0 MPa and 5572.5 MPa, respectively. Additionally, the bending strengths of the 940WQ, TriAC, and TriWQ samples are approximately 1773.75 MPa, 1267.5 MPa, and 5572.5 MPa, respectively. It is worth noting that the samples subjected to higher heat treatment temperatures (940WQ, TriAC, and TriWQ) exhibit a more significant decline in strength than those subjected to lower heat treatment temperatures (810AC and 810WQ).

Table 3. Bending performance test results of different specimens.

Sample Name	Slope of the Linear Elasticity m_E (MPa)	Maximum Bending Force F_{bb} (N)	Bending Strength R_{bb} (MPa)
None	466,811.41	2003.1	7511.25
810AC	59,630.48	1312.3	4920.0
810WQ	489,572.17	1486.6	5572.5
940WQ	364,184.23	473.2	1773.75
TriAC	360,134.39	338.3	1267.5
TriWQ	253,875.59	214.5	802.53

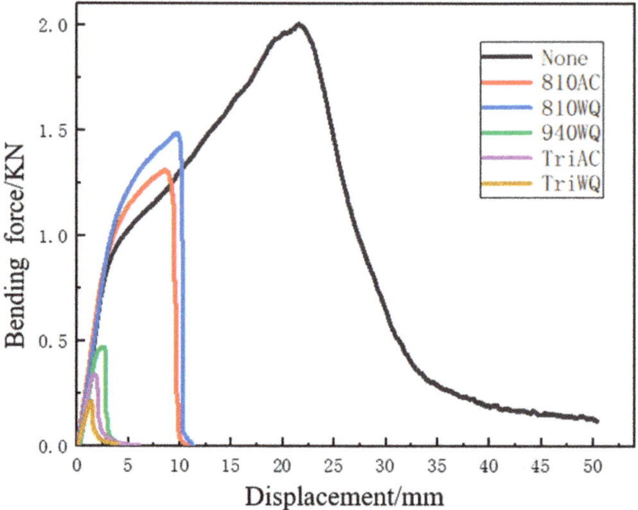

Figure 7. Effects of different heat treatment methods on bending properties of the TA15 titanium alloy.

3.5. Tensile Properties

Figure 8 illustrates the impact of different heat treatments on the tensile properties of TA15 titanium alloys. The data in Table 3 suggests that heat treatment results in a reduction in elongation (ε), with the highest elongation of approximately 27.92% observed in the initial sample. The yield strength (YS) of 810AC and 810WQ samples decreases slightly and insignificantly, while the ultimate tensile strength (UTS) increases significantly, with values of 987 MPa and 984 MPa, respectively. Notably, the tensile strengths are similar to the bending strengths observed in Figure 7. However, for the samples subjected to higher heat treatment temperatures (940WQ, TriAC, and TriWQ), YS, UTS, and elongation all decrease significantly. Specifically, the YS of TriAC and TriWQ samples are 529 MPa and 364 MPa, respectively, while their UTS are 272 MPa and 325 MPa, respectively. Furthermore, the very low elongation of TriAC and TriWQ samples suggests that the fracture mechanism is close to a brittle fracture.

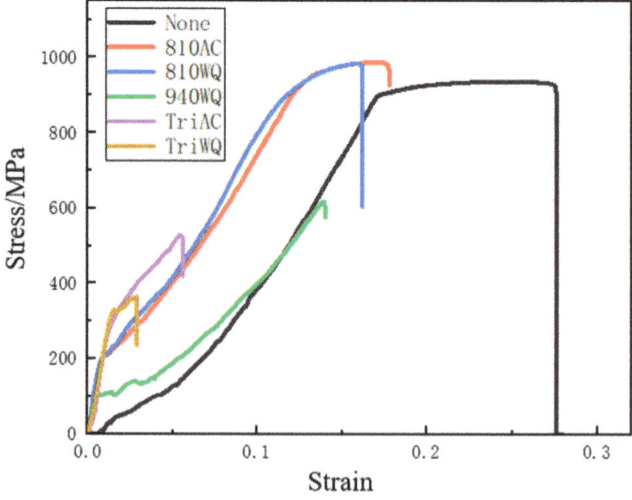

Figure 8. Effects of different heat treatments on tensile properties of the TA15 titanium alloy.

3.6. Fracture Analysis

To gain a thorough understanding of the fracture mechanism during a tensile test, an analysis was performed on the fracture characteristics of TA15 tensile samples, as presented in Figure 9. The fracture surface in Figure 9a exhibits numerous evenly distributed dimples, indicative of the sample's strong toughness and typical ductile fracture. In Figure 9b, a ductile fracture with numerous large depressions is shown. The size and depth of the dimples have increased significantly compared to the initial sample, and many shallow and small depressions are present, indicating that the ductile fracture is the primary fracture mechanism of the 810WQ sample. Lastly, Figure 9c depicts the tensile fracture morphology of the 810WQ sample. Compared to the None sample, the fracture surface undulation of the 810WQ TA15 titanium alloy is larger and contains many deep dimples, with a large number of small dimples distributed around them, which are indicative of ductile fractures [37]. As shown in Figure 9d, the fracture morphology of the 940WQ sample exhibits an obvious river-like tearing ridge, and the fracture is similar to a quasi-cleavage fracture. These results indicate that the plasticity of the TA15 titanium alloy after 940WQ treatment is significantly reduced, as the α_{Sr} strongly hinders slip and makes it more difficult to deform the sample, with the dislocation almost unable to cross the α/β boundary. Furthermore, Figure 9e shows that the cleavage steps, cleavage plane, and tear edges of the TriAC sample cover most of the fracture plane, with no dimple observed, and an unstable fracture tear edge existing on the cleavage plane. A portion of the fracture direction is nearly perpendicular to the tensile direction, resulting in a uniform fracture surface. Brittle fracture is the predominant mechanism in the fracture process, which aligns with the elongation value of 5.59%. As illustrated in Figure 9f, the TriWQ sample's fracture mainly exhibits fluvial tearing ridges, with micropores present at the grain boundary, representing a prominent characteristic of brittle intergranular fractures. These results indicate that the TriWQ sample exhibits low toughness and is prone to breakage, consistent with its ultimate tensile strength of 364 MPa and an elongation of 2.9% (refer to Table 4). Therefore, it can be inferred that the ductility of the TA15 titanium alloy is significantly reduced after triple heat treatment, as a result of a large number of αS and layered $\alpha+\beta$ structures that hinder slip.

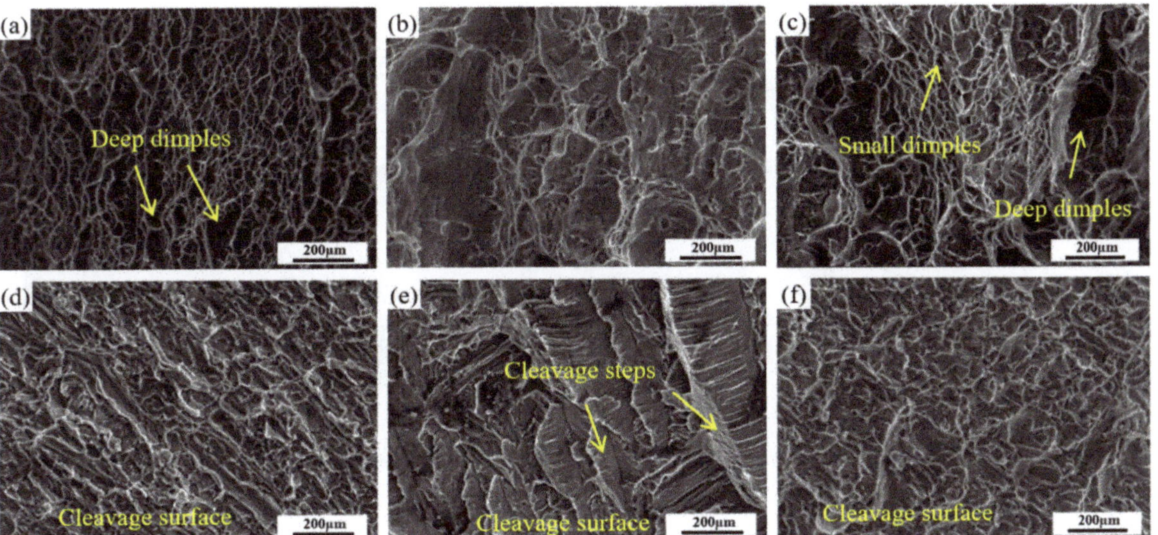

Figure 9. Tensile fracture morphology of the TA15 titanium alloy under different heat treatment methods: (**a**) None, (**b**) 810AC, (**c**) 810WQ, (**d**) 940WQ, (**e**) TriAC, (**f**) TriWQ.

Table 4. Measured mechanical properties of TA15 under different conditions.

Sample Name	Ultimate Tensile Strength σu (MPa)	Yield Strength σy (MPa)	Elongation ε (%)
None	936 ± 21	891 ± 18	27.92 ± 1.2
810AC	987 ± 25	886 ± 17	17.78 ± 1.3
810WQ	984 ± 40	805 ± 32	16.2 ± 2.3
940WQ	618 ± 84	105 ± 32	14.02 ± 3.2
TriAC	529 ± 36	272 ± 25	5.59 ± 0.8
TriWQ	364 ± 41	325 ± 38	2.9 ± 0.2

The impact fracture surfaces displayed in Figure 10a–c demonstrate distinct ductile fracture characteristics. The surface of the fracture exhibits multiple depressions of varying sizes with minimal surface undulations. The region situated between the deeper dimples is characterized by small dimples, which is a classic feature of ductile fractures. This type of fracture is a result of extensive plastic deformation before the sample's eventual rupture. In contrast, Figure 10d–f indicate that the impact fracture morphology displays a prominent cleavage plane, indicating a brittle fracture. However, some regions of the cleavage plane are overlaid with shallow dimples, representing a brittle ductile mixed fracture. After annealing at 810 °C, the dimple size of the sample increases significantly compared to the original sample, indicating excellent ductility and impact toughness. Conversely, at temperatures exceeding 940 °C, the area of the fracture cleavage plane of the sample increases significantly, and the size and number of dimples decrease, leading to a considerable reduction in impact toughness, as illustrated in Figure 6.

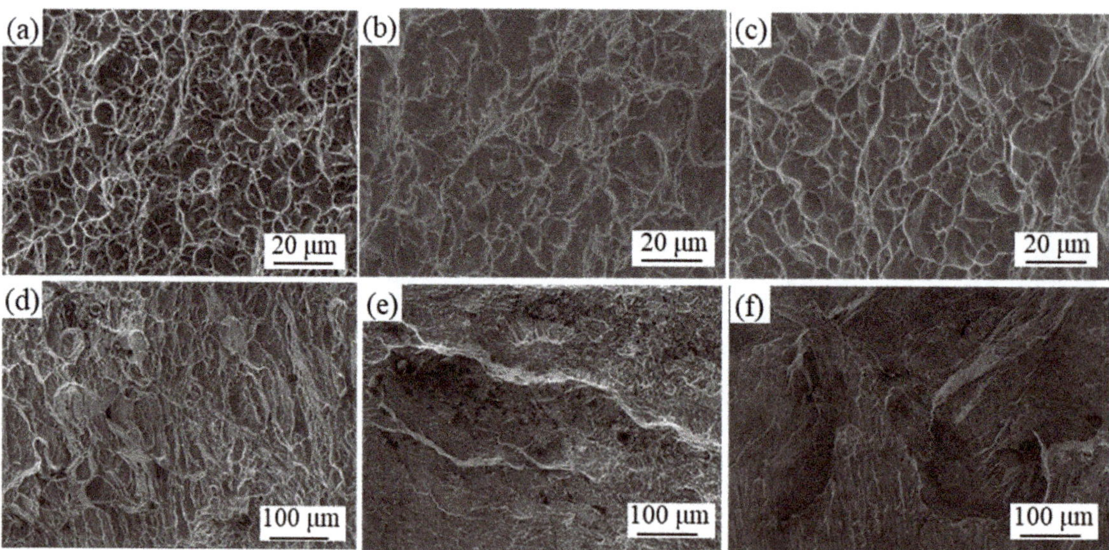

Figure 10. Impact fracture morphology of TA15 titanium alloys under different heat treatment methods: (**a**) None, (**b**) 810AC, (**c**) 810WQ, (**d**) 940WQ, (**e**) TriAC, (**f**) TriWQ.

Figure 11 illustrates the mechanism and schematic diagram of the impact of post-heat treatment on the microstructures and mechanical properties of TA15 titanium alloys. A lower heat treatment temperature has minimal effect on the mechanical properties. The transformed α phase and acicular α phase lead to a slight increase in microhardness (Figure 5). However, the impact toughness (Figure 6), tensile elongation (Figure 8), and bending strength (Figure 7) experience a significant decrease. Despite the reduction in toughness and plasticity, both the tensile fracture surface and impact test fracture surface's dimples indicate that it is a typical ductile fracture. However, a higher heat treatment temperature induces a complete transformation in the sample's microstructure morphology. The transformed α phase leads to a significant increase in microhardness (Figure 5) but a substantial reduction in plasticity (Figure 6). Consequently, the mechanical properties, including impact toughness (Figure 6), tensile elongation, tensile strength (Figure 8), and bending strength (Figure 7), undergo a considerable decrease due to the reduction in plasticity. Moreover, the fracture mode changes from a ductile fracture with lower heat treatment temperatures to a brittle fracture with higher heat treatment temperatures.

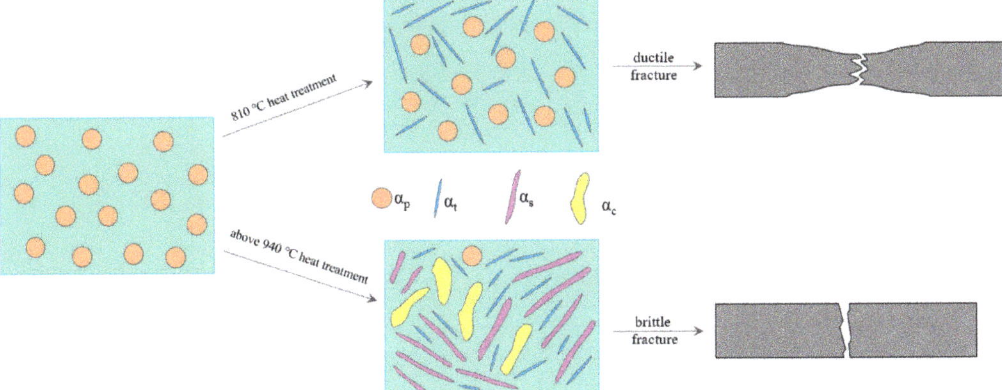

Figure 11. Mechanism and schematic diagram of the effect of post heat treatment on the microstructures and mechanical properties of TA15 titanium alloys.

4. Conclusions

This paper investigates the effects of heat treatment on the microstructures and mechanical properties of forged TA15 titanium alloy sheets using SEM, XRD, and microhardness characterization. Bending, impact, and tensile tests were performed to analyze the impact of different heat treatments on the strength and toughness of the alloy. Based on the results, the following conclusions are drawn:

(1) At a lower heat treatment temperature of 810 °C, cooling methods have minimal impact on the mechanical properties of the forged TA15 titanium alloys. The microstructure, microhardness, tensile strength, and impact toughness are similar to the initial samples.

(2) At a higher heat treatment temperature above 940 °C, all cooling methods significantly alter the microstructures and mechanical properties. Transformed α phase and acicular α phase result in increased microhardness but decreased strength and plasticity. Potentially, oxidation may be one of the main factors leading to performance degradation after high-temperature heat treatment.

(3) The tensile fracture morphology of the forged TA15 titanium alloys after different heat treatments is significantly different. The fracture morphology of the 940WQ sample is similar to quasi-cleavage fracture, and the tensile fracture of the TriAC sample is mainly brittle transgranular fracture, while the 810WQ and 810WQ samples have better tensile properties and exhibit deeper and more pronounced dimples than the initial sample.

Author Contributions: Conceptualization, L.L.; methodology, L.L., X.P., B.L. (Biao Liu) and B.L. (Bin Liu); software, X.P.; validation, L.L. and P.L.; formal analysis, P.L.; investigation, P.L. and Z.L.; resources, L.L.; data curation, L.L.; writing—original draft preparation, L.L.; writing—review and editing, L.L., X.P., B.L. (Biao Liu), B.L. (Bin Liu), P.L. and Z.L.; visualization, X.P. and B.L. (Bin Liu); supervision, P.L. and Z.L.; project administration, P.L.; funding acquisition, L.L. All authors have read and agreed to the published version of the manuscript.

Funding: This work was supported by the National Science and Technology Major Project (2017ZX04001001), the Key Laboratory of Vibration and Control of Aero-Propulsion System, Ministry of Education, Northeastern University (No. VCAME202208), the National Science Foundation of Jiangsu Province (No. BK20210758), China Postdoctoral Science Foundation Funded Project (No. 2022M710060) and Postgraduate Research and Practice Innovation Program of Jiangsu Province (Nos. KYCX22_3626 and SJCX22_1849).

Conflicts of Interest: The authors declare no conflict of interest..

References

1. Arab, A.; Chen, P.; Guo, Y. Effects of microstructure on the dynamic properties of TA15 titanium alloy. *Mech. Mater.* **2019**, *137*, 103121. [CrossRef]
2. Sun, Z.; Guo, S.; Yang, H. Nucleation and growth mechanism of α-lamellae of Ti alloy TA15 cooling from an α+ β phase field. *Acta Mater.* **2013**, *61*, 2057–2064. [CrossRef]
3. Napoli, G.; Paura, M.; Vela, T.; Schino, A. Di Colouring titanium alloys by anodic oxidation. *Metalurgija* **2018**, *57*, 111–113.
4. Yu, W.; Li, M.Q.; Luo, J. Effect of deformation parameters on the precipitation mechanism of secondary α phase under high temperature isothermal compression of Ti-6Al-4V alloy. *Mater. Sci. Eng. A* **2010**, *527*, 4210–4217. [CrossRef]
5. Li, P.; Liu, J.; Liu, B.; Li, L.; Zhou, J.; Meng, X.; Lu, J. Microstructure and mechanical properties of in-situ synthesized Ti (N, C) strengthen IN718/1040 steel laminate by directed energy deposition. *Mater. Sci. Eng. A* **2022**, *846*, 143247. [CrossRef]
6. Li, P.; Yang, Q.; Li, L.; Gong, Y.; Zhou, J.; Lu, J. Microstructure evolution and mechanical properties of in situ synthesized ceramic reinforced 316L/IN718 matrix composites. *J. Manuf. Process.* **2023**, *93*, 214–224. [CrossRef]
7. Majorell, A.; Srivatsa, S.; Picu, R.C. Mechanical behavior of Ti-6Al-4V at high and moderate temperatures-Part I: Experimental results. *Mater. Sci. Eng. A* **2002**, *326*, 297–305. [CrossRef]
8. Uhlmann, E.; Kersting, R.; Klein, T.B.; Cruz, M.F.; Borille, A.V. Additive Manufacturing of Titanium Alloy for Aircraft Components. *Procedia CIRP* **2015**, *35*, 55–60. [CrossRef]
9. Lütjering, G. Influence of processing on microstructure and mechanical properties of (α + β) titanium alloys. *Mater. Sci. Eng. A* **1998**, *243*, 32–45. [CrossRef]
10. Sun, Z.; Wu, H.; Wang, M.; Cao, J. Tri-Modal Microstructure Evolution in Near-β and Two Phase Field Heat Treatments of Conventionally Forged TA15 Ti-Alloy. *Adv. Eng. Mater.* **2017**, *19*, 1600796. [CrossRef]
11. Sun, Z.; Wu, H.; Sun, J.; Cao, J. Evolution of lamellar α phase during two-phase field heat treatment in TA15 alloy. *Int. J. Hydrogen Energy* **2017**, *42*, 20849–20856. [CrossRef]
12. Li, Y.; Gao, P.; Yu, J.; Jin, S.; Chen, S.; Zhan, M. Mesoscale deformation mechanisms in relation with slip and grain boundary sliding in TA15 titanium alloy during tensile deformation. *J. Mater. Sci. Technol.* **2022**, *98*, 72–86. [CrossRef]
13. Sun, Q.J.; Xie, X. Microstructure and mechanical properties of TA15 alloy after thermo-mechanical processing. *Mater. Sci. Eng. A* **2018**, *724*, 493–501. [CrossRef]
14. Lei, Z.; Gao, P.; Li, H.; Cai, Y.; Zhan, M. On the fracture behavior and toughness of TA15 titanium alloy with tri-modal microstructure. *Mater. Sci. Eng. A* **2019**, *753*, 238–246. [CrossRef]
15. Xu, X.; Liu, Q.; Wang, J.; Ren, X.; Hou, H. The heat treatment improving the mechanical and fatigue property of TA15 alloy joint by friction stir welding. *Mater. Charact.* **2021**, *180*, 111399. [CrossRef]
16. Wu, H.; Sun, Z.; Cao, J.; Yin, Z. Formation and evolution of tri-modal microstructure during dual heat treatment for TA15 Ti-alloy. *J. Alloys Compd.* **2019**, *786*, 894–905. [CrossRef]
17. Zhao, J.; Wang, K.; Huang, K.; Liu, G. Recrystallization behavior during hot tensile deformation of TA15 titanium alloy sheet with substantial prior deformed substructures. *Mater. Charact.* **2019**, *151*, 429–435. [CrossRef]
18. Wu, H.; Sun, Z.; Cao, J.; Yin, Z. Microstructure and Mechanical Behavior of Heat-Treated and Thermomechanically Processed TA15 Ti Alloy Composites. *J. Mater. Eng. Perform.* **2019**, *28*, 788–799. [CrossRef]
19. Zhong, Y.U. TA15 Effect of annealing process on three directions hardness and microstructure. *Heat Treat. Met.* **2011**, *36*, 73–75. (In Chinese)
20. Zhao, H.J.; Wang, B.Y.; Liu, G.; Yang, L.; Xiao, W.C. Effect of vacuum annealing on microstructure and mechanical properties of TA15 titanium alloy sheets. *Trans. Nonferrous Met. Soc. China* **2015**, *25*, 1881–1888. [CrossRef]
21. Wang, Y.; Xue, X.; Kou, H.; Chang, J.; Yin, Z.; Li, J. Improvement of microstructure homogenous and tensile properties of powder hot isostatic pressed TA15 titanium alloy via heat treatment. *Mater. Lett.* **2022**, *311*, 131585. [CrossRef]
22. Li, P.; Wang, Y.; Li, L.; Gong, Y.; Zhou, J.; Lu, J. Ablation oxidation and surface quality during laser polishing of TA15 aviation titanium alloy. *J. Mater. Res. Technol.* **2023**, *23*, 6101–6114. [CrossRef]
23. Wang, X.; Gao, P.; Zhan, M.; Yang, K.; Domg, Y.; Li, Y. Development of microstructural inhomogeneity in multi-pass flow forming of TA15 alloy cylindrical parts. *Chin. J. Aeronaut.* **2020**, *33*, 2088–2097. [CrossRef]
24. Wang, K.; Liu, G.; Tao, W.; Zhao, J.; Huang, K. Study on the mixed dynamic recrystallization mechanism during the globularization process of laser-welded TA15 Ti-alloy joint under hot tensile deformation. *Mater. Charact.* **2017**, *126*, 57–63. [CrossRef]
25. Vo, P.; Jahazi, M.; Yue, S.; Bocher, P. Flow stress prediction during hot working of near-α titanium alloys. *Mater. Sci. Eng. A* **2007**, *447*, 99–110. [CrossRef]

26. Yang, L.; Wang, B.; Liu, G.; Zhao, H.; Zhou, J. Hot Tensile Behavior and Self-consistent Constitutive Modeling of TA15 Titanium Alloy Sheets. *J. Mater. Eng. Perform.* **2015**, *24*, 4647–4655. [CrossRef]

27. Hao, F.; Xiao, J.; Feng, Y.; Wang, Y.; Ju, J.; Du, Y.; Wang, K.; Xue, L.; Nie, Z.; Tan, C. Tensile deformation behavior of a near-titanium alloy Ti-6Al-2Zr-1Mo-1V under a wide temperature range. *J. Mater. Res. Technol.* **2020**, *9*, 2818–2831. [CrossRef]

28. Zhao, H.-J.; Wang, B.-Y.; Ju, D.-Y.; Chen, G.-J. Hot tensile deformation behavior and globularization mechanism of bimodal microstructured Ti−6Al−2Zr−1Mo−1V alloy. *Trans. Nonferrous Met. Soc. China* **2018**, *28*, 2449–2459. [CrossRef]

29. Li, J.; Li, F.; Cai, J. Constitutive model prediction and flow behavior considering strain response in the thermal processing for the TA15 titanium alloy. *Materials* **2018**, *11*, 1985. [CrossRef]

30. Feng, Y.; Cui, G.; Zhang, W.; Chen, W.; Yu, Y. High temperature tensile fracture characteristics of the oriented TiB whisker reinforced TA15 matrix composites fabricated by pre-sintering and canned extrusion. *J. Alloys Compd.* **2018**, *738*, 164–172. [CrossRef]

31. Feng, Y.; Zhang, W.; Zeng, L.; Cui, G.; Chen, W. Room-temperature and high-temperature tensile mechanical properties of TA15 titanium alloy and TiB whisker-reinforced TA15 matrix composites fabricated by vacuum hot-pressing sintering. *Materials* **2017**, *10*, 424. [CrossRef] [PubMed]

32. Zhao, J.; Lv, L.; Liu, G.; Wang, K. Analysis of deformation inhomogeneity and slip mode of TA15 titanium alloy sheets during the hot tensile process based on crystal plasticity model. *Mater. Sci. Eng. A* **2017**, *707*, 30–39. [CrossRef]

33. Li, J.; Wang, B.; Huang, H.; Fang, S.; Chen, P.; Zhao, J.; Qin, Y. Behaviour and constitutive modelling of ductile damage of Ti-6Al-1.5Cr-2.5Mo-0.5Fe-0.3Si alloy under hot tensile deformation. *J. Alloys Compd.* **2019**, *780*, 284–292. [CrossRef]

34. Zhao, J.; Lv, L.; Liu, G. Experimental and simulated analysis of texture evolution of TA15 titanium alloy sheet during hot tensile deformation at 750 °C. *Procedia Eng.* **2017**, *207*, 2179–2184. [CrossRef]

35. Liu, G.; Wang, K.; He, B.; Huang, M.; Yuan, S. Mechanism of saturated flow stress during hot tensile deformation of a TA15 Ti alloy. *Mater. Des.* **2015**, *86*, 146–151. [CrossRef]

36. Ji, Z.; Shen, C.; Wei, F.; Li, H. Dependence of macro- and micro-properties on α plates in Ti-6Al-2Zr-1Mo-1V alloy with tri-modal microstructure. *Metals* **2018**, *8*, 299. [CrossRef]

37. Li, P.; Zhou, J.; Li, L.; Gong, Y.; Lu, J.; Meng, X. Influence of depositing sequence and materials on interfacial characteristics and mechanical properties of laminated composites. *Mater. Sci. Eng. A* **2021**, *827*, 142092. [CrossRef]

metals

Article

Increasing Hardness and Wear Resistance of Austenitic Stainless Steel Surface by Anodic Plasma Electrolytic Treatment

Sergei Kusmanov [1,*], Tatiana Mukhacheva [1], Ivan Tambovskiy [1], Alexander Naumov [2], Roman Belov [1], Ekaterina Sokova [1] and Irina Kusmanova [1]

[1]   Department of Mathematical and Natural Sciences, Kostroma State University, 156005 Kostroma, Russia
[2]   Department of Physics and Technology of Nanostructures, Saint Petersburg Academic University, 194021 Saint Petersburg, Russia
*   Correspondence: sakusmanov@yandex.ru; Tel.: +7-920-6473-090

Abstract: The results of modifying the surface of austenitic stainless steel by anodic plasma electrolytic treatment are presented. Surface treatment was carried out in aqueous electrolytes based on ammonium chloride (10%) with the addition of ammonia (5%) as a source of nitrogen (for nitriding), boric acid (3%) as a source of boron (for boriding) or glycerin (10%) as a carbon source (for carburizing). Morphology, surface roughness, phase composition and microhardness of the diffusion layers in addition to the tribological properties were studied. The influence of physicochemical processes during the anodic treatment of the features of the formation of the modified surface and its operational properties are shown. The study revealed the smoothing of irregularities and the reduction in surface roughness during anodic plasma electrolytic treatment due to electrochemical dissolution. An increase in the hardness of the nitrided layers to 1450 HV with a thickness of up to 20–25 μm was found due to the formation of iron nitrides and iron-chromium carbides with a 3.7-fold decrease in roughness accompanied by an increase in wear resistance by 2 orders. The carburizing of the steel surface leads to a smaller increase in hardness (up to 700 HV) but a greater thickness of the hardened layer (up to 80 μm) due to the formation of chromium carbides and a solid solution of carbon. The roughness and wear resistance of the carburized surface change are approximately the same values as after nitriding. As a result of the boriding of the austenitic stainless steel, there is no hardening of the surface, but, at the same time, there is a decrease in roughness and an increase in wear resistance on the surface. It has been established that frictional bonds in the friction process are destroyed after all types of processing as a result of the plastic displacement of the counter body material. The type of wear can be characterized as fatigue wear with boundary friction and plastic contact. The correlation of the friction coefficient with the Kragelsky–Kombalov criterion, a generalized dimensionless criterion of surface roughness, is shown.

Keywords: plasma electrolytic treatment; nitriding; boriding; carburizing; austenitic stainless steel; surface roughness; microhardness; wear resistance

Citation: Kusmanov, S.; Mukhacheva, T.; Tambovskiy, I.; Naumov, A.; Belov, R.; Sokova, E.; Kusmanova, I. Increasing Hardness and Wear Resistance of Austenitic Stainless Steel Surface by Anodic Plasma Electrolytic Treatment. *Metals* **2023**, *13*, 872. https://doi.org/10.3390/met13050872

Academic Editor: Francesca Borgioli

Received: 21 March 2023
Revised: 27 April 2023
Accepted: 28 April 2023
Published: 30 April 2023

1. Introduction

Stainless steels are used as structural and functional materials in products operating under aggressive conditions in the food, chemical, and thermal power engineering industries, among others. Products made of this material are characterized by high ductility, toughness, heat resistance and corrosion resistance but low strength and hardness. To harden stainless steels, surface plastic deformation technologies aimed at creating grain boundaries and substructural hardening are effectively used, for example, in ultrasonic strain engineering technology [1] and ultrasonic shot peening [2,3]. The disadvantage of deformation surface treatments is the significant increase in roughness and the need for subsequent finishing [1–3]. Laser shock peening increases the thickness of the hardened layer to 2 mm and have been proven to solve this problem [4–6].

Alternatives to the mechanical treatment of the metal surface include physical and chemical methods of surface hardening, such as plasma electrolytic treatment, which has a complex effect on the operational properties of metal products [7–9]. The majority of studies cover the cathodic variant of plasma electrolytic diffusion saturation variety. Thus, during the cathodic plasma electrolytic nitriding (PEN) of 316L steel in a solution of ammonium nitrate and potassium hydroxide, the formation of $FeN_{0.076}$ and Fe_3O_4 in the surface layer was revealed [10]. After PEN, the 10–15 μm thick oxide layer containing 39–41% oxygen is formed on the surface of the steel in a carbamide electrolyte. The maximum nitrogen content in steel of 0.68% is no longer detected at a depth of 30 μm. The cathodic nitriding of austenitic stainless steel in solutions of ammonium carbonate leads to the formation of nitrides (Fe_2N, Fe_4N, CrN and Cr_2N) and oxides (Fe_3O_4 and Cr_2O_3) of iron and chromium, and the microstructure of the nitrided layer contains a nitride zone and an internal nitriding zone [11]. As the temperature increases, more high-nitrogen nitride, Fe_2N, is formed. The nitriding of stainless steel has shown a positive result for increasing wear and corrosion resistance. After the PEN of 316L steel in solutions of carbamide or ammonium nitrate with the addition of potassium hydroxide, the dry friction coefficient with a corundum ball decreases from 0.19 in the untreated sample to 0.13 with an increase in wear resistance of 4.4–10 times [10]. The PEN of 304, 316L and 430 stainless steels in sodium nitrite solution proved to be an effective method of inhibiting pitting corrosion in 0.5M sodium chloride solution [12].

After the cathodic plasma electrolytic carburizing (PEC) of 12Cr18Ni10Ti steel, FeO iron oxides were detected in the glycerin electrolyte [13]. An increase in the applied voltage increased the degree of grain grinding to an extent to which nanoscale crystals formed alongside the increase in surface roughness [14]. After the carburizing of 12Cr18Ni10Ti steel, the compression of ferrite and austenite crystal lattices was observed, which occurs due to the displacement of the lines (110) α-Fe and (111) γ-Fe. Additional phases of Fe_3O_4, $(Cr,Fe)_7C_3$, $Fe_{15}Cr_4Ni_2$, CrN and $CrFe$ were detected in the PEC of 304 austenitic steel in a chloride-glycerin electrolyte [15] and 403 stainless martensitic steel, including Fe_3O_4, $CrFe$, FeO and CrC [16]. Cathodic carburizing during pulse treatment (250~600 V; 1500 Hz) in an electrolyte of glycerin and sodium chloride on 1Cr18Ni9Ti steel forms a hardened layer with a thickness of 0.2 mm and a microhardness up to 513 HV for 3–5 min [17]. After the cathodic carburizing of 304 steel in an electrolyte of glycerin (80%) and potassium chloride at a voltage of 350 V for 3 min, the thickness of the layer hardened to 762 HV reaches 0.085 mm [18].

In the cathodic plasma electrolytic nitrocarburizing (PENC) of 316L stainless steel in a carbamide electrolyte, the main phase appears as austenite nitrogen [19]. Additionally, oxides and oxygen-containing phases, including $NiFe_2O_4$, $FeCr_2O_4$ [20], Fe_2O_3, Fe_3O_4, Cr_2O_4 [21] and $Fe(Fe,Cr)_2O_4$ [22,23], nitrides, including Fe_3N [21], CrN and Cr_2N [23], carbides, including Cr_3C_2 and Cr_7C_3 [23], and silicon dioxide [21] are detected. The cathodic PENC of 304 steel in carbamide electrolytes increased microhardness up to 1380 HV with an increase in surface roughness from 0.025 to 0.14 μm and a 4.2-fold increase in wear resistance [19]. In one study, 316L steel after cathodic PENC in carbamide electrolytes with various additives had an increase in microhardness up to 1200 HV and 50 times the wear resistance [22] and, in another, an increase up to 1600 HV and 4.5 times the wear resistance [23].

The disadvantage of cathode plasma electrolytic treatment is the low controllability of the technological process and consequent properties. In addition, the increase in surface roughness that accompanies the cathodic treatment option requires additional finishing. The anodic version of plasma electrolytic saturation as a way to increase the operational properties of stainless steel products has not been practically considered in the literature. This option of plasma electrolytic treatment, however, in addition to hardening and increasing wear and corrosion resistance allows us to reduce surface roughness and exclude subsequent finishing treatment [24–27]. In this paper, the possibility of increasing hardness and wear resistance of the stainless steel surface with various types of anodic plasma electrolytic diffusion saturation (nitriding, boriding and carburizing) is considered.

2. Materials and Methods

2.1. Samples Processing

Cylindrical samples ($\varnothing$ 11 mm $\times$ 15 mm) of austenitic stainless steel (wt.%: 18 Cr; 10 Ni; 2 Mn; 0.8 Ti; 0.8 Si; 0.5 Mo; 0.3 Cu; 0.2 V; 0.2 W; 0.12 C; 0.03 P; 0.02 S and balanced Fe) were ground with SiC abrasive paper to a grit size of P100 to Ra~0.75 μm and ultrasonically cleaned with acetone.

These samples were subjected to anodic plasma electrolytic diffusion saturation with nitrogen (nitriding), boron (boriding) or carbon (carburizing). Plasma electrolytic treatment was carried out in a cylindrical electrolyzer with an axially symmetric electrolyte flow supplied through a nozzle located at the bottom of the electrolyzer (Figure 1) [28].

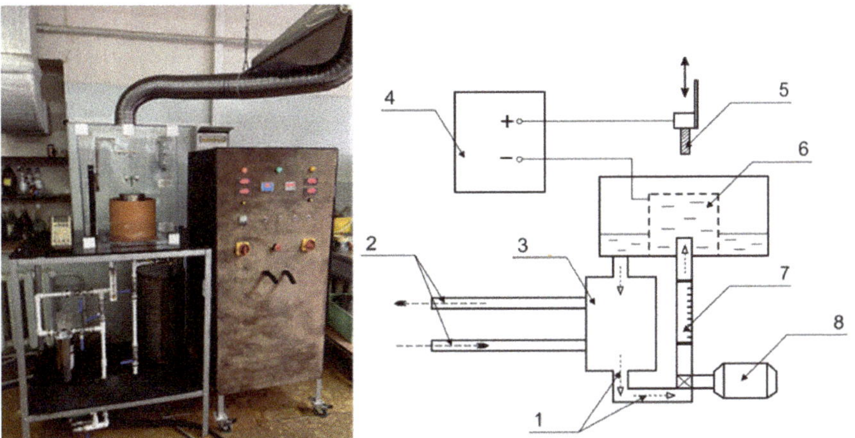

Figure 1. Setup for anodic plasma electrolytic treatment and schematic diagram: 1—electrolyte; 2—cold water; 3—heat exchanger; 4—power supply; 5—treated sample; 6—electrolytic cell; 7—flowmeter; 8—pump.

In the upper part of the electrolyzer, electrolyte was overflowing into the sump and was further pumped through a heat exchanger at a rate of 2.5 L/min, which was measured with a 0.4–4 LPM flowmeter (accuracy of ±2.5%) (Pribormarket, Arzamas, Russia). This scheme provides stabilization of the processing conditions. Solution temperature was measured using a a K-type thermocouple (Termoelement, Moscow, Russia) placed at the bottom of the chamber and maintained at 30 ± 2 °C. The samples were connected as the positive output, and the electrolyzer (Figure 1) was connected as the negative output of the 15 kW DC power supply.

The treatment was carried out in aqueous solutions of electrolyte based on ammonium chloride (10 wt.%) with the addition of ammonia (5 wt.%) for PEN, boric acid (3 wt.%) for plasma electrolytic boriding (PEB) or glycerin (10 wt.%) for PEC.

After switching the voltage to 200 V, the samples were immersed in the electrolyte at a speed of 1–2 mm/s. If the rate of immersion was slow, a vapor-gaseous envelope was easily formed on an initially small surface area of the sample near the electrolyte surface and extended further across the sample as it submerged. Once the sample was immersed at a depth equal to its height, the voltage was changed to the value in Tables 1–3 in order to reach the prescribed treatment temperature. The sample temperature was measured with a K-type thermocouple (Termoelement, Moscow, Russia) and a multimeter APPA109N (accuracy up to 3% over a temperature range of 400–1000 °C) (APPA TECHNOLOGY CORPORATION, Taipei, Taiwan (China)). The thermocouple was fixed in a hole made in the samples at a distance of 2 mm from the sample's bottom. The treatment continued for 5 min, and, after diffusion saturation, the samples were quenched in electrolyte (hardening).

Table 1. Conditions of PEN and results of sample testing.

Temperature (°C)	Voltage (V)	Current (A)	Weight Loss of Samples during PEN (mg)	Surface Roughness Ra [1] (μm)
650	160	7.2	51.0 ± 0.3	0.24 ± 0.03
700	166	7.5	53.5 ± 0.4	0.28 ± 0.03
750	182	7.1	53.1 ± 0.1	0.22 ± 0.03
800	188	7.3	61.1 ± 0.3	0.20 ± 0.02
850	191	7.7	76.1 ± 0.2	0.21 ± 0.01

[1] Untreated sample Ra is 0.75 ± 0.05 μm.

Table 2. Conditions of PEB and results of sample testing.

Temperature (°C)	Voltage (V)	Current (A)	Weight Loss of Samples during PEB (mg)	Surface Roughness Ra [1] (μm)
800	149	11.7	50.1 ± 0.3	0.33 ± 0.04
850	155	11.3	38.0 ± 0.2	0.38 ± 0.04
900	170	11.5	45.5 ± 0.2	0.49 ± 0.03
950	194	10.5	50.8 ± 0.3	0.62 ± 0.06

[1] Untreated sample Ra is 0.75 ± 0.05 μm.

Table 3. Conditions of PEC and results of sample testing.

Temperature (°C)	Voltage (V)	Current (A)	Weight Loss of Samples during PEC (mg)	Surface Roughness Ra [1] (μm)
750	138	10.2	90.9 ± 0.5	0.33 ± 0.04
800	157	8.8	64.1 ± 0.4	0.17 ± 0.04
850	175	8.2	52.5 ± 0.3	0.23 ± 0.03
900	195	6.9	41.6 ± 0.2	0.24 ± 0.01

[1] Untreated sample Ra is 0.75 ± 0.05 μm.

2.2. Study of the Surface Morphology and Microstructure

The Micromed MET (Micromed, St. Petersburg, Russia) optical metallographic microscope with digital image visualization served to study the surface morphology and microstructure of the cross-section of the austenitic stainless steel samples.

2.3. The Microhardness Measurement

The microhardness of the cross-sections of the treatment sample was measured using a Vickers microhardness tester (Falcon 503, Innovatest Europe BV, Maastricht, The Netherlands) under a 0.1 N load. According to 5 measurements, the average value of microhardness was found.

2.4. Surface Roughness and Weight of Samples Measurement

The surface roughness was measured with a TR-200 profilometer (Beijing TIME High Technology Ltd., Beijing, China). According to 10 measurements, the average value of roughness indicators was found. The change in the weight of the samples was determined on a CitizonCY224C electronic analytical balance (ACZET (Citizen Scale), Mumbai, India) with an accuracy of ± 0.0001 g after removing traces of salts by washing the samples in distilled water and subsequently drying.

2.5. Study of Phase Composition

X-ray diffraction (XRD) analysis was used to determine the phase composition of the samples. The XRD patterns were obtained by PANalytical Empyrean X-ray diffractometer (Malvern Panalytical, Malvern, UK) with CoKα radiation by a simple scanning mechanism in the theta-2theta-mode with a step of $0.026°$ and a scanning rate of $4.5°/\text{min}$. Phase composition analysis was performed using the PANalytical High Score Plus software [29] and the ICCD PDF-2 and COD databases [30].

2.6. Study of Tribological Properties

The friction scheme "shaft-block" was used in friction tests (Figure 2) [31,32].

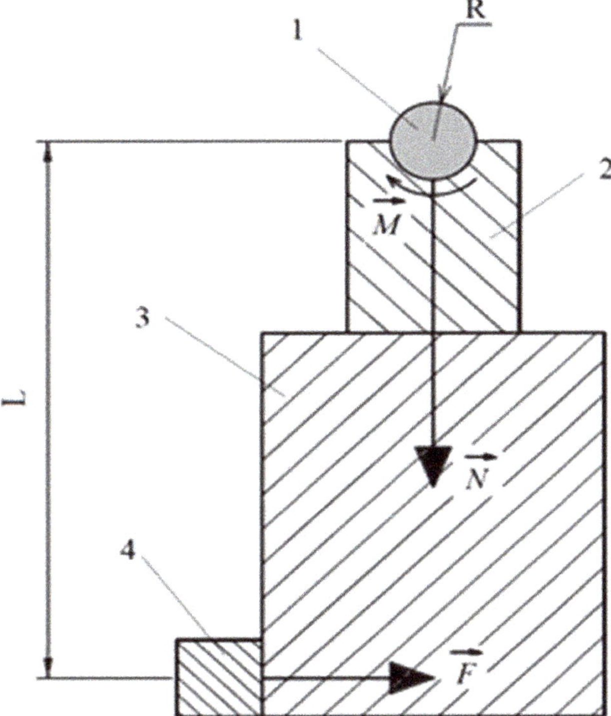

Figure 2. Friction scheme. 1—sample; 2—counter body; 3—pendulum; 4—strain gauge. *M*—frictional moment; *N*—force acting on the counter body and the pendulum from the side of the sample; *F*—force acting on the pendulum from the strain gauge; *L*—distance from the axis of rotation to the axis of symmetry of the strain gauge, *R* – radius of the sample.

The counter body was made of tool alloy steel (wt.%: 0.9–1.2 Cr, 1.2–1.6 W, 0.8–1.1 Mn and 0.9–1.05 C) in the form of a plate with a semicircular notch 10 mm in diameter enclosing the surface of the sample. The sample was mounted on a shaft driven by an electric motor. The counter body was mounted on a platform sliding along cylindrical guides. The platform was moved using a pneumatic cylinder. The cylinder, guides and the platform were able to rotate with the pendulum. The pendulum shaft was located coaxially with the sample. Such a scheme makes it possible to preserve the common rotation axis for the sample and the counter body as they exhaust and to avoid the influence of misalignment on the results of measurements of the frictional moment. Friction tests were carried out in dry friction mode under a load of 10 N. The sliding speed of the sample along the counter body was 1.555 m/s. The friction path was 1000 m. The parameters of the microgeometry of the friction tracks' surface were measured using the TR200 profilometer (Beijing TIME High Technology Ltd., Beijing, China). The temperature of the friction contact was measured on the friction track directly at the exit from the contact area using the MLX90614 digital infrared thermometer (Melexis Electronic Technology, Shanghai, China).

2.7. Wear Mechanism Calculation

Wear resistance is an important operational property of machines and mechanisms. Largely, wear resistance is caused by the contact interaction in tribo-conjugations, which is based on the properties of the surface layers. The processes implemented in tribo-conjugation depend on roughness. The rough surface model makes it possible to determine the type of tribo-tension stress state (elastic contact or plastic contact) and the wear mechanism and to evaluate the relationship of plasma electrolytic treatment conditions and the microgeometry parameters of the surface layer.

The model is based on experimental profilograms of friction tracks. The calculation is performed in relation to the contact of a rough surface with a smooth solid surface since the roughness of the sample surface significantly exceeds the roughness of the counterbody. This calculation is faster and easier to perform than the contact of two rough surfaces [32,33].

To describe the microgeometry of the surface of friction tracks, it is necessary to know the function of vertical material distribution throughout the rough layer and the function of the vertical material distribution throughout the single micro-roughness of the rough surface. The distribution of the material over the height of the rough layer is described by the curve of the support surface (Abbott curve). Curves are taken directly from the profiler by at least 15 pieces per 1 friction track on each ring.

To approximate the experimental reference curve, the Demkin function is selected [34].

$$\eta(\varepsilon) = l_m \left(\frac{z}{R_p} \right)^v = b \left(\frac{z}{R_{\max}} \right)^v = b\varepsilon_{\max}^v = \frac{A_r}{A_c} = \frac{P_c}{P_r} = \frac{n_r}{n_c}, \tag{1}$$

where z is the profile cross-section level measured from the protrusions line; A_r is the actual contact area; A_c is the contour contact area; P_r is the average actual pressure on the friction contact; P_c is the contour pressure; n_r is the number of contacting protrusions; n_c is the number of all protrusions on the contour area.

It is more convenient to calculate the heights of the experimental profile in relative terms:

$$\varepsilon = \frac{z}{R_p}, \ \varepsilon_{\max} = \frac{z}{R_{\max}}, \tag{2}$$

where R_p is the smoothing height (the distance from the protrusions line to the midline in terms of the base length); $R_{\max}$ is the maximum height of irregularities; z is the profile cross-section level measured from the protrusions line; ε and $\varepsilon_{\max}$ are relative dimensionless profile heights relative to the midline.

The parameters of the reference curve v and b are determined experimentally from the results of measurements of the parameters of the rough body profile:

$$v = 2l_m \left(\frac{R_p}{R_a} \right) - 1, \tag{3}$$

$$b = l_m \left(\frac{R_{\max}}{R_p} \right)^v, \tag{4}$$

where R_a is the arithmetic mean deviation of the profile. Additionally, l_m is the relative reference length of the profile on the midline:

$$l_m = \frac{\sum\limits_{1}^{n} \Delta l_i}{l}, \tag{5}$$

where l is the base length; Δl_i is the length of the segments cut off by the middle line in the profile. The l_m parameter is determined by direct measurements of the profilometer on the friction tracks.

To calculate the actual pressure P_r at the tops of the micro-steps, it is necessary to determine the type of deformations on the friction contact: elastic or plastic. The evaluation is made using the Greenwood–Williamson criterion:

$$K_p = \frac{\Theta}{HB} \sqrt{\frac{R_p}{r}}, \tag{6}$$

Θ is the reduced modulus of elasticity:

$$\Theta = \left(\frac{1 - \mu_1^2}{E_1} + \frac{1 - \mu_2^2}{E_2} \right)^{-1}, \tag{7}$$

where μ_i and E_i are Poisson's coefficients and the elastic modulus of interacting bodies, respectively.

The dimensionless parameter K_p takes into account the roughness and physical properties of the material at the same time. It describes the deformation properties of a rough surface. If the value of this parameter is lower than 3, then the deformations of the irregularities in contact with a flat surface will be completely elastic; if K_p exceeds 3, then the deformations will be predominantly plastic.

The average radius of a single micrometer, which characterizes the shape of the protrusion, is calculated directly from the profilometer data:

$$r = \frac{S_m^2}{8R_a} \cdot \frac{\gamma_1}{\gamma_2^2}, \tag{8}$$

where S_m is the average step of the irregularities; γ_1 is the vertical increase in the profiler; γ_2 is the horizontal increase in the profiler.

With elastic contact, the deformation of individual protrusions can be calculated according to the classical Hertz contact problem. Then, the average actual pressure at the contact is determined by the following expression:

$$P_r = (0.43\Theta)^{\frac{2\nu}{2\nu+1}} \left(\frac{2N}{\eta A_c} \right)^{\frac{1}{2\nu+1}} \left(\frac{R_p}{r} \right)^{\frac{2\nu}{2\nu+1}}, \tag{9}$$

where N is the normal load. With plastic contact, the average voltage at the contact is numerically equal to the microhardness $P_r \approx HB$.

The actual contact area is determined by the ratio of the normal load to the actual pressure:

$$A_r = \frac{N}{P_r}. \tag{10}$$

Substituting the reference curve (1) $z = h$ into the equation leads to the following expression of the absolute convergence of the surfaces:

$$h = R_{\max} \left(\frac{P_c}{bP_r} \right)^{\frac{1}{\nu}}. \tag{11}$$

Then, the ratio of the number of contacting protrusions n_r to the number of all protrusions on the contour area n_c can be determined using the Demkin function (1):

$$\frac{n_r}{n_c} = \left(\frac{N}{P_r \cdot l_m} \right)^{\frac{\nu-1}{\nu}}, \tag{12}$$

where l_m is the relative reference length of the profile at the midline level.

The relative embedding of the sample and the counterbody is the ratio of the absolute embedding to the average radius of the micronerosity:

$$\frac{h}{r} = \frac{R_{\max}}{r} \cdot \left(\frac{N}{b \cdot P_r} \right)^{\frac{1}{\nu}} = \frac{8 R_a R_{\max}}{S_m^2} \cdot \left(\frac{N}{b \cdot P_r} \right)^{\frac{1}{\nu}} \cdot \frac{\gamma_2^2}{\gamma_1}. \tag{13}$$

Relative convergence (13) characterizes the type of violation of frictional bonds in tribocontact. For steel surfaces, in the case when $h/r < 0.01$, destruction occurs because of friction fatigue, and friction bonds are broken due to elastic displacement of the sample material. At a value of $h/r < 0.1$, low-cycle friction fatigue develops, and the friction surfaces are destroyed due to plastic displacement of the sample material with residual deformation of the friction track after the passage of micro-steps along it.

To assess the bearing capacity of roughness, the dimensionless Kragelsky–Kombalov criterion is calculated [35]:

$$\Delta = \left(\frac{100}{l_m} \right)^{\frac{1}{\nu}} \cdot \left(\frac{R_p}{r} \right). \tag{14}$$

The complex (14) represents the most complete roughness assessment, including not only geometric but also statistical characteristics of the height distribution of the protrusions as well as the average radius of the rounding of the micro-protrusions. On a friction-worn surface, Δ shows how much its bearing capacity has been preserved. The lower the calculated value of Δ on the friction track, the higher the bearing capacity of the rough profile and the more favorable conditions for friction with the minimum possible wear.

The mathematical expression for the approximation of the Abbott curve (1) allows us to give a theoretical estimate of the wear value of the sample. If only the volume of embedded irregularities dV with density ρ is involved in the deformation, then

$$dm = \rho \cdot dV = \rho A_r dz, \tag{15}$$

where the contour area is defined by expression (10), and dz is the depth of the embedding of irregularities, which can vary from zero to the full value of the relative embedding of h, as defined by expression (11). Thus, weight wear is determined by volume V and is directly proportional to the area of the actual contact:

$$\Delta m = \rho V = \rho \int_0^z A_r dz = \rho \int_0^\varepsilon b \varepsilon^\nu d\varepsilon = \frac{\rho A_r h}{\nu + 1}. \tag{16}$$

3. Results

3.1. Morphology and Roughness of the Surface

After processing, under all varying conditions, there is a decrease in the weight of samples and a decrease in the surface roughness (Tables 1–3). At the same time, each type of diffusion saturation has its own characteristics. Thus, when nitriding with an increase in temperature from 650 to 850 °C, despite the practically constant value of the current strength, the surface roughness decreases with an increase in the intensity of anodic dissolution (the loss of sample weight) (Table 1). The surface morphology becomes more homogeneous with an increasing temperature when the textures of the untreated surface are completely removed (Figure 3).

The boriding of the steel surface was carried out at higher temperatures and, despite higher values of current and released power compared to those during nitriding, there is a smaller decrease in the weight of samples and roughness (Table 2). With an increase in temperature from 800 to 950 °C, the roughness increases but does not exceed the initial value. Pores are visible on the borided surface under all processing conditions, which determine its heterogeneity and the higher roughness compared to during nitriding (Figure 4).

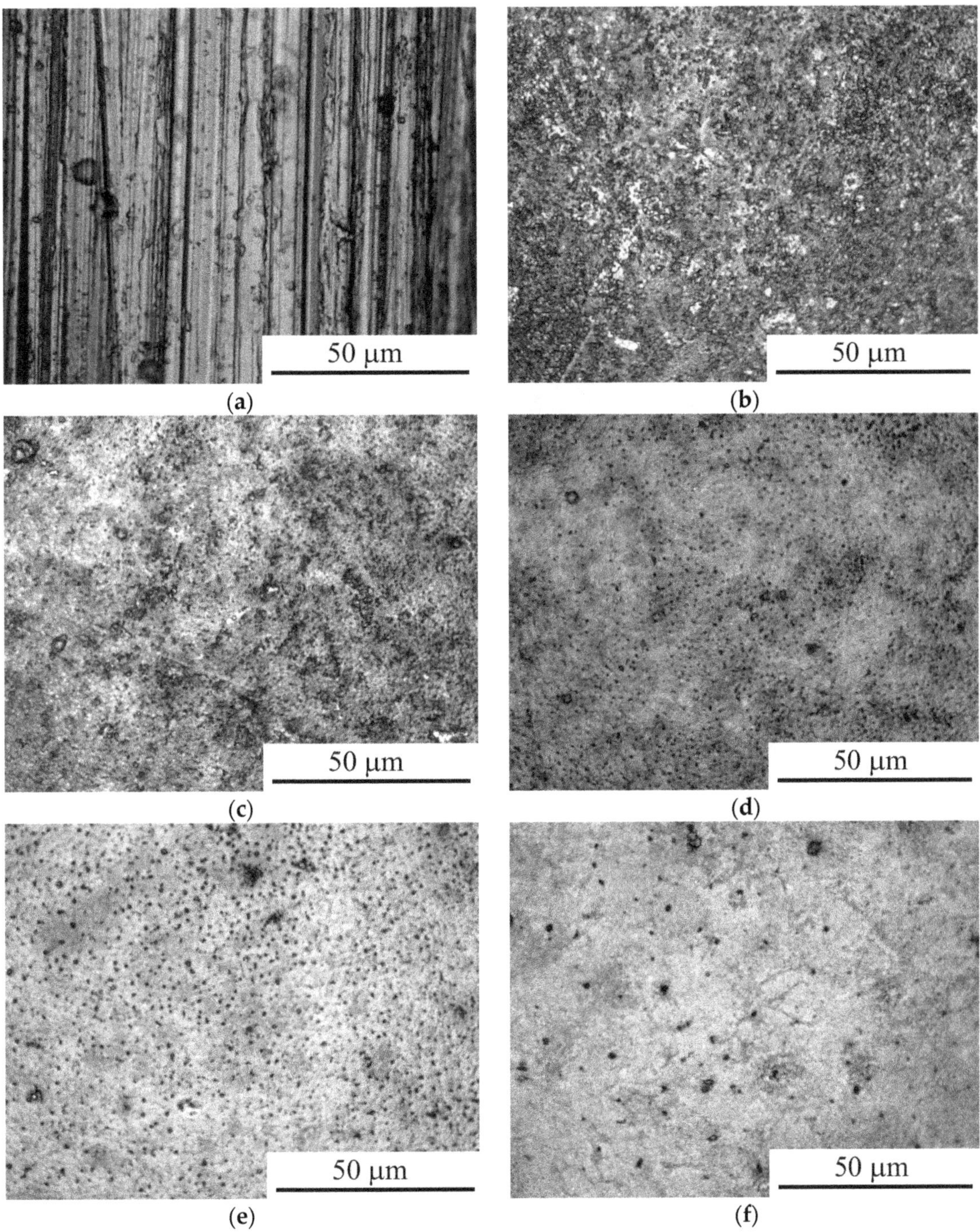

Figure 3. Morphology of the steel surface before (**a**) and after PEN at different treatment temperatures: (**b**) 650 °C; (**c**) 700 °C; (**d**) 750 °C; (**e**) 800 °C; (**f**) 850 °C.

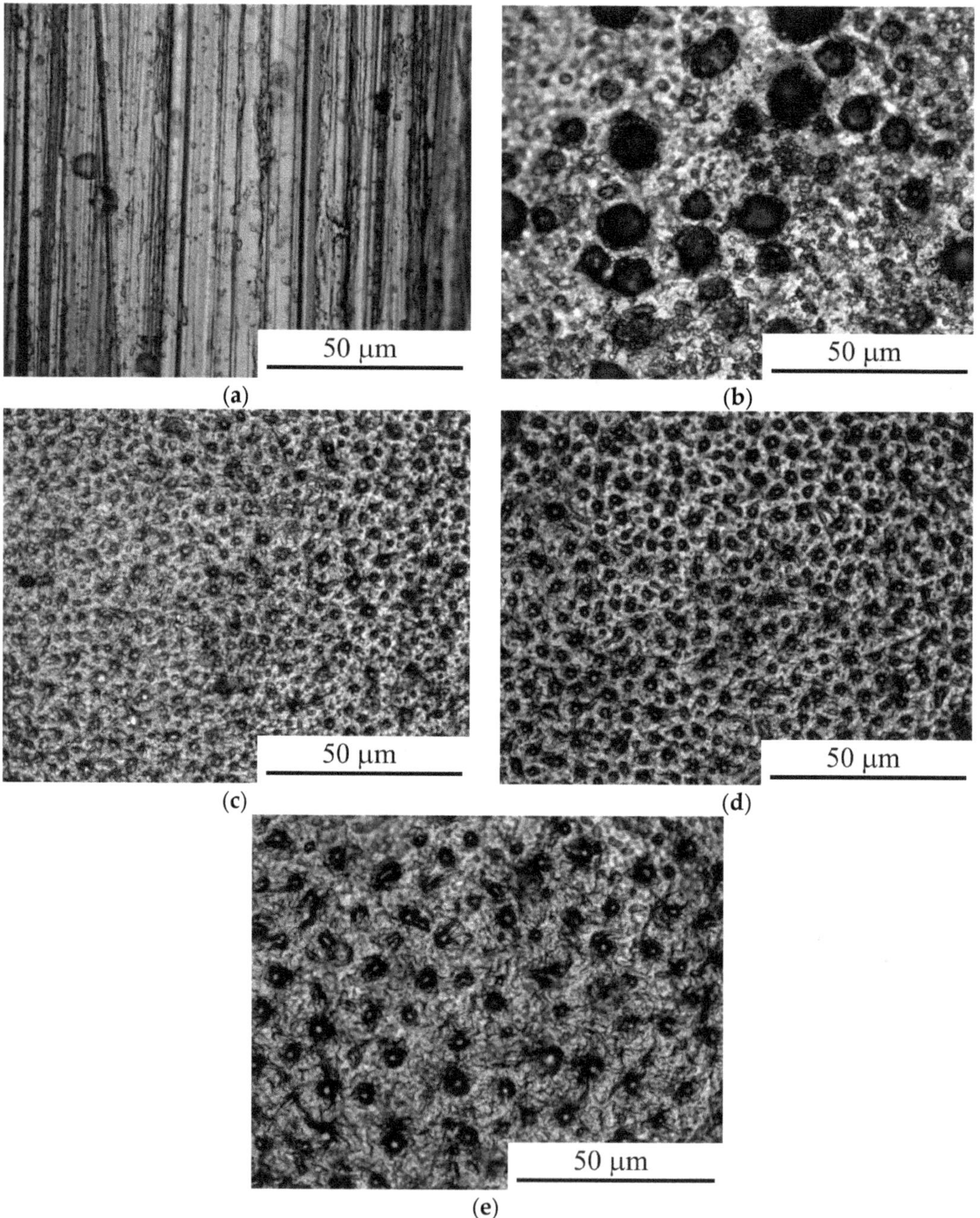

Figure 4. Morphology of the steel surface before (**a**) and after PEB at different treatment temperatures: (**b**) 800 °C; (**c**) 850 °C; (**d**) 900 °C; (**e**) 950 °C.

For the carburizing process, with an increase in temperature from 750 to 900 °C, there is a linear decrease in the current and weight of the samples (Table 3). At the same time, the surface roughness changes non-linearly: with an increase in the PEC temperature from 750 to 800 °C, it decreases by a factor of 4.4 compared to the initial value, and with a subsequent increase in temperature to 900 °C, a slight increase is observed (Table 3). The surface morphology after PEC is more uniform compared to after PEB, but after treatment at 900 °C, the treated material (steel) become visible (Figure 5).

Figure 5. Morphology of the steel surface before (**a**) and after PEC at different treatment temperatures: (**b**) 750 °C; (**c**) 800 °C; (**d**) 850 °C; (**e**) 900 °C.

3.2. Phase Composition, Structure and Microhardness of the Surface Layer

According to the metallographic analysis, because of the plasma electrolytic treatment, the diffusion saturation of the surface occurs with the formation of modified layers detected under the surface oxide layer (Figures 6–8).

Figure 6. Microstructure of cross-section of the steel surface after PEN at 850 °C. 1—oxide layer; 2—modified layer; 3—initial structure.

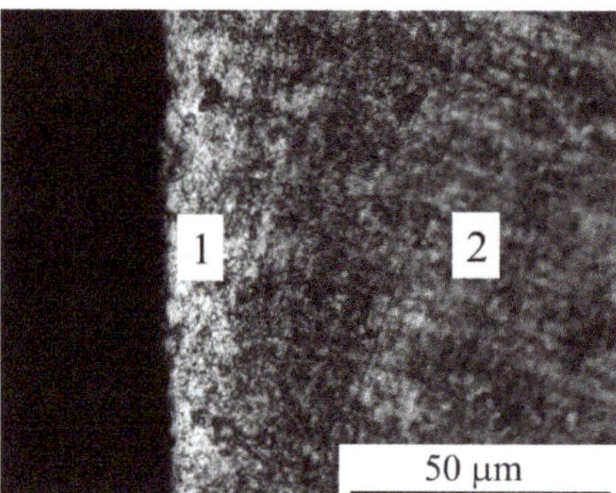

Figure 7. Microstructure of cross-section of the steel surface after PEB at 850 °C. 1—modified layer; 2—initial structure.

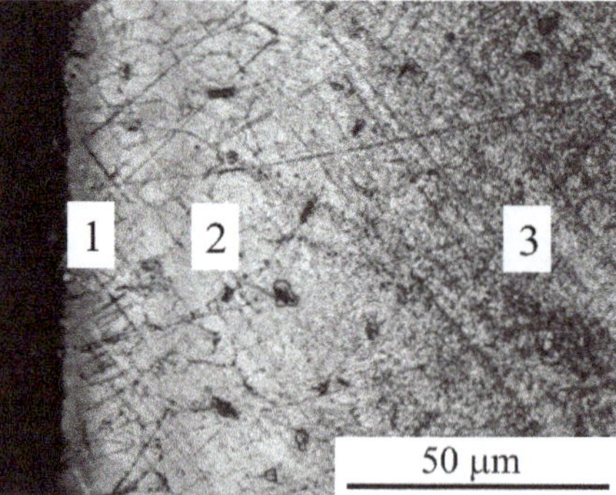

Figure 8. Microstructure of cross-section of the steel surface after PEC at 850 °C. 1—modified layer; 2—diffusion layer (N and C solid solution); 3—initial structure.

Depending on the type of processing, the modified layers will have a different phase composition and structure. According to X-ray analysis, after PEN, the oxide layer consists of α-Fe_2O_3 and γ-Fe_2O_3, and the modified layer includes phases $FeN_{0.0939}$ and $CrFe_7C_{0.45}$, which form due to diffusion processes, and intermetallides of the initial components of the alloy (Figure 9).

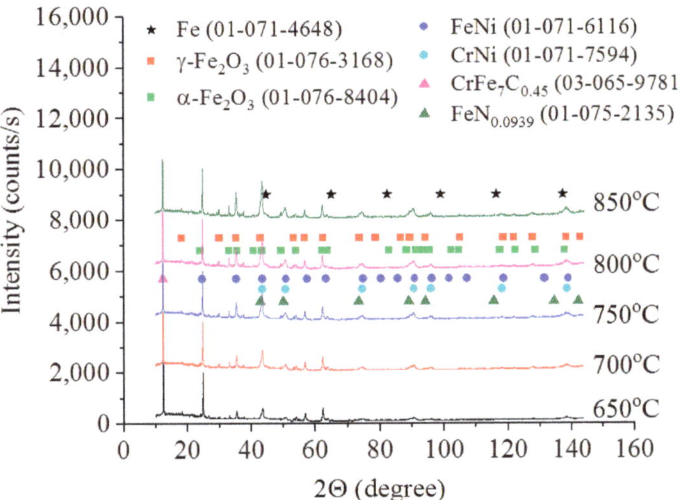

Figure 9. X-ray diffraction patterns of the steel surface layer after PEN at different treatment temperatures with the indication of ICDD card number.

After PEB, a porous FeO oxide is detected in the oxide layer, except for γ-Fe_2O_3, and no inclusion compounds were detected in the modified layer, except for intermetallides (Figure 10).

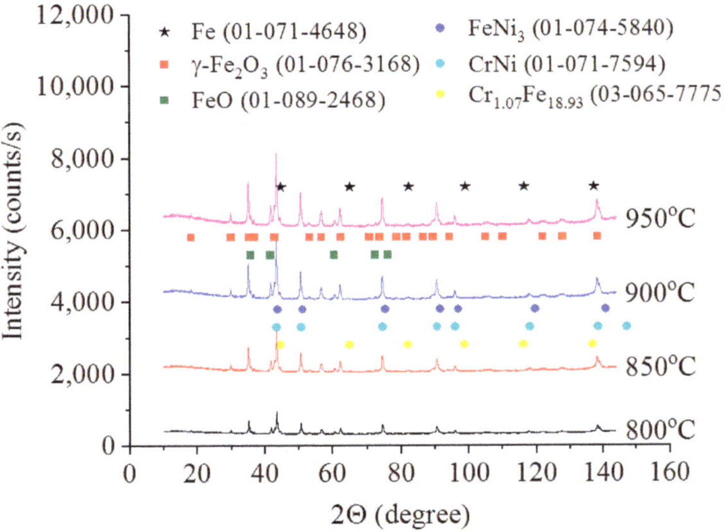

Figure 10. X-ray diffraction patterns of the steel surface layer after PEB at different treatment temperatures with the indication of ICDD card number.

As a result of the PEC, a more complex structure comprising three subsequent layers is formed: a surface oxide layer consisting of Fe_2O_3 and Fe_3O_4 phases; an outer modified layer, including intermetallides and chromium carbide; and a diffusion layer (a solid solution of diffusion atoms in the initial matrix) (Figure 11).

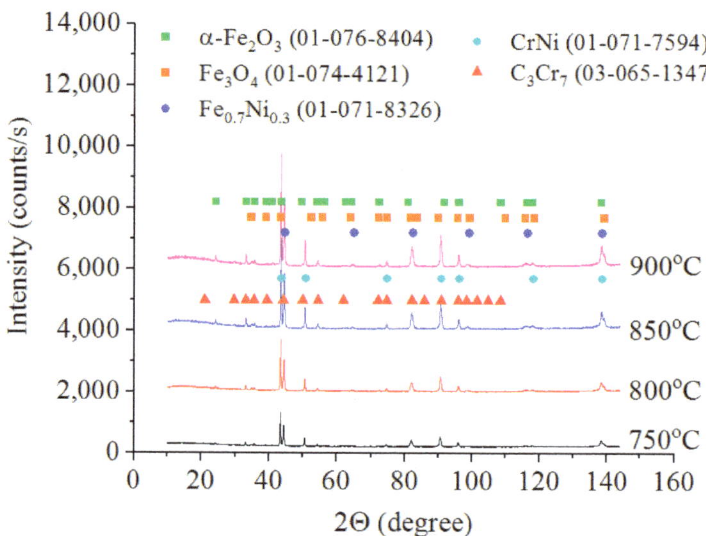

Figure 11. X-ray diffraction patterns of the steel surface layer after PEC at different treatment temperatures with the indication of ICDD card number.

Measurements of the microhardness of the surface layers showed that after PEN the surface is hardened to the depth of its modification, reaching 1400–1450 HV after nitriding at 650–700 °C (Figure 12). With an increase in the PEN temperature, the hardness of the nitrided layer decreases, which is associated with the coagulation of nitride particles and the breakdown of coherence. PEN at the maximum temperature of 850 °C nearly doubles the hardness on the surface compared to that in the core.

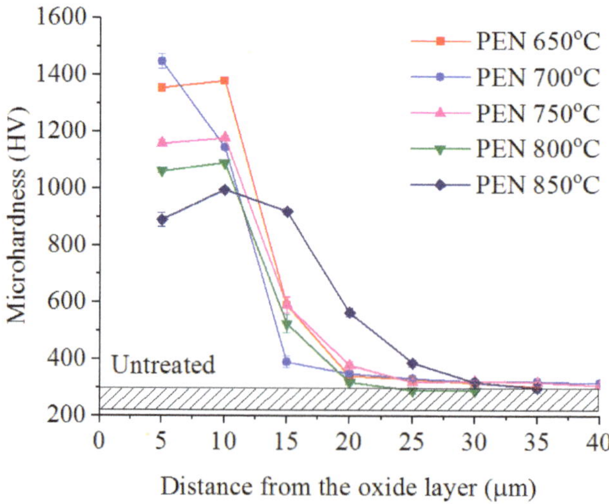

Figure 12. Microhardness distribution in the surface layer after PEN at different treatment temperatures.

After PEB, surface hardening does not occur (Figure 13).

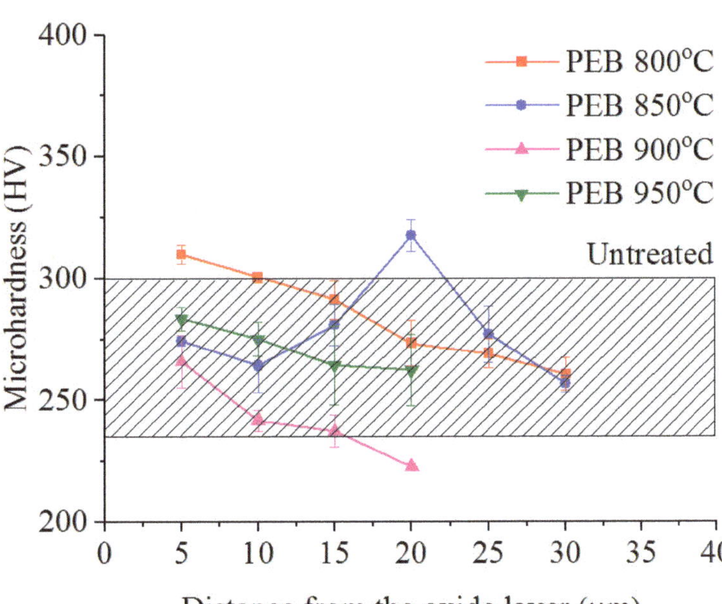

Figure 13. Microhardness distribution in the surface layer after PEB at different treatment temperatures.

After PEC, the microhardness increases with the rise in saturation temperature, similarly to during the carburizing of carbon steels [36,37], reaching 700 HV, and the thickness of the hardened layer correlates with the thickness of the modified and diffused layers (Figure 14).

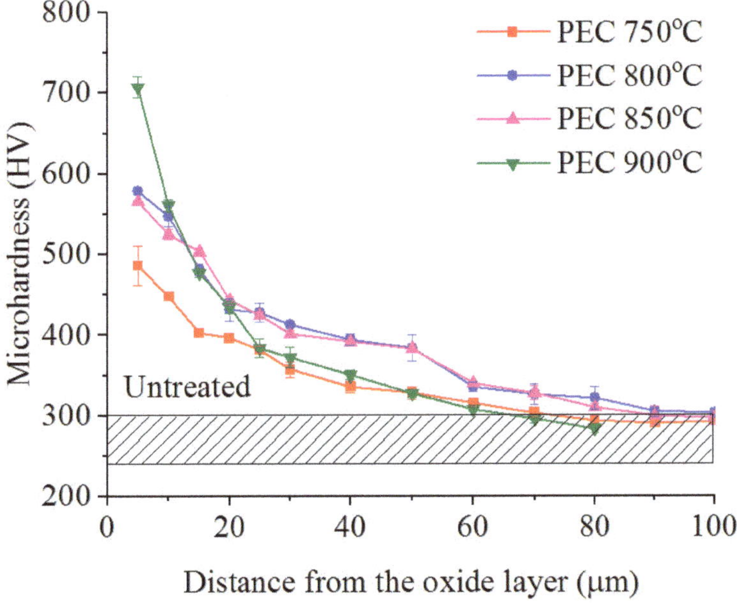

Figure 14. Microhardness distribution in the surface layer after PEC at different treatment temperatures.

3.3. Tribological Properties of Treated Surface

The results of tribological tests and calculations of the microgeometry parameters of the friction track surface are presented in Tables 4–6. According to the data presented in Table 4, the PEN of the steel surface under all varying treatment modes leads to a reduction in weight wear by two orders. At the same time, there is an increase of 10–18 degrees in the temperature in the friction contact area, and the friction coefficient reached 1.5–1.9 of the previous value with a tendency to decrease the latter with an increase in the PEN temperature. The dynamics of the change in the friction coefficient as the contact surfaces slide show their rapid stabilization (Figure 15).

Table 4. Friction parameters and microgeometry of the worn surface after PEN. Kragelsky–Kombalov criterion Δ; magnitude of the absolute penetration in the tribocontact h; average radius of rounding r; relative penetration of the deformed surfaces of the tribocontact h/r; Greenwood-Williamson criterion K_p; actual contact area A_r; the ratio of the actual and normal contact area A_r/A_a (relative error under 3.5%); the ratio of the number of vertices on the contour area to the number of micronerities that entered tribocontact n_c/n_r; friction track temperature over the last 100 m of the path at friction T_{fr} per 1 km; average friction coefficient over the last 100 m of the path with friction μ per 1 km; weight loss during friction at 1 km of the path Δm_{fr}.

T (°C)	Δ	h (MCM)	r (MCM)	h/r	K_p	A_r (MCM²)	A_r/A_a	n_c/n_r	T_{fr} (°C)	μ	Δm_{fr} (Mг)
untreated	0.989 ± 0.017	1.44 ± 0.02	14.38 ± 0.24	0.100 ± 0.002	22.5 ± 0.4	8.22 ± 0.14	0.24	202 ± 5	68	0.401 ± 0.004	23.2 ± 0.3
650	0.379 ± 0.006	0.48 ± 0.01	6.09 ± 0.11	0.079 ± 0.001	17.2 ± 0.4	2.07 ± 0.04	0.06	71 ± 1	79	0.698 ± 0.008	0.4 ± 0.1
700	0.408 ± 0.007	0.46 ± 0.01	5.54 ± 0.09	0.083 ± 0.001	21.3 ± 0.5	1.95 ± 0.03	0.06	78 ± 2	81	0.773 ± 0.009	0.5 ± 0.1
750	0.402 ± 0.007	0.48 ± 0.01	5.98 ± 0.10	0.080 ± 0.001	19.2 ± 0.4	2.62 ± 0.04	0.08	70 ± 1	86	0.615 ± 0.007	0.4 ± 0.1
800	0.398 ± 0.007	0.47 ± 0.01	5.95 ± 0.11	0.079 ± 0.001	18.8 ± 0.4	0.31 ± 0.01	0.01	69 ± 1	78	0.586 ± 0.007	0.4 ± 0.1
850	0.387 ± 0.007	0.45 ± 0.01	6.03 ± 0.10	0.074 ± 0.001	20.0 ± 0.5	2.87 ± 0.05	0.08	73 ± 1	81	0.606 ± 0.007	0.4 ± 0.1

Table 5. Friction parameters and microgeometry of the worn surface after PEB. Kragelsky–Kombalov criterion, Δ; magnitude of the absolute penetration in the tribocontact, h; average radius of rounding, r; relative penetration of the deformed surfaces of the tribocontact, h/r; Greenwood–Williamson criterion, K_p; actual contact area, A_r; the ratio of the actual and normal contact area, A_r/A_a (relative error under 3.5%); the ratio of the number of vertices on the contour area to the number of micronerities that entered tribocontact, n_c/n_r; friction track temperature over the last 100 m of the path at friction, T_{fr}, per 1 km; average friction coefficient over the last 100 m of the path with friction, μ, per 1 km; weight loss during friction at 1 km of the path, Δm_{fr}.

T (°C)	Δ	h (MCM)	r (MCM)	h/r	K_p	A_r (MCM²)	A_r/A_a	n_c/n_r	T_{fr} (°C)	μ	Δm_{fr} (Mг)
untreated	0.989 ± 0.017	1.44 ± 0.02	14.38 ± 0.24	0.100 ± 0.002	22.5 ± 0.4	8.22 ± 0.14	0.24	202 ± 5	68	0.401 ± 0.004	23.2 ± 0.3
800	0.451 ± 0.008	0.45 ± 0.01	5.33 ± 0.09	0.084 ± 0.001	22.6 ± 0.4	6.82 ± 0.12	0.20	115 ± 3	62	0.478 ± 0.005	24.7 ± 0.2
850	0.415 ± 0.007	0.54 ± 0.01	6.73 ± 0.11	0.080 ± 0.001	16.3 ± 0.3	6.51 ± 0.11	0.19	112 ± 3	74	0.388 ± 0.004	3.3 ± 0.1
900	0.430 ± 0.007	0.52 ± 0.01	6.54 ± 0.11	0.079 ± 0.001	17.2 ± 0.3	6.43 ± 0.11	0.19	118 ± 3	83	0.386 ± 0.004	1.8 ± 0.1
950	0.462 ± 0.008	0.53 ± 0.01	5.98 ± 0.10	0.089 ± 0.002	24.4 ± 0.4	6.72 ± 0.11	0.19	120 ± 3	76	0.564 ± 0.006	26.2 ± 0.1

Table 6. Friction parameters and microgeometry of the worn surface after PEC. Kragelsky–Kombalov criterion, Δ; magnitude of the absolute penetration in the tribocontact, h; average radius of rounding, r; relative penetration of the deformed surfaces of the tribocontact, h/r; Greenwood–Williamson criterion, K_p; actual contact area, A_r; the ratio of the actual and normal contact area, A_r/A_a (relative error under 3.5%); the ratio of the number of vertices on the contour area to the number of micronerities that entered tribocontact, n_c/n_r; friction track temperature over the last 100 m of the path at friction, T_{fr},per 1 km; average friction coefficient over the last 100 m of the path with friction, μ, per 1 km; weight loss during friction at 1 km of the path, Δm_{fr}.

T (°C)	Δ	h (MCM)	r (MCM)	h/r	K_p	A_r (MCM^2)	A_r/A_a	n_c/n_r	T_{fr} (°C)	μ	Δm_{fr} (Mг)
untreated	0.989 ± 0.017	1.44 ± 0.02	14.38 ± 0.24	0.100 ± 0.002	22.5 ± 0.4	8.22 ± 0.14	0.24	202 ± 5	68	0.401 ± 0.004	23.2 ± 0.3
750	0.362 ± 0.006	0.43 ± 0.01	7.22 ± 0.12	0.059 ± 0.001	16.8 ± 0.3	4.43 ± 0.08	0.13	87 ± 2	45	0.313 ± 0.003	0.3 ± 0.1
800	0.412 ± 0.007	0.46 ± 0.01	6.54 ± 0.11	0.071 ± 0.001	23.5 ± 0.4	3.76 ± 0.06	0.11	120 ± 3	51	0.452 ± 0.005	0.5 ± 0.1
850	0.387 ± 0.007	0.42 ± 0.01	7.06 ± 0.12	0.061 ± 0.001	17.3 ± 0.3	3.81 ± 0.06	0.11	94 ± 2	79	0.354 ± 0.004	0.3 ± 0.1
900	0.423 ± 0.007	0.49 ± 0.01	6.32 ± 0.11	0.078 ± 0.001	24.1 ± 0.4	3.27 ± 0.06	0.09	125 ± 3	76	0.497 ± 0.005	0.1 ± 0.1

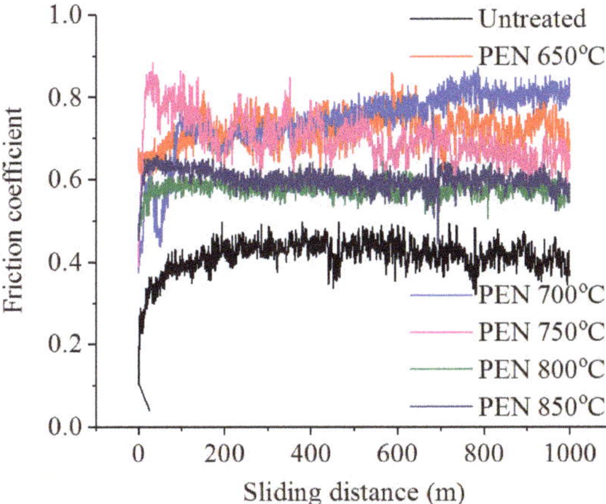

Figure 15. Dependence of friction coefficient on sliding distance of the untreated and PEN samples.

It is shown that as a result of PEN, the value of Kragelsky-Kambalov criterion becomes 2.5 times lower regardless of processing temperature. Calculations of microgeometry of worn surface parameters before and after PEN allowed us to determine the deformation properties of the rough surface (according to the Greenwood-Williamson criterion), which are predominantly plastic. According to the calculated h/r indicator, it can be stated that the destruction of friction bonds occurs due to the development of low-cycle friction fatigue and friction surfaces are destroyed due to plastic displacement of the sample material with residual deformation of the friction track.

Test results of PEB samples presented in Table 5 showed, that only after processing under 850 and 900 °C weight wear value was 7 and 12.9 times, respectively, as low as before. Under these conditions, friction coefficient decreased insignificantly compared to the untreated surface (Figure 16). The greatest increase in temperature in the friction contact area occurs with the greatest reduction in weight wear. The values of Kragelsky–Kombalov criterion of PEB samples after friction do not notably exceed similar parameters of the nitrided surfaces. The calculation showed the presence of plastic deformation properties of a rough surface according to Greenwood-Williamson criterion, and the destruction of friction bonds occurs in the same way as for nitrided surfaces.

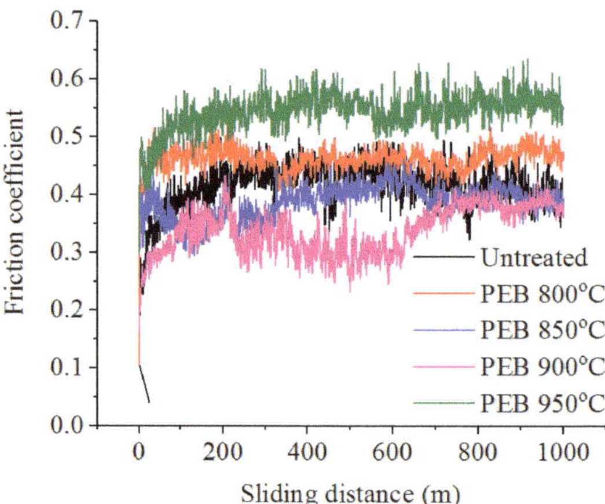

Figure 16. Dependence of friction coefficient on sliding distance of the untreated and PEB samples.

Evidently, as a result of PEC, there is a significant decrease in weight wear corresponding to the values obtained after nitriding, while the friction coefficient can both increase (after PEC at 800 and 900 °C) and decrease (after PEC at 750 and 850 °C) compared to the untreated surface with a rapid stabilization of its values (Table 6, Figure 17). A decrease in temperature in the friction contact area at low nitriding temperatures and an increase after treatment at 850 and 900 °C were revealed. The values of the Kragelsky–Kombalov criterion of the PEC samples after friction correlate with those for the nitrided surfaces in addition to those for the deformation properties of the surface and the mechanical destruction of friction bonds, according to the calculated parameters of worn surface microgeometry.

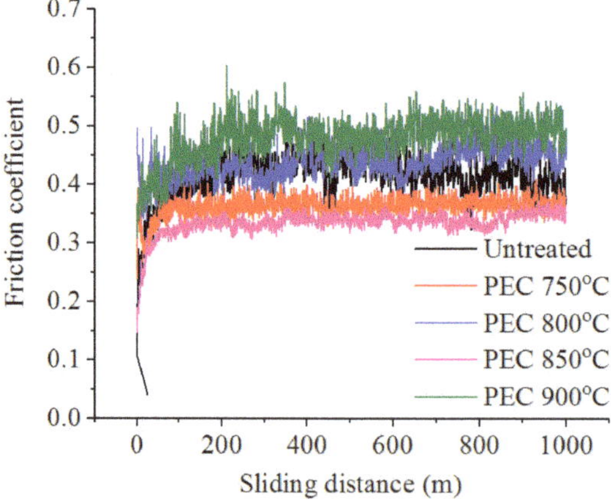

Figure 17. Dependence of friction coefficient on sliding distance of the untreated and PEC samples.

4. Discussion

Surface morphology during anodic plasma electrolytic treatment in aqueous electrolytes is determined by the competition of the processes of high-temperature oxidation, which leads to the formation of an oxide layer with a growth of roughness on the surface, and the anodic dissolution of the treated material, which leads to the alignment of the surface profile and a decrease in roughness [38–40]. In the considered cases of the anodic diffusion saturation of austenitic stainless steel samples, in general, the prevalence of anodic dissolution is observed, alongside a decrease in surface roughness. This determines the fundamental difference between the result of the anodic and cathodic treatment shown in the Introduction. At the same time, at high saturation temperatures, the intensive oxidation of the surface is observed, partially compensating for the decrease in the weight of the samples and the roughness during anodic dissolution. Under these conditions, the surface morphology will be determined by the structural features of the oxide layers—pores (after PEB) and areas with traces of the detachment of the fragile oxidized material (after PEC at high temperatures) are visually observed. Similar morphological features were observed after the PEN [27,41], PEB [42,43] and PEC [36,37] of carbon steels, which determine the general mechanism of the processes of the high-temperature oxidation and the anodic dissolution of the surfaces of carbon and high-alloy steels.

In contrast to the plasma electrolytic treatment of carbon steels that do not contain alloying additives, the surface hardening of austenitic stainless steel with a low carbon content develops due to the formation of inclusion compounds in the form of carbides and nitrides. In this case, an increase in microhardness reaches the depth of phase transformations up to 20–25 microns, which is observed during PEN (Figure 12). During PEC, when quenching with the formation of martensite or the consolidation of the crystal lattice is possible, due to the presence of carbon diffusion at a greater depth than nitrogen, the surface layer hardens up to 80 μm (Figure 14). According to this mechanism, the hardening of low-carbon unalloyed steels is possible during anodic PEC [36,37] and PENC [39,40,44]. The results of PEB clearly showed that the absence of inclusion compounds in the surface layer does not lead to an increase in microhardness (Figure 13), while the PEB of medium carbon steel (0.45 wt.% C) makes it possible to harden the surface to 1800 HV [42,43].

Despite the decrease in the hardness of samples after PEN with an increase in the processing temperature, their wear resistance does not decrease and the relationship between hardness and wear resistance in the function of nitriding temperature is not one-digit. Wear resistance is affected by the formation of nitride particles and the formation of a low level of micro-deformations in the lattice. Nitrides formed in the diffusion layer can be incoherent, coherent or semi-coherent. Coherent and semi-coherent nitrides lead to the greater deformation of the matrix than incoherent ones. Plastic deformation plays a leading role in the wear process. At a higher nitriding temperature, the matrix of the diffusion layer apparently has greater plasticity, which significantly reduces the level of micro-deformations of the crystal lattice of the iron matrix. Therefore, wear resistance does not decrease following a decrease in hardness. The level of weight wear within the margin of error does not change.

Samples after PEB show the maximum level of weight wear and the highest friction coefficient after treatment at temperatures of 800 and 950 °C, which corresponds to the cases of the most developed pores on the surface (Figure 4). The X-ray analysis of the sample after saturation (Figure 10) shows the presence of FeO on the surface, which can lead to both an increase in the friction coefficient and friction weight losses.

The minimum weight wear after PEC is achieved by processing at 900 °C. It is influenced by two factors, including the maximum hardness (Figure 14) and a large number of oxides (Figure 5c), among which Fe_3O_4—a highly effective lubricant—appears, according to X-ray analysis (Figure 11) [45].

All samples after PEN, PEB and PEC show a correlation of the friction coefficient with the Kragelsky–Kombalov criterion, which is a generalized dimensionless criterion for surface roughness. With a predominance of plastic deformations in the tribo-conjugation, the molecular component of the external friction coefficient does not depend on the micro-geometry of the surface. In addition, the deformation component of the friction coefficient increases with an increase in complex Δ. The friction cumulative coefficient also increases with an increase in Kragelsky–Kombalov criterion. In all plasma electrolytic treatment sessions, the maximum Kragelsky-Kombalov criterion on the friction track correlates with the highest friction coefficient of this sample.

The Kragelsky–Kombalov criterion determines the bearing capacity of the roughness profile. The smaller Δ is, the higher the bearing capacity of the roughness profile. Tables 4–6 show that the loss in weight during friction is smaller in samples with lower values of the

Kragelsky–Kombalov criterion. The maximum value Δ on the friction track of an untreated sample (0.989) corresponds to weight loss during friction of 23.2 ± 0.3 g.

PEN at all temperatures reduces Δ 2.4 to 2.6 times, and the weight loss due to friction at 1 km falls 46–58 times. PEB at 850 and 900 °C shows a 2.3–2.4 times lower value of Δ, and weight decreases 12.9 and 7.0 times, respectively. At 800 and 950 °C, values of Δ increase; in addition, a pronounced porosity of the oxide surface, leading to a strong increase in weight losses, becomes of great importance. After PEC, the friction on the track is less than double compared to an untreated sample, and such a roughness profile provides weight losses per 1 km of friction 232 times as small as in an untreated sample.

The relief of a rough surface also influences the friction coefficient via the distribution of material along the height of a single protrusion, that is, the shape and size of the protrusion. With a decrease in the radii of the curvature of the vertices of the microfoils, their deeper penetration into the volume of the material occurs in absolute magnitude, and the friction coefficient (the deformation part) increases, which is confirmed by Tables 4–6.

Friction bonds in the process of friction after treatment are broken as a result of the plastic displacement of the counter body material, as indicated by the value of the relative insertion $h/r < 0.1$ in all cases. The type of wear can be characterized as fatigue wear with boundary friction and plastic contact for samples after all the described types of processing.

The assumption about the type of wear and the nature of the destruction of friction bonds is confirmed by the values of the Greenwood–Williamson complex parameter, which are greater than 3 for all the described types of processing (Tables 4–6).

The actual contact area differs significantly from the nominal (geometric contact area of counterbody with the sample). The actual contact area has a minimum value of 1% (PEN at 800 °C, Table 1) of the nominal value and a maximum of 20% (PEB at 800 °C, Table 5) for treated samples versus 24% for untreated.

In all cases of processing, an unsaturated plastic contact is realized during the friction process. With this type of contact, the deformation of micro-dimensions does not influence the load increase and the number of protrusions increases with the load increase. As can be seen from Tables 4–6, the number of protrusions that come into contact in the tribo-connection is always smaller than the number of protrusions on the contour area.

The highest values of the friction coefficients after plasma electrolytic treatment are demonstrated by samples with PEN. Moreover, the values of their friction coefficients after all nitriding temperatures are greater than those of the untreated sample, but the weight wear is two orders smaller than that of the untreated one. Equation (16) serves to explain this fact. The actual contact area A_r of samples with PEN is the smallest of all experimental series of samples and varies from 1 to 8% of the nominal depending on the PEN temperature (Table 4). For comparison, the actual contact area of samples after PEB is 19–20% of the nominal (Table 5) samples with that of PEC being 9–13% of the nominal (Table 6) samples at different temperatures. A strong decrease in the actual contact area of PEN samples at values of absolute penetration h, which is comparable to that of other series, leads to low values of friction losses of weight per 1 km at fairly high values of the coefficient of friction.

Thus, the study showed a number of fundamental differences between the results of anodic plasma electrolytic saturation with light elements of austenitic stainless steel from the cathodic treatment option. In particular, with smaller prolonged saturation, as with the anodic treatment, there is no accumulation of high concentrations of diffusant atoms in the surface layer or a formation of inclusion compounds with a high content of nitrogen and carbon, such as $Fe_{2-3}N$ during the cathodic nitriding of steel; 12Cr18Ni10Ti [6], Fe_2N, Fe_4N, CrN and Cr_2N during the cathodic nitriding of 316L steel [9]; and $(Cr,Fe)_7C_3$ and CrN during the cathodic cementation of 304 steel [15]. Nevertheless, after anodic PEC, the microhardness value of the diffusion layers (700 HV) exceeds that obtained by cathodic carburizing (513 HV [17]), and the results of the tribological tests showed an improvement in the wear resistance index by two orders of magnitude, significantly exceeding the results on the friction of nitrided 316L steel by the cathodic method [8]. The analysis of friction track microtopology showed the correlation of the Kragelsky–Kombalov criterion and the friction coefficient. All this testifies to the complex influence of the hardness of the reinforced layer and the composition and morphology of the surface. Thus, the effectiveness of the use of anodic plasma electrolytic treatment to increase the hardness and wear resistance of austenitic stainless steel is shown.

5. Conclusions

(a) The paper shows the possibility of increasing the hardness and wear resistance of an austenitic stainless steel surface using anodic plasma electrolytic treatment. The positive effect of anodic dissolution on the reduction in surface roughness, as well as the complex effect of high-temperature oxidation and anodic dissolution on the morphology of the surface, has been confirmed. The structural features and phase composition of the modified surfaces are revealed, which determine the hardness of the surface layers and, together with the morphology of the surface, the tribological properties of the processed products.

(b) The hardness of the nitrided layers increases to 1450 HV with a thickness of up to 20–25 μm due to the formation of iron nitrides and iron-chromium carbides with a decrease in roughness of 3.7 times after PEN. This leads to an increase in wear resistance by two orders of magnitude.

(c) The PEC of the steel surface leads to an increase in hardness up to 800 HV and a thickness of the hardened layer up to 80 μm due to the formation of chromium carbides and a solid solution of carbon. The roughness and wear resistance of the treated surface change by approximately the same values as after PEN.

(d) It was revealed that when PEB is conducted on austenitic stainless steel, there is no hardening of the surface, but, at the same time, there is a decrease in roughness and an increase in the wear resistance of the surface.

(e) The frictional bonds in the friction process of each type of treatment are broken because of the plastic displacement of the counterbody material. The type of wear can be characterized as fatigue wear with boundary friction and plastic contact. The correlation of the friction coefficient and the Kragelsky–Kombalov criterion, a generalized dimensionless criterion of surface roughness, is shown.

Author Contributions: Conceptualization, S.K. and A.N.; methodology, T.M. and I.T.; validation, S.K., I.T. and T.M.; formal analysis, A.N.; investigation, T.M., I.T., R.B., E.S. and I.K.; resources, S.K.; writing—original draft preparation, S.K.; writing—review and editing, A.N. and I.K.; visualization, T.M. and I.T.; supervision, S.K.; project administration, S.K.; funding acquisition, S.K. All authors have read and agreed to the published version of the manuscript.

Funding: This research was funded by the Russian Science Foundation, grant number 18-79-10094.

Institutional Review Board Statement: Not applicable.

Informed Consent Statement: Not applicable.

Data Availability Statement: Not applicable.

Conflicts of Interest: The authors declare no conflict of interest.

References

1. Yin, F.; Hu, S.; Xu, R.; Han, X.; Qian, D.; Wei, W.; Hua, L.; Zhao, K. Strain rate sensitivity of the ultrastrong gradient nanocrystalline 316L stainless steel and its rate-dependent modeling at nanoscale. *Int. J. Plast.* **2020**, *129*, 102696. [CrossRef]
2. Li, P.; Hu, S.; Liu, Y.; Hua, L.; Yin, F. Surface Nanocrystallization and Numerical Modeling of 316L Stainless Steel during Ultrasonic Shot Peening Process. *Metals* **2022**, *12*, 1673. [CrossRef]
3. Chen, Y.; Deng, S.; Zhu, C.; Hu, K.; Yin, F. The effect of ultrasonic shot peening on the fatigue life of alloy materials: A review. *Int. J. Comput. Mater. Sci. Surf. Eng.* **2021**, *10*, 209–234. [CrossRef]
4. Liu, S.; Kim, Y.; Jung, J.; Bae, S.; Jeong, S.; Shin, K. Effect of Ultrasonic Shot Peening and Laser Shock Peening on the Microstructure and Microhardness of IN738LC Alloys. *Materials* **2023**, *16*, 1802. [CrossRef] [PubMed]
5. Shi, X.; Feng, X.; Teng, J.; Zhang, K.; Zhou, L. Effect of laser shock peening on microstructure and fatigue properties of thin-wall welded Ti-6A1-4V alloy. *Vacuum* **2021**, *184*, 109986. [CrossRef]

6. Wang, P.; Cao, Q.; Liu, S.; Peng, Q. Surface strengthening of stainless steels by nondestructive laser peening. *Mater. Des.* **2021**, *205*, 109754. [CrossRef]

7. Mora-Sanchez, H.; Pixner, F.; Buzolin, R.; Mohedano, M.; Arrabal, R.; Warchomicka, F.; Matykina, E. Combination of Electron Beam Surface Structuring and Plasma Electrolytic Oxidation for Advanced Surface Modification of Ti6Al4V Alloy. *Coatings* **2022**, *12*, 1573. [CrossRef]

8. Perez, H.; Vargas, G.; Magdaleno, C.; Silva, R. Article: Oxy-Nitriding AISI 304 Stainless Steel by Plasma Electrolytic Surface Saturation to Increase Wear Resistance. *Metals* **2023**, *13*, 309. [CrossRef]

9. Marcuz, N.; Ribeiro, R.P.; Rangel, E.C.; Cristino da Cruz, N.; Correa, D.R.N. The Effect of PEO Treatment in a Ta-Rich Electrolyte on the Surface and Corrosion Properties of Low-Carbon Steel for Potential Use as a Biomedical Material. *Metals* **2023**, *13*, 520. [CrossRef]

10. Kumruoglu, L.C.; Ozel, A. Plasma electrolytic saturation of 316L stainless steel in an aqueous electrolyte containing urea and ammonium nitrate. *Mater. Technol.* **2013**, *47*, 307–310.

11. Roy, A.; Tewari, R.K.; Sharma, R.C.; Sherhar, R. Feasibility study of aqueous electrolyte plasma nitriding. *Surf. Eng.* **2007**, *23*, 243–246. [CrossRef]

12. Aliev, M.K.; Sabour, A.; Taheri, P. Study of corrosion protection of different stainless steels by nanocrystalline plasma electrolysis. *Prot. Met. Phys. Chem. Surf.* **2008**, *44*, 402–407. [CrossRef]

13. Skakov, M.; Kurbanbekov, S.; Scheffler, M. Influence of Regimes Electrolytic Plasma Cementation on the Mechanical Properties of Steel 12Cr18Ni10Ti. *Key Eng. Mater.* **2013**, *531–532*, 173–177. [CrossRef]

14. Aliofkhazraei, M.; Sabour Rouhaghdam, A. Relationship Study among Nanocrystallite Distribution and Roughness of a Nanostructured Hard Layer. *Defect Diffus. Forum* **2009**, *293*, 83–90. [CrossRef]

15. Andrei, A.; Lungu, C.; Oncioiu, G.; Diaconu, C. Plasma processing for improvements of structural materials properties. In Proceedings of the 35th EPS Conference on Plasma Physics, Hersonissos, Greece, 9–13 June 2008.

16. Andrei, V.; Vlaicu, G.; Fulger, M.; Ducu, C.; Diaconu, C.; Oncioiu, G.; Andrei, E.; Bahrim, M.; Gheboianu, A. Chemical and structural modifications induced in structural materials by electrolchemical process. *Roman. Report. Phys.* **2009**, *61*, 95–104.

17. Hao, J.; Liu, B.; Chen, H. Research on liquid plasma surface strengthening of stainless steel. *Hot Work. Technol.* **2007**, *12*, 58–59. Available online: http://en.cnki.com.cn/KCMS/detail/detail.aspx?dbcode=CJFD&dbname=CJFD2007&filename=SJGY200712019&uniplatform=OVERSEA&v=UmKJpFeqc4eDOKZqgJD0JOIw42o5SJBWWBr09sbbr-MvLaU0wx3gItmBOAIoBDNU (accessed on 15 January 2023).

18. Xue, W.; Jin, Q.; Zhu, Q.; Wu, X. Characterization of Plasma Electrolytic Carburized Stainless Steel in Glycerin Aqueous Solution. *J. Aeronaut. Mater.* **2010**, *30*, 38–42. [CrossRef]

19. Yerokhin, A.L.; Leyland, A.; Tsotsos, C.; Wilson, A.D.; Nie, X.; Matthews, A. Duplex surface treatments combining plasma electrolytic nitrocarburising and plasma-immersion ion-assisted deposition. *Surf. Coat. Technol.* **2001**, *142–144*, 1129–1136. [CrossRef]

20. Taheri, P.; Dehghanian, C. A phenomenological model of nanocrystalline coating production using plasma electrolytic saturation (PES) technique. *Sci. Iran.* **2009**, *16*, 87–91.

21. Kazerooniy, N.A.; Bahrololoom, M.E.; Shariat, M.H.; Mahzoon, F.; Jozaghi, T. Effect of Ringer's solution on wear and friction of stainless steel 316L after plasma electrolytic nitrocarburising at low voltages. *J. Mater. Sci. Technol.* **2011**, *27*, 906–912. [CrossRef]

22. Nie, X.; Tsotsos, C.; Wilson, A.; Yerokhin, A.L.; Leyland, A.; Matthews, A. Characteristics of a plasma electrolytic nitrocarburising treatment for stainless steels. *Surf. Coat. Technol.* **2001**, *139*, 135–142. [CrossRef]

23. Taheri, P.; Dehghanian, C. Wear and corrosion properties of nanocrystalline coatings on stainless steel produced by plasma electrolytic nitrocarburising. *Int. J. Mat. Res.* **2008**, *99*, 92–100. [CrossRef]

24. Belkin, P.N.; Kusmanov, S.A.; Zhirov, A.V.; Belkin, V.S.; Parfenyuk, V.I. Anode Plasma Electrolytic Saturation of Titanium Alloys with Nitrogen and Oxygen. *J. Mat. Sci. Tech.* **2016**, *32*, 1027–1032. [CrossRef]

25. Kusmanov, S.A.; Dyakov, I.G.; Belkin, P.N.; Gracheva, I.A.; Belkin, V.S. Plasma electrolytic modification of the VT1-0 titanium alloy surface. *J. Surf. Investig. X-ray Synchrotron Neutron Tech.* **2015**, *9*, 98–104. [CrossRef]

26. Belkin, P.N.; Borisov, A.M.; Kusmanov, S.A. Plasma Electrolytic Saturation of Titanium and Its Alloys with Light Elements. *J. Surf. Investig. X-ray Synchrotron Neutron Tech.* **2016**, *10*, 516–535. [CrossRef]

27. Belkin, P.N.; Kusmanov, S.A. Plasma electrolytic nitriding of steels. *J. Surf. Investig. X-ray Synchrotron Neutron Tech.* **2017**, *11*, 767–789. [CrossRef]

28. Kusmanov, S.; Tambovskiy, I.; Silkin, S.; Nikiforov, R.; Belov, R. Increasing the Hardness and Corrosion Resistance of the Surface of CP-Ti by Plasma Electrolytic Nitrocarburising and Polishing. *Materials* **2023**, *16*, 1102. [CrossRef]

29. Shelekhov, E.V.; Sviridova, T.A. Programs for X-ray analysis of polycrystals. *Metal Sci. Heat Treat.* **2000**, *42*, 309–313 . [CrossRef]

30. Grazulis, S.; Chateigner, D.; Downs, R.T.; Yokochi, A.T.; Le Bail, A. Crystallography open database—An open-access collection of crystal structures. *J. Appl. Cryst.* **2009**, *42*, 726–729. [CrossRef]

31. Tambovskiy, I.; Mukhacheva, T.; Gorokhov, I.; Suminov, I.; Silkin, S.; Dyakov, I.; Kusmanov, S.; Grigoriev, S. Features of Cathodic Plasma Electrolytic Nitrocarburizing of Low-Carbon Steel in an Aqueous Electrolyte of Ammonium Nitrate and Glycerin. *Metals* **2022**, *12*, 1773. [CrossRef]

32. Mukhacheva, T.; Kusmanov, S.; Suminov, I.; Podrabinnik, P.; Khmyrov, R.; Grigoriev, S. Increasing Wear Resistance of Low-Carbon Steel by Anodic Plasma Electrolytic Sulfiding. *Metals* **2022**, *12*, 1641. [CrossRef]

33. Mukhacheva, T.L.; Belkin, P.N.; Dyakov, I.G.; Kusmanov, S.A. Wear mechanism of medium carbon steel after its plasma electrolytic nitrocarburising. *Wear* **2020**, *462–463*, 203516. [CrossRef]
34. Demkin, N.B.; Izmailov, V.V. Surface topography and properties frictional contacts. *Trib. Int.* **1991**, *24*, 21–24. [CrossRef]
35. Kragelsky, I.V.; Dobychin, M.N.; Kombalov, V.S. *Friction and Wear Calculation Methods*; Pergamon Press Ltd.: Oxford, UK, 1982; Available online: https://books.google.ru/books?id=QLcgBQAAQBAJ&hl=ru (accessed on 24 February 2023).
36. Belkin, P.N.; Kusmanov, S.A. Plasma Electrolytic Carburising of Metals and Alloys. *Surf. Eng. Appl. Electrochem.* **2021**, *57*, 19–50. [CrossRef]
37. Shadrin, S.Y.; Belkin, P.N.; Tambovskiy, I.V.; Kusmanov, S.A. Physical Features of Anodic Plasma Electrolytic Carburising of Low-Carbon Steels. *Plasma Chem. Plasma Process.* **2020**, *40*, 549–570. [CrossRef]
38. Kusmanov, S.A.; Silkin, S.A.; Smirnov, A.A.; Belkin, P.N. Possibilities of increasing wear resistance of steel surface by plasma electrolytic treatment. *Wear* **2017**, *386–387*, 239–246. [CrossRef]
39. Kusmanov, S.A.; Dyakov, I.G.; Kusmanova, Y.V.; Belkin, P.N. Surface Modification of Low-Carbon Steels by Plasma Electrolytic Nitrocarburising. *Plasma Chem. Plasma Process.* **2016**, *36*, 1271–1286. [CrossRef]
40. Kusmanov, S.A.; Kusmanova, Y.V.; Naumov, A.R.; Belkin, P.N. Formation of diffusion layers by anode plasma electrolytic nitrocarburising of low carbon steel. *J. Mater. Eng. Perform.* **2015**, *24*, 3187–3193. [CrossRef]
41. Kusmanov, S.A.; Smirnov, A.A.; Silkin, S.A.; Belkin, P.N. Increasing wear and corrosion resistance of low-alloy steel by anode plasma electrolytic nitriding. *Surf. Coat. Technol.* **2016**, *307*, 1350–1356. [CrossRef]
42. Kusmanov, S.A.; Tambovskiy, I.V.; Sevostyanova, V.S.; Savushkina, S.V.; Belkin, P.N. Anode plasma electrolytic boriding of medium carbon steel. *Surf. Coat. Technol.* **2016**, *291*, 334–341. [CrossRef]
43. Belkin, P.N.; Kusmanov, S.A. Plasma Electrolytic Boriding of Steels and Titanium Alloys. *Surf. Eng. Appl. Electrochem.* **2019**, *55*, 1–30. [CrossRef]
44. Kusmanov, S.A.; Kusmanova, Y.V.; Smirnov, A.A.; Belkin, P.N. Modification of steel surface by plasma electrolytic saturation with nitrogen and carbon. *Mater. Chem. Phys.* **2016**, *175*, 164–171. [CrossRef]
45. Song, X.; Qiu, Z.; Yang, X.; Gong, H.; Zheng, S.; Cao, B.; Wang, H.; Möhwald, H.; Shchukin, D. Submicron-Lubricant Based on Crystallized Fe_3O_4 Spheres for Enhanced Tribology Performance. *Chem. Mater.* **2014**, *26*, 5113–5119. [CrossRef]

Article

Effect of Li Content on the Microstructure and Mechanical Properties of as-Homogenized Mg-Li-Al-Zn-Zr Alloys

Yuehua Sun [1,2], **Fan Zhang** [1], **Jian Ren** [1,2,*] **and Guangsheng Song** [1,2]

[1] Key Laboratory of Green Fabrication and Surface Technology of Advanced Metal Materials (Anhui University of Technology), Ministry of Education, Maanshan 243002, China; sunyuehua1008@126.com (Y.S.)
[2] School of Materials Science and Engineering, Anhui University of Technology, Maanshan 243002, China
* Correspondence: renjianahut@126.com

Abstract: The microstructure and mechanical properties of as-homogenized Mg-xLi-3Al-2Zn-0.2Zr alloys (x = 5, 7, 8, 9, 11 wt.%) were studied. As the Li content increased from 5 wt.% to 11 wt.%, the alloy matrix changed from the α-Mg single-phase to α-Mg+β-Li dual-phase and then to the β-Li single-phase. Homogenized With the increase in Li content, the alloy strength decreased while the elongation increased, and the corresponding fracture mechanism changed from cleavage fracture to microvoid coalescence fracture. This is mainly attributed to the matrix changing from α-Mg with hcp structure to β-Li with bcc structure. Additionally, the increase in the AlLi softening phase led to the reduction of Al and Zn dissolved in the alloy matrix with increasing Li content, which is one of the reasons for the decrease in alloy strength.

Keywords: Mg-Li alloy; microstructure; mechanical property; fracture mechanism

1. Introduction

Alloying Mg with Li can produce the lightest Mg-Li alloy; its density (1.25–1.65 g/cm^3) is close to that of plastic, with only 3/5–3/4 of that of conventional Mg alloy and 1/2–2/3 of that of Al alloy. The Mg-Li alloy exhibits a unique phase transition with the change in Li content, i.e., the alloy is an α-Mg single-phase alloy with an hcp structure when the Li content is lower than 5.7 wt.%, the alloy is a β-Li single-phase with a bcc structure when the Li content is higher than 10.3 wt.%, and the alloy is an α-Mg+β-Li dual-phase alloy with an hcp+bcc structure when the Li content is between 5.7 wt.% and 10.3 wt.% [1]. The addition of Li can reduce the c/a axial ratio of the Mg lattice and even change the crystal structure from an hcp structure to a bcc structure, which makes the Mg-Li alloy exhibit incomparable plastic deformation ability compared to a conventional Mg alloy [1,2]. In addition, the Mg-Li alloy also shows the advantages of high specific strength, excellent damping, good electromagnetic shielding performance, and weak anisotropy, being lightweight and having good prospects in the fields of energy saving, and emission reduction, especially in aerospace and electron industries [3,4]. Taking aerospace satellites as an example, the application of the Mg-Li alloy can achieve a weight reduction effect of about 20–50%, significantly improving the payload of the satellite, and thus generating huge economic benefits.

However, since the binary Mg-Li alloy exhibits low strength, it is necessary to add alloying elements to strengthen the binary Mg-Li alloy. Al and Zn are the common alloying elements for a Mg-Li alloy; they can improve the strength of the Mg-Li alloy by solid solution strengthening and second phase strengthening, owing to their high solid solubility in the Mg matrix and ease of combining with other elements to form compounds [5]. Previous research has shown that Al and Zn have similar effects on Mg-Li alloy; that is, the alloy strength increases while the elongation decreases with the increase in Al content or Zn content [5,6]. In the Mg-Li-Al alloy, the Al content is generally less than 5 wt.%–6 wt.%, above which the alloy strength does not increase significantly and the elongation loss is

Citation: Sun, Y.; Zhang, F.; Ren, J.; Song, G. Effect of Li Content on the Microstructure and Mechanical Properties of as-Homogenized Mg-Li-Al-Zn-Zr Alloys. *Alloys* **2023**, *2*, 89–99. https://doi.org/10.3390/alloys2020006

Academic Editor: Andrea Di Schino

Received: 14 March 2023
Revised: 12 April 2023
Accepted: 20 April 2023
Published: 28 April 2023

obvious. When the Zn content exceeds 2.5 wt.%, it has a negative effect on the corrosion resistance of the Mg-Li alloy, so the Zn content is usually controlled within 2.5 wt.% [7]. Zr is often used as a grain refiner to improve the microstructure and mechanical properties of an alloy, and its addition amount is generally 0.1 wt.%–0.3 wt.%. In recent years, due to the excellent mechanical properties of Mg-Li-Al-Zn (LAZ) alloys, much research has have focused on improving the mechanical properties of LAZ532 and LAZ832 alloys through alloying with rare earth elements [8–11]. Cui et al. [8] revealed that the addition of Y element into the LAZ532 alloy could form an Al_2Y strengthening phase, reduce the AlLi softening phase, and refine grains, and the tensile strength was increased by 32.66% when the Y content was 0.8 wt.%. Zhu et al. [9] indicated that Y and Nd had a synergistic strengthening effect on the LAZ532 alloy, and the grain refinement effect was the best and the mechanical properties reached their maximum values (UTS = 231 MPa, δ = 16%) when adding 1.2 wt.% Y and 0.8 wt.% Nd simultaneously. Zhao et al. [10] showed that the LAZ832-0.5Y alloy had the best tensile strength of 218.5 MPa and elongation of 16.9%, which was mainly attributed to the grain refinement and second phase strengthening effect caused by the addition of Y. However, there are few systematic studies on the microstructure and mechanical properties of Mg-Li-Al-Zn alloys with different Li contents.

In this work, the microstructure and mechanical properties of as-homogenized Mg-xLi-3Al-2Zn-0.2Zr alloys (x = 5, 7, 8, 9, 11 wt.%) were investigated, and the fundamental reason for the variation of mechanical properties was explained.

2. Experimental Procedures

The alloys used in this work were prepared by melting pure Mg (>99.9 wt.%), pure Li (>99.9 wt.%), pure Al (>99.9 wt.%), pure Zn (>99.9 wt.%), and the Mg-Zr (30 wt.%) master alloy in a vacuum induction melting furnace under pure argon atmosphere. The melts were held at 730 °C for 10 min, and then poured into a stainless steel mold with a diameter of 80 mm and cooled to room temperature. After that, the cast ingots were homogenized at 280 °C for 24 h to obtain the experimental alloys with uniform composition and microstructure. Choosing the homogenization temperature of 280 °C not only ensured that the alloys could be completely homogenized, but also ensured that the alloys were kept warm for a long time without any combustion hazards. The densities of as-homogenized alloys were determined with an electronic density balance (AEL-200, Mettler Toledo, Greifensee, Switzerland). The phase composition was analyzed using an X-ray diffractometer (XRD, D/Max 2500, Rigaku, Tokyo, Japan) using monochromatic Cu Kα radiation, and the 2θ was in the range of 20° to 80° with a step size of 0.02° and a scan rate of 4°/min. The microstructure was observed with a scanning electron microscope (SEM, Quanta-200, FEI, Eindhoven, The Netherlands), an electron probe microanalyser (EPMA, JXA-8230, JEOL, Tokyo, Japan), and a transmission electron microscope (TEM, Tecnai G20ST, FEI, Eindhoven, The Netherlands). The samples for TEM observation were firstly ground to 80 μm, then thinned to perforation with a twin-jet electropolisher, and finally thinned for 20 min using an ion-beam thinner at a small angle. The electrolyte for twin-jet electropolishing was a mixed solution containing 8 vol.% perchloric acid and 92 vol.% alcohol, and the specific experimental parameters were a temperature of about −30 °C, working voltage of 12–20 V, and working current of 12–16 mA. The tensile tests were carried out with a tensile tester (MTS-858, MTS, Costa Mesa, CA, USA) at a tensile rate of 2 mm/min, and the dimensions of samples for tensile tests were 30 mm in gauge length and 6 mm in diameter. The hardness was measured with Vickers hardness tester (HV-50AP, Mitutoyo, Kanagawa, Japan) with a loading force of 98 N and a holding time of 10 s.

3. Results and Discussion

3.1. Microstructure

Figure 1 shows the XRD patterns of as-homogenized Mg-xLi-3Al-2Zn-0.2Zr alloys. Among these alloys, the LAZ532-0.2Zr alloy is a typical α-Mg single-phase alloy with an hcp structure, while the LAZ1132-0.2Zr alloy is a typical β-Li single-phase alloy with a bcc structure. When the Li contents were 7 wt.%, 8 wt.%, and 9 wt.%, the alloys were α-Mg+β-Li dual-phase alloys with an hcp+bcc mixed structure. Obviously, as the Li content increased from 5 wt.% to 11 wt.%, the matrix phase of the alloy changed from α-Mg single-phase to α-Mg+β-Li dual-phase and then to β-Li single-phase, and the crystal structure transformed from hcp structure to hcp+bcc structure and then to bcc structure, which are results consistent with the phase diagram of the Mg-Li binary alloy [12]. Except for the matrix phase, all alloys contained the AlLi phase (fcc structure), and the corresponding phase peaks become more and more obvious with increasing Li content. No phases containing Zn and Zr elements were detected in the alloys, which indicates that Zn and Zr exist in the alloy matrix in the form of solid solutions and do not form intermetallic compounds.

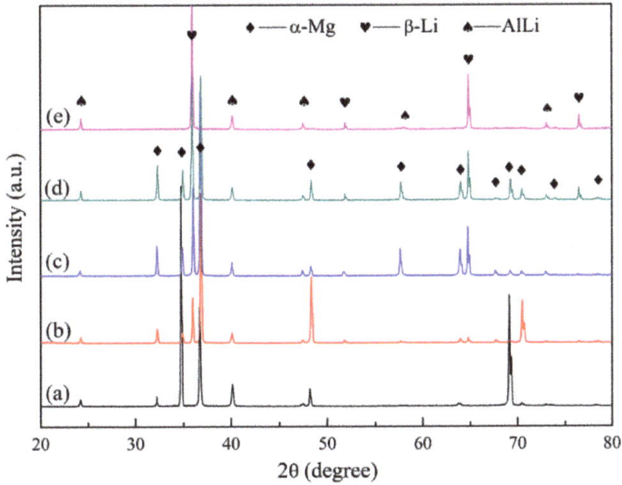

Figure 1. XRD patterns of as-homogenized Mg-*x*Li-3Al-2Zn-0.2Zr alloys: (**a**) *x* = 5; (**b**) *x* = 7; (**c**) *x* = 8; (**d**) *x* = 9; (**e**) *x* = 11.

Figure 2 displays the SEM micrographs of as-homogenized Mg-*x*Li-3Al-2Zn-0.2Zr alloys. The as-homogenized LAZ532-0.2Zr alloy was composed of a light gray α-Mg matrix phase, with a kind of white fibrous phase in the α-Mg matrix. According to the XRD results (Figure 1) and works in the literature [13,14], this fibrous phase can be identified as the AlLi phase, and it exists in the α-Mg matrix in the form of eutectic. As-homogenized LAZ732-0.2Zr, LAZ832-0.2Zr, and LAZ932-0.2Zr alloys consisted of a light gray α-Mg matrix and a dark gray β-Li matrix. With the increase in Li content, the α-Mg phase changes from a smooth block to long strip, and its content decreases gradually. It can be seen from the local enlarged view in the lower left corner of Figure 2c that a white fibrous phase similar to that in Figure 2a is distributed in the α-Mg matrix and a white granular phase is distributed in the β-Li matrix. According to the XRD results (Figure 1) and the literature [15,16], both the fibrous phase in the α-Mg matrix and the granular phase in the β-Li matrix were the AlLi phase. The as-homogenized LAZ1132-0.2Zr alloy consisted entirely of a dark gray β-Li matrix phase, and a large number of AlLi particles were uniformly distributed in the β-Li matrix.

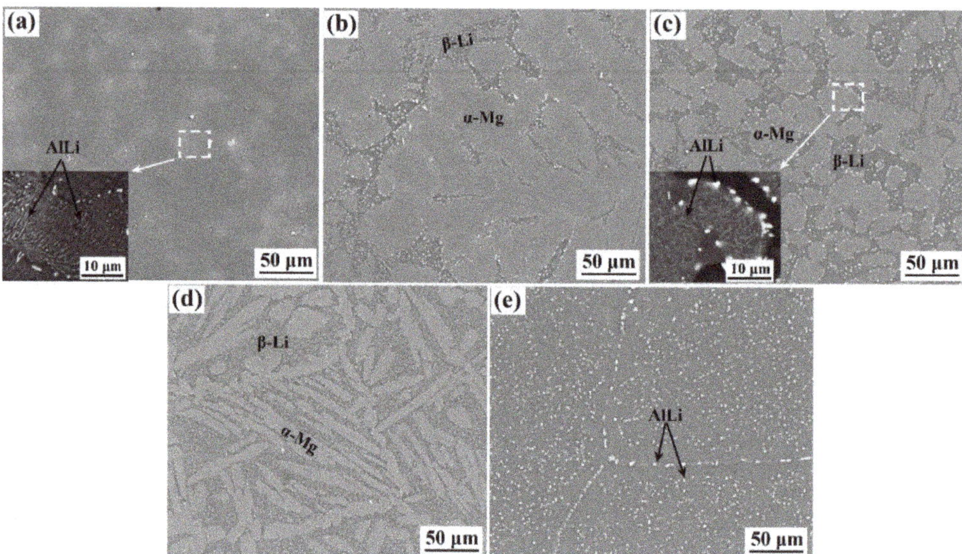

Figure 2. SEM micrographs of as-homogenized Mg-*x*Li-3Al-2Zn-0.2Zr alloys: (**a**) *x* = 5; (**b**) *x* = 7; (**c**) *x* = 8; (**d**) *x* = 9; (**e**) *x* = 11.

Figure 3 depicts the element distribution of as-homogenized LAZ532-0.2Zr, LAZ832-0.2Zr, and LAZ1132-0.2Zr alloys. Clearly, both the fibrous AlLi phase and the granular AlLi phase were rich in Al and Zn, and poor in Mg. It is worth noting that Li is too light to detect. The enrichment of Zn in the AlLi phase was mainly due to the high solid solubility of Zn in Al (83.1 wt.%) [17]. A part of the added Al element was combined with Li to form the AlLi phase, and the remainder was dissolved in the alloy matrix. Because the solid solubility of Al in Mg (12.7 wt.%) is much higher than that of Al in Li (extremely limited), the dissolved Al element was mainly distributed in the α-Mg matrix. A part of the added Zn element was enriched in the AlLi phase, and the remainder was dissolved in the alloy matrix. Because the solid solubility of Zn in Mg (6.2 wt.%) is lower than that of Zn in Li (12.5 wt.%), the solid solution concentration of Zn in the β-Li matrix was higher than that in the α-Mg matrix, but the difference was not great.

Figure 4 displays the TEM micrographs of the AlLi phase in the as-homogenized LAZ832-0.2Zr alloy. The AlLi phase distributed in the α-Mg matrix and β-Li matrix shows quite different morphologies. Fibrous AlLi phases with a width in the range of 0.02–0.11 μm are found in the α-Mg matrix, and round-like AlLi particles with a diameter in the range of 0.42–0.92 μm are observed in the β-Li matrix, which is consistent with the microstructure shown in Figures 2 and 3.

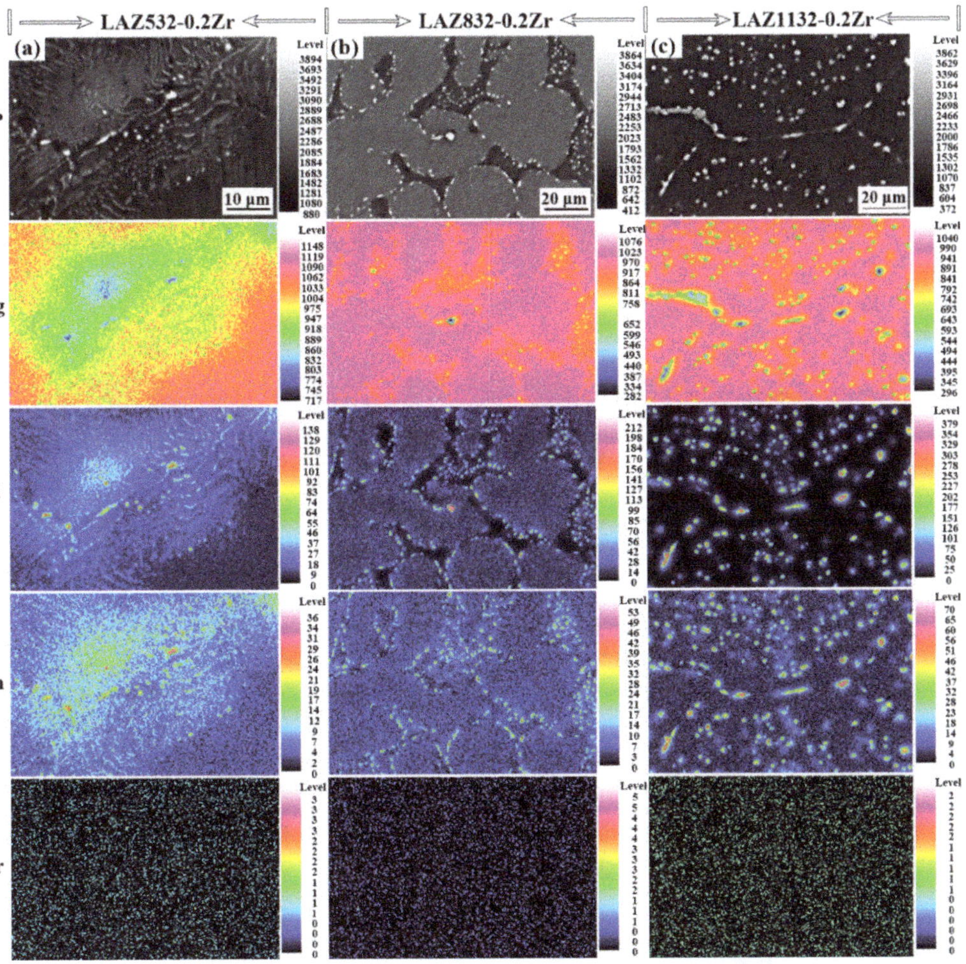

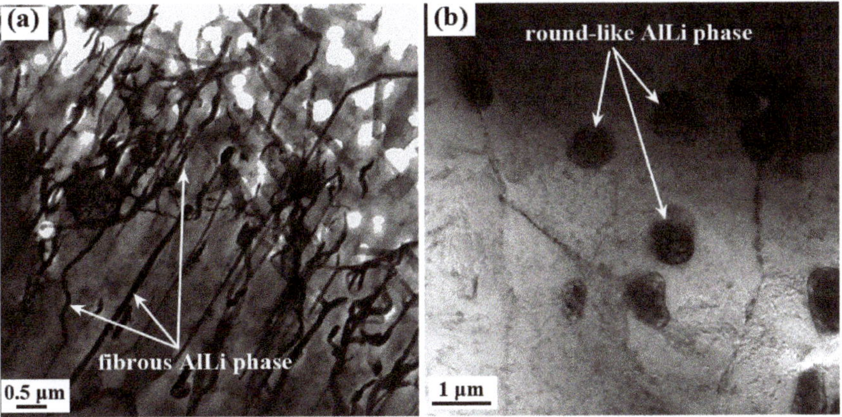

Figure 3. EPMA area analysis of (**a**) LAZ532-0.2Zr, (**b**) LAZ832-0.2Zr, and (**c**) LAZ1132-0.2Zr alloys.

Figure 4. TEM micrographs of (**a**) fibrous AlLi phase and (**b**) round-like AlLi phase in as-homogenized LAZ832-0.2Zr alloy.

3.2. Mechanical Properties

Figure 5 shows the stress–strain curves and hardness of as-homogenized Mg-xLi-3Al-2Zn-0.2Zr alloys, and the densities and mechanical properties of the alloys are listed in Table 1. The ultimate tensile strengths (UTS) of as-homogenized LAZ532-0.2Zr, LAZ732-0.2Zr, LAZ832-0.2Zr, LAZ932-0.2Zr, and LAZ1132-0.2Zr alloys were consecutively 197.2, 185.4, 181.4, 168.7, and 143.7 MPa; their elongations (δ) were consecutively 10.1%, 11.6%, 21.5%, 24.7%, and 27.3%; and their hardness was consecutively 68.6, 64.1, 61.3, 58.9, and 56.2 HV. In addition, the specific strength decreased from 124.8 MPa/(g/cm^3) to 100.5 MPa/(g/cm^3) as the Li content increased from 5 wt.% to 11 wt.%, exhibiting excellent specific strength. Clearly, as the Li content increases, the strength and hardness of the alloy decreases while the elongation increases, which is consistent with the variation law reported in previous research [18,19]. This result is mainly attributed to the transformation of the crystal structure of the Mg-Li alloy from the hcp structure to the bcc structure. It has been reported that the hardness of the α-Mg matrix is higher than that of the β-Li matrix [20]. With the increase in Li content, the content of the soft β-Li matrix in the alloy increased gradually, and thus the hardness of the alloy decreased gradually. In the Mg-Li alloy, the hard α-Mg matrix with hcp structure showed higher strength and worse ductility, while the soft β-Li matrix with bcc structure showed lower strength and excellent ductility. For the dual-phase Mg-Li alloy, plastic deformation occurred preferentially in the soft β-Li matrix during the deformation process, and then the hard α-Mg matrix began to undergo plastic deformation when the stress transmitted from the β-Li matrix to the α-Mg matrix was greater than the elastic limit of the α-Mg matrix. This is because the β-Li matrix with bcc structure is softer and has more independent slip systems than the α-Mg matrix. Table 2 lists the independent slip systems and the critical resolved shear stress (CRSS) in Mg and Li at room temperature [21–24]. The common slip systems in Mg are $\langle \vec{a} \rangle$ basal slip, $\langle \vec{a} \rangle$ prismatic slip, $\langle \vec{a} \rangle$ pyramidal slip, and $\langle \vec{c} + \vec{a} \rangle$ pyramidal slip. The slip direction of these $\langle \vec{a} \rangle$ dislocation slips is $\langle 11\bar{2}0 \rangle$ direction perpendicular to the c axis, which cannot coordinate the strain in the c axis, while $\langle \vec{c} + \vec{a} \rangle$ dislocation slip along $\langle 11\bar{2}3 \rangle$ direction can coordinate the strain in the c axis. At room temperature, the CRSS required for the basal slip initiation of Mg is about 0.45–0.81 MPa, and the CRSS required for prismatic slip and pyramidal slip is about 100 times that required for basal slip. Therefore, the basal slip is the easiest to initiate during the deformation process of Mg at room temperature, but the prismatic slip and pyramidal slip are difficult to initiate. However, the basal slip of Mg can only provide two independent slip systems, which fails to meet the requirements of five independent slip systems in Von Mises criterion, and thus Mg has poor plasticity and is difficult to deform. Due to the limited slip systems of Mg, twinning often occurs to coordinate the c-axis strain during deformation at room temperature. Compared with compression twins ($\{10\bar{1}1\}$ and $\{10\bar{1}2\}$), tension twin ($\{10\bar{1}2\}$) is the most easily initiated twinning mode in Mg at room temperature because of its lower CRSS (2.0–2.8 MPa) and smaller shear displacement [25]. For Li with a bcc structure, the primary slip system is the $\{110\}\langle 111 \rangle$ with the secondary slip systems of $\{112\}\langle 111 \rangle$ and $\{123\}\langle 111 \rangle$. It has been reported that the $\{110\}\langle 111 \rangle$ slip system of Li has a lower CRSS (0.54–0.57 MPa), and the corresponding number of the independent slip system is 12, so Li exhibits good ductility during the deformation process [23]. Moreover, it has been considered that the excellent plasticity of the β-Li matrix is attributed to the nearly equal CRSS values of the primary slip system ($\{110\}\langle 111 \rangle$) and secondary slip systems ($\{112\}\langle 111 \rangle$ and $\{123\}\langle 111 \rangle$) [24]. Therefore, as the Li content increases, the content of the β-Li matrix with bcc structure increases gradually, and accordingly the tensile strength decreases while the elongation increases. In addition, as the Li content increases, the AlLi softening phase increases gradually, as shown in Figure 1, and the contents of Al and Zn dissolved in the alloy matrix decrease, which is one of the reasons for the gradual decrease in the alloy strength, although the effect is not great.

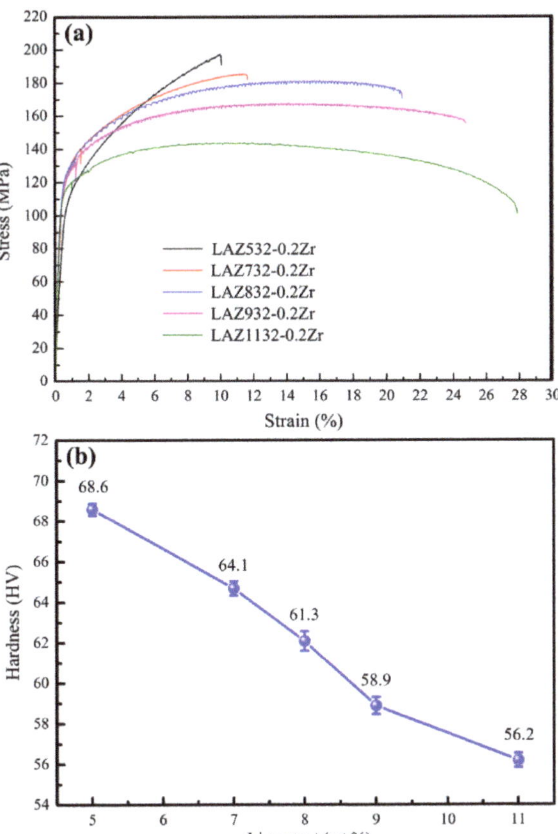

Figure 5. (**a**) Stress–strain curves and (**b**) hardness of as-homogenized Mg-*x*Li-3Al-2Zn-0.2Zr alloys.

Figure 6 depicts the fracture morphologies of as-homogenized Mg-*x*Li-3Al-2Zn-0.2Zr alloys. The fracture surface of the as-homogenized LAZ532-0.2Zr alloy was full of cleavage planes and cleavage steps, showing typical cleavage fracture characteristics. Cleavage fracture is caused by the stacking of dislocation at the grain boundary or twin boundary, which usually occurs along the cleavage plane, and the main cleavage planes of the α-Mg matrix with hcp structure were {0001} and {$1\bar{1}00$} with low index. In addition to cleavage planes and cleavage steps, many dimples could be observed on the fracture surfaces of as-homogenized dual-phase alloys (LAZ732-0.2Zr, LAZ832-0.2Zr, and LAZ932-0.2Zr), and the cleavage steps decreased while the dimples increased with the increase in Li content. Because a dimple is the basic feature of the microvoid coalescence fracture, the fracture modes of as-homogenized dual-phase alloys are mixture fractures of cleavage fracture and microvoid coalescence fracture. In the early stage of microvoid coalescence fracture, the microvoids are formed by the fracture of the second phase (AlLi), the separation of the second phase from the matrix, and the separation of the phase interface, owing to the elastic and plastic difference between the second phase and the matrix. The fracture surface of the as-homogenized LAZ1132-0.2Zr alloy was full of dimples with almost no cleavage steps, so its fracture mode was a microvoid coalescence fracture. The fracture mode of the as-homogenized alloys changed from cleavage fracture to microvoid coalescence fracture with increasing Li content, indicating that the elongation of the alloy gradually became better, which is consistent with the variation trend of elongation in Figure 5a.

Table 1. Density and mechanical properties of as-homogenized Mg-xLi-3Al-2Zn-0.2Zr alloys.

Alloy	Density (g/cm^3)	UTS (MPa)	δ (%)	Hardness (HV)	Specific Strength (MPa/(g/cm^3))
LAZ532-0.2Zr	1.58	197.2	10.1	68.6	124.8
LAZ732-0.2Zr	1.53	185.4	11.6	64.1	121.2
LAZ832-0.2Zr	1.51	181.4	21.5	61.3	120.1
LAZ932-0.2Zr	1.48	168.7	24.7	58.9	114.0
LAZ1132-0.2Zr	1.43	143.7	27.3	56.2	100.5

Table 2. Independent slip systems and the CRSS in Mg and Li at room temperature [21–24].

Metal	Type of Slip Plane	Type of Slip Direction	Slip Plane	Slip Direction	CRSS (MPa)	Number of Independent Slip Systems
Mg (hcp)	basal slip	$\langle \vec{a} \rangle$	$\{0001\}$	$\langle 11\bar{2}0 \rangle$	0.45–0.81	2
	prismatic slip	$\langle \vec{a} \rangle$	$\{1\bar{1}00\}$	$\langle 11\bar{2}0 \rangle$	39.2–40	2
		$\langle \vec{a} \rangle$	$\{11\bar{2}0\}$	$\langle 11\bar{2}0 \rangle$		
	pyramidal slip	$\langle \vec{a} \rangle$	$\{10\bar{1}1\}$	$\langle 11\bar{2}0 \rangle$	45–81	4
		$\langle \vec{c} + \vec{a} \rangle$	$\{11\bar{2}1\}$	$\langle 11\bar{2}3 \rangle$		5
		$\langle \vec{c} + \vec{a} \rangle$	$\{11\bar{2}2\}$	$\langle 11\bar{2}3 \rangle$		
Li (bcc)	prismatic slip	$\langle \vec{c} + \vec{a} \rangle$	$\{110\}$	$\langle 111 \rangle$	0.54–0.57	12
	pyramidal slip	$\langle \vec{c} + \vec{a} \rangle$	$\{112\}$	$\langle 111 \rangle$	—	12
		$\langle \vec{c} + \vec{a} \rangle$	$\{123\}$	$\langle 111 \rangle$	—	24

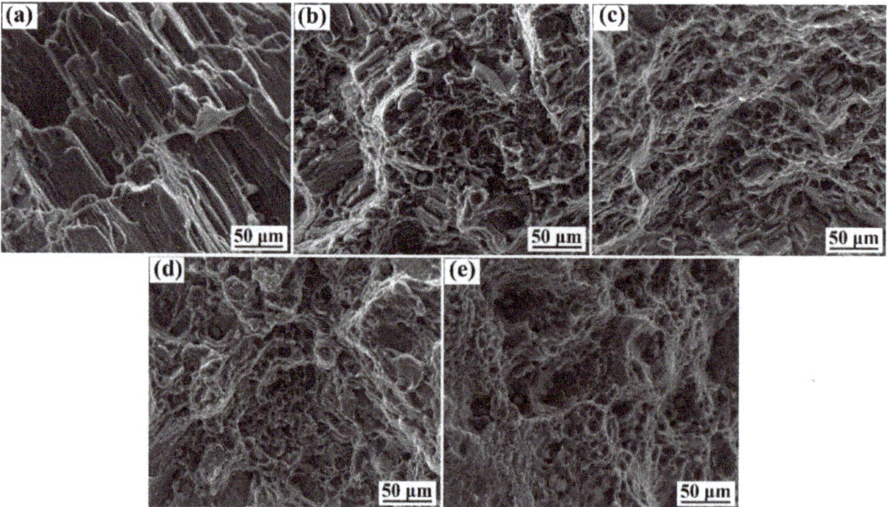

Figure 6. Fracture morphologies of as-homogenized Mg-xLi-3Al-2Zn-0.2Zr alloys: (**a**) $x = 5$; (**b**) $x = 7$; (**c**) $x = 8$; (**d**) $x = 9$; (**e**) $x = 11$.

Figure 7 depicts the side surfaces of the fracture for as-homogenized Mg-xLi-3Al-2Zn-0.2Zr alloys. In as-homogenized LAZ532-0.2Zr and LAZ732-0.2Zr alloys, the microvoids mainly came from the rupture of AlLi particles with larger sizes (marked by a red oval), and large microcracks could even be observed in the α-Mg matrix (marked by a yellow oval), while no microvoids were found around the fibrous AlLi phase and tiny AlLi particles. In the as-homogenized LAZ832-0.2Zr alloy, the microvoids mainly came from the separation of AlLi particles from the matrix, and in addition the separation of the matrix phase interface (marked by a blue oval) due to the AlLi particles was extremely tiny. In as-homogenized LAZ932-0.2Zr and LAZ1132-0.2Zr alloys, microvoids caused by the rupture of AlLi particles and the separation of AlLi particles from the matrix were observed. Since the fracture mechanism of the α-Mg matrix is mainly cleavage fracture, large and straight microcracks could be observed in the α-Mg matrix. These microcracks grew through the grains and continued to expand, and finally showed the cleavage steps. The fracture mechanism of the β-Li matrix was a microvoid coalescence fracture with the basic feature of dimples, and the microvoids were formed by the rupture of the AlLi phase and the separation of the phase interface. According to statistics, AlLi particles with a diameter greater than 1.55 μm break easily during the tensile test, the smaller particles are prone to separate from the matrix, and extremely tiny particles cannot produce microvoids easily. The formation of microvoids at tiny particles is more difficult than at large particles, because the tensile stress required for the formation of microvoids is inversely proportional to the square root of particle size. Therefore, except for the fact that the increase of the β-Li matrix greatly improves the alloy elongation, tiny AlLi particles inhibit the formation of microvoids and are also conducive to the improvement of elongation.

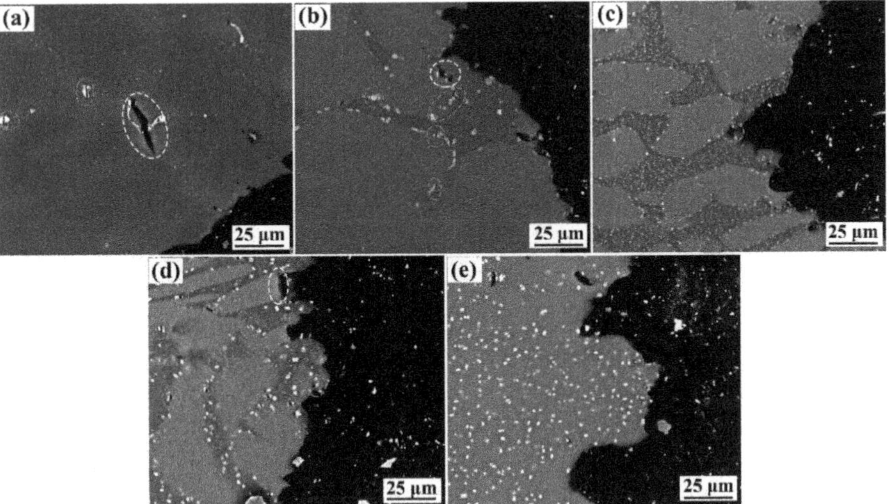

Figure 7. Side surfaces of the fracture for as-homogenized Mg-xLi-3Al-2Zn-0.2Zr alloys: (**a**) x = 5; (**b**) x = 7; (**c**) x = 8; (**d**) x = 9; (**e**) x = 11.

4. Conclusions

(1) When the Li content increases from 5 wt.% to 11 wt.%, the alloy matrix changes from the α-Mg single-phase to the α-Mg+β-Li dual-phase, and then to the β-Li single-phase. All alloys contain the AlLi phase, and its content gradually increases with the increase in Li content;

(2) Part of the Al element in the alloys forms the AlLi phase, and the remainder is mainly dissolved in the α-Mg matrix. Zn element in the alloys does not form compounds; part of Zn is enriched in the AlLi phase, and the rest is dissolved in the α-Mg matrix and β-Li matrix;

(3)   As the Li content increases, the tensile strength and hardness of as-homogenized Mg-*x*Li-3Al-2Zn-0.2Zr alloys decrease while the elongation increases, and the corresponding fracture mode changes from cleavage fracture to microvoid coalescence fracture. This is mainly ascribed to the matrix of alloys changing from α-Mg with an hcp structure to β-Li with a bcc structure.

Author Contributions: Conceptualization, G.S. and Y.S.; methodology, Y.S.; validation, F.Z.; formal analysis, Y.S. and J.R.; investigation, Y.S., F.Z. and Y.S.; resources, J.R.; data curation, F.Z.; writing—original draft preparation, Y.S. and J.R.; writing—review and editing, J.R. and G.S.; supervision, G.S.; funding acquisition, Y.S. and J.R. All authors have read and agreed to the published version of the manuscript.

Funding: The authors are grateful for the financial support from the open project of the Key Laboratory of Green Fabrication and Surface Technology of Advanced Metal Materials (No. GFST2021KF04 and No. GFST2021KF09), the University Natural Science Research Project of Anhui Province (No. KJ2021A0394 and No. KJ2021A0395), and Anhui Provincial Natural Science Foundation (No. 2208085QE124).

Data Availability Statement: The raw data required to reproduce these findings cannot be shared at this time as the data also forms part of an ongoing study.

Conflicts of Interest: The authors declare that they have no known competing financial interest or personal relationships that could have appeared to influence the work reported in this paper.

References

1. Al-Samman, T. Comparative study of the deformation behavior of hexagonal magnesium–lithium alloys and a conventional magnesium AZ31 alloy. *Acta Mater.* **2009**, *57*, 2229–2242. [CrossRef]
2. Xu, W.; Birbilis, N.; Sha, G.; Wang, Y.; Daniels, J.E.; Xiao, Y. A high-specific-strength and corrosion-resistant magnesium alloy. *Nat. Mater.* **2015**, *14*, 1229–1235. [CrossRef] [PubMed]
3. Zhao, Z.; Xing, X.; Ma, J.; Bian, L.; Liang, W.; Wang, Y. Effect of addition of Al-Si eutectic alloy on microstructure and mechanical properties of Mg-12wt% Li alloy. *J. Mater. Sci. Technol.* **2018**, *34*, 1564–1569. [CrossRef]
4. Wu, R.; Yan, Y.; Wang, G.; Murr, L.E.; Han, W.; Zhang, Z.; Zhang, M. Recent progress in magnesium–lithium alloys. *Int. Mater. Rev.* **2014**, *60*, 65–100. [CrossRef]
5. Wu, R.Z.; Qu, Z.K.; Zhang, M.L. Reviews on the influences of alloying elements on the microstructure and mechanical properties of Mg–Li base alloys. *Rev. Adv. Mater. Sci.* **2010**, *24*, 35–43.
6. Jackson, J.H.; Frost, P.D.; Loonam, A.C.; Eastwood, L.W.; Lorig, C.H. Magnesium-lithium base alloys—Preparation, fabrication, and general characteristics. *JOM* **1949**, *1*, 149–168. [CrossRef]
7. Gusieva, K.; Davies, C.H.J.; Scully, J.R.; Birbilis, N. Corrosion of magnesium alloys: The role of alloying. *Int. Mater. Rev.* **2015**, *60*, 169–194. [CrossRef]
8. Cui, C.; Wu, L.; Wu, R.; Zhang, J.; Zhang, M. Influence of yttrium on microstructure and mechanical properties of as-cast Mg–5Li–3Al–2Zn alloy. *J. Alloys Compd.* **2011**, *509*, 9045–9049. [CrossRef]
9. Zhu, T.; Cui, C.; Zhang, T.; Wu, R.; Betsofen, S.; Leng, Z.; Zhang, J.; Zhang, M. Influence of the combined addition of Y and Nd on the microstructure and mechanical properties of Mg–Li alloy. *Mater. Des.* **2014**, *57*, 245–249. [CrossRef]
10. Zhao, J.; Zhang, J.; Liu, W.; Wu, G.; Zhang, L. Effect of Y content on microstructure and mechanical properties of as-cast Mg–8Li–3Al–2Zn alloy with duplex structure. *Mater. Sci. Eng. A* **2016**, *650*, 240–247. [CrossRef]
11. Cao, F.; Xue, G.; Xu, G. Superplasticity of a dual-phase-dominated Mg-Li-Al-Zn-Sr alloy processed by multidirectional forging and rolling. *Mater. Sci. Eng. A* **2017**, *704*, 360–374. [CrossRef]
12. Wang, J.; Wu, R.; Feng, J.; Zhang, J.; Hou, L.; Zhang, M. Influence of rolling strain on electromagnetic shielding property and mechanical properties of dual-phase Mg-9Li alloy. *Mater. Charact.* **2019**, *157*, 109924. [CrossRef]
13. Li, J.; Qu, Z.; Wu, R.; Zhang, M. Effects of Cu addition on the microstructure and hardness of Mg–5Li–3Al–2Zn alloy. *Mater. Sci. Eng. A* **2010**, *527*, 2780–2783. [CrossRef]
14. Sun, Y.; Wang, R.; Peng, C.; Feng, Y. Effects of Sn and Y on the microstructure, texture, and mechanical properties of as-extruded Mg-5Li-3Al-2Zn alloy. *Mater. Sci. Eng. A* **2018**, *733*, 429–439. [CrossRef]
15. Fei, P.; Qu, Z.; Wu, R. Microstructure and hardness of Mg-9Li-6Al-xLa (x = 0, 2, 5) alloys during solid solution treatment. *Mater. Sci. Eng. A* **2015**, *625*, 169–176. [CrossRef]
16. Sun, Y.; Wang, R.; Peng, C.; Wang, X. Microstructure and corrosion behavior of as-homogenized Mg-xLi-3Al-2Zn-0.2 Zr alloys (x = 5, 8, 11 wt%). *Mater. Charact.* **2020**, *159*, 110031. [CrossRef]
17. Tang, R.; Tian, R. *Binary Alloy Phase Diagrams and Crystal Structure of Intermediate Phase*; Central South University Press: Changsha, China, 2009; p. 87. (In Chinese)

18. Zhang, M.; Elkin, F.M. *Magnesium-Lithium Superlight Alloy*; Science Press: Beijing, China, 2010; p. 173. (In Chinese)
19. Zhang, C.; Wu, L.; Zhao, Z.; Xie, Z.; Huang, G. Effect of Li content on microstructure and mechanical property of Mg−xLi−3 (Al−Si) alloys. *Trans. Nonferrous Met. Soc. China* **2019**, *29*, 2506–2513. [CrossRef]
20. Cao, F.; Zhou, B.; Ding, X.; Zhang, J.; Xu, G. Mechanical properties and microstructural evolution in a superlight Mg-7.28 Li-2.19 Al-0.091 Y alloy fabricated by rolling. *J. Alloys Compd.* **2018**, *745*, 436–445. [CrossRef]
21. Koike, J. Enhanced deformation mechanisms by anisotropic plasticity in polycrystalline Mg alloys at room temperature. *Metall. Mater. Trans. A* **2005**, *36*, 1689–1696. [CrossRef]
22. Syed, B.; Geng, J.; Mishra, R.K.; Kumar, K.S. [0 0 0 1] Compression response at room temperature of single-crystal magnesium. *Scr. Mater.* **2012**, *67*, 700–703. [CrossRef]
23. Mushtaq, N.; Butt, M.Z. Rate Process of Yielding in Some Body-Centered Cubic Alkali Metals. *Int. J. Mod. Phys. B* **2010**, *24*, 4233–4242. [CrossRef]
24. Russell, A.M.; Chumbley, L.S.; Gantovnik, V.B.; Xu, K.; Tian, Y.; Laabs, F.C. Anomalously high impact fracture toughness in BCC Mg-Li between 4.2K and 77K. *Scr. Mater.* **1998**, *39*, 1663–1667. [CrossRef]
25. Matsuda, M.; Ii, S.; Kawamura, Y.; Ikuhara, Y.; Nishida, M. Interaction between long period stacking order phase and deformation twin in rapidly solidified Mg97Zn1Y2 alloy. *Mater. Sci. Eng. A* **2004**, *386*, 447–452. [CrossRef]

aerospace

Article

Experimental Evaluation of Fatigue Strength of AlSi10Mg Lattice Structures Fabricated by AM

Carlo Giovanni Ferro *, Sara Varetti and Paolo Maggiore

Department of Mechanical Engineering and Aerospace (DIMEAS), Polytechnic University of Turin, 10129 Turin, Italy
* Correspondence: carlo.ferro@polito.it; Tel.: +39-011-090-6850

Abstract: There is evidence that Additive Manufacturing (AM) plays a crucial role in the fourth industrial revolution. The design freedom provided by this technology is disrupting limits and rules from the past, enabling engineers to produce new products that are otherwise unfeasible. Recent developments in the field of Selective Laser Melting (SLM) have led to a renewed interest in lattice structures that can be produced non-stochastically in previously unfeasible dimensional scales. One of the primary applications is aerospace engineering where the need for light weights and performance is urgent to reduce the carbon footprint of civil transport around the globe. Of particular concern is fatigue strength. Being able to predict fatigue life in both LCF (Low Cycle Fatigue) and HCF (High Cycle Fatigue) is crucial for a safe and reliable design in aerospace systems and structures. In the present work, an experimental evaluation of compressive–compressive fatigue behavior has been performed to evaluate the fatigue curves of different cells, varying sizes and relative densities. A Design of Experiment (DOE) approach has been adopted in order to maximize the information extractable in a reliable form.

Keywords: additive manufacturing; lattice structures; fatigue strength; trabecular structures

1. Introduction

Additive Manufacturing (AM) has emerged as a game-changer in the field of aerospace engineering, providing new opportunities for the design and manufacture of complex structures and systems with improved thermo-mechanical properties [1–5]. Currently, one of the key challenges in aerospace engineering is the development of low power effective anti-ice systems, which are essential for ensuring safe and efficient operations in cold and icy conditions [6–8]. State-of-the-art anti-icing systems use hot air spilled from the turbine engine compressors [9]. These systems present low thermal efficiencies and high losses [10–13]. In recent years, lattice structures made of AlSi10Mg alloy have emerged as a promising candidate for the development of innovative anti-ice systems [14–16]. Lattice structures indeed present interesting thermal and mechanical properties for heat and structure components making them the ideal candidate for a novel anti-icing material to integrate into the leading edge structure (Figure 1). The emerging benefit could be a sensible reduction of the air spilled and therefore, a reduction of the fuel consumption. Unfortunately, lattice structures in the core of the sandwich panel are subjected to cyclic loads. Being able to predict the fatigue strength under cyclic loading is currently one of the greatest challenges [17,18]. A search of the literature revealed a few studies, which experimentally evaluated the fatigue curves of lattice structures made of AlSi10Mg fabricated by AM [19,20]. In order to fill the gap between the requirements for a safe design of a critical component and the available literature, an extensive analysis campaign has been developed. This paper presents results aimed at investigating the fatigue strength of lattice structures made of AlSi10Mg alloy fabricated by AM in order to provides insights into the design of innovative systems for aerospace applications [21].

Citation: Ferro, C.G.; Varetti, S.; Maggiore, P. Experimental Evaluation of Fatigue Strength of AlSi10Mg Lattice Structures Fabricated by AM. *Aerospace* **2023**, *10*, 400. https://doi.org/10.3390/aerospace10050400

Academic Editor: Andrea Di Schino

Received: 8 March 2023
Revised: 28 March 2023
Accepted: 21 April 2023
Published: 25 April 2023

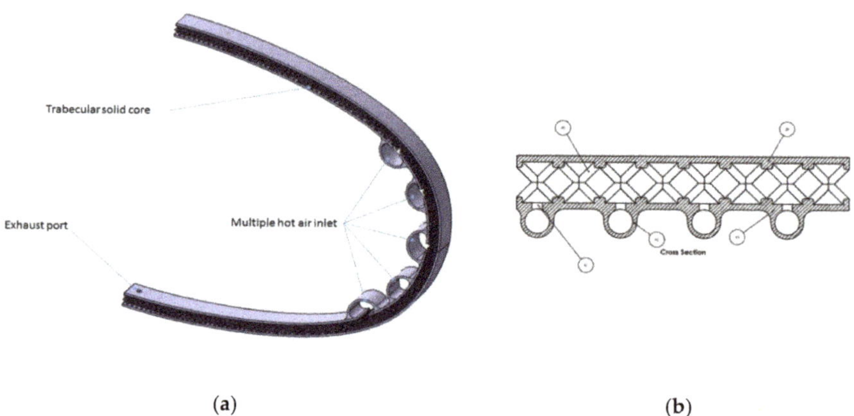

(a)   (b)

Figure 1. Integrated Anti-Ice Panel. (**a**): Isometrical View of the proposed novel anti-ice panel (**b**): Cross Section of the patented solutions.

Fatigue strength is a crucial parameter in the design of aerospace components, especially those that are subjected to cyclic loading. Lattice structures, also known as cellular or honeycomb structures, are highly desirable due to their lightweight nature and high stiffness-to-weight ratio [22–24]. They have been used in various applications, including heat exchangers, sandwich panels, and energy absorption systems [25–28]. AlSi10Mg is an aluminum alloy that has been extensively used in AM due to its excellent mechanical properties and good weldability [29]. In the case of anti-ice systems, lattice structures made of AlSi10Mg have been proposed as a potential solution due to their ability to provide excellent anti-ice performance by allowing a continuous flow of hot air to pass through the structure, thus melting the ice on its surface [21,30]. However, the fatigue strength of these structures is not well understood, which poses a significant challenge in their design and optimization. The experimental study presented in this paper involved the testing of AlSi10Mg lattices fabricated by AM under cyclic loading to determine their fatigue limits. The results of this study will provide new insights into the behavior of these structures under cyclic loading and will help in understanding their failure mechanisms. The findings will also aid in the development of design guidelines for lattice structures and will enable the optimization of innovative anti-ice systems for aerospace applications [31]. In summary, the experimental evaluation of the fatigue strength of lattice structures is of strong significant interest to the aerospace industry.

2. Materials and Methods

2.1. Design of Experiment (DOE)

The fatigue tests on trabecular lattices had the primary objective to experimentally characterize the fatigue life curve of different types of non-stochastic foams built with AlSi10Mg by AM. The data for this study were collected using a DOE approach [32] in order to provide novel information regarding the effects of design parameters (cell type, cell size and relative density) on lattice fatigue performance [33]. Although some research has been carried out on fatigue with this approach [33], none of the previous studies have reported such a wide investigation for AlSi10Mg.

Three different trusses layout have been selected: Body-Centered Cubic with vertical beams (Bccz), Octet and Rhombic Dodecahedron. For each cell type, two cells sizes, 5 mm and 7 mm, were adopted with two relative densities: 25% and 30%. In order to limit the number of specimens and the cost of the experimental campaign, the DOE with three factors and hybrid levels for each factor was executed in fractional factorial form. The details of the specimens produced for the fatigue test are reported in Table 1.

Table 1. Factors selected to design specimens for fatigue tests.

Cell Type	Cell Size [mm]	Relative Density [%]
Bccz	5	25
Bccz	5	30
Bccz	7	30
Rhombic Dodecahedron	5	25
Rhombic Dodecahedron	5	30
Rhombic Dodecahedron	7	30
Octet Truss	5	25
Octet Truss	5	30
Octet Truss	7	30

2.2. Specimen Design

The trabecular specimens presented a height/base side ratio of 2:1, similar to the one proposed previously [7]. Each specimen had a square bottom and a longer height to prevent the effect of borders and specimen shape. A picture of Bccz is shown in Figure 2 while detailed drawings with dimensions are reported in Appendix C. All specimens have two flat skins 1 mm thick at the square bases. This skin has been added in order to uniformly distribute the stress on first layer cells and to avoid peaks caused by manufacturing imperfections.

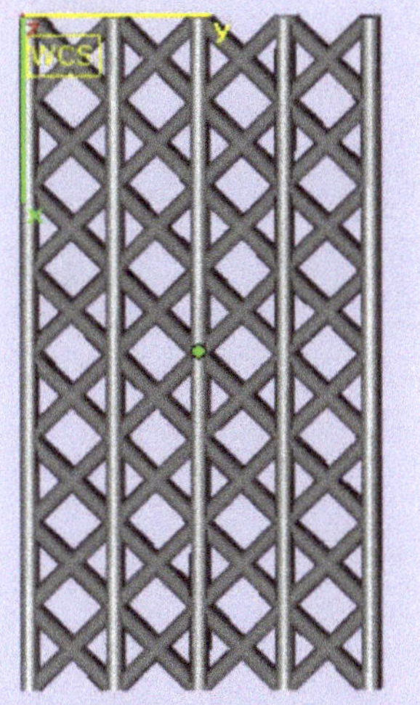

Figure 2. CAD Design of a Specimen.

2.3. Fatigue Test Setup

Compression–compression fatigue tests were carried out according to protocols in the scientific literature [34–37].

A variable maximum load of 80%, 60%, 40% and 20% of the static yield (σ_{02} calculated according to ASTM E9) was imposed in order to obtain the Wohler curves. The ratio parameter of *R = 0.1* implies that the compressive load on the specimens oscillates with a sinusoidal from 10% to 100% of the imposed load with a frequency speed of 50 cyclic loads per second (50 Hz).

The maximum value of the compression is defined as σ_M and can be obtained from Equation (1); the parameter x corresponds to a linear multiplying value of 80%, 60%, 40% and 20%; σ_{02} is the average yield load of the static uniaxial compression tests on the corresponding trabecular specimens. This value has been separately calculated by previous experimental static compressive test on analogous specimens [38].

The minimum value of the compression is reported as σ_m, obtained from Equation (2), where the R parameter is maintained equal to 0.1. The median value between σ_M and σ_m is σ_a, calculated using Equation (3). In order to evaluate the Wohler curve, it is then necessary to record the Number of cycles (N) corresponding to the specimen failure. The imposed upper limit for the cycles is N equal to 1.5×10^7, suitable for airframe structures and the systems safe-life approach [39].

$$\sigma_M = x\sigma_{02} \tag{1}$$

$$\sigma_m = R\sigma_M \tag{2}$$

$$\sigma_a = (\sigma_M - \sigma_m)/2 \tag{3}$$

The values of σ_M, σ_m, σ_a and N are tabulated in Appendix A. All the specimens, three for each type, were processed with an Instron machine. The test machine setup has been prepared with a flat attachment, a load cell of 50 kN and a constant displacement of 1 mm/min, as shown in Figure 3.

Figure 3. Testing Machine Setup.

The same AlSi10Mg material as [7] was used. The specimen's size was maintained as much as possible to be coherent to the 20 mm × 20 mm × 40 mm proposed in [7] in order to preserve the comparability of the data.

3. Results

In this section, the results obtained from the fatigue tests will be disclosed. All the numerical evidence are reported in Appendix A subdivided by cell type. For each cell type, Wohler curves will be provided and discussed. The Wohler curves relate the amplitude of the load, σ_a, for each sample and the number of cycles at which the specimen breakdown occurs. Subsequently, images of broken specimens will be shown to provide insights on the failure mechanism.

3.1. Rhombic Dodecahedron

The graph in Figure 4. reports the Wohler curves for all the rhombic specimens tested. At equal σ_a (load amplitude), the greatest fatigue life was that of specimens with a smaller cell size, 5 mm, and higher density, 30%. The second longest endurance was produced by cells with a 7 mm cell size and 30% relative density. The lowest fatigue life was witnessed from specimens with a 5 mm cell size and 25% relative density. The most relevant aspect of this graph was the effect of cell size. Increasing cell size, while maintaining fixed the relative density of the specimen (i.e., same amount material per cubic centimeter), noticeably reduced the fatigue life as reported by the comparison between 7_30 and 5_30 cells.

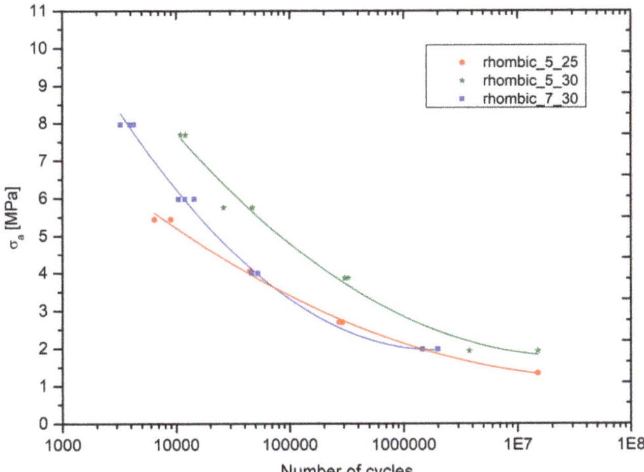

Figure 4. Wohler curves for rhombic dodecahedron specimens.

The effect of relative density was less stable in all the tested stress amplitudes. As reported in Appendix B, in the main effect graph in Figure A1, the effect of the relative density was not monotonous. While at higher loads the increase of the relative density seemed to provide a higher fatigue life, this was not true for lower loads (60% and 20% of the yield stress). This aspect will require further dedicated analysis in the future.

The photos of the rhombic specimens before the fatigue test are shown in Figures 5 and 6. while pictures of specimens after the test are shown in Figure 6.

The failure occurred along a 45° plane with respect to the Z-axis and started from a corner of the specimen. The deformation evidenced before the failure was almost absent; the connections broke internally one by one along the plane until the resistant section became insufficient and caused the collapse of the specimen. From the images, it was possible to observe that the breaking of the specimens subjected to higher loads (80% and 60% of σ_{02}) was clearer and more similar to the one from the static compression tests. On the other hand, the rupture of the specimens subjected to lower loads (40% and 20% of σ_{02}) was more irregular and acted through several fractures' planes simultaneously, causing the final separation of the specimen into more than two pieces. This difference was particularly marked for the 7 mm cells, as visible in Figure 6. For cells with a lower relative density, such as 5-25, irregular fractures were already present for higher loads, such as 60% of σ_{02}.

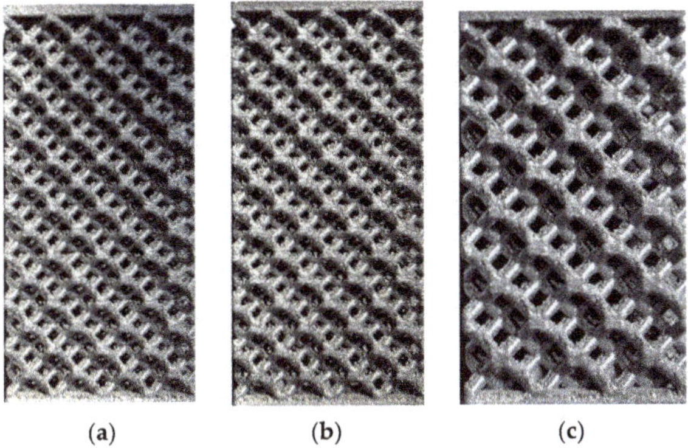

Figure 5. Lateral view of the Rhombic dodecahedron AlSi10Mg specimens for fatigue test: (**a**) Rhom-5-25, (**b**) Rhom-5-30 and (**c**) Rhom-7-30.

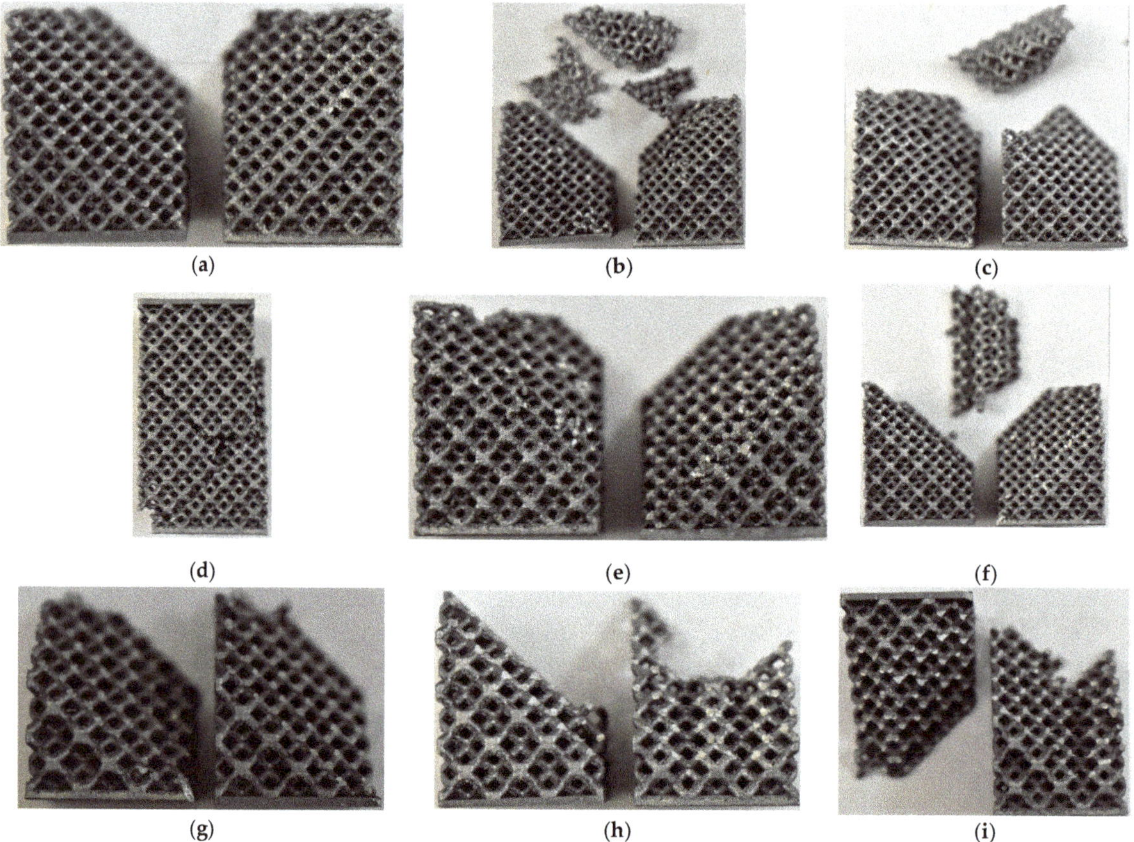

Figure 6. Rhombic cells subjected to fatigue test: (**a**) cell size 5 mm, relative density 25%, max load 80% σ_{02}; (**b**) cell size 5 mm, relative density 25%, max load 60% σ_{02}; (**c**) cell size 5 mm, relative density 25%, max load 40% σ_{02}; (**d**) cell size 5 mm, relative density 30%, max load 80% σ_{02}; (**e**) cell size, 5 mm relative density 30%, max load 60% σ_{02}; (**f**) cell size 5 mm, relative density 30%, max load 40% σ_{02}; (**g**) cell size 7 mm, relative density 30%, max load 80% σ_{02}; (**h**) cell size 7 mm, relative density 30%, max load 60% σ_{02}; (**i**) cell size 7 mm, relative density 30%, max load 40% σ_{02}.

3.2. Octet Truss

The graph in Figure 7 reports the Wohler curves for all the tested Octet truss specimens. In accordance with the present results, the longest fatigue life was shown by the 5-30 specimens, followed by the specimens 7-30 and 5-25 with intersecting behaviors.

Similar to the case with Rhombic cells, Octet truss curves for 5-30 and 7-30 were almost parallel to each other, with the 7-30 case shifted to lower cycles. This behavior suggests a clear negative effect of increasing of the cell size with same relative density.

As for the 7-30 vs. 5-25 trend, relative density played a beneficial effect at higher loads (80% and 60% of σ_{02}) and reported a dissimilar trend at lower loads (40% and 20% of yield stress). Additionally, in this scenario, further analyses are required to deeply investigate the effect of this parameter. It is important to note that specimens 5-30 and 5-25 reached the imposed fatigue limit (1.5×10^{7} cycles) without breaking with a σ_{M} equal to the 20% of yield stress.

The photos of the Octet specimens before the fatigue test are reported in Figure 8 while pictures of specimens after the test are in Figure 9.

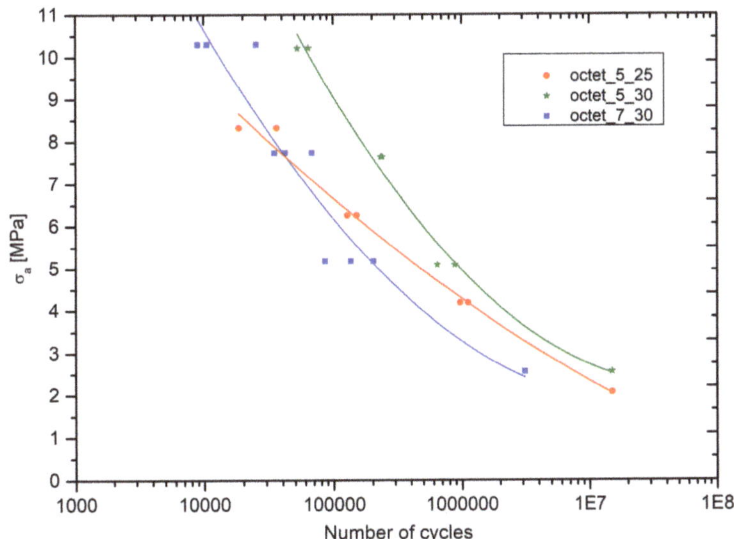

Figure 7. Wohler curves for Octet truss specimens.

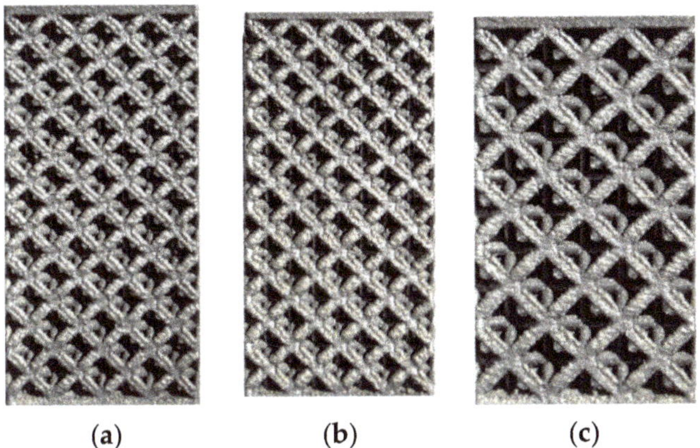

Figure 8. Lateral view of the Octet truss AlSi10Mg specimens for fatigue test: (**a**) Octet truss 5-25, (**b**) Octet truss 5-30 and (**c**) Octet truss 7-30.

The fracture of the specimen was, in part, similar to that evidenced during compression tests. The failure occurred along a 45° plane with respect to the Z-axis and started from a corner of the specimen. Deformation before the failure was almost absent; the connections broke internally one by one along the plane until the resistant section became insufficient and caused the collapse of the specimen. The failure mode of the 5-25 specimens subjected to all loads was more irregular compared to the others and acted simultaneously on several planes causing the final separation of the specimen into more than two pieces. Similarly, the 7-30 specimens subjected to higher loads (80% and 60% of σ_{02}) also failed in a similar manner to that seen for the compression tests; on the other hand, the failure mode of the same cells when subjected to lower loads (40% and 20% of σ_{02}) was more irregular and acted on several planes simultaneously, causing final separation of the specimen into more than two pieces.

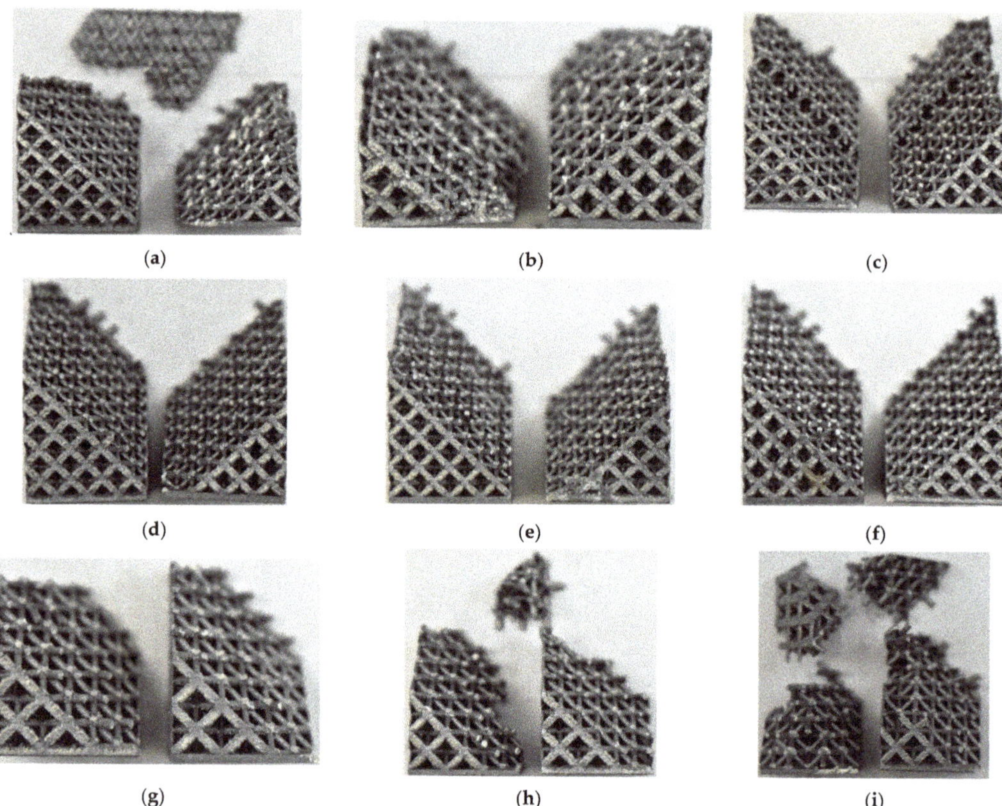

Figure 9. Octet cells subjected to fatigue: (**a**) cell size 5 mm, relative density 25%, max load 80% σ_{02}; (**b**) cell size 5 mm, relative density 25%, max load 60% σ_{02}; (**c**) cell size 5 mm, relative density 25%, max load 40% σ_{02}; (**d**) cell size 5 mm, relative density 30%, max load 80% σ_{02}; (**e**) cell size 5 mm, relative density 30%, max load 60% σ_{02}; (**f**) cell size 5 mm, relative density 30%, max load 40% σ_{02}; (**g**) cell size 7 mm, relative density 30%, max load 80% σ_{02}; (**h**) cell size 7 mm, relative density 30%, max load 60% σ_{02}; (**i**) cell size 7 mm, relative density 30%, max load 40% σ_{02}.

3.3. Bccz

The fatigue Wohler curves for the Bccz specimens, presented in Figure 10, remained partially completed due to the different resistance behaviors found. The specimens already reached the preset limit of 1.5×10^7 cycles at 60% of the load for the 5 mm cells and 40% for the 7 mm cells; for this reason, it was impossible to obtain complete Wohler curves. Further improvements will enhance the span of loads tested by introducing a setup to produce between 80% and 100% of the yield load. The greatest fatigue resistance of the Bccz specimens was given by the vertical struts. During the static compression tests, the vertical struts deformed and then broke with buckling-like deformation. The photos of the Bccz specimens before the fatigue test are shown Figure 11 while pictures of specimens after the test are shown in Figure 12.

In the fatigue tests, the imposed load was not sufficient to reach the buckling limit for the vertical struts. As visible in Figure 12, the failure of the specimen, when it occurs, was due to the breaking of the struts at the central nodes, as seen for Rhombic and Octet cells. The fracture then followed the plane at 45° only in some points, but in most cases, there was a sparse and irregular separation of the cells. Further analysis will be dedicated to a higher σ_M.

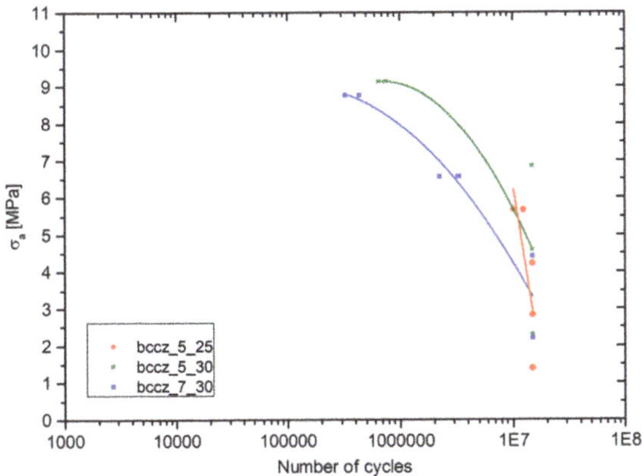

Figure 10. Wohler curves for Bccz specimens.

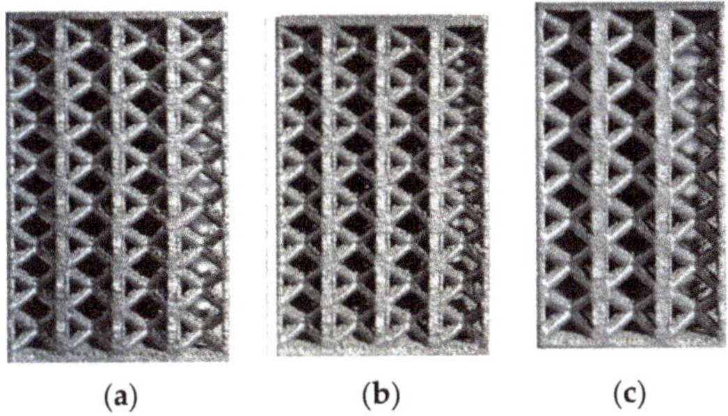

Figure 11. Lateral view of the Bccz AlSi10Mg specimens for fatigue test: (**a**) Bccz 5-25, (**b**) Bccz 5-30 and (**c**) Bccz 7-30.

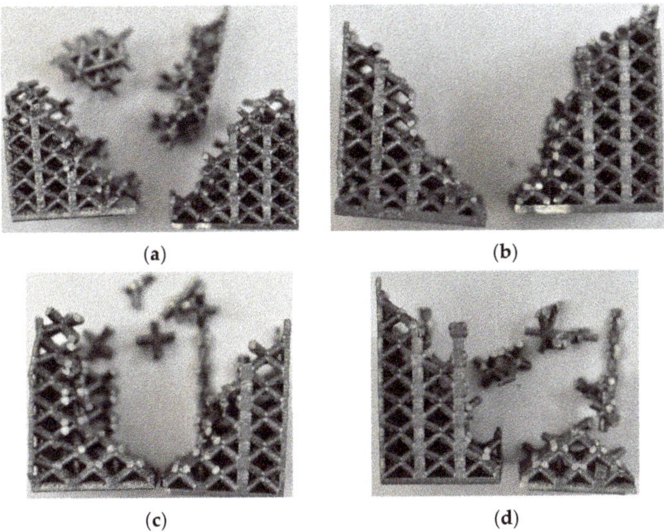

Figure 12. Bccz cells subjected to fatigue: (**a**) cell size 5 mm, relative density 25%, max load 80% σ_{02}; (**b**) cell size 5 mm, relative density 30%, max load 80% σ_{02}; (**c**) cell size 7 mm, relative density 30%, max load 80% σ_{02}; (**d**) cell size 7 mm, relative density 30%, max load 60% σ_{02}.

4. Discussion

Several reports have shown that different cell topologies provide different fatigue life behaviors [40,41]. This section will provide a comparison of the three architected cells subjected to the same load path.

What is striking about the graphs reported in Figure 13 is the homogeneous trend. Bccz cells, as commented above, always performed better, followed by Octet cells and Rhombic cells. The last two cells' topologies presented similar curves, which showed different trends than that of the Bccz cells. A possible explanation for this might be that Rhombic and Octet cells are bending-dominated cells [42] while Bccz cells are stretch-dominated cells that benefit enormously when the compression load on the vertical cells does not reach the buckling limit.

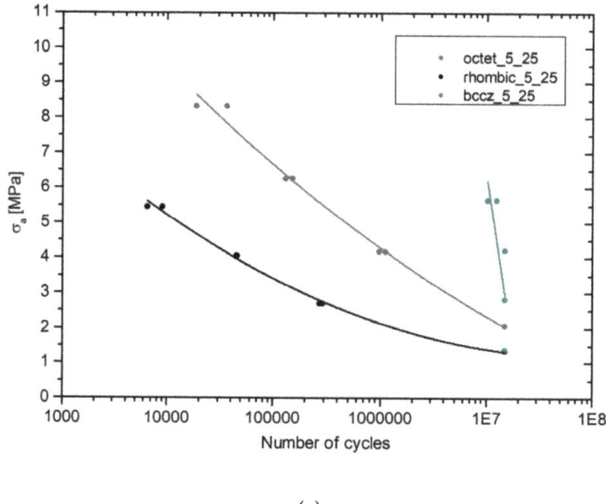

(a)

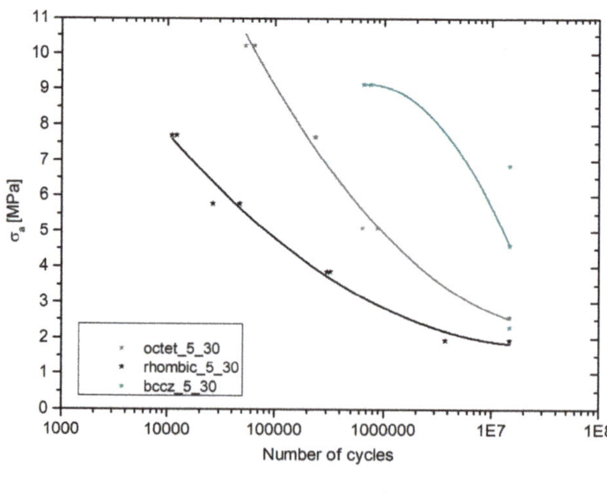

(b)

Figure 13. *Cont.*

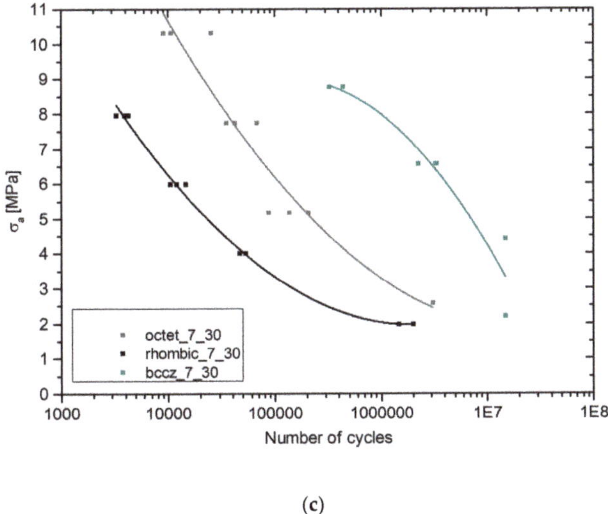

(**c**)

Figure 13. Wohler curves: (**a**) cell size 5 mm and 25% relative density; (**b**) cell size 5 mm and 30% relative density; (**c**) cell size 7 mm and 30% relative density.

The results provided are significant in at least two major respects. Firstly, they provide reliable encouraging data on the fatigue life of different cells with various cell sizes and different relative densities. Secondly, they establish some useful correlations between design parameters and fatigue life.

Despite these promising results, further work is still required to establish the real fatigue limit of Bccz cells and deeply evaluate the effect of relative density on the fatigue performance for bending-dominated cells type.

5. Conclusions

Wohler fatigue curves for different types of trabecular structures have been evaluated for the first time. This study has identified clear trends and correlations for the cell type and cell size effects on the number of cycles at rupture during compression–compression fatigue. Further investigations are needed to deepen our understanding of some nonlinear phenomena on the effect of the relative density. Overall, this study has strengthened the idea that lattice structures are a valid resource for aerospace structures due to their light weight and their high specific structural performances. In spite of its limitations, this study certainly adds important novel findings to aid our understanding of fatigue life prediction.

6. Patents

In this paper, research on the fatigue life of trabecular structures was performed. This work is part of effort towards a patent for a novel anti-icing system for aircraft use [8].

Author Contributions: Conceptualization, C.G.F.; methodology, C.G.F.; software, C.G.F. and S.V.; validation, C.G.F. and S.V.; formal analysis, C.G.F. and S.V.; investigation, C.G.F. and S.V.; resources, P.M.; data curation, S.V.; writing—original draft preparation, C.G.F.; writing—review and editing, S.V.; visualization, P.M.; supervision, P.M.; project administration, P.M.; funding acquisition, P.M. All authors have read and agreed to the published version of the manuscript.

Funding: This research received no external funding.

Data Availability Statement: Data are already all disclosed in Appendices A–C.

Conflicts of Interest: The authors declare no conflict of interest.

Appendix A

AlSi10Mg specimens for fatigue test with Bccz cell. The identification code of the specimens is: Celle shape-Cell size-Solid volume fraction (nominal)-Serial number.

Table A1. BCC-Z Fatigue Results.

Id Code	σ_M [MPa]	σ_m [MPa]	σ_a [MPa]	N
Bccz-5-25-1	12.6	1.26	5.67	10,178,091
Bccz-5-25-2	12.6	1.26	5.67	12,441,038
Bccz-5-25-3	12.6	1.26	5.67	-
Bccz-5-25-4	9.4	0.94	4.23	>15,000,000
Bccz-5-25-5	9.4	0.94	4.23	>15,000,000
Bccz-5-25-6	9.4	0.94	4.23	-
Bccz-5-25-7	6.3	0.63	2.84	>15,000,000
Bccz-5-25-8	6.3	0.63	2.84	>15,000,000
Bccz-5-25-9	6.3	0.63	2.84	-
Bccz-5-25-10	3.1	0.31	1.40	>15,000,000
Bccz-5-25-11	3.1	0.31	1.40	>15,000,000
Bccz-5-25-12	3.1	0.31	1.40	-
Bccz-5-30-1	20.3	2.03	9.14	654,563
Bccz-5-30-2	20.3	2.03	9.14	762,810
Bccz-5-30-3	20.3	2.03	9.14	-
Bccz-5-30-4	15.2	1.52	6.84	>15,000,000
Bccz-5-30-5	15.2	1.52	6.84	>15,000,000
Bccz-5-30-6	15.2	1.52	6.84	-
Bccz-5-30-7	10.2	1.02	4.59	>15,000,000
Bccz-5-30-8	10.2	1.02	4.59	>15,000,000
Bccz-5-30-9	10.2	1.02	4.59	-
Bccz-5-30-10	5.1	0.51	2.30	>15,000,000
Bccz-5-30-11	5.1	0.51	2.30	>15,000,000
Bccz-5-30-12	5.1	0.51	2.30	-
Bccz-7-30-1	19.5	1.95	8.78	326,837
Bccz-7-30-2	19.5	1.95	8.78	440,166
Bccz-7-30-3	19.5	1.95	8.78	326,161
Bccz-7-30-4	14.6	1.46	6.57	3,362,847
Bccz-7-30-5	14.6	1.46	6.57	3,331,678
Bccz-7-30-6	14.6	1.46	6.57	2,249,586
Bccz-7-30-7	9.8	0.98	4.41	>15,000,000
Bccz-7-30-8	9.8	0.98	4.41	>15,000,000
Bccz-7-30-9	9.8	0.98	4.41	>15,000,000
Bccz-7-30-10	4.9	0.49	2.21	>15,000,000
Bccz-7-30-11	4.9	0.49	2.21	>15,000,000
Bccz-7-30-12	4.9	0.49	2.21	>15,000,000

AlSi10Mg specimens for fatigue test with Rhombic dodecahedron cell. The identification code of the specimens is: Cell shape-Cell size-Solid volume fraction (nominal)-Serial number.

Table A2. Rhombic Dodecahedron Fatigue Results.

Id Code	σ_M [MPa]	σ_m [MPa]	σ_a [MPa]	N
Rhom-5-25-1	12.1	1.21	5.45	9044
Rhom-5-25-2	12.1	1.21	5.45	6480
Rhom-5-25-3	12.1	1.21	5.45	-
Rhom-5-25-4	9	0.9	4.05	45,544
Rhom-5-25-5	9	0.9	4.05	45,031
Rhom-5-25-6	9	0.9	4.05	-
Rhom-5-25-7	6	0.6	2.70	291,522
Rhom-5-25-8	6	0.6	2.70	274,062
Rhom-5-25-9	6	0.6	2.70	-
Rhom-5-25-10	3	0.3	1.35	>15,000,000
Rhom-5-25-11	3	0.3	1.35	>15,000,000
Rhom-5-25-12	3	0.3	1.35	-
Rhom-5-30-1	17.1	1.71	7.70	12,100

Table A2. *Cont.*

Id Code	σ_M [MPa]	σ_m [MPa]	σ_a [MPa]	N
Rhom-5-30-2	17.1	1.71	7.70	10,947
Rhom-5-30-3	17.1	1.71	7.70	-
Rhom-5-30-4	12.8	1.28	5.76	47,132
Rhom-5-30-5	12.8	1.28	5.76	26,329
Rhom-5-30-6	12.8	1.28	5.76	-
Rhom-5-30-7	8.6	0.86	3.87	327,084
Rhom-5-30-8	8.6	0.86	3.87	307,454
Rhom-5-30-9	8.6	0.86	3.87	-
Rhom-5-30-10	4.3	0.43	1.94	>15,000,000
Rhom-5-30-11	4.3	0.43	1.94	3,755,825
Rhom-5-30-12	4.3	0.43	1.94	-
Rhom-7-30-1	17.7	1.77	7.97	4281
Rhom-7-30-2	17.7	1.77	7.97	3281
Rhom-7-30-3	17.7	1.77	7.97	3970
Rhom-7-30-4	13.3	1.33	5.99	14,484
Rhom-7-30-5	13.3	1.33	5.99	12,010
Rhom-7-30-6	13.3	1.33	5.99	10,550
Rhom-7-30-7	8.9	0.89	4.01	53,097
Rhom-7-30-8	8.9	0.89	4.01	46,449
Rhom-7-30-9	8.9	0.89	4.01	46,723
Rhom-7-30-10	4.4	0.44	1.98	1,471,143
Rhom-7-30-11	4.4	0.44	1.98	2,004,674
Rhom-7-30-12	4.4	0.44	1.98	-

AlSi10Mg specimens for fatigue test with Octet truss cell. The identification code of the specimens is: Cell shape-Cell size-Solid volume fraction (nominal)-Serial number.

Table A3. Octet Truss Fatigue Results.

Id Code	σ_M [MPa]	σ_m [MPa]	σ_a [MPa]	N
Oct-5-25-1	18.5	1.85	8.33	18,568
Oct-5-25-2	18.5	1.85	8.33	36,394
Oct-5-25-3	18.5	1.85	8.33	-
Oct-5-25-4	13.9	1.39	6.26	128,641
Oct-5-25-5	13.9	1.39	6.26	151,236
Oct-5-25-6	13.9	1.39	6.26	-
Oct-5-25-7	9.3	0.93	4.19	971,841
Oct-5-25-8	9.3	0.93	4.19	1,119,612
Oct-5-25-9	9.3	0.93	4.19	-
Oct-5-25-10	4.6	0.46	2.07	>15,000,000
Oct-5-25-11	4.6	0.46	2.07	15,000,000
Oct-5-25-12	4.6	0.46	2.07	-
Oct-5-30-1	22.7	2.27	10.22	64,099
Oct-5-30-2	22.7	2.27	10.22	52,684
Oct-5-30-3	22.7	2.27	10.22	-
Oct-5-30-4	17	1.7	7.65	239,878
Oct-5-30-5	17	1.7	7.65	234,151
Oct-5-30-6	17	1.7	7.65	-
Oct-5-30-7	11.3	1.13	5.09	880,174
Oct-5-30-8	11.3	1.13	5.09	645,908
Oct-5-30-9	11.3	1.13	5.09	-
Oct-5-30-10	5.7	0.57	2.57	>15,000,000
Oct-5-30-11	5.7	0.57	2.57	>15,000,000
Oct-5-30-12	5.7	0.57	2.57	-
Oct-7-30-1	22.9	2.29	10.31	10,607
Oct-7-30-2	22.9	2.29	10.31	9009
Oct-7-30-3	22.9	2.29	10.31	25,343
Oct-7-30-4	17.2	1.72	7.74	42,201
Oct-7-30-5	17.2	1.72	7.74	68,178
Oct-7-30-6	17.2	1.72	7.74	35,084
Oct-7-30-7	11.5	1.15	5.18	136,536
Oct-7-30-8	11.5	1.15	5.18	86,464
Oct-7-30-9	11.5	1.15	5.18	204,722
Oct-7-30-10	5.7	0.57	2.57	3,108,586
Oct-7-30-11	5.7	0.57	2.57	-
Oct-7-30-12	5.7	0.57	2.57	-

Appendix B

This section reports the quality analysis of the outcomes provided by the fatigue experiments.

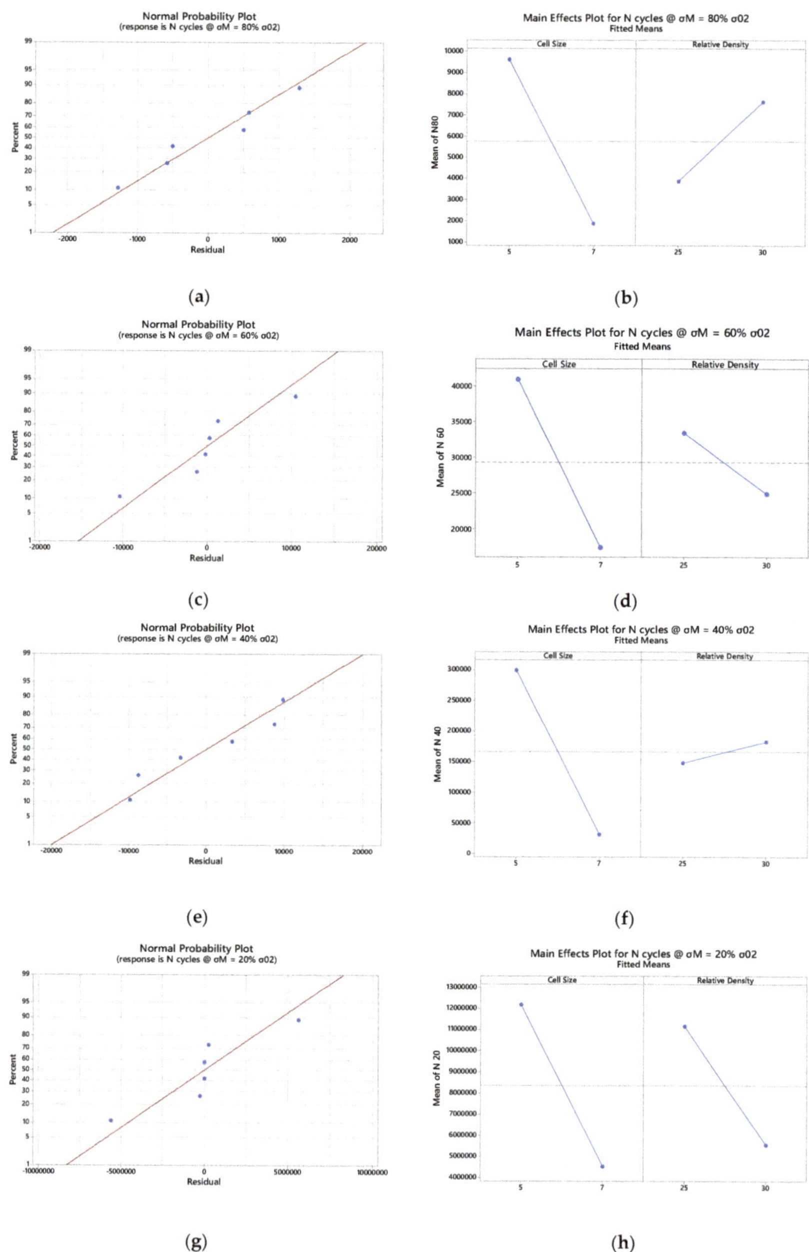

Figure A1. Rhombic Dodecahedron Cells Fatigue Outcome Statistical Analysis: (**a**) normal probability plot, max load 80% σ_{02}; (**b**) main effect plot, max load 80% σ_{02}; (**c**) normal probability plot, max load 60% σ_{02}; (**d**) main effect plot, max load 60% σ_{02}; (**e**) normal probability plot, max load 40% σ_{02}; (**f**) main effect plot, max load 40% σ_{02}; (**g**) normal probability plot, max load 20% σ_{02}; (**h**) main effect plot, max load 20% σ_{02}.

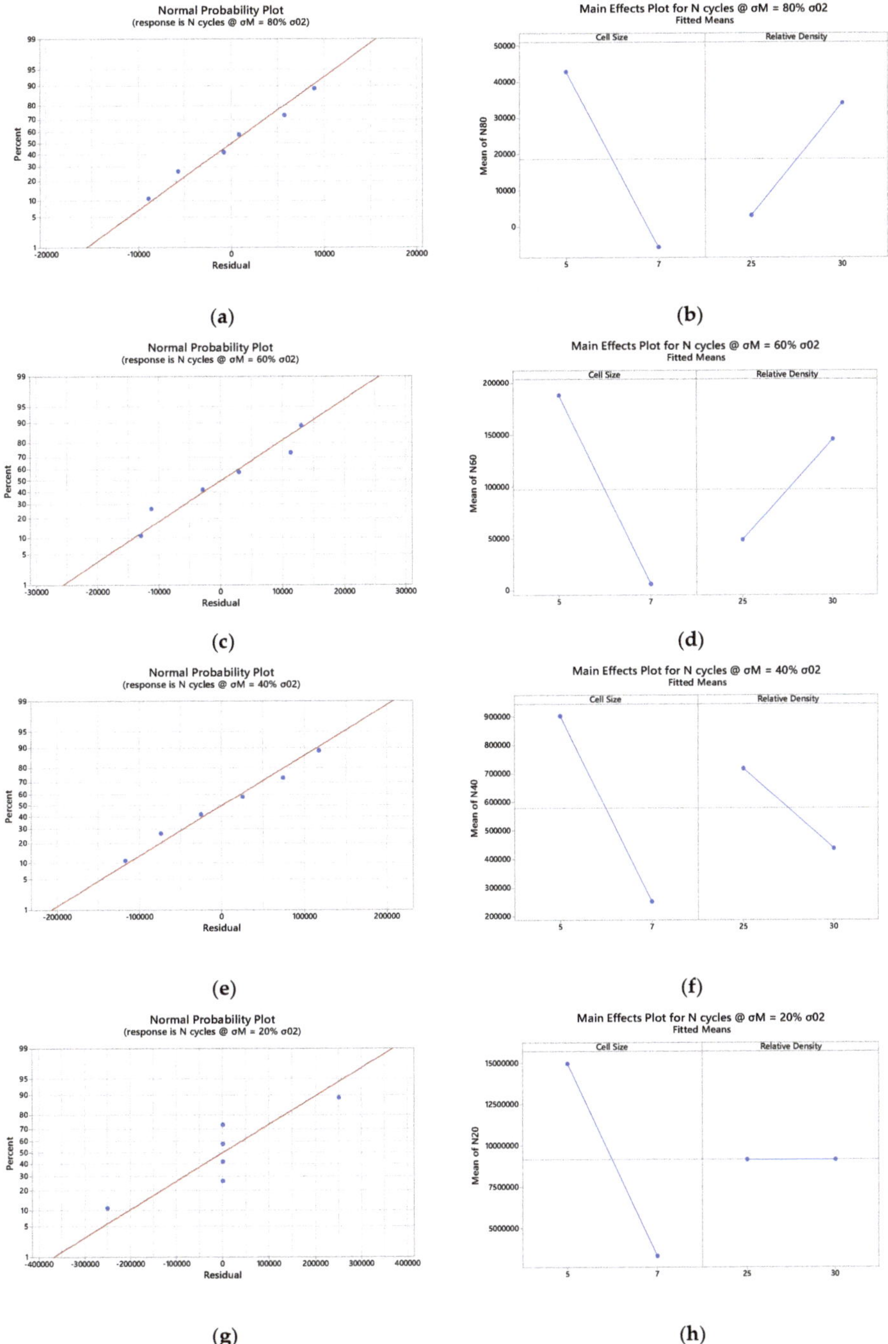

Figure A2. Octet Truss Cell Fatigue Outcome Statistical Analysis: (**a**) normal probability plot, max load 80% σ_{02}; (**b**) main effect plot, max load 80% σ_{02}; (**c**) normal probability plot, max load 60% σ_{02}; (**d**) main effect plot, max load 60% σ_{02}; (**e**) normal probability plot, max load 40% σ_{02}; (**f**) main effect plot, max load 40% σ_{02}; (**g**) normal probability plot, max load 20% σ_{02}; (**h**) main effect plot, max load 20% σ_{02}.

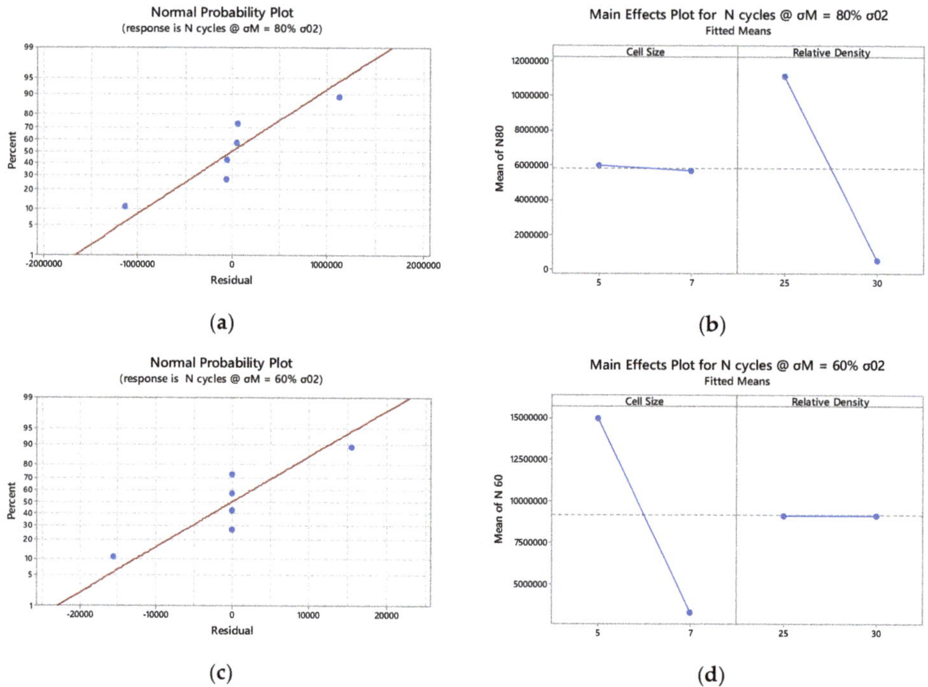

(a)

(b)

(c)

(d)

Figure A3. Bccz Truss Cells Fatigue Outcome Statistical Analysis: (**a**) normal probability plot, max load 80% σ_{02}; (**b**) main effect plot, max load 80% σ_{02}; (**c**) normal probability plot, max load 60% σ_{02}; (**d**) main effect plot, max load 60% σ_{02}.

Appendix C

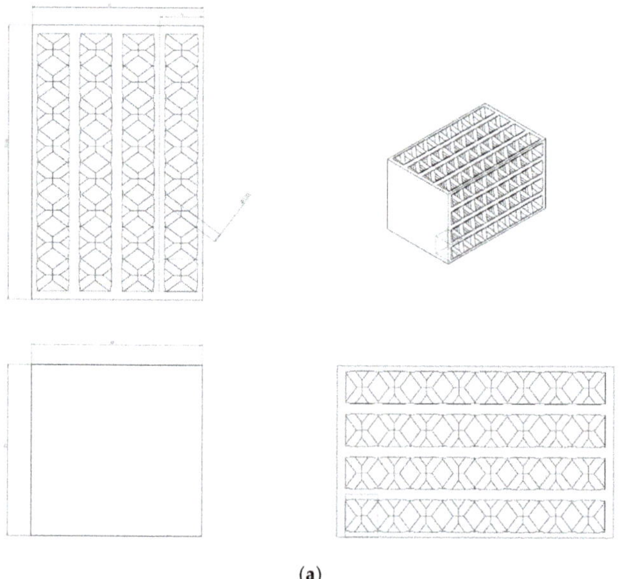

(a)

Figure A4. *Cont.*

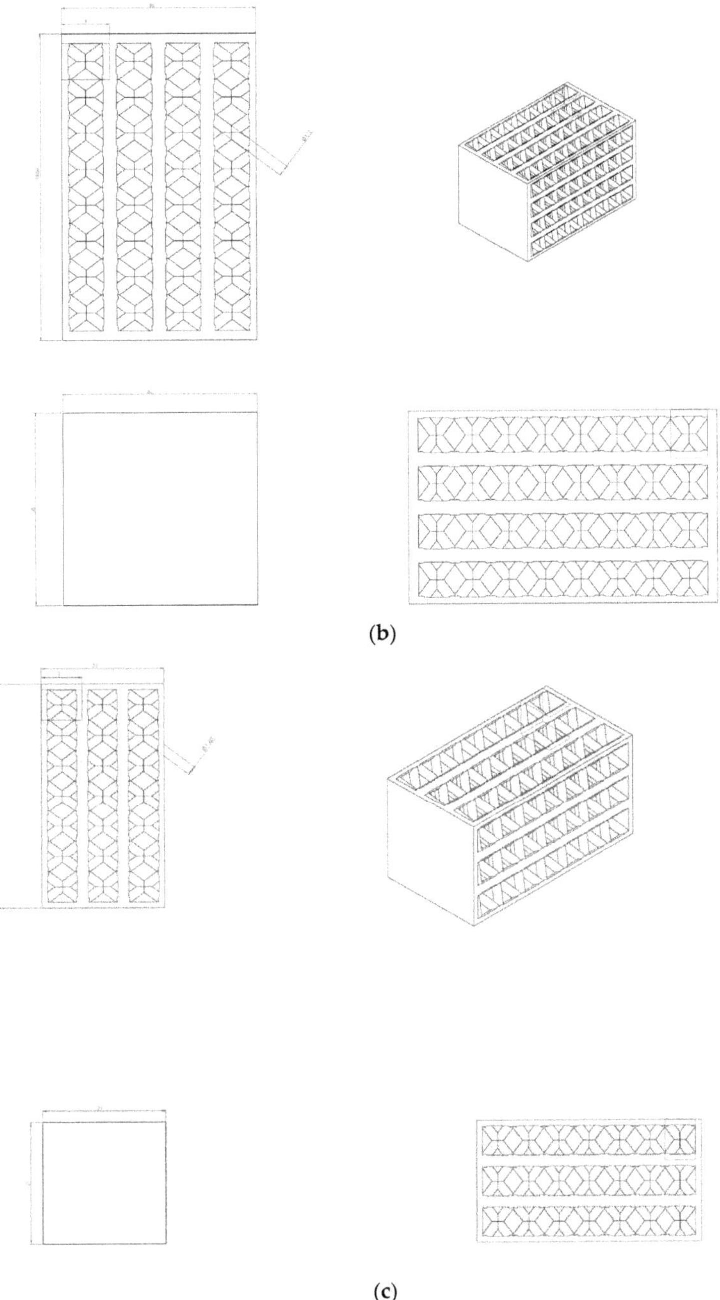

Figure A4. Bccz Cells Design: (**a**) 5 mm cells, 25% relative density; (**b**) 5 mm cells, 30% relative density; (**c**) 7 mm cells, 25% relative density.

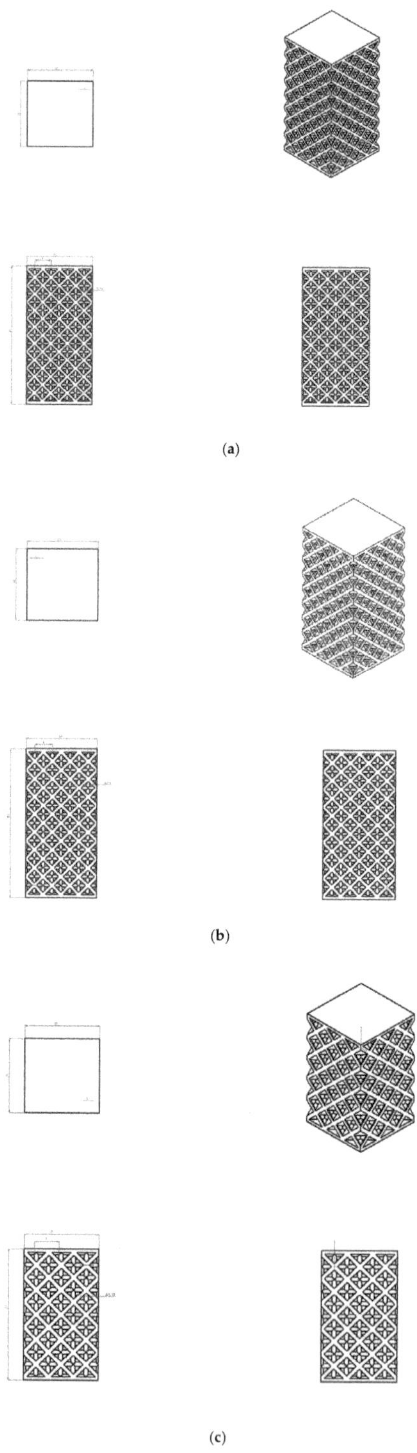

Figure A5. Octet Cells Design: (**a**) 5 mm cells, 25% relative density; (**b**) 5 mm cells, 30% relative density; (**c**) 7 mm cells, 25% relative density.

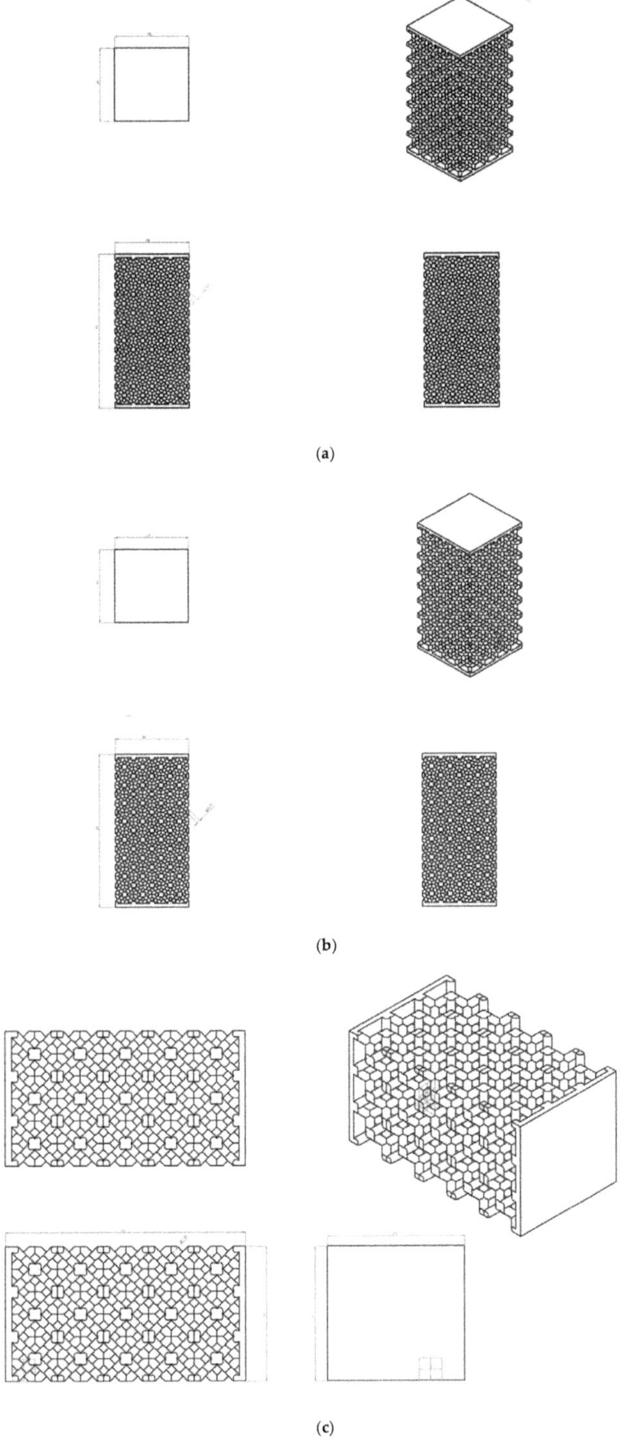

Figure A6. Rhombic Cells Design: (**a**) 5 mm cells, 25% relative density; (**b**) 5 mm cells, 30% relative density; (**c**) 7 mm cells, 25% relative density.

References

1. Ferro, C.; Grassi, R.; Seclì, C.; Maggiore, P. Additive Manufacturing Offers New Opportunities in UAV Research. *Procedia CIRP* **2016**, *41*, 1004–1010. [CrossRef]
2. Stornelli, G.; Gaggia, D.; Rallini, M.; Di Schino, A. Heat Treatment Effect on Maraging Steel Manufactured by Laser Powder Bed Fusion Technology: Microstructure and Mechanical Properties. *Acta Metall. Slovaca* **2021**, *27*, 122–126. [CrossRef]
3. Benedetti, M.; du Plessis, A.; Ritchie, R.; Dallago, M.; Razavi, S.; Berto, F. Architected cellular materials: A review on their mechanical properties towards fatigue-tolerant design and fabrication. *Mater. Sci. Eng. R Rep.* **2021**, *144*, 100606. [CrossRef]
4. Blakey-Milner, B.; Gradl, P.; Snedden, G.; Brooks, M.; Pitot, J.; Lopez, E.; Leary, M.; Berto, F.; du Plessis, A. Metal additive manufacturing in aerospace: A review. *Mater. Des.* **2021**, *209*, 110008. [CrossRef]
5. Korkmaz, M.E.; Gupta, M.K.; Robak, G.; Moj, K.; Krolczyk, G.M.; Kuntoğlu, M. Development of lattice structure with selective laser melting process: A state of the art on properties, future trends and challenges. *J. Manuf. Process.* **2022**, *81*, 1040–1063. [CrossRef]
6. Cao, Y.; Tan, W.; Wu, Z. Aircraft icing: An ongoing threat to aviation safety. *Aerosp. Sci. Technol.* **2018**, *75*, 353–385. [CrossRef]
7. Ferro, C.G.; Varetti, S.; Maggiore, P.; Lombardi, M.; Biamino, S.; Manfredi, D.; Calignano, F. Design and characterization of trabecular structures for an anti-icing sandwich panel produced by additive manufacturing. *J. Sandw. Struct. Mater.* **2020**, *22*, 1111–1131. [CrossRef]
8. Maggiore, P.; Vitti, F.; Ferro, C.G.; Sara, V. Thermal Anti Ice System Integrated in the Structure and Method for Its Fabrication. U.S. Patent 102016000098196, 2016.
9. Goraj, Z. An Overview of the Deicing and Anti-Icing Technologies with Prospects for the Future. In Proceedings of the 24th International Congress of the Aeronautical Sciences, Yokohama, Japan, 29 August–3 September 2004.
10. Pellissier, M.P.C.; Habashi, W.G.; Pueyo, A. Optimization via FENSAP-ICE of aircraft hot-air anti-icing systems. *J. Aircr.* **2011**, *48*, 265–276. [CrossRef]
11. Al-Khalil, K.M.; Keith, T.G.; DeWitt, K.J.; Nathman, J.K.; Dietrich, D.A. Thermal analysis of engine inlet anti-icing systems. *J. Propuls. Power* **1990**, *6*, 628–634. [CrossRef]
12. Papadakis, M.; Wong, S.-H.; Yeong, H.-W.; Vu, G. Icing Tunnel Experiments with a Hot Air Anti-Icing System. In Proceedings of the 46th AIAA Aerospace Sciences Meeting and Exhibit, Reno, NV, USA, 7–10 January 2008. [CrossRef]
13. Morency, F.; Tezok, F.; Paraschivoiu, I. Heat and mass transfer in the case of anti-icing system simulation. *J. Aircr.* **2000**, *37*, 245–252. [CrossRef]
14. Manogharan, G. *Analysis of Non-Stochastic Lattice Structure Design for Heat Exchanger*; NC State University: Raleigh, NC, USA, 2009.
15. Tancogne-Dejean, T.; Spierings, A.B.; Mohr, D. Additively-manufactured metallic micro-lattice materials for high specific energy absorption under static and dynamic loading. *Acta Mater.* **2016**, *116*, 14–28. [CrossRef]
16. Hao, L.; Raymont, D.; Yan, C.; Hussein, A.; Young, P. Design and Additive Manufacturing of Cellular Lattice Structures. In Proceedings of the The International Conference on Advanced Research in Virtual and Rapid Prototyping (VRAP), Leiria, Portugal, 28 September–1 October 2011; pp. 249–254. [CrossRef]
17. Zargarian, A.; Esfahanian, M.; Kadkhodapour, J.; Ziaei-Rad, S.; Zamani, D. On the fatigue behavior of additive manufactured lattice structures. *Theor. Appl. Fract. Mech.* **2019**, *100*, 225–232. [CrossRef]
18. Burr, A.; Persenot, T.; Doutre, P.-T.; Buffiere, J.-Y.; Lhuissier, P.; Martin, G.; Dendievel, R. A numerical framework to predict the fatigue life of lattice structures built by additive manufacturing. *Int. J. Fatigue* **2020**, *139*, 105769. [CrossRef]
19. Agenbag, N.; McDuling, C. Fatigue Life Testing of Locally Additive Manufactured AlSi1OMg Test Specimens. *R&D J.* **2021**, *37*, 19–25. [CrossRef]
20. Tommasi, A.; Maillol, N.; Bertinetti, A.; Penchev, P.; Bajolet, J.; Gili, F.; Pullini, D.; Mataix, D.B. Influence of surface preparation and heat treatment on mechanical behavior of hybrid aluminum parts manufactured by a combination of laser powder bed fusion and conventional manufacturing processes. *Metals* **2021**, *11*, 522. [CrossRef]
21. Bici, M.; Brischetto, S.; Campana, F.; Ferro, C.G.; Seclì, C.; Varetti, S.; Maggiore, P.; Mazza, A. Development of a multifunctional panel for aerospace use through SLM additive manufacturing. *Procedia CIRP* **2018**, *67*, 215–220. [CrossRef]
22. Yan, C.; Hao, L.; Hussein, A.; Bubb, S.L.; Young, P.; Raymont, D. Evaluation of light-weight AlSi10Mg periodic cellular lattice structures fabricated via direct metal laser sintering. *J. Mater. Process. Technol.* **2014**, *214*, 856–864. [CrossRef]
23. Perello, M. Numerical Simulation and Experimental Validation of Lattice Structures for an Innovative Anti-Ice Leading Edge. 2018. Available online: http://webthesis.biblio.polito.it/id/eprint/9219 (accessed on 1 March 2023).
24. Hussein, A.; Hao, L.; Yan, C.; Everson, R.; Young, P. Advanced lattice support structures for metal additive manufacturing. *J. Mater. Process. Technol.* **2013**, *213*, 1019–1026. [CrossRef]
25. Maconachie, T.; Leary, M.; Lozanovski, B.; Zhang, X.; Qian, M.; Faruque, O.; Brandt, M. SLM lattice structures: Properties, performance, applications and challenges. *Mater. Des.* **2019**, *183*, 108137. [CrossRef]
26. du Plessis, A.; Razavi, S.M.J.; Benedetti, M.; Murchio, S.; Leary, M.; Watson, M.; Bhate, D.; Berto, F. Properties and applications of additively manufactured metallic cellular materials: A review. *Prog. Mater. Sci.* **2022**, *125*, 10918. [CrossRef]
27. Seharing, A.; Azman, A.H.; Abdullah, S. A review on integration of lightweight gradient lattice structures in additive manufacturing parts. *Adv. Mech. Eng.* **2020**, *12*, 1687814020916951. [CrossRef]

28. Mahmoud, D.; Elbestawi, M.A. Elbestawi, Lattice structures and functionally graded materials applications in additive manufacturing of orthopedic implants: A review. *J. Manuf. Mater. Process.* **2017**, *1*, 13. [CrossRef]

29. Muhammad, M.; Nezhadfar, P.; Thompson, S.; Saharan, A.; Phan, N.; Shamsaei, N. A comparative investigation on the microstructure and mechanical properties of additively manufactured aluminum alloys. *Int. J. Fatigue* **2021**, *146*, 106165. [CrossRef]

30. Ferro, C.G.; Varetti, S.; Vitti, F.; Maggiore, P.; Lombardi, M.; Biamino, S.; Manfredi, D.; Calignano, F. A robust multifunctional sandwich panel design with trabecular structures by the use ofadditive manufacturing technology for a new de-icing system. *Technologies* **2017**, *5*, 35. [CrossRef]

31. Zilio, C.; Patricelli, L. Aircraft anti-ice system: Evaluation of system performance with a new time dependent mathematical model. *Appl. Therm. Eng.* **2014**, *63*, 40–51. [CrossRef]

32. Montgomery, D.C. *Design and Analysis of Experiments*; John Wiley & Sons, Inc.: Scottsdale, AZ, USA, 2019; p. 684.

33. Lambert, D.; Adler, M. IN718 Additive Manufacturing Properties and Influences. In Proceedings of the JANNAF Propulsion Meeting, Nashville, TN, USA, 1–4 June 2014.

34. McCullough, K.Y.G.; Fleck, N.A.; Ashby, M.F. Stress-life fatigue behaviour of aluminum alloy foams. *Fatigue Fract. Eng. Mater. Struct.* **2000**, *23*, 199–208. [CrossRef]

35. Zhao, S.; Li, S.; Hou, W.; Hao, Y.; Yang, R.; Misra, R. The influence of cell morphology on the compressive fatigue behavior of Ti-6Al-4V meshes fabricated by electron beam melting. *J. Mech. Behav. Biomed. Mater.* **2016**, *59*, 251–264. [CrossRef] [PubMed]

36. Zenkert, D.; Burman, M. Tension, compression and shear fatigue of a closed cell polymer foam. *Compos. Sci. Technol.* **2009**, *69*, 785–792. [CrossRef]

37. Zhai, Y.; Galarraga, H.; Lados, D.A. Microstructure, static properties, and fatigue crack growth mechanisms in Ti-6Al-4V fabricated by additive manufacturing: LENS and EBM. *Eng. Fail. Anal.* **2016**, *69*, 3–14. [CrossRef]

38. Varetti, S. *Study and Development of an Innovative L-PBF Demonstrator and an Anti-Ice Solution Based on Trabecular Structures*; Politecnico di Torino: Turin, Italy, 2020.

39. Lazzeri, R. A comparison between safe life, damage tolerance and probabilistic approaches to aircraft structure fatigue design. *Aerotec. Missili Spaz.* **2002**, *81*.

40. Li, Y.; Pavier, M.; Coules, H. Compressive fatigue characteristics of octet-truss lattices in different orientations. *Mech. Adv. Mater. Struct.* **2022**, *29*, 6390–6402. [CrossRef]

41. Li, Y.; Pavier, M.; Coules, H. Experimental study on fatigue crack propagation of octet-truss lattice. *Procedia Struct. Integr.* **2021**, *37*, 41–48. [CrossRef]

42. Ashby, M.F. The properties of foams and lattices. *Philos. Trans. R. Soc. A Math. Phys. Eng. Sci.* **2006**, *364*, 15–30. [CrossRef] [PubMed]

 *metals*

Article

Design of Glass Fiber-Doped High-Resistivity Hot-Pressed Permanent Magnets for Reducing Eddy Current Loss

Yingjian Guo, Minggang Zhu *, Ziliang Wang, Qisong Sun, Yu Wang and Zhengxiao Li

Division of Function Materials, Central Iron and Steel Research Institute, Beijing 100081, China; yingjguo@126.com (Y.G.)
* Correspondence: mgzhu@126.com

Abstract: The Nd-Fe-B hot-deformation magnet with high resistivity was successfully prepared by hot-pressing and hot-deformation of Nd-Fe-B fast-quenched powder with amorphous glass fiber. After the process optimization, the resistivity of the magnet was increased from 0.383 mΩ·cm to 7.2 mΩ·cm. Therefore, the eddy current loss of magnets can be greatly reduced. The microstructure shows that the granular glass fiber forms a continuous isolation layer during hot deformation. At the same time, the boundary of Nd-Fe-B quick-quenched the flake and glass fiber from the transition layer, which improves the binding of the two, and which can effectively prevent the spalling of the isolation layer. In addition, adding glass fiber improves the orientation of the hot deformation magnet to a certain extent. The novel design concept of insulation materials provides new insights into the development and application of rare earth permanent magnet materials.

Keywords: Nd-Fe-B; composite magnet; glass fiber; high resistivity; thermorheological property

1. Introduction

Permanent magnet materials are important for industry, military, and information technology [1–4]. At present, in an environment where green and clean energy is promoted, the demand for permanent magnet materials in the fields of wind power, electric vehicles, electric motors, robots, and aerospace is continuously increasing [5]. Among the permanent magnets, the demand for Nd-Fe-B increased the most, reaching 2.1×10^4 t production in 2020. However, after 2011, the price of rare earth soared due to fluctuations in prices [1]. At the same time, Ce-containing dual-main phase (DMP) Nd-Fe-B magnets emerged and became popular due to their balanced utilization of high-abundance rare earth elements [6]. DMP magnets are made by adding Ce-Fe-B during the preparation of Nd-Fe-B magnets to form magnets with both Nd-Fe-B and Ce-Fe-B main phases, and they have excellent magnetic properties compared to single main phase Nd-Fe-B magnets [7,8].

As the temperature increases, all magnets inevitably face the problem of irreversible demagnetization, and the above means only alleviating the problem to a certain extent. After the high temperature demagnetization of the magnet, it still has enough magnetic properties for use. The ideal magnet should have a high magnetic energy product and a square demagnetization curve, which undoubtedly describes Nd-Fe-B magnets. However, the Curie temperature of 315 °C limits its usable condition, which has to be lower than 80 °C (N grade); otherwise, irreversible demagnetization will occur. Although the coercivity is increased by adding heavy rare earth elements such as Dy and Tb, and although the Curie temperature is increased by adding Co, the maximum service temperature is not more than 240 °C (AH grade) [3]. When the Nd-Fe-B magnet rises from 20 °C to 100 °C, the magnetic energy product will decrease by half, and the motor power will decrease [9], which will cause a waste of magnetic energy.

Because the resistivity of rare earth permanent magnets is very small, the eddy current generated in them is the main factor for the heating of magnets. Reducing the eddy current loss in magnets and thus reducing the temperature of magnets during operation is a

Citation: Guo, Y.; Zhu, M.; Wang, Z.; Sun, Q.; Wang, Y.; Li, Z. Design of Glass Fiber-Doped High-Resistivity Hot-Pressed Permanent Magnets for Reducing Eddy Current Loss. *Metals* **2023**, *13*, 808. https://doi.org/10.3390/met13040808

Academic Editors: Andrea Di Schino and Claudio Testani

Received: 11 March 2023
Revised: 12 April 2023
Accepted: 18 April 2023
Published: 20 April 2023

problem that needs to be considered. Through the optimization of motor structure designs, although a certain effect has been achieved [9–11], the magnet resistivity is often used as a fixed value, so the optimization results are not satisfactory. In addition to optimizing the design of the motor structure, Aoyama [12] noted that increasing the resistivity of the magnet can also effectively reduce the eddy current loss. Considering the skin effect of the eddy current and the insulation of the main phase grains or raw material particles in the magnet, the electron transport in the magnet can be reduced by the heat source, and the resistivity can be increased. Thus, the working temperature of the magnet can be decreased. Gabay [13] introduced CaS for insulation and isolation, and it was found that adding sulfur (phosphorus) compounds would form NdS, which would reduce the metallurgical bonding of the cladding layer. Although the introduction of SiO_2 increases the resistivity to a certain extent, Nd will combine with O to form NdO, resulting in the loss of the Nd-rich phase and the reduction of magnetic properties [14]. Fluoride is suitable for addition to permanent magnet materials because of its inertia. The addition of NdF increased the magnetic resistivity by 200% [15]. Komuro et al. [16] prepared a magnet with a fluoride coating with a resistivity of 1.4 $m\Omega \cdot mm$. The surface resistivity of the magnet increased by 10 times, and the rotor temperature decreased by 50%. Although these works have made some progress, some problems still occur. In the above methods, most of the materials involved are inorganic compounds. In the process of magnet insulation and coating, these inorganic compounds cannot deform with the magnet; thus, spalling occurs during the deformation process of the magnet, which cannot effectively isolate the magnet. In addition, it is often necessary to complete the coating process with the help of liquid re-rich phase flow. The participation of the rare-earth rich phase in the coating will not only reduce the insulating ability of the coating layer but also cause the loss of the rare-earth rich phase and reduce the magnetic properties.

The method of adding insulating materials and the form of insulating materials play an important role in the resistivity of composite materials. McLachlan [17] gave the general effective media equation and described the different wetting and coating conditions between the conductive phase and the insulating phase. For calculating the resistivity of composite materials with continuous isolation layers:

$$\rho_m = \rho_h (1 - \Phi)^{m_\Phi}$$

ρ_m is the resistivity of the composite, ρ_h is the resistivity of the low-conductivity phase, Φ is the volume fraction of the high-conductivity phase, and m_Φ is the exponent for randomly oriented high-conductivity ellipsoids.

Depending on the value range of m_Φ, a thin layer of insulation distributed continuously helps to isolate electron transport between adjacent particles, thus increasing the resistivity of the composite magnet. As seen from the metal binary phase diagram, most of the crystalline compound insulating materials have very high melting points, which cannot follow the deformation of the magnet during the particle coating process. Moreover, spalling occurs during the deformation process. At the same time, liquid re-rich phase flow is often needed to complete the coating process. The isolation layer formed by these crystalline compounds is still dispersed rather than continued under the non-wetting or intermediate wetting case discussed for the general effective media equation.

In this work, to overcome the shortcomings of the insulating layer, we designed a composite magnet doped with amorphous materials. In contrast to crystalline materials, amorphous materials, which can be deformed well with magnetic powder, will exhibit viscous flow with an increasing temperature. Through the systematic study of magnets with amorphous glass fibers, it is verified that this design can significantly improve the resistivity of the magnets and facilitate the preparation of magnets with oriented textures. The new design idea provides a new way to develop high-resistivity rare earth permanent magnets with oriented textures.

2. Materials and Methods

The magnetic powder was commercial melt-spun ribbon powder with a composition of $Nd_{30.7}Fe_{bal}(Co_{3.98}Ga_{0.48})B_{0.89}$ (wt.%), and the glass fiber was crushed through a 2000 mesh screen, with a composition of $CaO \cdot SiO_2$. Three different additions of mixed magnetic powder were prepared, and the additions were 2 wt.%, 5 wt.%, and 10 wt.%, respectively. The mixing powder was hot-pressed at temperatures of 550 °C under 130 MPa for 2 min. After that, the sample was heated to 850 °C and kept warm for 1 min with a deformation rate of approximately 60~70%. A simple hot-deformed magnet without the addition of glass fiber was also prepared. The resistivity was by a four-point probe resistivity. The microstructure was studied using scanning electron microscopy (SEM) JEOL JSM 7100F, and energy dispersion spectrum (EDS) was used for elemental analysis. X-ray diffraction (XRD) was used to examine the crystalline constituents and detected on PANalytical Empyrean Series 2. The orthogonal experiment in Table $L_9(3^4)$ was used to explore and optimize the magnet preparation experimental parameters. For the experimental parameters, the hot-pressing temperature was 550 °C, 575 °C, and 600 °C, and then the hot-deformed temperature was 850 °C, 870 °C, and 890 °C, respectively.

3. Results

Backscattered electron (BSE)-SEM image from the hot-pressed magnet is shown in Figure 1a. The black regions are the grainy glass fiber between the original flakes of Nd-Fe-B, and they have relatively rough interfaces. BSE-SEM image of the hot-deformed magnet sample (Figure 1b) shows that glass fiber particles aggregate to form a continuous isolation layer after being hot deformed, and the interface with Nd-Fe-B flakes becomes smoother. Figure 1c is a partial enlargement of Figure 1b. It can be seen that the glass fiber not only aggregates itself but also forms transitional layers between the glass fiber and the Nd-Fe-B flakes, as shown by the line segment in Figure 1c.

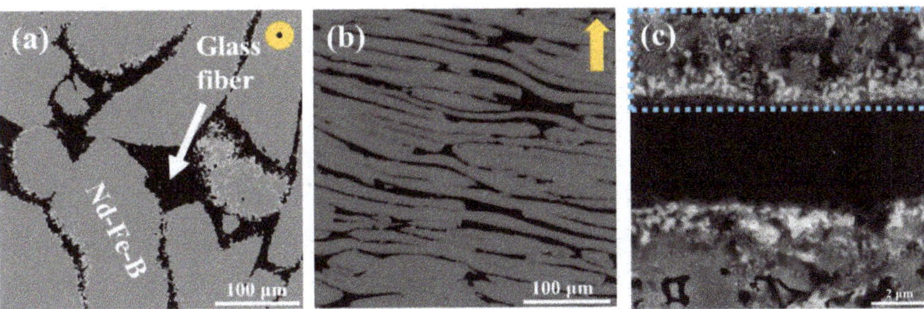

Figure 1. BSE-SEM images obtained from magnet with 10 wt.% addition (**a**) hot-pressed temperature at 550 °C, (**b**) hot-deformed temperature at 850 °C, and (**c**) partial enlargement of (**b**), yellow arrows and circles indicate different C-axis orientations of the magnet.

As shown in Figure 2, as the additions increase, the glass fiber tends to coat the main-phase grains. The distribution style of the magnetic flakes and the glass fiber was perpendicular to the c-axis direction, stacked in layers. Meanwhile, the distribution of the glass fiber made the grain boundary more clearly visible with increasing additions.

Figure 3 show the element maps in different directions of the hot deformed magnet. According to the role of element distribution, it is clear that only a small amount of Si atoms diffuse into the boundary of the Nd-Fe-B grains, and Fe atoms diffuse in the opposite direction; thus, the formation of the transitional layers is mainly caused by the mutual diffusion of Fe and Si elements.

Figure 4 shows the XRD of the hot deformed magnet with different glass fiber additions. There are the characteristic peaks of the calcium-containing compound, the peaks $(22\bar{1})$ and $(31\bar{1})$ in Figure 4b,e, and the peak (002) in Figure 4d of the $CaFe(Si_2O_6)$ phase. The diffracted intensity of Ca element and CaO increased obviously with the increase of glass fiber addition, indicating that the glass fiber decomposed during the hot deformation process. This also means that there is a diffusion reaction between the glass fiber and the magnetic powder, which effectively improves the resistivity.

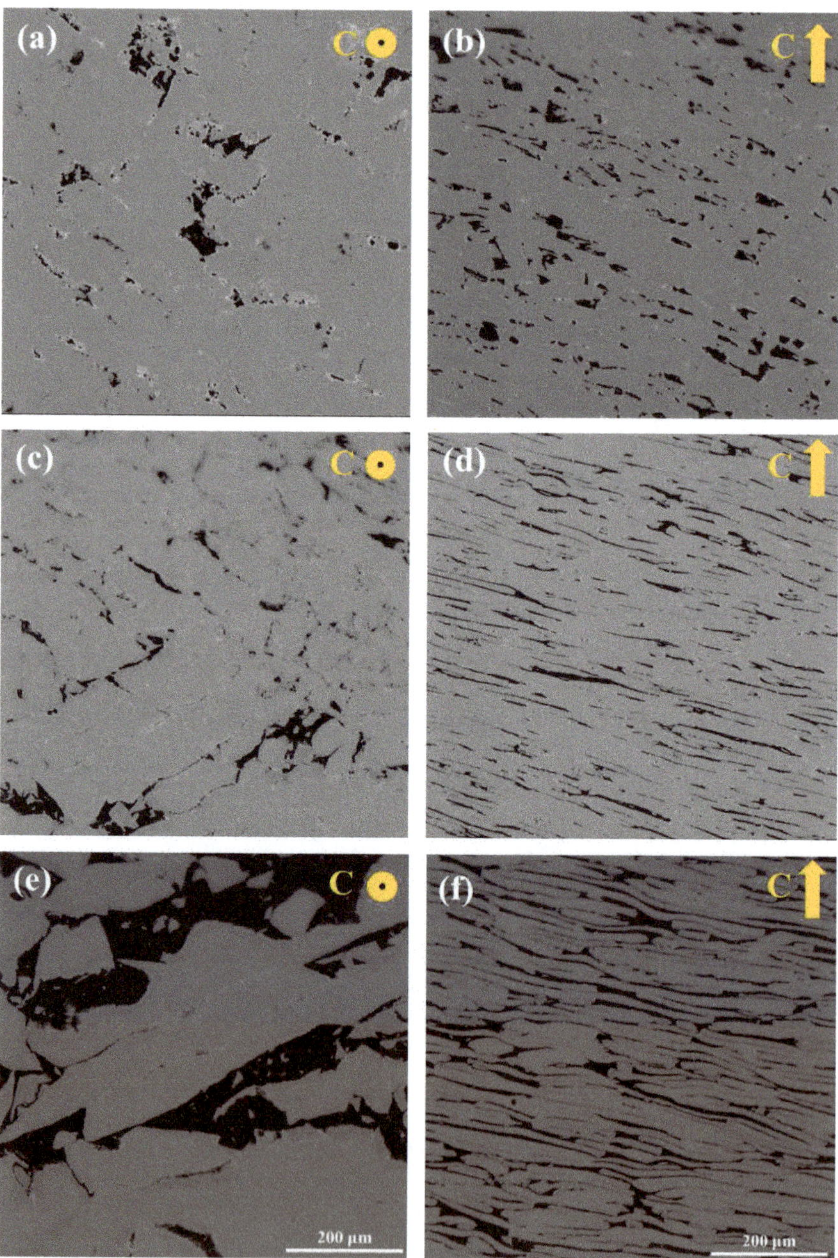

Figure 2. The same magnification BSE-SEM images obtained from (**a**,**b**) hot-deformed magnet with 2wt.% glass fiber, (**c**,**d**) with 5wt.% glass fiber, (**e**,**f**) with 10wt.% glass fiber. Furthermore, the direction of images (**a**,**c**,**e**) are parallel to the pressure, and (**b**,**d**,**f**) are perpendicular to the pressure, yellow arrows and circles indicate different C-axis orientations of the magnet.

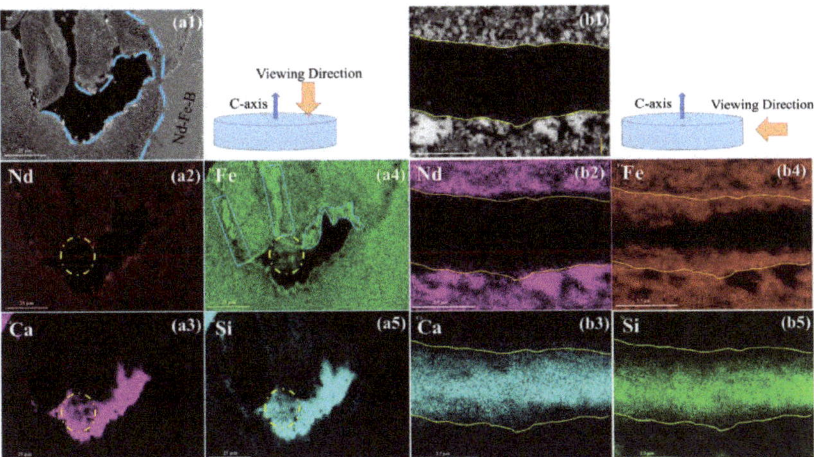

Figure 3. EDS mapping in a different direction of hot-deformed magnet with 10%wt glass fiber. (**a1**–**a5**) parallel to the pressure direction and (**b1**–**b5**) are perpendicular to the pressure direction.

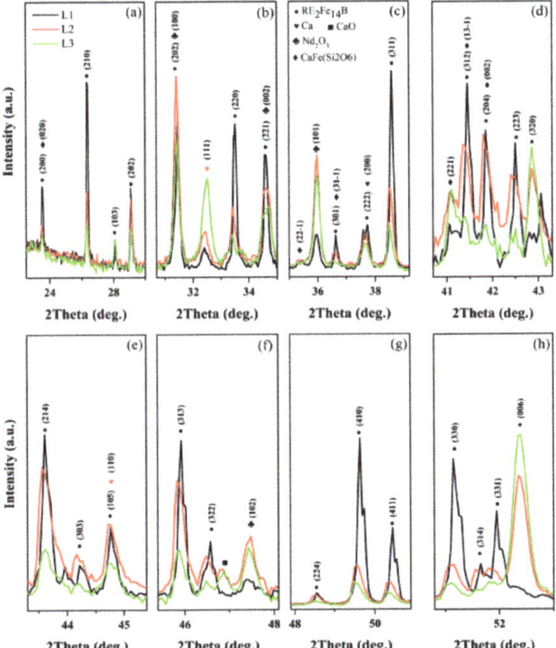

Figure 4. XRD patterns of the hot-deformed magnet with different glass fiber additions: (**a**) 2 wt.% (indexed as L1), (**b**) 5 wt.% (indexed as L2), and (**c**) 10 wt.% (indexed as L3). (**a**–**h**) are different main peak areas of XRD, and the height of the peaks was scaled to the same height.

The results of the orthogonal experiment are shown in Tables 1 and 2. Where Xi is the different measurement, Q is the variation sum of squares, and EOV is the estimator of variance. As shown in Table 2, all three factors are significantly affected. Furthermore, in the variance analysis results, the relationship of the three factors should be ABC, and the optimal analysis condition is A3B1C3.

Table 1. Resistivity of hot-pressed permanent magnets with different mass fractions of glass fiber-doped.

Mass Fraction m (wt.%)	Hot-Pressed Temperature T1 (°C)	Hot-Deformed Temperature T2 (°C)	Xi (mΩ·cm)		
			1	2	3
A1(2)	B1(550)	C1(850)	0.99	0.97	0.97
A1(2)	B2(575)	C2(870)	1.05	1.08	1.03
A1(2)	B3(600)	C3(890)	1.18	1.05	1.02
A2(5)	B1(550)	C2(870)	0.86	0.86	0.86
A2(5)	B2(575)	C3(890)	0.90	0.92	0.90
A2(5)	B3(600)	C1(850)	1.00	1.02	1.18
A3(10)	B1(550)	C3(890)	1.84	2.35	3.01
A3(10)	B2(575)	C1(850)	0.81	0.78	0.84
A3(10)	B3(600)	C2(870)	0.88	0.90	0.92

Table 2. Analysis of Variance.

Source of Variation	Q	DOF	EOV	F	f	Significance Level (*, **, ***)	The Optimal Level
m (wt.%)	2.12	2	1.06	10.47	3.49	**	A3
T_1 (°C)	1.21	2	0.61	6.00	3.49	**	B1
T_2 (°C)	1.15	2	0.58	5.71	3.49	**	C3
error	2.02	20	0.10				
total	6.50	26					

Significance level: * insignificant, ** significant, *** highly significant.

Figure 5 shows the confirmatory experiment results carried out under the explored conditions. The electrical resistivities of the magnets are 0.532, 2.35, and 7.2 mΩ·cm when the amount of glass fiber added is 2 wt.%, 5 wt.%, and 10 wt.%, respectively. Compared to the 0.383 mΩ·cm of the magnet without glass fibers, the resistivities of the magnets are increased by approximately 39%, 513%, and 1780%. It is more than 25 times the resistivity of the magnet with 10 wt.%, in contrast to the nano-silica layer coating magnets [14].

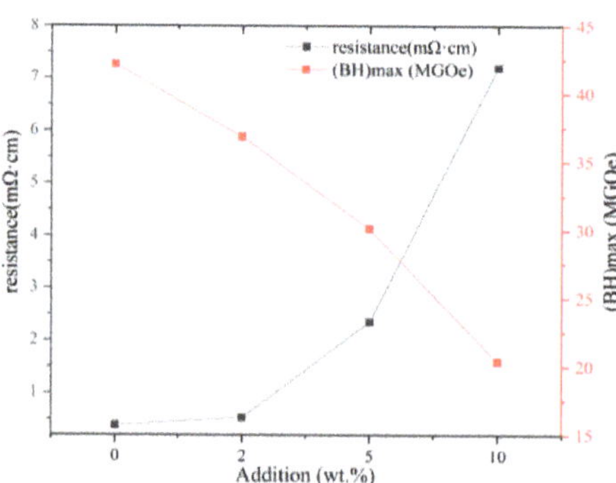

Figure 5. Electrical resistivity and magnetic energy products of the hot-deformed ones in confirmatory experiment samples with different additions.

4. Discussion

Figure 6a shows the microstructure of the magnet without glass fiber. The gray part in the image is the main Nd-Fe-B phase, and the highlight is the rare-earth rich phase between Nd-Fe-B layers. Nd-Fe-B particles change into sheets due to endure along the direction of pressure (arrow in Figure 6a). Figure 6b shows the microstructure of the magnet with 10wt% glass fiber added. The black part is the glass fiber, and the gray part is the Nd-Fe-B main phase. The dark area between the glass fiber and the Nd-Fe-B sheet is the transition layer (dashed blue area in the image).

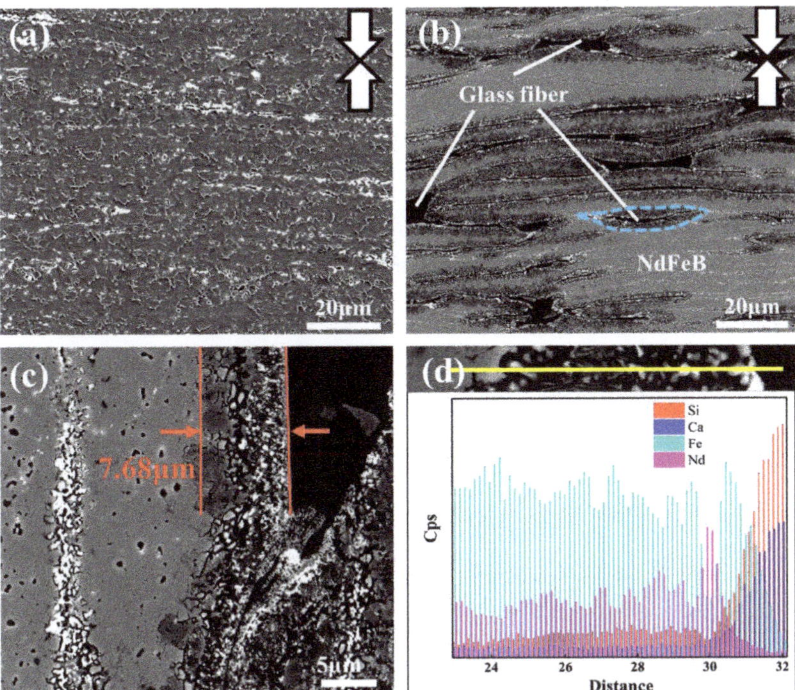

Figure 6. SEM image of hot-deformed magnets, (**a**) without glass fiber, (**b**) with 5wt% glass fiber; (**c**) partial enlarged drawing of (**b**,**d**) EDS line scanning near the transition layer.

EDS line scanning was carried out for the distribution of interface elements between magnetic powder and glass fiber, and the results are drawn in Figure 6c,d. The Fe element kept a high element concentration in the Nd-Fe-B phase, and the concentration decreases step-by-step as the distance approaches the magnetic powder interface. The variation tendency of Nd element concentration is consistent with that of the Fe element. Nevertheless, the mutation occurs in eutectoid tissue, where the Nd element concentration is more than that of Fe element at this mutation position. The Si element, on the contrary, whose concentration emerges gradually, decreases from glass fiber to the Nd-Fe-B phase until the Nd element is in the position of the mutation. The Ca element concentration is almost the same as Si element concentration, with only a small change, simultaneously, and simultaneously disappeared until the Nd element reached the position of the mutation.

According to energy spectrum analysis, the Si element is present in all interface reaction regions, indicating that the Si element is capable of a wide range of diffusion phenomena. Xu et al. [18] investigated the corrosion of Ni-based alloys with various elements added by molten glass and discovered that the metal elements in the alloy could reduce the Si element from glass. Its reduction reaction is:

$$Si^{4+} + 4e \rightarrow Si^0 \tag{1}$$

The Fe element is enriched at the forefront of the interfacial reaction, indicating that it was the first to participate in the reaction process. In their studies on the interface reaction between Kovar alloy/Invar alloy and glass, Luo et al. [19] and Khachatryan et al. [20] discovered that FeO, Fe_3O_4, and Fe_2O_3 can be formed sequentially as the O concentration changes. These products can react with the glass indefinitely, eventually forming the Fe_2SiO_4 phase. As a result, Fe oxidation occurs because:

$$Fe^0 - 2e \rightarrow Fe^{2+} \tag{2}$$

$$Fe^{2+} - e \rightarrow Fe^{3+} \tag{3}$$

Finally, the interface reaction process at the forefront of the interface between magnetic particle and glass fiber should begin with the contact of the surface of the magnetic particle and glass fiber, and the elements on both sides should begin to diffuse. Because the Fe element is the first to participate in the reaction, FeO, Fe_3O_4, and Fe_2O_3 are produced sequentially. Furthermore, the $CaFe(Si_2O_6)$ phase is formed in the glass fiber by combining with the CaO, and the reduced Si element diffuses into the magnetic particles. The precise equation is:

$$Fe^0 - 2e \rightarrow FeO \tag{4}$$

$$Fe^{2+} - e \rightarrow Fe_2O_3 \tag{5}$$

$$Fe_2O_3 + FeO \rightarrow Fe_3O_4 \tag{6}$$

$$Fe_3O_4 + SiO_2 \rightarrow Fe_2SiO_4 + O_2 \tag{7}$$

$$Fe_2SiO_4 + CaO + SiO_2 \rightarrow CaFe(Si_2O_6) \tag{8}$$

$$Si^{4+} + 4e \rightarrow Si^0 \tag{9}$$

It should be pointed out that in this reaction, the O element will be released, so the O element will also diffuse into the interior of the magnetic particle.

As previously stated, the O element was released during the reaction process. Because of the higher activity of RE elements, the Nd element began to combine with the O element at this point to form the Nd_2O_3 RE-rich phase. This reaction consumes the released O element, and the Fe element that has diffused with it must be diffused further into the glass fiber to keep the reaction going. This explains why there is a mutation in the concentration of Nd elements in this region. The precise equation is:

$$Nd + O \rightarrow Nd_2O_3 \tag{10}$$

In the preparation process of the thermal deformation magnet, the glass fiber was added to the magnetic particle interface reaction, forming a variety of reaction products, and through analysis of the three sections, there was no doubt that the interface reaction between the magnetic particle and the glass fiber was controlled by the diffusion process. At the forefront of the interface, the Fe element substituted the Si element in the glass fiber, forming a series of oxides, for instance, that eventually reacted with each other and formed the $CaFe(Si_2O_6)$ phase. Analysis of the energy spectrum near the interface indicates that the concentration of the Nd element increases abnormally before the Fe-rich zone, and the Nd element can hardly diffuse through the $CaFe(Si_2O_6)$ phase, just like a wall blocking the diffusion direction of the RE element. Similarly, the Ca element does not spread beyond the Fe-rich zone into the interior of the magnetic powder. It is inferred that the glass fiber that was added has an important role to form a "Fe-rich wall", which can effectively block the "leakage" of RE elements from the main-phase grains. This phenomenon can be applied in the process of the diffusion of heavy RE elements in the magnet, or to regulate the diffusion of heavy RE elements into the magnet.

In the above process, CaO and SiO$_2$ decomposed from glass fiber will be pushed into the narrow grain boundary with the softening flow of glass fiber, more fully isolating the ferromagnetic coupling between the main grains, and oxides have a good insulation effect, so the diffusion reaction between the glass fiber and magnetic powder can lead to an increase in resistivity.

The structure of the above experimental magnets is shown in Figure 3. The black area is glass fiber, the bright white area is the rare-earth rich phase precipitated during the preparation of the magnets, and the gray part is the Nd-Fe-B phase. With the increase in glass fiber addition, the distribution of black glass fiber gradually changed from sporadic distribution to continuous distribution. According to the design of the orthogonal experiment, Figure 7a–c show the morphology of glass fiber with 2 wt.% addition. Due to the small amount of addition, glass fiber is less distributed among Nd-Fe-B grain lamellas and scattered in the matrix structure. Figure 7d–f show the morphology of the glass fiber with the additional amount of 5 wt.%. It can be seen that with the increase in the addition, the distribution of the glass fiber between Nd-Fe-B laminates increases, and, thus, the resistivity of the magnet increases. However, it can be seen that due to the insufficient addition, there is still a large area uncoated between the Nd-Fe-B lamellae. When the addition amount reaches 10 wt.%, as shown in Figure 7g–i, the glass fiber has formed a relatively complete coating layer, and so the resistivity of this group of samples is, relatively, the highest.

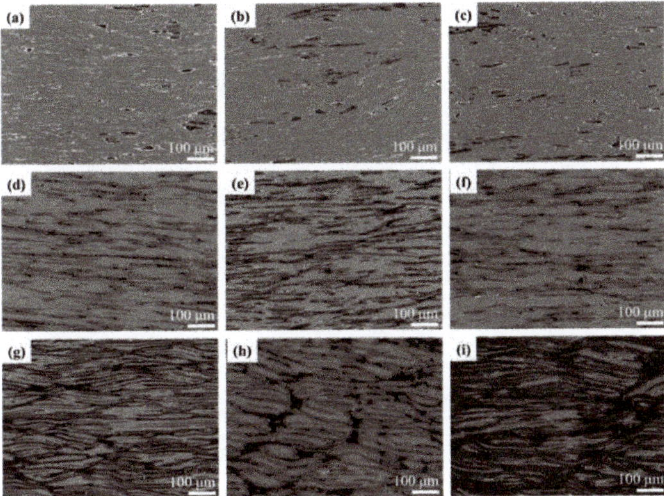

Figure 7. Sample morphology of orthogonal experiment, (**a–i**) experiments 1–9, respectively.

By comparing different deformation temperatures, Figure 7a,f,h is 850 °C. Due to the low temperature, the magnetic particle deformation degree is small, the glass fiber deformation degree is also insufficient, and the glass fiber aggregates in the magnet. With the increase in deformation temperature, that is, Figure 7b,d,i are the morphologies at 870 °C, and Figure 7c,e,g are the morphologies at 890 °C. The analysis shows that the deformation of the grain increases gradually, while the aggregation of the glass fiber decreases, indicating that the glass fiber is also deformed in the process of thermal deformation and is squeezed into the grains to form the coating layer. Compared with 9 groups of experimental samples, sample no. 7 (Figure 7g) has the best coating effect, a large deformation degree, uniform distribution of tissue, and, therefore, the highest resistivity.

Figure 8 shows the schematic of the preparation process of the magnets with glass fiber. Based on the microstructure, glass fibers are fully embedded around the magnetic particles and maintain the original particle shape during the hot-pressed process. After being hot-deformed, the magnetic flakes change into thin sheets and distribute in parallel to the direction of pressure. The glass fiber particles aggregate with each other and form a thin sheet distributed among the magnetic sheets, effectively separating the magnetic flakes. Moreover, there is a transitional layer at the contact surface between the glass fiber and the magnetic sheets, indicating that there is a mutual diffusion of elements between the two, forming a new phase. Therefore, the magnetic flakes and the glass fiber can be firmly bonded together, thus effectively avoiding the isolation failure caused by peeling, which is beneficial for improving resistivity.

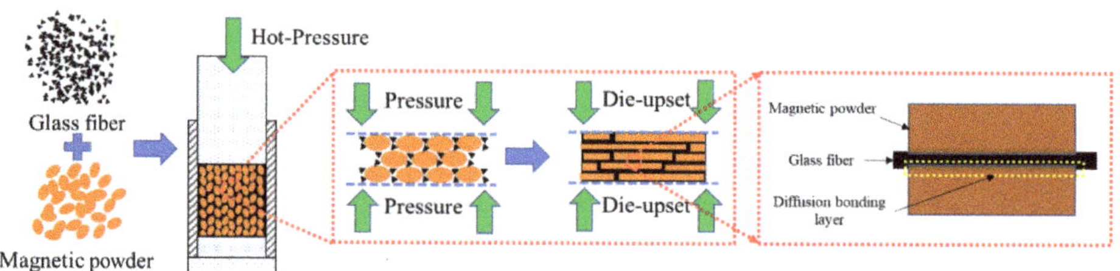

Figure 8. Schematic of the preparation process of the magnets with glass fiber.

With increasing additions of the glass fiber, the intensity of the main-phase peaks decreases; nevertheless, the intensity of the main-phase peak (006) is abnormal in Figure 4h. There is no doubt that the orientation degree (I(006)/I(105)) increased dramatically, which means that the orientation of the magnet is improved. This is due to the glass fiber collaborative deformation with Nd-Fe-B powders during the hot deformation process. According to Zhu's theory [21], the grains of magnetic powders slide and rotate in the thermorheological process. Softened glass fiber acts as a lubricant in the gaps of the Nd-Fe-B powders, and the grains of the Nd-Fe-B powders are more likely to slide and rotate under the action of external force, as shown in Figure 9.

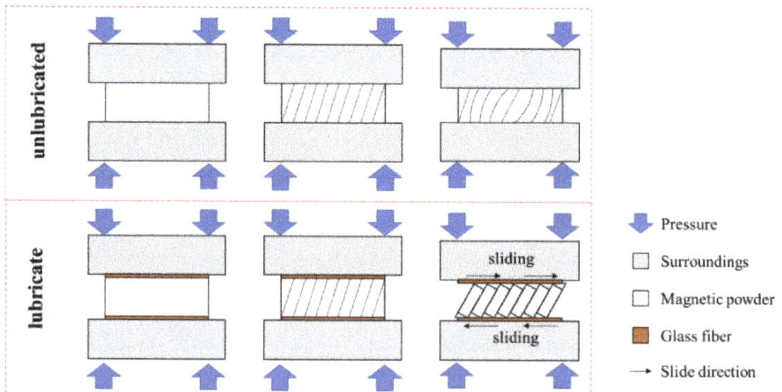

Figure 9. Diagram of the deformation with and without lubrication.

In contrast to other insulating materials, adding amorphous glass fiber to the magnet greatly improves the resistivity. Shen [22] calculated and simulated the relationship between eddy current loss and the resistivity of magnets by FEA models. As shown in Figure 10, the relationship between the resistivity of a magnet and the reduction of the eddy current losses in magnets was shown. It should be pointed out that *pu* (per unit) is the resistivity of sintered Sm_2Co_{17} magnets (8.503×10^{-2} mΩ·cm). The resistivity of the magnet prepared in this work is 85 times that of the one in Figure 10. This means that eddy current losses are reduced by at least two thirds, relative to the magnets that have not been treated with high resistivity. Therefore, the addition of the amorphous glass fiber magnet can reduce eddy current loss very effectively.

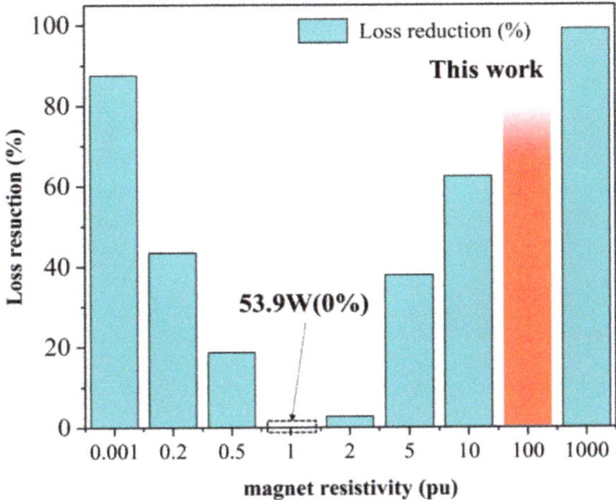

Figure 10. Magnet resistivity vs. reduction of eddy losses in magnets, where *pu* (per unit) is the resistivity of sintered Sm_2Co_{17} magnets (8.503×10^{-2} mΩ·cm), and take the loss of one-time unit as 100%. Data from [22].

5. Conclusions

We have shown that the highest resistivity of the obtained magnet was 7.2 mΩ·cm, which is almost 18 times as high as that of the magnet without additions and 25 times higher than that of the nano-silica layer coating Nd-Fe-B magnets. Thus, the eddy current loss of the magnets can be effectively reduced. The continuous coating between the glass fibers and magnet particles is diffusion controlled, which can effectively prevent the incomplete coating phenomenon caused by the peeling of the coating during thermal deformation. At the same time, the analysis results show that the addition of glass fibers can improve the deformation ability of the magnet and thus improve the orientation of the magnet. This work proves that adding amorphous glass fiber is an effective means for improving the resistivity of magnets. The novel design concept of insulation materials provides new insight into the development and application of rare earth permanent magnet materials. Furthermore, the search for additives that can substantially improve the resistivity with an enhanced magnetic energy product should be considered in future work.

Author Contributions: Y.G.: Carrying out measurements and manuscript composition, Writing—Original draft preparation, Writing—Reviewing, and Editing. M.Z.: Conception, Experimental design, Methodology, Investigation, Validation, Project administration, Funding acquisition. Z.W.: Carrying out measurements and manuscript composition, Data curation. Q.S.: Carrying out measurements and manuscript composition, Writing—Reviewing, and Editing. Y.W. and Z.L.: Carrying out measurements and manuscript composition, Formal analysis. All authors have read and agreed to the published version of the manuscript.

Funding: The work was supported by the National Key Research and Development Program of China (Grant No. 2022YFB3505600; 2022YFB3503300; 2022YFB3505200), the National Natural Science Foundation of China (Grant No. 51871063), and Major Projects in Inner Mongolia Autonomous Region (Grant No. 2021ZD0035).

Data Availability Statement: Data sharing not applicable. No new data were created or analyzed in this study. Data sharing is not applicable to this article.

Conflicts of Interest: The authors declare that they have no known competing financial interests or personal relationships that could have appeared to influence the work reported in this paper.

References

1. Coey, J.M.D. Perspective and Prospects for Rare Earth Permanent Magnets. *Engineering* **2020**, *6*, 119–131. [CrossRef]
2. Skokov, K.P.; Gutfleisch, O. Heavy rare earth free, free rare earth and rare earth free magnets—Vision and reality. *Scr. Mater.* **2018**, *154*, 289–294. [CrossRef]
3. Herbst, J.F. R2Fe14B materials: Intrinsic properties and technological aspects. *Rev. Mod. Phys.* **1991**, *63*, 819–898. [CrossRef]
4. Sugimoto, S. Current status and recent topics of rare-earth permanent magnets. *J. Phys. D Appl. Phys.* **2011**, *44*, 064001. Available online: https://iopscience.iop.org/article/10.1088/0022-3727/44/6/064001 (accessed on 10 March 2023). [CrossRef]
5. Xia, W.; He, Y.; Huang, H.; Wang, H.; Shi, X.; Zhang, T.; Liu, J.; Stamenov, P.; Chen, L.; Coey, J.M.D.; et al. Initial Irreversible Losses and Enhanced High-Temperature Performance of Rare-Earth Permanent Magnets. *Adv. Funct. Mater.* **2019**, *29*, 1900690. [CrossRef]
6. Zhu, M.; Li, W.; Wang, J.; Zheng, L.; Li, Y.; Zhang, K.; Feng, H.; Liu, T. Influence of Ce Content on the Rectangularity of Demagnetization Curves and Magnetic Properties of Re-Fe-B Magnets Sintered by Double Main Phase Alloy Method. *IEEE Trans. Magn.* **2014**, *50*, 1–4. [CrossRef]
7. Zhang, L.; Zhu, M.; Guo, Y.; Song, L.; Li, W. Grains orientation and restructure mechanism of Ce-contained magnets processed by reduction diffusion. *J. Alloys Compd.* **2022**, *891*, 161921. [CrossRef]
8. Zhu, M.; Li, Y.; Li, W.; Zheng, L.; Zhou, D.; Feng, H.; Chen, L.; Du, A. Relation between microstructure and magnetic properties of shock wave-compressed Nd–Fe–B magnets. *Rare Met.* **2022**, *41*, 2353–2356. [CrossRef]
9. Haavisto, M.; Tuominen, S.; Santa-nokki, T.; Kankaanpää, H.; Paju, M.; Ruuskanen, P. Magnetic Behavior of Sintered NdFeB Magnets on a Long-Term Timescale. *Adv. Mater. Sci. Eng.* **2014**, *7*, 760584. [CrossRef]
10. Prakht, V.; Dmitrievskii, V. Optimal Design of a High-Speed Flux Reversal Motor with Bonded Rare-Earth Permanent Magnets. *Mathematics* **2021**, *9*, 256. [CrossRef]
11. Ding, H.; Gong, X.; Gong, Y. Estimation of Rotor Temperature of Permanent Magnet Synchronous Motor Based on Model Reference Fuzzy Adaptive Control. *Math. Probl. Eng.* **2020**, *16*, 4183706. [CrossRef]
12. Aoyama, Y.; Miyata, K.; Ohashi, K. Simulations and experiments on eddy current in Nd-Fe-B magnet. *IEEE Trans. Magn.* **2005**, *41*, 3790–3792. [CrossRef]
13. Gabay, A.M.; Marinescu-Jasinski, M.; Liu, J.; Hadjipanayis, G.C. Internally segmented Nd-Fe-B/CaF2 sintered magnets. *IEEE Trans. Magn.* **2013**, *49*, 558–561. [CrossRef]
14. Zheng, L.; Li, W.; Zhu, M.; Ye, L.; Bi, W. Microstructure, magnetic and electrical properties of the composite magnets of Nd-Fe-B powders coated with silica layer. *J. Alloys Compd.* **2013**, *560*, 80–83. [CrossRef]
15. Marinescu, M.; Gabay, A.M.; Liu, J.F.; Hadjipanayis, G.C. Fluoride-added Pr-Fe-B die-upset magnets with increased electrical resistivity. *J. Appl. Phys.* **2009**, *105*, 07A711. [CrossRef]
16. Komuro, M.; Satsu, Y.; Enomoto, Y.; Koharagi, H. High electrical resistance hot-pressed NdFeB magnet for low loss motors. *Appl. Phys. Lett.* **2007**, *91*, 102503. [CrossRef]
17. McLachlan, D.S.; Blaszkiewicz, M.; Newnham, R.E. Electrical Resistivity of Composites. *J. Am. Ceram. Soc.* **1990**, *73*, 2187–2203. [CrossRef]
18. Xu, Y.; Yan, J.; Sun, F.; Gu, Y. Effect of alloyed Al on the corrosion behaviour of Ni-base alloys in molten glass under static condition. *Corros. Sci.* **2016**, *112*, 635–646. [CrossRef]
19. Khachatryan, H.; Baek, S.-H.; Lee, S.-N.; Kim, H.K.; Kim, M.; Kim, K.B. Metal to glass sealing using glass powder: Iron induced crystallization of glass. *Mater. Chem. Phys.* **2019**, *226*, 331–337. [CrossRef]
20. Yang, W.; Zha, L.; Lai, Y.; Qiao, G.; Du, H.; Liu, S.; Wang, C.; Han, J.; Yang, Y.; Hou, Y.; et al. Structural and magnetic properties of the R10Fe90-xSix alloys with R=Y, Ce, Pr, Nd, Sm, Gd, Tb, Dy, Ho, and Er. *Intermetallics* **2018**, *99*, 8–17. [CrossRef]
21. Zhu, M.; Li, W. Texture formation mechanism and constitutive equation for anisotropic thermorheological rare-earth permanent magnets. *AIP Adv.* **2017**, *7*, 056236. [CrossRef]
22. Shen, J.; Han, T.; Yao, L.; Wang, Y. Is Higher Resistivity of Magnet Beneficial to Reduce Rotor Eddy Current Loss in High-Speed Permanent Magnets AC Machines? Trans. *CHINA Electrotech. Soc.* **2020**, *35*, 2074–2078. [CrossRef]

 *materials*

Article

Influence of Vanadium Micro-Alloying on the Microstructure of Structural High Strength Steels Welded Joints

Giulia Stornelli [1,*], Anastasiya Tselikova [2], Daniele Mirabile Gattia [3], Michelangelo Mortello [4], Rolf Schmidt [2], Mirko Sgambetterra [5], Claudio Testani [6], Guido Zucca [5] and Andrea Di Schino [1]

[1] Dipartimento di Ingegneria, Università degli Studi di Perugia, Via G. Duranti 93, 06125 Perugia, Italy; andrea.dischino@unipg.it

[2] Vantage Alloys AG, 6300 Zug, Switzerland; anastasiya.tselikova@vantage-alloys.com (A.T.); rolf.schmidt@vantage-alloys.com (R.S.)

[3] Dipartimento Sostenibilità dei Sistemi Produttivi e Territoriali, ENEA—CR Casaccia, 00123 Rome, Italy; daniele.mirabile@enea.it

[4] Sede di Genova, Istituto Italiano della Saldatura, Lungobisagno Istria 15, 16141 Genova, Italy; michelangelo.mortello@iis.it

[5] Aeronautical and Space Test Division, Italian Air Force, Via Pratica di Mare 45, 00040 Pomezia, Italy; mirko.sgambetterra@gmail.com (M.S.); zucca.guido@gmail.com (G.Z.)

[6] CALEF-ENEA CR Casaccia, Via Anguillarese 301, Santa Maria di Galeria, 00123 Rome, Italy; claudio.testani@consorziocalef.it

* Correspondence: giulia.stornelli@unipg.it

Abstract: The inter-critically reheated grain coarsened heat affected zone (IC GC HAZ) has been reported as one of the most brittle section of high-strength low-alloy (HSLA) steels welds. The presence of micro-alloying elements in HSLA steels induces the formation of microstructural constituents, capable to improve the mechanical performance of welded joints. Following double welding thermal cycle, with second peak temperature in the range between Ac1 and Ac3, the IC GC HAZ undergoes a strong loss of toughness and fatigue resistance, mainly caused by the formation of residual austenite (RA). The present study aims to investigate the behavior of IC GC HAZ of a S355 steel grade, with the addition of different vanadium contents. The influence of vanadium micro-alloying on the microstructural variation, RA fraction formation and precipitation state of samples subjected to thermal cycles experienced during double-pass welding was reported. Double-pass welding thermal cycles were reproduced by heat treatment using a dilatometer at five different maximum temperatures of the secondary peak in the inter-critical area, from 720 °C to 790 °C. Although after the heat treatment it appears that the addition of V favors the formation of residual austenite, the amount of residual austenite formed is not significant for inducing detrimental effects (from the EBSD analysis the values are always less than 0.6%). Moreover, the precipitation state for the variant with 0.1 wt.% of V (high content) showed the presence of vanadium rich precipitates with size smaller than 60 nm of which, more than 50% are smaller than 15 nm.

Keywords: vanadium micro-alloying; microstructure; residual austenite; precipitation

Citation: Stornelli, G.; Tselikova, A.; Mirabile Gattia, D.; Mortello, M.; Schmidt, R.; Sgambetterra, M.; Testani, C.; Zucca, G.; Di Schino, A. Influence of Vanadium Micro-Alloying on the Microstructure of Structural High Strength Steels Welded Joints. *Materials* **2023**, *16*, 2897. https://doi.org/10.3390/ma16072897

Academic Editor: Abdollah Saboori

Received: 21 March 2023
Revised: 30 March 2023
Accepted: 3 April 2023
Published: 5 April 2023

1. Introduction

Recent developments of high strength low alloyed steels (HSLA) for energy sector focused on the need to target optimized combinations of strength, toughness, weldability on industrial scale at affordable prices [1–5]. A similar scenario also applies over other application sectors (e.g., offshore structural application and shipbuilding) with different specific requirements, which are set as a function of technological and exercise needs [6]. Vanadium, due to its own thermodynamic and kinetic ability to precipitate in form of carbide and nitride, is considered a key element in the metallurgical design of modern HSLA steels [7–9], as it enables efficient and cost-effective solution across a broad range of applications [10–13]. For example, in high strength-high toughness steels for pipelines, the increase in the available strength level up to X80–X100 grade of line pipe steels, promoted in the latest decades, has produced economic advantages estimated in the billion-dollar range [14].

The evolution of the effect of microalloying on the microstructure and the properties of a girth weld is challenging as this depends on the number of inter-correlated metallurgical phenomena correlated to the steel chemical composition and the welding processing conditions [15–17].

Despite the importance of microalloying for the development of high strength steels with increased toughness, a decay of material properties in girth welded joints is reported in literature [18] which has limited an excessive use of microalloying.

Thermal cycles experienced during welding have a large impact on equilibrium set between high strength and high toughness in HSLA steels as these cycles are the main cause of toughness loss in the heat affected zone (HAZ).

Welds and heat-affected zones are critical when considering structural integrity, specifically, fracture and fatigue properties [19,20]. Weld design in thicker gauge plates requires consideration of the time required to perform the weld, which is partially controlled through the heat input. With the aim of saving time, reducing component manufacturing costs, and improving efficiency, high heat input welding technology has been widely used.

Historically, the lowest toughness was expected in the grain coarsened heat affected zone (GC HAZ), which is the part of the HAZ closest to the welding fusion line [21–24]. During welding, the GC HAZ experiences peak temperatures up to the melting point, followed by rapid cooling. The high temperatures can lead to significant austenite grain coarsening [25], the combination of a coarse austenite grain size and rapid cooling promotes brittle microstructures, which contain high proportions of ferrite side-plates and bainite [26].

In recent years, it has been found that the most degraded part in the HAZ is the inter-critically reheated grain coarsened HAZ (IC GC HAZ), which is the region of the GC HAZ reheated to temperatures between the Ac1 and Ac3 by subsequent welding passes [27]. During the inter-critical thermal cycle, partial transformation to austenite occurs, particularly where austenite stabilizers, such as carbon or manganese, are segregated in the initial microstructure [28]. These areas include pearlite/bainite colonies. When cooling, these high carbon regions transform into pearlite/bainite or residual austenite (RA) depending on the hardenability of the austenite and cooling rate [29]. The presence of RA phase is generally regarded as the major factor which reduces the HAZ toughness [30,31].

However, it is also reported that the loss in toughness is not just due to the presence of RA phase, but is related to the distribution and morphology of the RA constituent, and the matrix microstructure. Cui et al. [32] reported that block residual austenite significantly deteriorates impact toughness of super-critical reheated coarse grain heat affected zone (SC CGHAZ).

Niobium is commonly added to enhance the strength of HSLA steels. However, under welding conditions, niobium adoption shows detrimental effect on the HAZ toughness, although its effect is strongly dependent on heat input [33]. At medium to high heat input despite a precipitation hardening effect via Nb(C,N), niobium has a detrimental influence on the fracture toughness of GC HAZ. Niobium reduces the grain boundary ferrite formation and promotes the nucleation of coarse structure of ferrite with aligned RA resulting in increased mechanical properties. A small addition of niobium suppresses ferrite nucleation at prior austenite grain boundaries and increase the volume fraction of either martensite or bainite. Previous studies reported that the major advantages of a niobium addition, i.e., the grain refinement and the resultant improvement of base metal mechanical properties, appear to be outweighed by the detrimental effects of martensite formation, when the steel plates are welded [34].

On the other hand, vanadium leads to grain refinement and precipitation strengthening to HSLA steels. The effect of vanadium on the GC HAZ microstructure is quite different from that of niobium. Vanadium has a beneficial effect on the toughness of the GC HAZ, because it reduces the bainitic colony size and, due to the low misfit between vanadium nitrides (VN) and ferrite in comparison with other types of inclusions, promotes intragranular nucleation of acicular ferrite [20,35].

Moreover, alloying associated with the precipitate formation is an important consideration in weld design to achieve desirable microstructures in the HAZ [36]. For instance, Zajac et al. [37] performed HAZ simulations on 25 mm-thick HSLA plates (Fe-0.09%C-1.4%Mn-0.08%V-0.010%Ti-xN) with low (0.003%N) and high (0.013%N) nitrogen contents. They showed that for the high nitrogen steel, intragranular forms for a wider range of cooling rates as compared to the low nitrogen steel. In contrast, in high nitrogen steel for the high heat input (slowly cooled) conditions, a significant fraction of coarse ferrite grains forms at the austenite grain boundaries, which leads to poorer toughness compared to the low nitrogen steel. Zajac et al. [37] interpreted that the increased fraction of coarse grain boundary ferrite was associated with V(C,N), the amount of which would likely be increased during slow cooling; it should be noted that the mechanism for V(C,N) to accelerate the formation of grain boundary ferrite is not clearly discussed in the study. Zajac et al. [37] also pointed out that the precipitation status, which depends on the peak temperature in HAZ simulation, influences austenite grain size and amount of 'free' nitrogen, both of which affect phase transformations upon cooling.

Hu et al. [38] and Wu et al. [39] showed that V(C,N) with sizes between 20–30 nm were not detrimental on impact toughness because of their small sizes.

Following the above mentioned latest developments, the influence of vanadium on the toughness and fatigue resistance of the IC GC HAZ is not fully understood and requires further investigation. To understand the effective resistive behavior due to the addition of vanadium in the IC GC ZTA of a welded joint, it is first of all considered appropriate to investigate the real effect of the alloying elements on the microstructural characteristics.

In this regard, this study aims to assess the effect of vanadium alloying on material properties (in terms of microstructural constituent variation, RA formation and precipitation state) of an S355 steel (EN10025-2), when subjected to welding-representative thermal cycles in the IC GC HAZ.

2. Materials and Methods

S355 steel grade (EN10025-2) plates for structural application were manufactured by Vacuum Induction Melting (VIM) plant in the form of three 80 kg ingots (diameter 120 mm) in four variants, including its base reference. The nominal chemical composition of the considered steels are reported in Table 1.

Table 1. Nominal chemical composition of the considered steels (wt.%) (Fe to balance).

	C	Mn	V	Si	Nb
Reference material	0.16	1.45	-	0.03	-
Variant I	0.16	1.45	0.05	0.03	-
Variant II	0.16	1.45	0.10	0.03	-
Variant III	0.16	1.45	0.03	0.03	0.02

The ingots have been hot rolled down to 16 mm thickness in 10 passes. The steels chemical compositions to be investigated were designed in order to have a Carbon Equivalent Content (Ceq) value lower than 0.42%, according to International Institute of Welding (IIW) Equation (1), as a function of weight percentage (%) [40]:

$$Ceq = \%\,C + \frac{\%\,Mn}{6} + \frac{\%\,Cr + \%\,Mo + \%\,V}{5} + \frac{\%\,Cu + \%\,Ni}{15} \tag{1}$$

The hot rolled microstructures are reported in Figure 1 after 2% Nital etching. Starting from the hot rolled material, cylindrical specimens (10 mm in length, 4 mm in diameter) were machined to be heat treated in controlled conditions by using a dilatometer. The IC GC HAZ thermal cycles, in accordance with Figure 2, were designed to simulate a double pass submerged arc welding process with heat input of 2.5 kJ/mm in 16 mm thick plate [41]. The initial temperature of the first pass was assumed at the room temperature (25 °C) while, for the second pass, the value was set to 150 °C. Because of the technological limitations of the dilatometer, the samples were heated up to 1100 °C with a heating rate of 100 °C/s whilst the holding time was set at 3 s. The cooling profile was set in order to guarantee the cooling time between 800 °C and 500 °C (t8/5) of about 25 s [41,42]. The second peak of weld conditions, with a peak temperature in the inter-critical zone, was selected by considering the values of critical temperature Ac1 and Ac3. In fact, these temperatures, obtained by dilatometric test, were reported in Table 2 and they are dependent on steel variants. Ac1 and Ac3 can be estimated through empirical equations taking into account the alloy elements [43]. However, in this study Ac1 and Ac3 were assumed equal to 715 °C and 815 °C respectively so that the peak temperature of the second pass was in the inter-critical zone for all steel variants. Therefore, the heat treatments were designed with the aim to reproduce different microstructures corresponding to different positions of the HAZ in a welded joint. In particular, the inter-critical zone of the second welding pass has been simulated with five different peak temperatures: 720, 735, 750, 775 and 790 °C (see Figure 2).

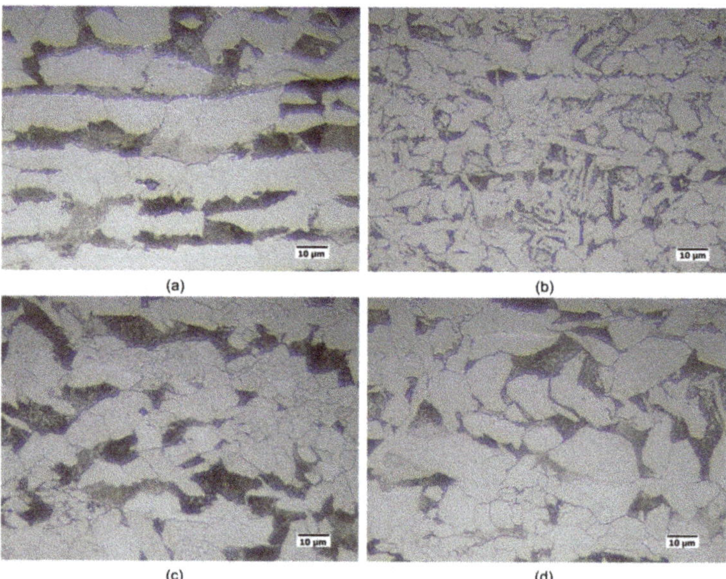

Figure 1. Hot rolled material (2% Nital etching) ((**a**) Reference material, (**b**) Variant I, (**c**) Variant II, (**d**) Variant III).

In order to investigate the presence of residual austenite (RA) and to define the most suitable methodology to assess the RA presence in the considered steels after the heat treatment, three different methods have been applied: X-ray Diffraction (XRD), Electron Backscattered Diffraction (EBSD), and LePerà selective etching. XRD analysis was carried out by using a Smartlab Rigaku diffractometer equipped with Cu kα source radiation and a D/teX Ultra 250 SL detector, operated at 40 kV and 30 mA in continuous mode in the angular range 30–110 2θ degree. The automated sample alignment routine has been used. EBSD measurements were performed with the aim to detect the presence and position of RA islands, by means of a field emission gun scanning electron microscope (FEG-SEM) (Ultra-Plus Carl-Zeiss-Oberkochen, Jena, Germany) equipped with an EBSD detector (C Nano Oxford Instruments, Stockholm, Sweden), using a 0.1 µm scanning step size. RA was revealed by building up phase maps, taking into account both face-centered cube (fcc) and body-centered cube (bcc) phases: automatic image analysis of such maps allowed to determine RA volume fraction.

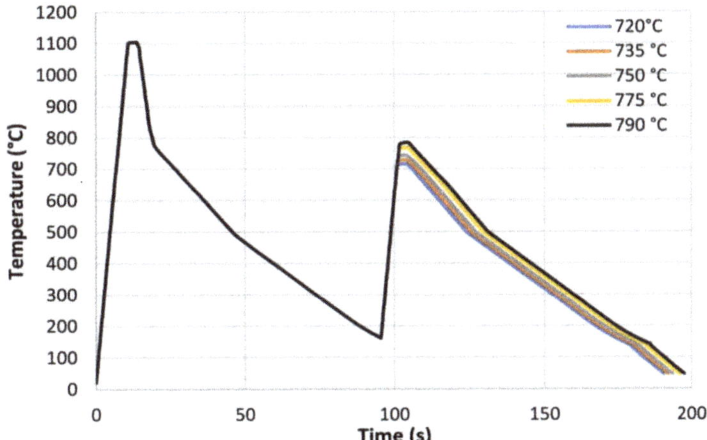

Figure 2. Experimental thermal profiles as acquired by thermocouples as obtained by dilatometry.

Table 2. Critical temperature Ac1 and Ac3 evaluated by means of dilatometric test.

	Ac1 (°C)	Ac3 (°C)
Reference material	715	825
Variant I	712	820
Variant II	715	845
Variant III	700	815

LePerà solution (1 g $H_2S_2O_5$ + 100 mL H_2O + 4 g $C_6H_3N_3O_7$ + 100 mL C_2H_5OH) for about 60–90 s was conducted for selective etching. The microstructure was then analyzed by optical microscopy (OM) (Eclipse LV150 NL, Nikon, Tokyo, Japan) whilst the image analysis was performed using dedicated software (AlexaSoft, X-Plus, serial number: 6308919690486393, Florence, Italy), in order to determine the RA fraction. The procedures were performed on low magnification image and on three different fields: the RA % reported refers to the average of three values. Vickers hardness tests were made by means of a HV50 (Remet, Bologna, Italy) instrument by using a load of 10 kg. Three hardness tests were performed on each sample. Precipitation state was analyzed by transmission electron microscope (TEM) on extraction replica specimens. The observations were performed with a JEOL 200CX transmission electron microscopy (JEOL Ltd., Tokyo, Japan). The analysis was carried out over a significant area, evaluating the chemical composition (by means of EDX analysis) and the average size of the precipitates, within a limit of 50 precipitates for each sample analysed.

3. Results and Discussions

3.1. Microstructure and Hardness

The microstructural evolution detected by SEM for all the considered steels subjected to heat treatments (as shown in Figure 2) is reported in Figures 3–6.

Moving from 790 °C to 720 °C the microstructure changes from ferrite-perlite to bainite, regardless the chemical composition, as confirmed by EBSD pole figure maps, reported in Figures 7–10. This demonstrated that the formation of certain microstructural constituent, after welding thermal cycles, is not sensitive to micro-alloying addition in the selected ranges (V up to 0.10% and V-Nb addition up to 0.05%). Moreover, it is visible that Variant

II shows the smallest and more uniform grain size among all the other variant for a peak temperature of 790 °C.

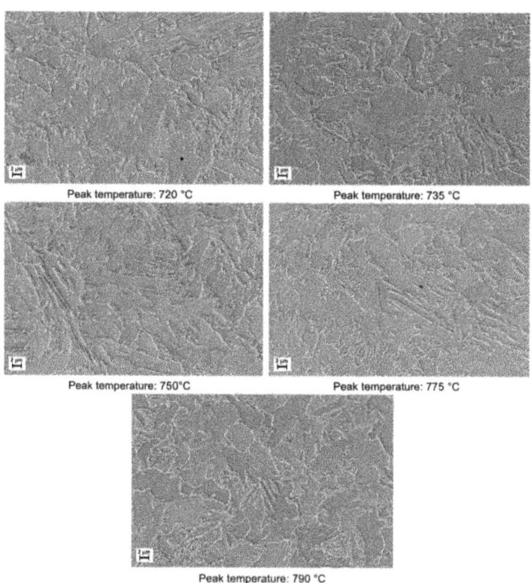

Figure 3. Microstructures as obtained by dilatometric cycles (reference material).

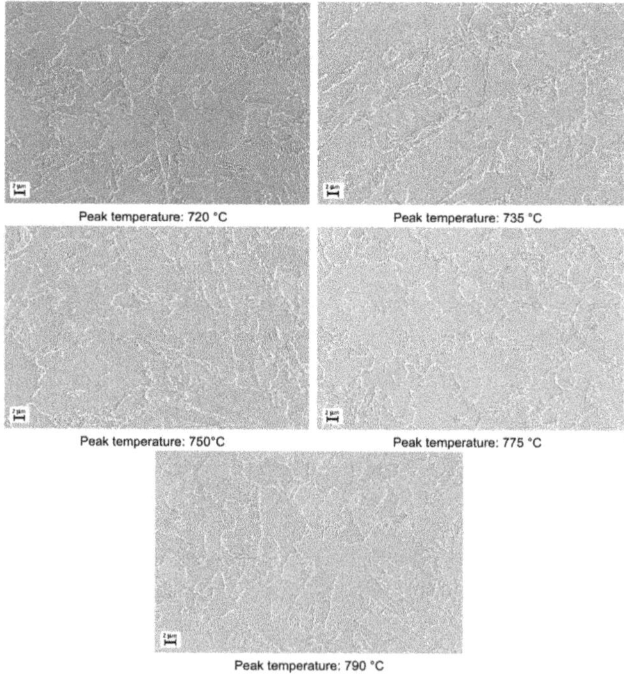

Figure 4. Microstructures as obtained by dilatometric cycles (variant I).

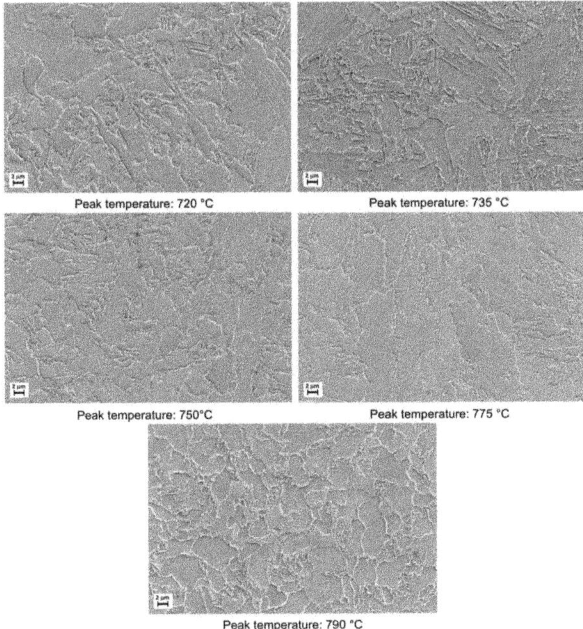

Figure 5. Microstructures as obtained by dilatometric cycles (variant II).

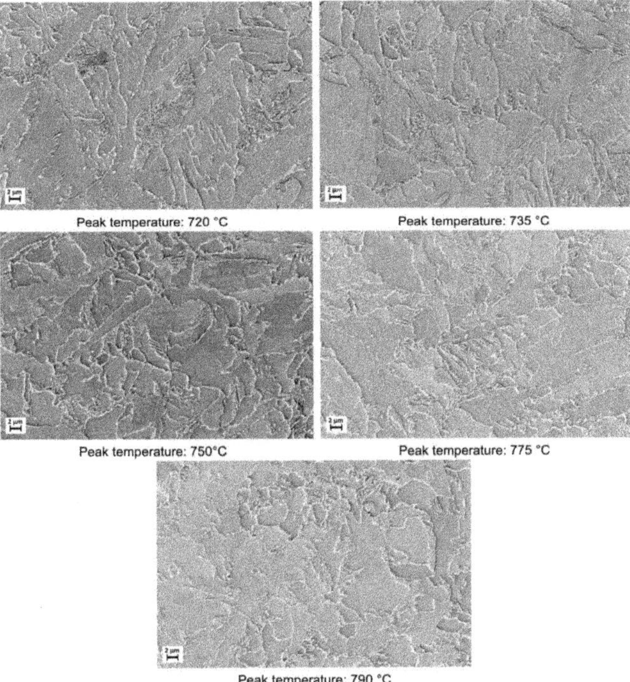

Figure 6. Microstructures as obtained by dilatometric cycles (variant III).

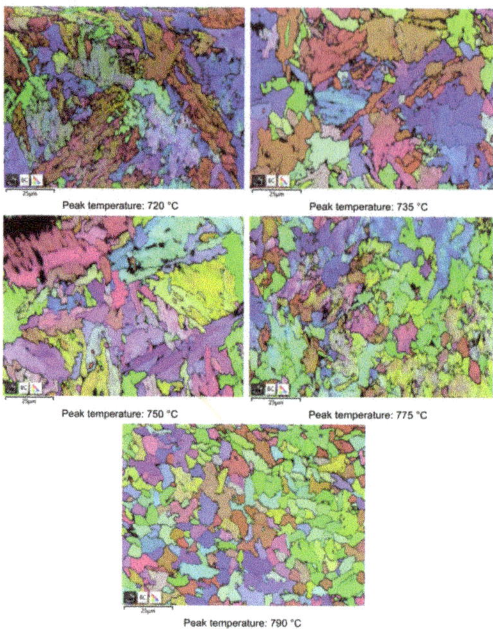

Figure 7. EBSD polar figure maps of specimens subjected to dilatometric cycles (reference material).

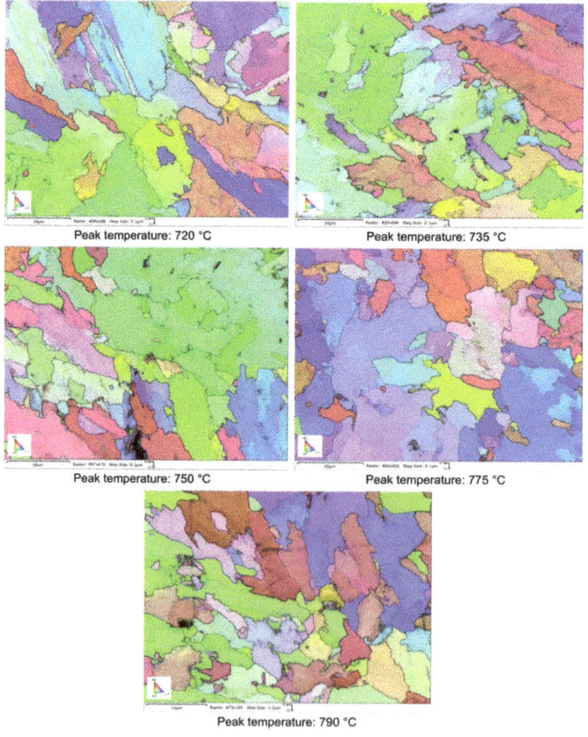

Figure 8. EBSD polar figure maps of specimens subjected to dilatometric cycles (variant I).

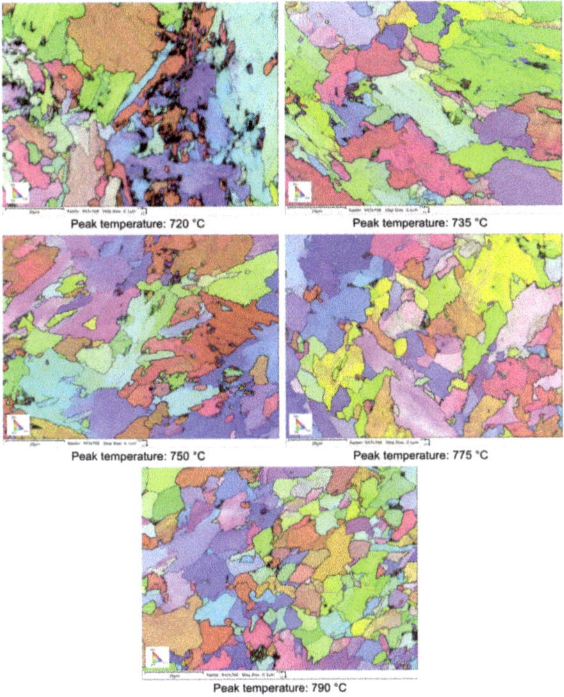

Figure 9. EBSD polar figure maps of specimens subjected to dilatometric cycles (variant II).

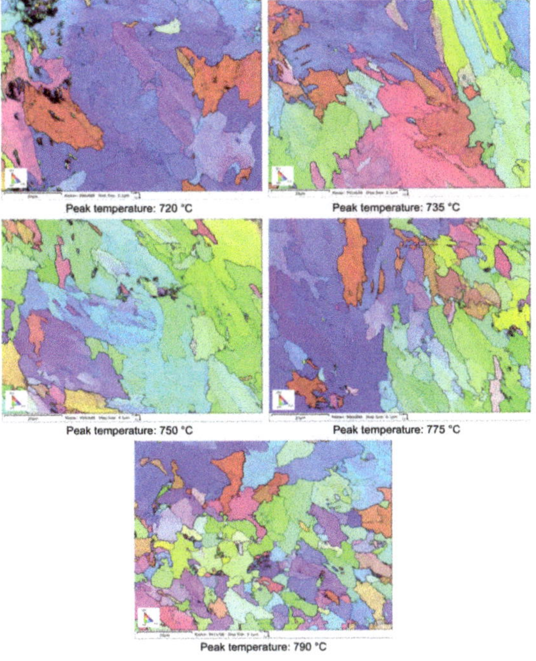

Figure 10. EBSD polar figure maps of specimens subjected to dilatometric cycles (variant III).

Figure 11 shows the quantification of high angle grain boundaries (HAGBs %) ($\phi > 10°$), obtained by EBSD analysis, for each condition. The tendency of all the variants, except for Reference material, is to increase HAGBs fraction with the increase of the inter-critical temperature. Variant II has the highest fraction of HAGBs in comparison to other variants. At the same time, Variant I shows the same trend of Variant II and exhibits higher fraction of HAGBs in comparison to Variant III, except for 735 °C and 790 °C. This suggests that, in regards to the HAGBs fraction, Variant II is expected to show the highest fatigue and toughness performance, since these grain boundaries type are responsible for a higher deflection of cracks during a fatigue cycle and an obstacle to cleavage propagation [44–46].

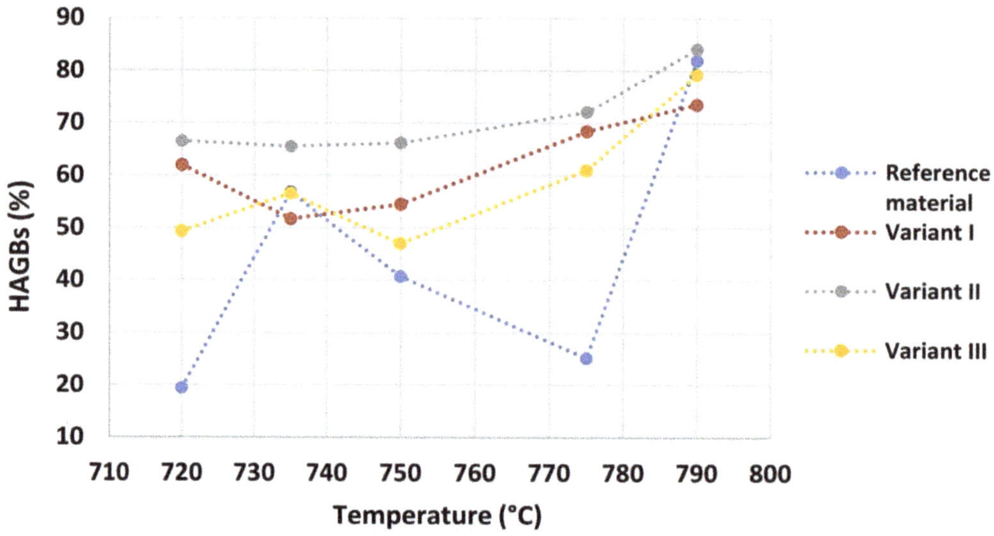

Figure 11. High-angle grain boundaries quantification (HAGBs %) ($\phi > 10°$) for each condition.

The hardness dependence of all the considered variants as a function of the inter-critical temperature is reported in Figure 12. As expected, after welding thermal cycle, for each steel variant there is an increase of hardness value compared to hot rolled state. Moreover, while the reference material appears to be independent on the tested temperature, Variant I is subjected to a hardness loss starting from a hardness value approximately similar to that of the reference material. A clear effect of micro-alloying is reported for Variant II (0.10% V) and Variant III (0.03% V–0.02% Nb). These results show that an increase in the inter-critical temperature leads to a decrease of hardness. Typically, at a fixed inter-critical temperature of 720 °C an increase by approximately 30 Vickers points of each variant in comparison to its initial value in hot rolled state is present. For a temperature of 790 °C, Variant I kept the same hardness, Reference material experienced an increase of 30 Vickers points, while Variant III and Variant II experience an increase of hardness by 20 Vickers points. These results are consistent with the ones published by [47,48].

In particular, both for Variant II and for Variant III, a peak of hardness at 735 °C is evident and the nature of this behavior can be attributed to both the presence of residual austenite and a different precipitation state. A desired strengthening of the steels due to formation of the residual austenite would be detrimental in terms of toughness [30,31]. Otherwise, an adequate state of precipitation (fine and homogeneously dispersed precipitates) would ensure the strengthening and, at the same time, a better fatigue behavior [49]. The investigation of these aspects (RA and precipitation state) has been conducted and is illustrated in the Sections 3.2 and 3.3.

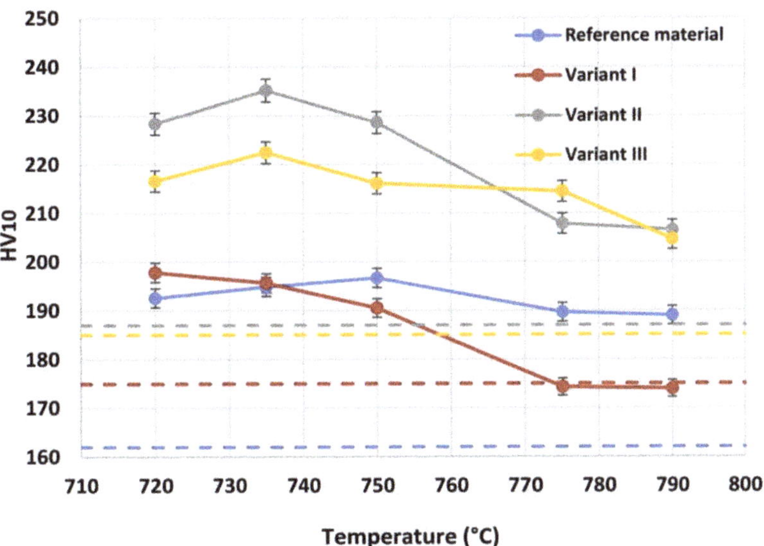

Figure 12. Hardness dependence on inter-critical temperature for the different considered materials.

3.2. Residual Austenite

In order to evaluate the RA content on the above considered specimens, XRD patterns have been acquired. Results reported in Figure 13 for the reference material and for Variant III show no evidence of variation between the different conditions, since the RA fraction is low enough to stand below the intrinsic sensibility threshold of the technique (equal to about 1 wt.%). Therefore, X-ray diffraction technique showed to be not suitable for RA determination in HAZ for the considered set of samples.

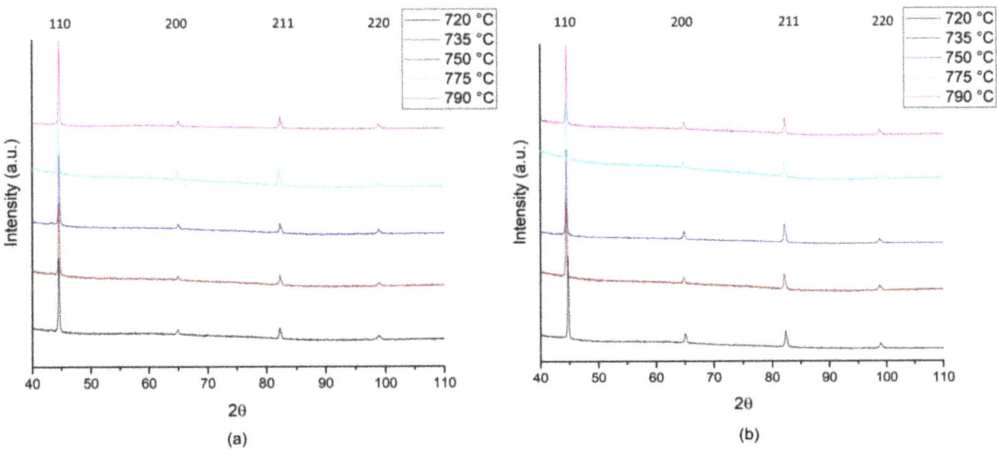

Figure 13. XRD spectra for reference material (**a**) and variant III (**b**) as a function of second temperature peak.

RA content has also been evaluated by the analysis of phase maps as obtained by EBSD (Figures 14–17).

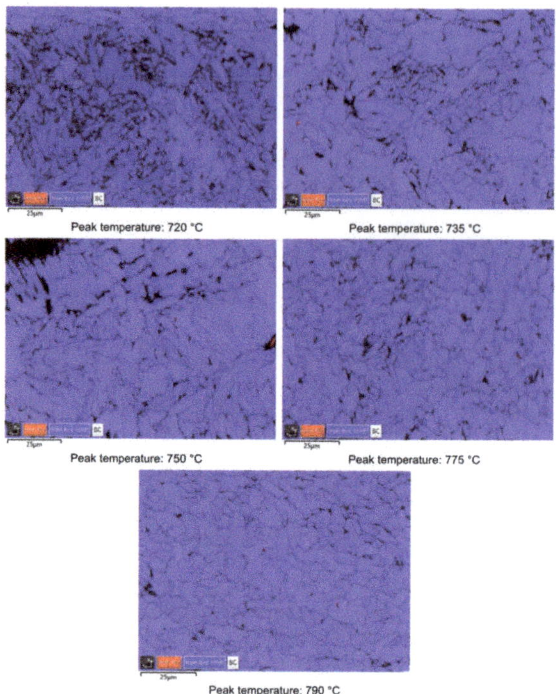

Figure 14. EBSD phase maps. Red zones: RA phase (reference material).

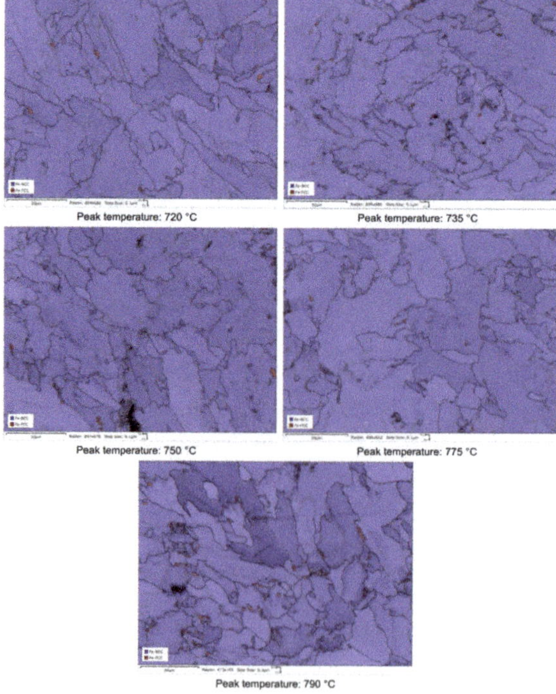

Figure 15. EBSD phase maps. Red zones: RA phase (variant I).

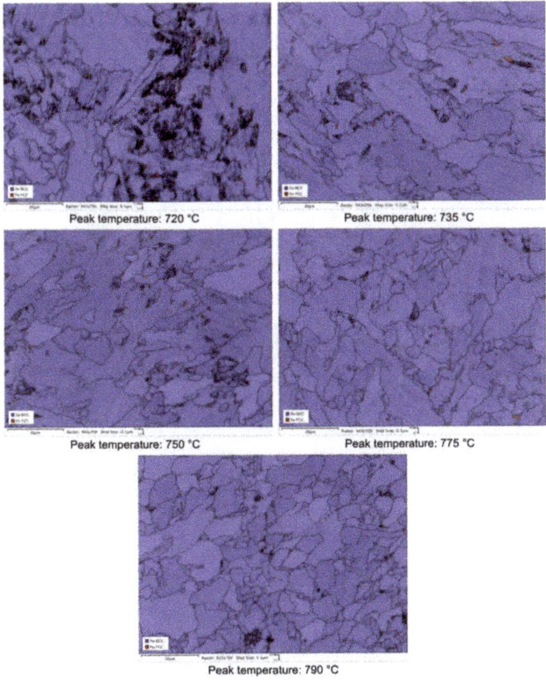

Figure 16. EBSD phase maps. Red zones: RA phase (variant II).

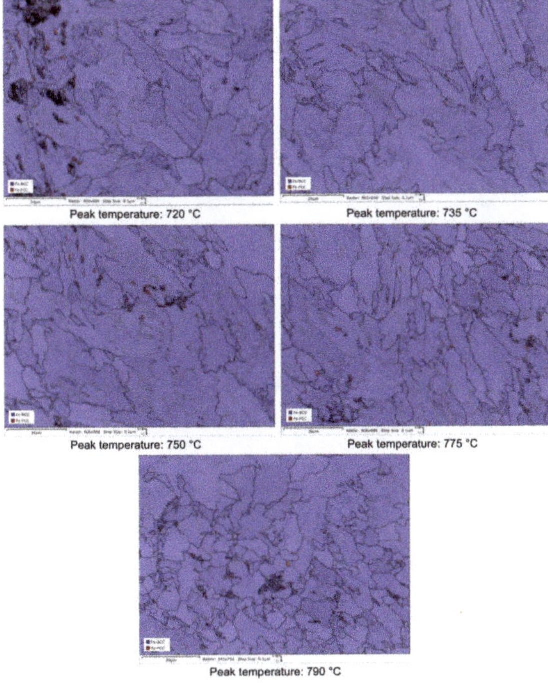

Figure 17. EBSD phase maps. Red zones: RA phase (variant III).

Quantitative optical microscopy metallography after LePerà selective etching has been applied to the same specimens and the micrographs referring to Variant I and Variant II are shown as an example in Figures 18 and 19.

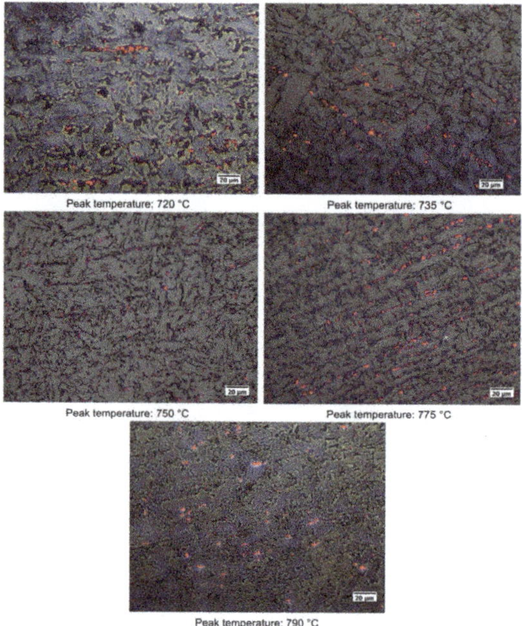

Figure 18. Microstructures of the considered specimens after LePerà etching. Red zones: RA phase (Variant I).

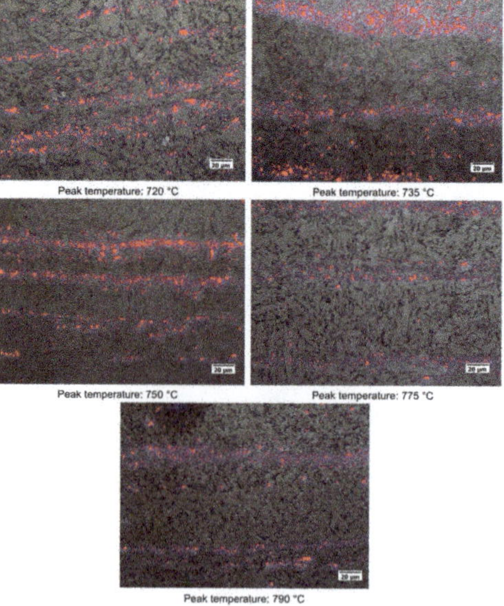

Figure 19. Microstructures of the considered specimens after LePerà etching. Red zones: RA phase (Variant II).

RA fractions as a function of second peak temperature for the considered steels are reported in Figure 20, as obtained by the analysis of EBSD phase maps (Figure 20a) and by optical metallography after selective etching (Figure 20b), giving scope for a comparison. RA % values as obtained by light microscopy analysis after selective etching appears to be higher in magnitude than those by crystallographic information given by EBSD. Moreover, the reference material shows more residual austenite than the rest of the variants in the selective etching technique. Such a result, not predicted, is already mentioned in literature (e.g., [18]) and it is related to the fact that the selective etching could partially affect some zones neighboring the austenite areas, together with some limitations in the measurements accuracy. Therefore, although the EBSD technique applies for smaller investigation areas, it is still more accurate for evaluating RA % values compared to selective etching. As a consequence, the values obtained by analyzing the EBSD phase maps were considered more reliable. However, the capability of observing an extended area by optical microscopy enables the estimation of localization and distribution of residual austenite. In fact, as shown in Figures 18 and 19, the constituent RA is arranged along bands, precisely in correspondence with the segregated areas of the original microstructure (Figure 1), where the stabilizing elements of austenite (such as C of Mn) are more concentrated [28]. It is also worth to mention that the RA values determined by means of EBSD technique are well below the X-ray diffraction threshold (equal to 1 wt.%), thus confirming the unsuitability of such a technique in the analysis of RA content in inter-critical zones of welded joints of considered steels. Furthermore, taking each steel variant as a separate reference, RA values shown in Figure 20 prove that the temperature of the second peak does not seem to influence the content of RA regardless of the technique used (either EBSD or LePerà selective etching). In this regard it is useful to consider these two methods as complementary, EBSD to determine the numerical value and LePerà selective etching method to give a localization overview of RA in HAZ of welded joint. Moreover, the experimental results of RA % obtained by EBSD technique (Figure 20a) show that V addition promotes the formation of residual austenite in agreement with [33]. However, there is no evidence that such low RA contents which were determined can be considered responsible for the hardness behavior reported in Figure 10. In this regard, in order to understand the hardness peak at 735 °C shown in Figure 10, the analysis of the precipitation state conducted by TEM is reported in the Section 3.3.

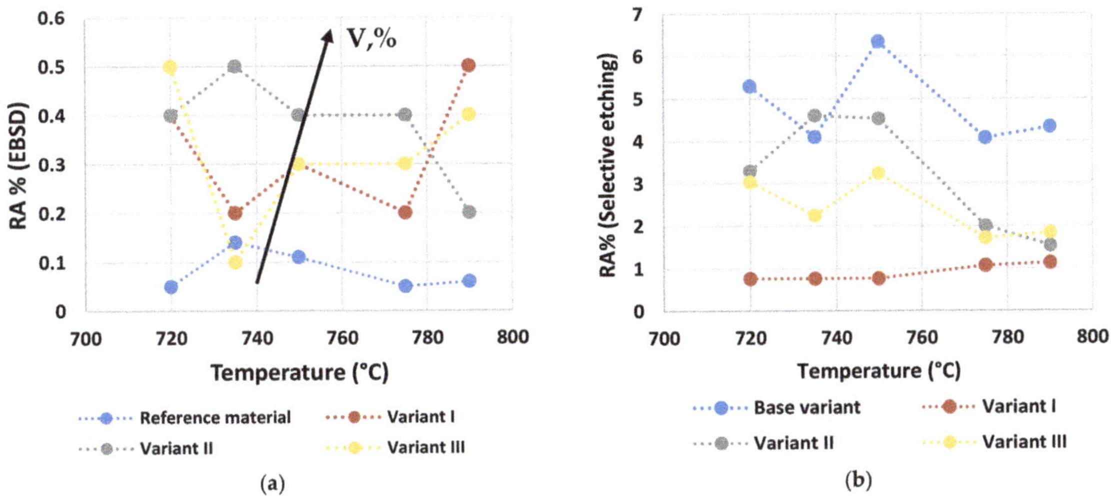

Figure 20. Quantified RA % with EBSD phase maps (**a**) and selective etching (**b**) as a function of the second peak temperature for the different considered steels.

3.3. Precipitation State

Precipitation state analysis has been performed on selected specimens corresponding to the highest hardness values, according to Figure 12, in order to focus on the most critical scenario on toughness and fatigue behavior perspective. In this paragraph, the analysis of the Variant II (0.10% V) and Variant III (0.03% V and 0.02% Nb) with second peak temperature in the inter-critical range at 735 °C is reported. Analysis of Variant II specimen shows the presence of cementite regions (as expected) (Figure 21) and others with very fine V-rich precipitates in the matrix (Figures 22 and 23). The precipitates size distribution is reported in Figure 24, taking into account the chemical composition of precipitates. Results show that V-rich precipitates range sizes is below 60 nm (vanadium content in largest precipitates is lower than 0.5% as compared to a value larger than 30% for the precipitates smaller than 60 nm). In addition, more than 50% of V-rich precipitates have a size below 15 nm. This means that in such a condition vanadium addition does not appear to be critical in terms of fatigue resistance, as it would be expected in the case of its presence in largest precipitates [50].

Similarly, the analysis of Variant III shows the presence of cementite (Figure 25) and areas with precipitates rich in Nb-V (Figure 26) and Nb (Figure 27). However, in this specific case, the distribution of the frequency of the size of the precipitates (Figure 28) shows a different behavior of the precipitation, compared to Variant II. V is always present combined with Nb, in precipitates smaller than 90 nm, whereas the larger precipitates, which size is up to 250 nm, are only rich in Nb. Furthermore, only 30% of the precipitates of Nb-V are smaller than 15 nm in size, evidencing that the combination of V and Nb micro-alloying could compromise the fatigue performance in the HAZ of a welded joint [51].

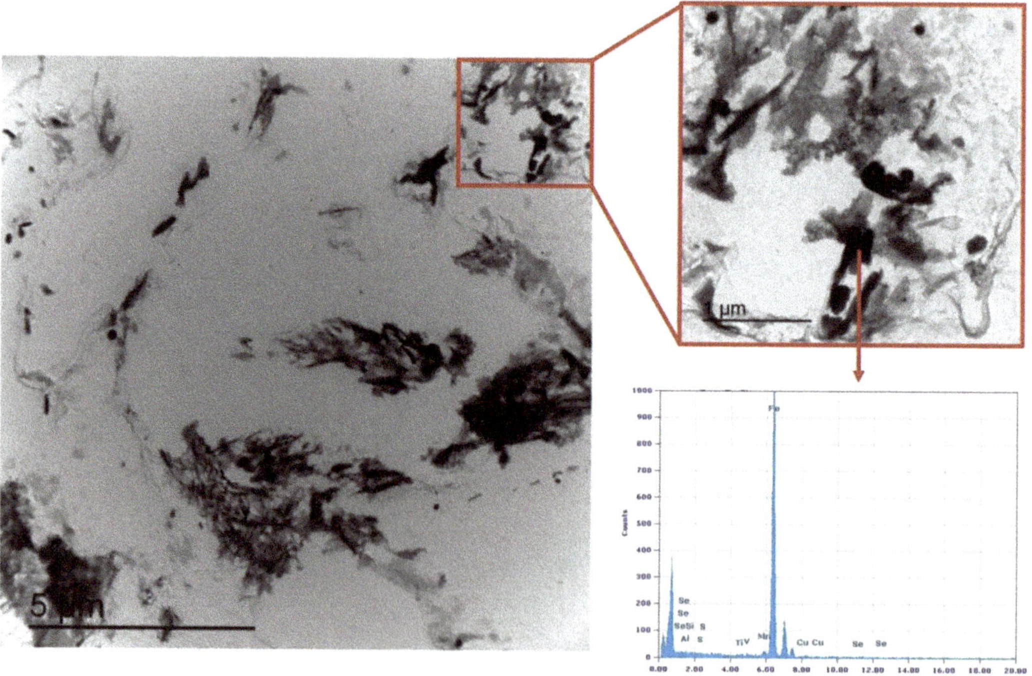

Figure 21. TEM micrograph of Variant II steel after inter-critical treatment with second peak temperature at 735 °C. Highlighted are the areas of cementite.

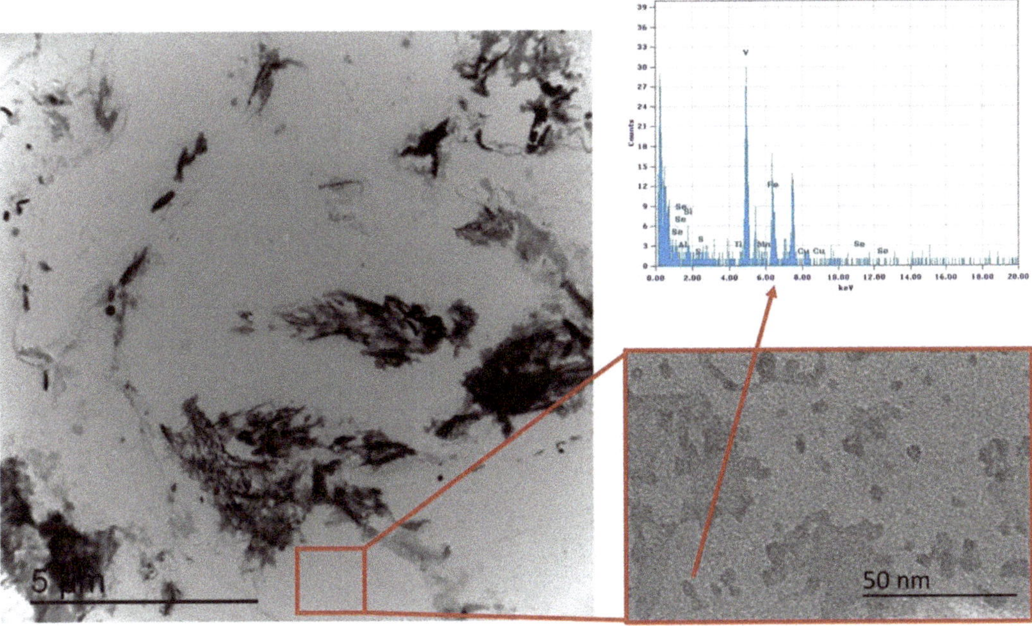

Figure 22. TEM micrograph of Variant II steel after inter-critical treatment with second peak temperature at 735 °C. Highlighted are the fine V-rich precipitates in the matrix.

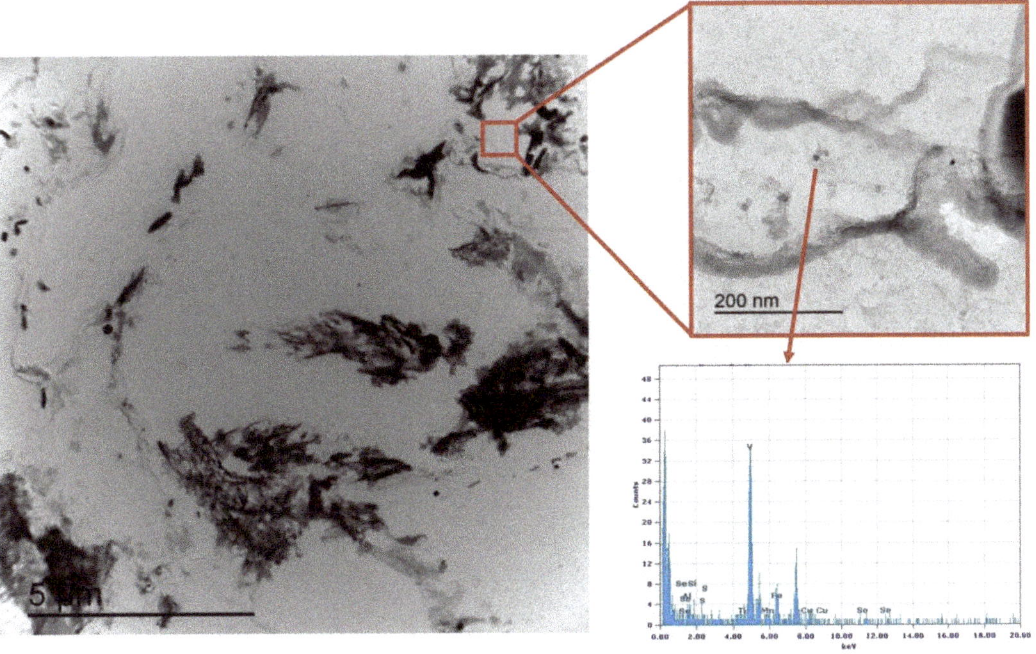

Figure 23. TEM micrograph detail of Variant II steel after inter-critical treatment with second peak temperature at 735 °C. Highlighted are the fine V-rich precipitates in the matrix.

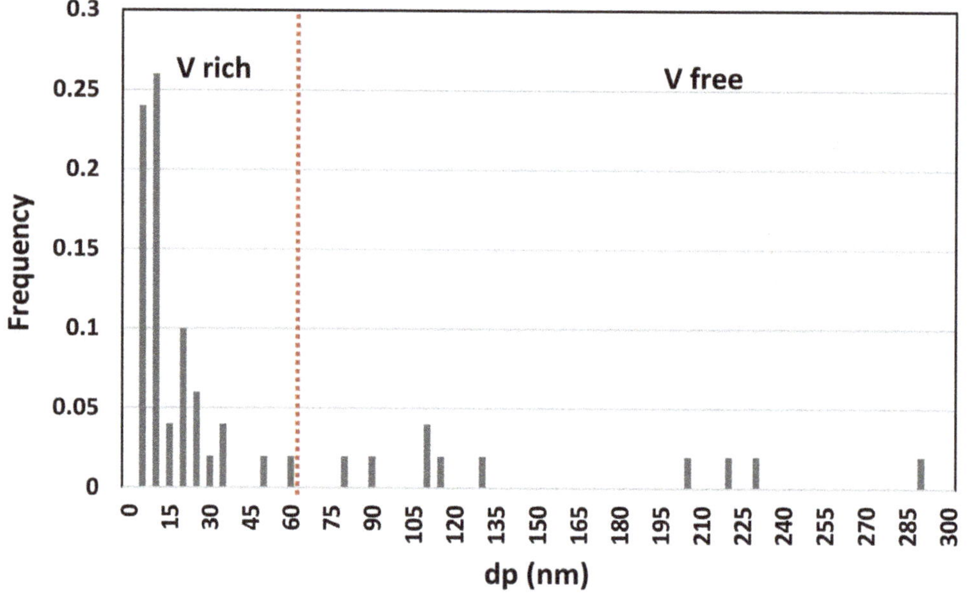

Figure 24. Precipitates size distribution (Variant II, 735 °C).

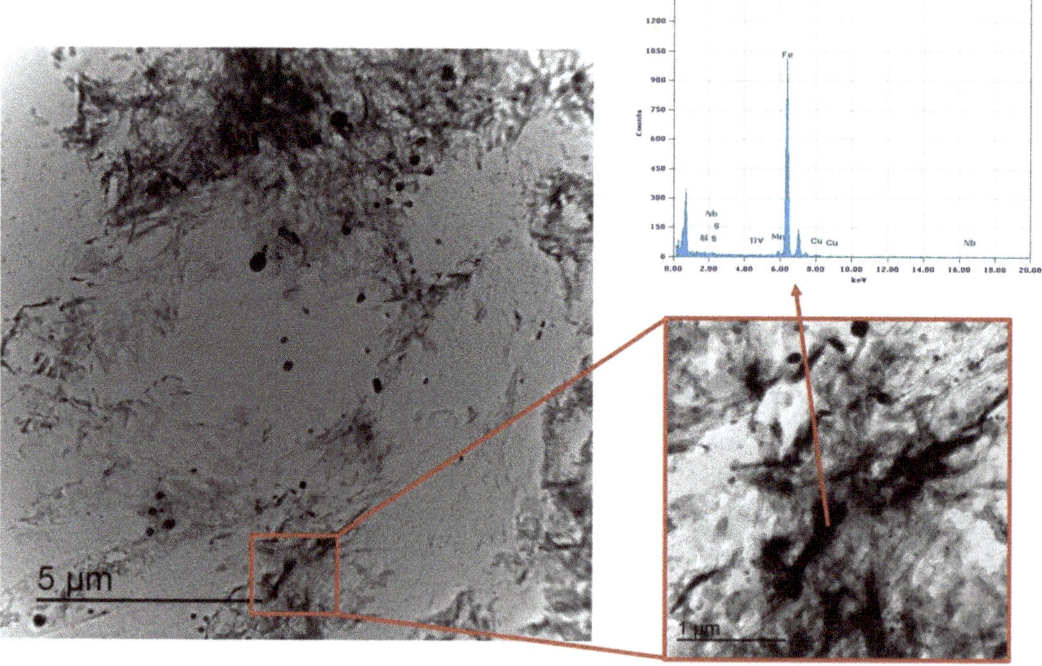

Figure 25. TEM micrograph of Variant III steel after inter-critical treatment with second peak temperature at 735 °C. Highlighted are the areas of cementite.

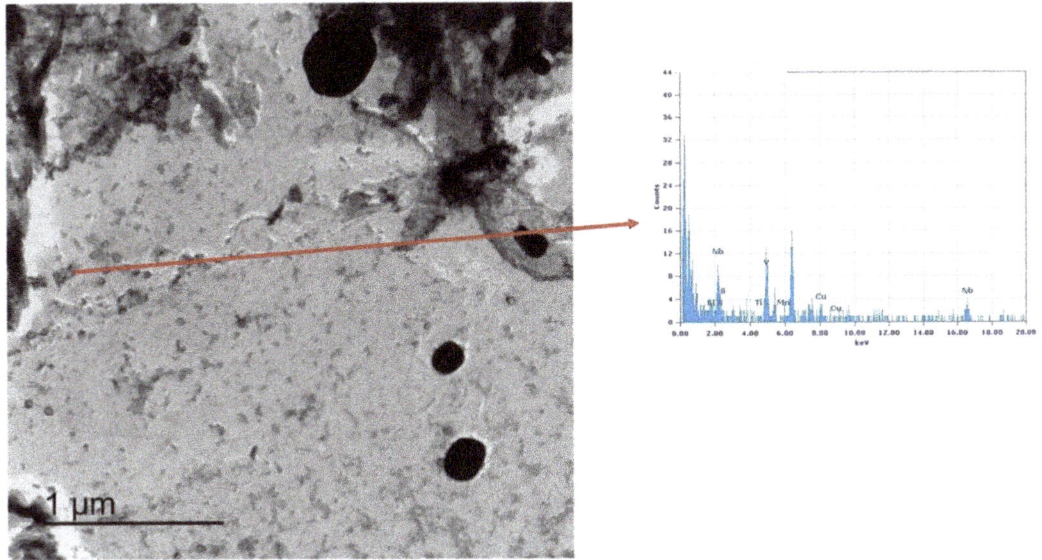

Figure 26. TEM micrograph of Variant III steel after inter-critical treatment with second peak temperature at 735 °C. Highlighted are the Nb-V rich precipitates in the matrix.

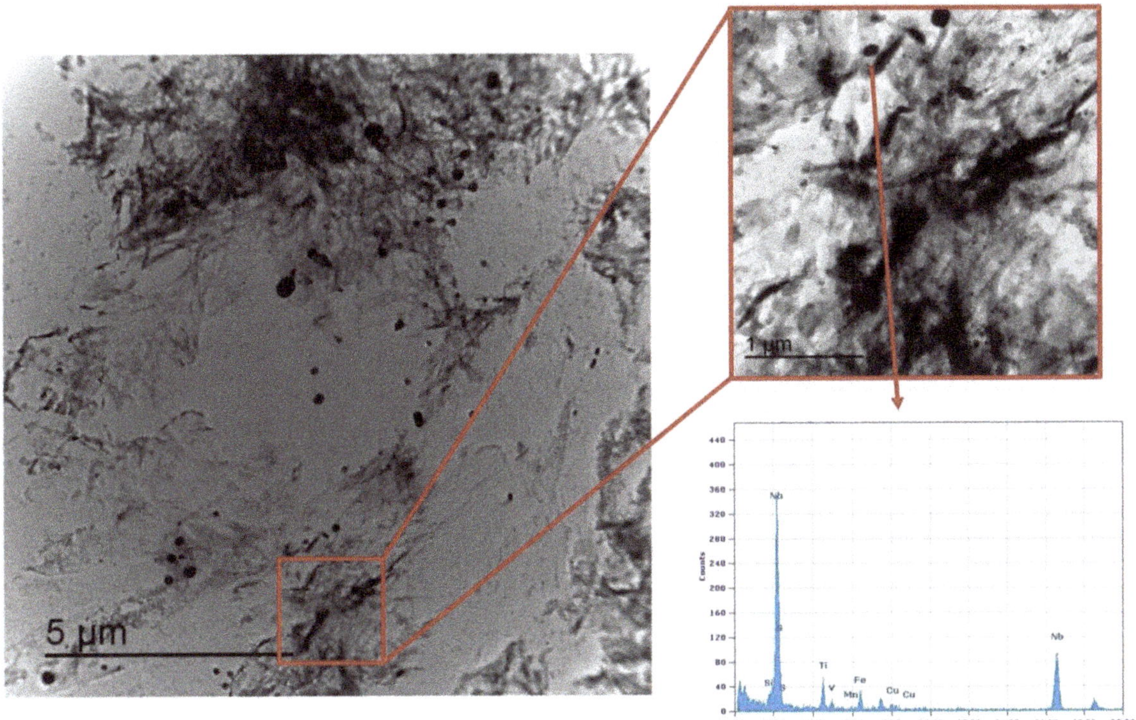

Figure 27. TEM micrograph of Variant III steel after inter-critical treatment with second peak temperature at 735 °C. Highlighted are the Nb rich precipitates in the matrix.

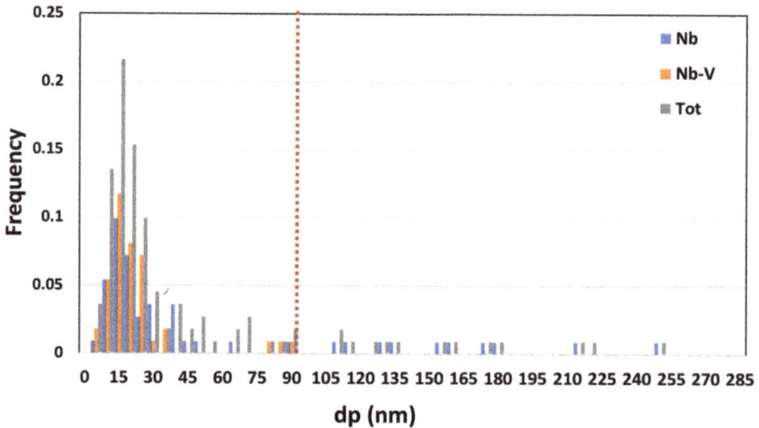

Figure 28. Precipitates size distribution (Variant III, 735 °C).

4. Conclusions

The behavior of the inter-critical region of a S355 grade steel with different vanadium content is reported in this paper. Double-pass welding thermal cycles were simulated using a dilatometer, with the maximum temperature of the secondary peak in the inter-critical area, in the range between 720 °C and 790 °C.

- Results show a negligible effect of vanadium addition on steel hardenability: this is consistent with a poor dependence of the microstructure type on steel chemical composition (in the considered variation range) taking into account the same inter-critical temperature.
- A low residual austenite value is found in combination with an increase of residual austenite content as the micro-alloying content increases. This result is independent on the inter-critical temperature and is consistent with the negligible effect on hardenability. EBSD technique has shown being more accurate in the quantification of RA fraction than selective etching in small areas, being consistent with them.
- HAGBs fraction, except for reference material, increase with the increase of inter-critical temperature for all the variants. The Variant II with high V content (0.1 wt.%) shows the highest HAGBs fraction for all the temperatures, which is expected to improve the fatigue and toughness behavior.
- Vanadium micro-alloying promotes the formation of very fine precipitates in the IC GC HAZ. Results also show that V-rich precipitates size is below 60 nm with more than 50% of V-rich precipitates having a size below 15 nm. This means that vanadium addition in such a condition does not appear to be critical in terms of fatigue resistance, as it would be expected in the case of its presence in largest precipitates. On the other hand, from the combination of V and Nb, the results of the analysis of the precipitates showed that V is present in precipitates smaller than 90 nm and combined with Nb and only 30% of the Nb-V precipitates have a size smaller than 15 nm. Furthermore, the precipitates with size larger than 90 nm are purely Nb, which is compromising for fatigue behavior.

In conclusion, in this work it has been demonstrated that vanadium addition in HSLA steel does not lead to the formation of a significant percentage of residual austenite in IC GC HAZ of a double-pass welding process. The combination of a more fine-grained microstructure, higher fraction of HAGBs and the formation of fine precipitates, can be promising for the improvement of fatigue and toughness behavior. This makes the adoption of high strength vanadium micro-alloyed steels very promising in structural applications, also enabling the use of a reduced quantity of raw-materials, hence mitigating the environmental impact of the resulting formulations.

Author Contributions: G.S.: Conceptualization, Methodology, Formal analysis, Investigation, Writing—original draft preparation, Writing—review and editing, A.T.: Conceptualization, Methodology, Writing—review and editing, Supervision, D.M.G.: Investigation, Writing—review and editing, M.M.: Investigation, Writing—review and editing, R.S.: Conceptualization, Methodology, Writing—review and editing, Supervision, M.S.: Investigation, Writing—review and editing, C.T.: Investigation, Writing—review and editing, G.Z.: Investigation, Writing—review and editin, Supervision, A.D.S.: Conceptualization, Methodology, Formal analysis, Writing—original draft preparation, Writing—review and editing, Supervision. All authors have read and agreed to the published version of the manuscript.

Funding: This research was funded by Vantage Alloys AG, Zug, Switzerland.

Institutional Review Board Statement: Not applicable.

Informed Consent Statement: Not applicable.

Data Availability Statement: The data presented in this study are available on request from the corresponding author.

Acknowledgments: Authors thank Dario Venditti (RINA Centro Sviluppo Materiali SpA) for TEM analysis.

Conflicts of Interest: The authors declare no conflict of interest.

References

1. Kim, D.W.; Yang, J.; Kim, Y.G.; Kim, W.K.; Lee, S.; Sohn, S.S. Effects of Granular Bainite and Polygonal Ferrite on Yield Strength Anisotropy in API X65 Linepipe Steel. *Mater. Sci. Eng. A* **2022**, *843*, 143151. [CrossRef]
2. Roy, S.; Romualdi, N.; Yamada, K.; Poole, W.; Militzer, M.; Collins, L. The Relationship Between Microstructure and Hardness in the Heat-Affected Zone of Line Pipe Steels. *JOM* **2022**, *74*, 2395–2401. [CrossRef]
3. Stornelli, G.; Di Schino, A.; Mancini, S.; Montanari, R.; Testani, C.; Varone, A. Grain Refinement and Improved Mechanical Properties of Eurofer97 by Thermo-Mechanical Treatments. *Appl. Sci.* **2021**, *11*, 10598. [CrossRef]
4. Stornelli, G.; Gaggia, D.; Rallini, M.; Di Schino, A. Heat Treatment Effect on Maraging Steel Manufactured by Laser Powder Bed Fusion Technology: Microstructure and Mechanical Properties. *Acta Metall. Slovaca* **2021**, *27*, 122–126. [CrossRef]
5. Stornelli, G.; Gaggiotti, M.; Mancini, S.; Napoli, G.; Rocchi, C.; Tirasso, C.; Di Schino, A. Recrystallization and Grain Growth of AISI 904L Super-Austenitic Stainless Steel: A Multivariate Regression Approach. *Metals* **2022**, *12*, 200. [CrossRef]
6. Lo, K.H.; Shek, C.H.; Lai, J.K.L. Recent Developments in Stainless Steels. *Mater. Sci. Eng. R Rep.* **2009**, *65*, 39–104. [CrossRef]
7. Benz, J. The Effect of Vanadium and Other Microalloying Elements on the Microstructure and Properties of Bainitic HSLA Steels. *Mater. Sci. Technol. Conf. Exhib.* **2017**, *1*, 490.
8. Fazeli, F.; Amirkhiz, B.S.; Scott, C.; Arafin, M.; Collins, L. Kinetics and Microstructural Change of Low-Carbon Bainite Due to Vanadium Microalloying. *Mater. Sci. Eng. A* **2018**, *720*, 248–256. [CrossRef]
9. Baker, T.N. Microalloyed Steels. *Ironmak. Steelmak.* **2016**, *43*, 264–307. [CrossRef]
10. Di Schino, A.; Gaggiotti, M.; Testani, C. Heat Treatment Effect on Microstructure Evolution in a 7% Cr Steel for Forging. *Metals* **2020**, *10*, 808. [CrossRef]
11. Tian, Y.; Zhao, M.C.; Zeng, Y.P.; Shi, X.B.; Yan, W.; Yang, K.; Zeng, T.Y. Elimination of Primary NbC Carbides in HSLA Steels for Oil Industry Tubular Goods. *JOM* **2022**, *74*, 2409–2419. [CrossRef]
12. Li, X.; Cai, Z.; Hu, M.; Li, K.; Hou, M.; Pan, J. Effect of NbC Precipitation on Toughness of X12CrMoWNbVN10-1-1 Martensitic Heat Resistant Steel for Steam Turbine Blade. *J. Mater. Res. Technol.* **2021**, *11*, 2092–2105. [CrossRef]
13. Mancini, S.; Langellotto, L.; Di Nunzio, P.E.; Zitelli, C.; Di Schino, A. Defect Reduction and Quality Optimization by Modeling Plastic Deformation and Metallurgical Evolution in Ferritic Stainless Steels. *Metals* **2020**, *10*, 186. [CrossRef]
14. Bay, Y.; Bhattacharyya, R.; Mc Cormick, M.E. Use of High Strength Steels. *Elsevier Ocean. Eng. Ser.* **2001**, *3*, 353.
15. Narimani, M.; Hajjari, E.; Eskandari, M.; Szpunar, J.A. Electron Backscattered Diffraction Characterization of S900 HSLA Steel Welded Joints and Evolution of Mechanical Properties. *J. Mater. Eng. Perform.* **2022**, *31*, 3985–3997. [CrossRef]
16. Geng, R.; Li, J.; Shi, C.; Zhi, J.; Lu, B. Effect of Ce on Microstructures, Carbides and Mechanical Properties in Simulated Coarse-Grained Heat-Affected Zone of 800-MPa High-Strength Low-Alloy Steel. *Mater. Sci. Eng. A* **2022**, *840*, 142919. [CrossRef]
17. Shi, S.C.; Wang, W.C.; Ko, D.K. Influence of Inclusions on Mechanical Properties in Flash Butt Welding Joint of High-Strength Low-Alloy Steel. *Metals* **2022**, *12*, 242. [CrossRef]
18. Lambert-Perlade, A.; Gourgues, A.F.; Besson, J.; Sturel, T.; Pineau, A. Mechanisms and Modeling of Cleavage Fracture in Simulated Heat-Affected Zone Microstructures of a High-Strength Low Alloy Steel. *Metall. Mater. Trans. A* **2004**, *35A*, 1039–1053. [CrossRef]
19. Miletić, I.; Ilić, A.; Nikolić, R.R.; Ulewicz, R.; Ivanović, L.; Sczygiol, N. Analysis of Selected Properties of Welded Joints of the HSLA Steels. *Materials* **2020**, *13*, 1301. [CrossRef]

20. Cho, L.; Tselikova, A.; Holtgrewe, K.; de Moor, E.; Schmidt, R.; Findley, K.O. Critical Assessment 42: Acicular Ferrite Formation and Its Influence on Weld Metal and Heat-Affected Zone Properties of Steels. *Mater. Sci. Technol.* **2022**, *38*, 1425–1433. [CrossRef]

21. Kim, B.C.; Lee, S.; Kim, N.J.; Lee, D.Y. Microstructure and Local Brittle Zone phenomena in High-Strength Low-Alloy Steel Welds. *Metall. Trans. A* **1991**, *22*, 139–149. [CrossRef]

22. Di Schino, A.; Mortello, M.; Schmidt, R.; Stornelli, g.; Tselikiva, A.; Zucca, G. Studio della microstruttura in zona termicamente alterata di un acciaio S355 microlegato al vanadio. *Riv. Ital. Della Saldatura* **2023**, *75*, 7–16.

23. Prasad, K.; Dwivedi, D.K. Some Investigations on Microstructure and Mechanical Properties of Submerged Arc Welded HSLA Steel Joints. *Int. J. Adv. Manuf. Technol.* **2008**, *36*, 475–483. [CrossRef]

24. Kvackaj, T.; Bidulská, J.; Bidulský, R. Overview of Hss Steel Grades Development and Study of Reheating Condition Effects on Austenite Grain Size Changes. *Materials* **2021**, *14*, 1988. [CrossRef] [PubMed]

25. Mengaroni, S.; Cianetti, F.; Calderini, M.; Evangelista, E.; Di Schino, A.; McQueen, H. Tool steels: Forging simulation and microstructure evolution of large-scale ingot. *Acta Phys. Pol. A* **2015**, *128*, 629–632. [CrossRef]

26. Ouchi, C. Advances in Physical Metallurgy and Processing of Steels. Development of Steel Plates by Intensive Use of TMCP and Direct Quenching Processes. *ISIJ Int.* **2001**, *41*, 542–553. [CrossRef]

27. Spanos, G.; Fonda, R.W.; Vandermeer, R.A.; Matuszeski, A. Microstructural Changes in HSLA-100 Steel Thermally Cycled to Simulate the Heat-Affected Zone during Welding. *Metall. Mater. Trans. A* **1995**, *26*, 3277–3293. [CrossRef]

28. Di Schino, A.; Testani, C. Corrosion behavior and mechanical properties of AISI 316 stainless steel clad Q235 plate. *Metals* **2020**, *10*, 552. [CrossRef]

29. Lee, S.; Kim, B.C.; Kwon, D. Correlation of Microstructure and Fracture Properties in Weld Heat- Affected Zones of Thermome-chanically Controlled Processed Steels. *Metall. Trans. A* **1992**, *23*, 2803–2816. [CrossRef]

30. Taillard, R.; Verrier, P.; Maurickx, T.; Foct, J. Effect of Silicon on CGHAZ Toughness and Microstructure of Microalloyed Steels. *Metall. Mater. Trans. A* **1995**, *26*, 447–457. [CrossRef]

31. Taillard, R.; Foct, J.; Verrier, P.; Maurickx, T. Residual Austenite and Its Effect on Fracture Toughness of Coarse-Grained Heat-Affected Zone of H.S.L.A. Steels. In Proceedings of the ESOMAT 1989—Ist European Symposium on Martensitic Transformations in Science and Technology, Bochum, Germany, 9–10 March 1989; EDP Sciences: Les Ulis, France, 1989; pp. 495–502.

32. Cui, J.; Zhu, W.; Chen, Z.; Chen, L. Microstructural Characteristics and Impact Fracture Behaviors of a Novel High-Strength Low-Carbon Bainitic Steel with Different Reheated Coarse-Grained Heat-Affected Zones. *Met. Mater. Trans. A Phys. Met. Mater. Sci.* **2020**, *51*, 6258–6268. [CrossRef]

33. Li, Y.; Crowther, D.N.; Green, M.J.W.; Mitchell, P.S.; Baker, T.N. The Effect of Vanadium and Niobium on the Properties and Microstructure of the Intercritically Reheated Coarse Grained Heat Affected Zone in Low Carbon Microalloyed Steels. *ISIJ Int.* **2001**, *41*, 46–55. [CrossRef]

34. Mitchell, P.S.; Hart, P.H.M.; Morrison, W.B. The Effect of Microalloying on HAZ Toughness. In Proceedings of the International Conference Microalloying, Pittsburgh, PA, USA, 11–14 June 1995; pp. 149–162.

35. Babu, S.S.; Bhadeshia, H.K.D.H. Stress and the Acicular Ferrite Transformation. *Mater. Sci. Eng. A* **1992**, *156*, 1–9. [CrossRef]

36. Stornelli, G.; Gaggiotti, M.; Gattia, D.M.; Schmidt, R.; Sgambetterra, M.; Tselikova, A.; Zucca, G.; Di Schino, A. Vanadium alloying in S355 structural steel: Effect on residual austenite formation in welded joints heat affected zone. *Acta Metall. Slovaca* **2022**, *28*, 127–132. [CrossRef]

37. Zajac, S.; Siwecki, T.; Hutchinson, B.; Svensson, L.E.; Attlegård, M. *Weldability of High Nitrogen Ti-V Microalloyed Steel Plates Processed via Thermomechanical Controlled Rolling*; Internal Report IM-2764; Swedish Institute for Metals Research: Stockholm, Sweden, 1991.

38. Hu, J.; Du, L.X.; Wang, J.J.; Xie, H.; Gao, C.R.; Misra, R.D.K. High Toughness in the Intercritically Reheated Coarse-Grained (ICRCG) Heat-Affected Zone (HAZ) of Low Carbon Microalloyed Steel. *Mater. Sci. Eng. A* **2014**, *590*, 323–328. [CrossRef]

39. Wu, H.; Xia, D.; Ma, H.; Du, Y.; Gao, C.; Gao, X.; Du, L. Study on Microstructure Characterization and Impact Toughness in the Reheated Coarse-Grained Heat Affected Zone of V-N Microalloyed Steel. *J. Mater. Eng. Perform.* **2022**, *31*, 376–382. [CrossRef]

40. Kasuya, T.; Yurioka, N. Carbon Equivalent and Multiplying Factor for Hardenability of Steel. *Weld. Res. Suppl.* **1993**, *72*, 263.

41. Li, Y.; Baker, T.N. Effect of Morphology of Martensite–Austenite Phase on Fracture of Weld Heat Affected Zone in Vanadium and Niobium Microalloyed Steels. *Mater. Sci. Technol.* **2010**, *26*, 1029–1040. [CrossRef]

42. Qi, X.; Di, H.; Wang, X.; Liu, Z.; Misra, R.D.K.; Huan, P.; Gao, Y. Effect of Secondary Peak Temperature on Microstructure and Toughness in ICCGHAZ of Laser-Arc Hybrid Welded X100 Pipeline Steel Joints. *J. Mater. Res. Technol.* **2020**, *9*, 7838–7849. [CrossRef]

43. Sun, J.; Hensel, J.; Klassen, J.; Nitschke-Pagel, T.; Dilger, K. Solid-State Phase Transformation and Strain Hardening on the Residual Stresses in S355 Steel Weldments. *J. Mater. Process. Technol.* **2019**, *265*, 173–184. [CrossRef]

44. Díaz-fuentes, M.; Iza-mendia, A.; Gutiérrez, I. Analysis of Different Acicular Ferrite Microstructures in Low-Carbon Steels by Electron Backscattered Diffraction. Study of Their Toughness Behavior. *Metall. Mater. Trans. A* **2003**, *34*, 2505–2516. [CrossRef]

45. Zhao, M.C.; Yang, K.; Shan, Y.Y. Comparison on Strength and Toughness Behaviors of Microalloyed Pipeline Steels with Acicular Ferrite and Ultrafine Ferrite. *Mater. Lett.* **2003**, *57*, 1496–1500. [CrossRef]

46. Rodriguez-Ibabe, J.M. The Role of Microstructure in Toughness Behaviour of Microalloyed Steels. *Mater. Sci. Forum* **1998**, *284–286*, 51–62. [CrossRef]

47. Zhu, Z.; Kuzmikova, L.; Li, H.; Barbaro, F. Effect of Inter-Critically Reheating Temperature on Microstructure and Properties of Simulated Inter-Critically Reheated Coarse Grained Heat Affected Zone in X70 Steel. *Mater. Sci. Eng. A* **2014**, *605*, 8–13. [CrossRef]
48. Moeinifar, S.; Kokabi, A.H.; Madaah Hosseini, H.R. Influence of Peak Temperature during Simulation and Real Thermal Cycles on Microstructure and Fracture Properties of the Reheated Zones. *Mater. Des.* **2010**, *31*, 2948–2955. [CrossRef]
49. Hajisafari, M.; Nategh, S.; Yoozbashizadeh, H.; Ekrami, A. Fatigue Properties of Heat-Treated 30MSV6 Vanadium Microalloyed Steel. *J. Mater. Eng. Perform.* **2013**, *22*, 830–839. [CrossRef]
50. Avtokratova, E.; Sitdikov, O.; Latypova, O.E.; Markushev, M.v.; Linderov, M.L.; Merson, D.L.; Vinogradov, Y.A. Effect of Precipitates on Static and Fatigue Strength of a Severely Forged Aluminum Alloy 1570C. In Proceedings of the IOP Conference Series: Materials Science and Engineering, Yekaterinburg, Russia, 13–16 November 2018; Institute of Physics Publishing: Bristol, UK, 2018; Volume 447.
51. Di Schino, A.; Barteri, M.; Kenny, J.M. Fatigue behavior of a high nitrogen austenitic stainless steel as a function of its grain size. *J. Mater. Sci. Lett.* **2003**, *22*, 1511–1513. [CrossRef]

Article

Modification of the Tensile Performance of an Extruded ZK60 Magnesium Alloy with the Addition of Rare Earth Elements

Soroush Najafi [1], Alireza Sheikhani [1], Mahdi Sabbaghian [1], Péter Nagy [2], Klaudia Fekete [3] and Jenő Gubicza [2,*]

[1] School of Metallurgy and Materials Engineering, College of Engineering, University of Tehran, Tehran P.O. Box 11155-4563, Iran; soroush.najafiran@gmail.com (S.N.); ali.ksa72@gmail.com (A.S.); msabbaghians@ut.ac.ir (M.S.)

[2] Department of Materials Physics, Eötvös Loránd University, P.O. Box 32, H-1518 Budapest, Hungary; nagyp@student.elte.hu

[3] Department of Physics of Materials, Faculty of Mathematics and Physics, Charles University, 121 16 Prague, Czech Republic; fekete@karlov.mff.cuni.cz

* Correspondence: jeno.gubicza@ttk.elte.hu

Abstract: The influence of rare earth (RE) elements on the microstructure and mechanical performance of an extruded ZK60 Mg alloy was studied. Two types of RE elements were added to a ZK60 material and then extruded at a ratio of 18:1. The first new alloy contained 2 wt% Y while the second one was produced using 2 wt% Ce-rich mischmetal. The microstructure, the texture, and the dislocation density in a base ZK60 alloy and two materials with RE additives were studied by scanning electron microscopy, electron backscattered diffraction, and X-ray line profile analysis, respectively. It was found that the addition of RE elements caused a finer grain size, the formation of new precipitates, and changes in the initial fiber texture. As a consequence, Y and Ce-rich RE elements increased the strength and reduced the ductility. The addition of these two types of RE elements to the ZK60 alloy decreased the work hardening capacity and the hardening exponent mainly due to grain refinement.

Keywords: ZK60 alloy; rare earth element; dislocation density; texture; work hardening

Citation: Najafi, S.; Sheikhani, A.; Sabbaghian, M.; Nagy, P.; Fekete, K.; Gubicza, J. Modification of the Tensile Performance of an Extruded ZK60 Magnesium Alloy with the Addition of Rare Earth Elements. *Materials* **2023**, *16*, 2828. https://doi.org/10.3390/ma16072828

Academic Editors: Andrea Di Schino and Claudio Testani

Received: 15 March 2023
Revised: 29 March 2023
Accepted: 31 March 2023
Published: 2 April 2023

1. Introduction

Magnesium (Mg) and its alloys are promising materials for automobile or aerospace applications or also as biodegradable implants. This extensive capability is due to properties such as low density, high specific strength, good castability, a controllable corrosion rate, and biocompatibility [1,2]. However, there are poor mechanical features of Mg, e.g., the low room temperature formability [2,3]. For obtaining the better mechanical behavior of Mg alloys, various strategies have been developed, such as hot deformation, dynamic recrystallization (DRX), and alloying with additional elements [3–5]. This study investigates the latter strategy in a ZK60 Mg alloy.

The ZK60 alloy is one of the most common Mg alloys with acceptable strength and ductility, in which the main alloying elements are Zn and Zr in concentrations of about 6 and 0.5 wt%, respectively. Hot extrusion of a ZK60 alloy helps to improve the strength due to grain refinement [4,6]. In hot deformation processes such as extrusion, DRX at grain boundaries and the formation of twins cause grain refinement [4,7]. Other methods such as severe plastic deformation (SPD) techniques can also refine the grain structure of a ZK60 alloy. For instance, Fakhar and Sabbaghian reported that applying five passes of repeated upsetting (RU), as a powerful SPD method, could reduce the grain size of a rolled and annealed ZK60 alloy from 40 μm to 2.8 μm [8].

Alloying is also an effective way to medicate the mechanical performance of Mg alloys. It has been reported that the addition of rare earth (RE) elements can improve the mechanical properties at ambient and high temperatures through the formation of thermally stable precipitates with high melting points at the grain boundaries and inside the grains [6,9,10]. RE elements also influence mechanical properties by changing the crystallographic texture. Due to the phenomenon of particle-stimulated nucleation (PSN), precipitates also influence grain size, especially if they have a high volume fraction [11]. Owing to the hexagonal close-packed (HCP) structure of Mg alloys, their mechanical behavior also depends on the activation of different dislocation slip systems, which is influenced by their texture [12–14]. Li et al. proposed that a Yb addition effectively refined recrystallized grains and yielded dense Mg–Zn–Yb nanoprecipitates that could inhibit grain boundary migration, and also could promote the formation of the RE texture component [15]. It was reported that the yield stress of the ZK60 alloy increased from 212 MPa to 308 MPa in a ZK60–3Ce alloy due to the development of a finer grain structure and a large volume fraction of secondary phase particles ($MgZn_2Ce$) [6].

The improvement of work hardening (WH) behavior is one of the most important ways to enhance both the strength and workability of Mg alloys [16]. The enhanced hardening capability leads to resistance against the creation of tensile mechanical instabilities and therefore improves ductility. It has been shown that RE elements have a significant effect on the strength and WH behavior of Mg alloys [17]. In our previous research, we studied the effects of 1 and 2 wt% RE elements on the shear strength of a ZK60 alloy extruded at a ratio of 12:1 [18]. It was reported by Zhou et al. that the WH in an extruded Mg–2Nd–0.5Zr alloy was low, most probably due to the high density of basal dislocations and the relatively low density of non-basal dislocations during deformation [19]. Shi et al. found that an improved WH capability was achieved for an extruded Mg-6Zn-1Gd-0.3Ca alloy with an increased volume fraction of recrystallized grains, which was attributed to the higher dislocation storage capacity in the recrystallized volumes [20]. In that study, the volume fractions of the recrystallized and the coarse deformed grains were tailored by rolling and subsequent annealing under different conditions. It was also observed that when the ratio of the volumes of recrystallized grains and coarse deformed grains increased, the WH rate decreased in stage III, while an opposite trend was detected in stage IV for a Mg–6Zn-1Gd-0.12Y alloy. These effects were closely related to the initial dislocation density in the rolled and annealed samples, as well as the dislocation evolution under tension [21].

In this study, the change in the tensile performance of an extruded ZK60 alloy (at an extrusion ratio of 18:1) due to the addition of RE elements is investigated, and its mechanical behavior (strength and WH) is related to its microstructure and crystallographic texture. In addition to the base ZK60 alloy, two other compositions were processed. One sample contained 2 wt% Y while to the other alloy a Ce-rich mixture of large RE elements of a total fraction of 2 wt% was added. This mixture of RE alloying elements contained Ce, La, Nd, and Pr. The dislocation density in the extruded samples with or without RE addition was studied by X-ray line profile analysis (XLPA). The grain structure and the crystallographic texture were analyzed by electron back-scattered diffraction (EBSD). The effect of the addition of RE elements on the microstructure and the mechanical performance of the ZK60 alloy was discussed. To the knowledge of the authors, such a careful study of the WH behavior of ZK60 alloys containing RE elements is missing from the literature.

2. Materials and Methods

Three alloys were investigated in this study: a ZK60 Mg-based alloy with the alloying element concentrations of 6 wt% Zn and 0.5 wt%, Zr, and two other materials in which 2 wt% Y or 2 wt% Ce-rich RE elements were added to the ZK60 alloy (the relative fractions of the constituents in wt% were 50.6% Ce, 23.2% La, 20.3% Nd, and 5.9% Pr). Similar to our previous work [18], all three materials were prepared by melting them in an induction furnace at 740 °C. After casting and homogenization at 440 °C for 12 h, extrusion was performed on the samples at 180 °C at the extrusion ratio of 18:1.

The phase composition of the alloys was analyzed by X-ray diffraction (XRD) using a Smartlab diffractometer (manufacturer: Rigaku, Tokyo, Japan). In the XRD experiments, a divergent beam was applied in Bragg–Brenatno (θ–θ) diffraction geometry using CuKα radiation (wavelength: 0.15418 nm).

The microstructure was studied by scanning electron microscopy (SEM) using a ZEISS SIGMA VP microscope (manufacturer: Carl Zeiss AG, Jena, Germany). During the surface preparation, the samples were polished with an alumina suspension (particle size: 0.5 μm) and then etched in a picral solution at room temperature. The same instrument was used for energy-dispersive X-ray spectroscopy (EDS) in order to determine the chemical composition of the secondary phases.

The crystallite size and the dislocation density were investigated by XLPA [22]. This analysis was carried out on diffractograms measured by a high-resolution rotating anode diffractometer (type: RA-MultiMax9, manufacturer: Rigaku, Tokyo, Japan) using CuKα_1 radiation (wavelength: 0.15406 nm). The measured XRD patterns were evaluated by the convolutional multiple whole profile (CMWP) fitting method which yielded the crystallite size and the dislocation density in the studied samples. In this procedure, all peaks in the measured pattern are fitted simultaneously using theoretical profile functions calculated on the basis of the kinematical theory of XRD [23]. These calculated functions contained unknown parameters of the microstructure, such as the median and the variance of the crystallite size distribution and the dislocation density, which were then changed during the CMWP procedure in order to obtain the best fit between the measured and the calculated patterns. It should be noted that the calculated XRD pattern contains not only the diffraction peaks but also the background under the peaks which was approximated by a spline. The strongest 16 reflections of the main Mg phase in the diffraction angle (2θ) range between 30 and 125° were used in the evaluation. An example for CMWP fitting will be shown in Section 3.1. It should be noted that the recent development in XLPA enables the determination of the microstructural parameters with the help of artificial intelligence [24]. This novel machine learning-based method is much easier to use and faster than the conventional pattern-fitting procedure. However, the present Mg alloys cannot be evaluated by this new method since it is available only for face-centered cubic (fcc) structures as yet.

The EBSD technique was used to investigate the crystallographic texture and to characterize the grain boundaries in the samples. First, the surface of the specimens was mechanically polished with a diamond paste down to 1 μm, and then refined by ion beam polishing using a Leica EM RES102 system (manufacturer: Leica Mikrosysteme, Wetzlar, Germany). The EBSD was carried out with Zeiss Cross Beam Auriga SEM (manufacturer: Carl Zeiss AG, Jena, Germany) with a step size of 0.8 μm. The misorientation distribution was determined by the EDAX OIM software (manufacturer: EDAX Inc., Mahwah, NJ, USA). In addition, the OIM software was also used for the determination of the area-weighted mean grain size from the EBSD images. In this evaluation, the regions containing at least 4 pixels and bounded by high-angle grain boundaries (HAGBs) with misorientation angles higher than 15° were considered grains.

An electrodischarge machine (EDM) was used to prepare tensile specimens parallel to the extrusion direction (ED) according to the ASTM E8-04 standard. The tensile tests were carried out on at least three samples for each condition using a SANTAM tensile machine with a capacity of 2 tons and at an initial strain rate of 10^{-3} s^{-1}. The samples containing Y and the specimens with Ce-rich rare earth elements are denoted as ZK60–2Y and ZK60–2RE, respectively.

3. Results and Discussion

3.1. Microstructure and Texture of the Extruded ZK60 Alloys with and without RE Addition

The SEM micrographs in Figure 1a–c revealed that the addition of Y and Ce-rich elements to the ZK60 alloy increased the volume fraction of precipitates from 2% to 14% and 15% in the ZK60–2Y and ZK60–2RE alloys, respectively (see also Table 1). Spherical and network-like precipitates were observed in both the Y- and RE-containing alloys; however, the ZK60–2RE alloy contained more network-like secondary phase particles than the ZK60–2Y alloy did. Figure 1d,e shows color-coded EDS elemental maps for the ZK60–2Y and

ZK60–2RE alloys (Mg: orange, Zn: red, Y and Ce: turquoise). The EDS analysis suggests that the precipitates were enriched in Y and Zn in the ZK60–2Y alloy, and in Ce and Zn in the ZK60–2RE sample. Namely, the secondary phase particles indicated by the red arrows in Figure 1a–c had the following composition, as obtained by EDS (in at.%): 66.1% Mg and 33.9% Zn for the ZK60 alloy, 29.6% Mg, 33.1% Zn and 27.3% Y for sample ZK60–2Y, and 89.2% Mg and 10.8% Ce for the ZK60–2RE alloy. The corresponding EDS spectra are shown in Figure 2.

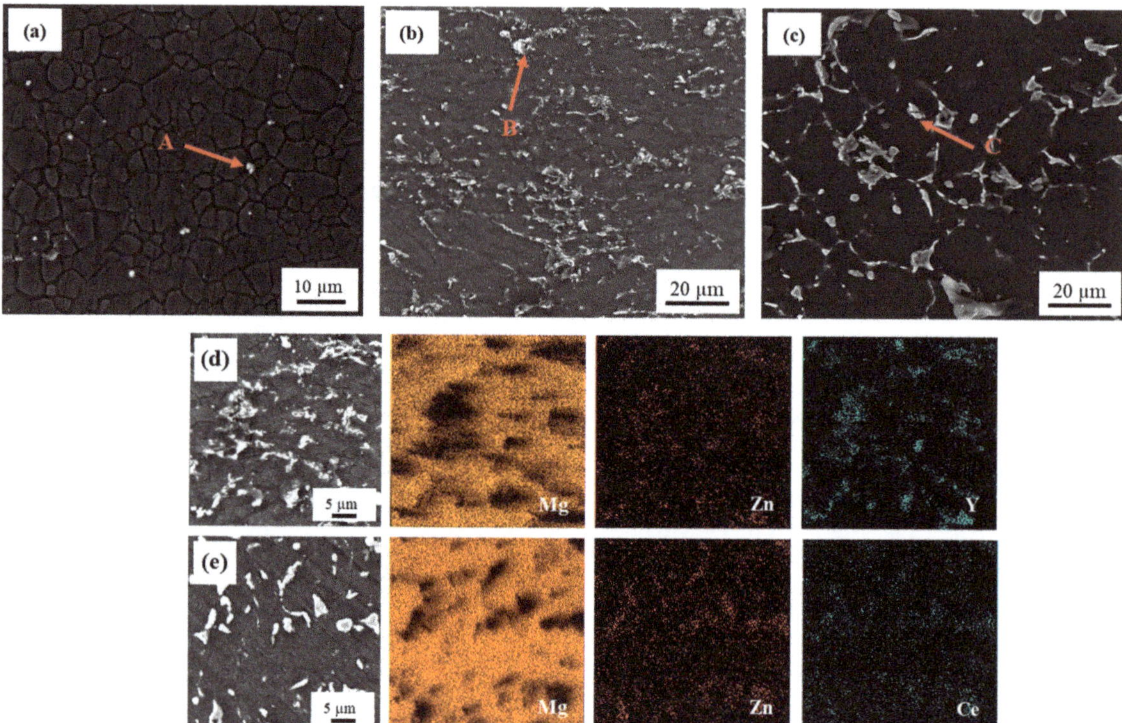

Figure 1. SEM images for (**a**) ZK60, (**b**) ZK60–2Y, and (**c**) ZK60–2RE alloys. The bright areas in the SEM images correspond to the secondary phase particles. The red arrows and the letters A, B and C show the locations where the EDS analysis was performed. The corresponding EDS spectra are shown in Figure 2. Color-coded EDS elemental maps for (**d**) ZK60–2Y and (**e**) ZK60–2RE alloys (Mg: orange, Zn: red, Y and Ce: turquoise).

Table 1. Microstructural characteristics and mechanical properties of the three studied alloys (f_P: fraction of secondary phase particles as obtained by SEM; f_{DRX}: fraction of recrystallized volumes determined by EBSD; f_{HAGB}: HAGB fraction obtained by EBSD; ρ: dislocation density determined by XLPA; UTS: ultimate tensile strength).

Alloy	Grain Size (µm)	f_P (%)	f_{DRX} (%)	f_{HAGB} (%)	ρ (10^{14} m^{-2})	Crystallite Size (nm)	σ_y (MPa)	UTS (MPa)
ZK60	6.5 ± 0.3	2 ± 0.5	96.1	95.2	1.2 ± 0.2	119 ± 14	212 ± 9	299 ± 11
ZK60–2Y	2.1 ± 0.2	14 ± 1	63.4	92.5	1.0 ± 0.2	96 ± 10	303 ± 6	348 ± 8
ZK60–2RE	2.8 ± 0.1	15 ± 1	61.2	92.7	2.3 ± 0.2	74 ± 10	299 ± 7	337 ± 5

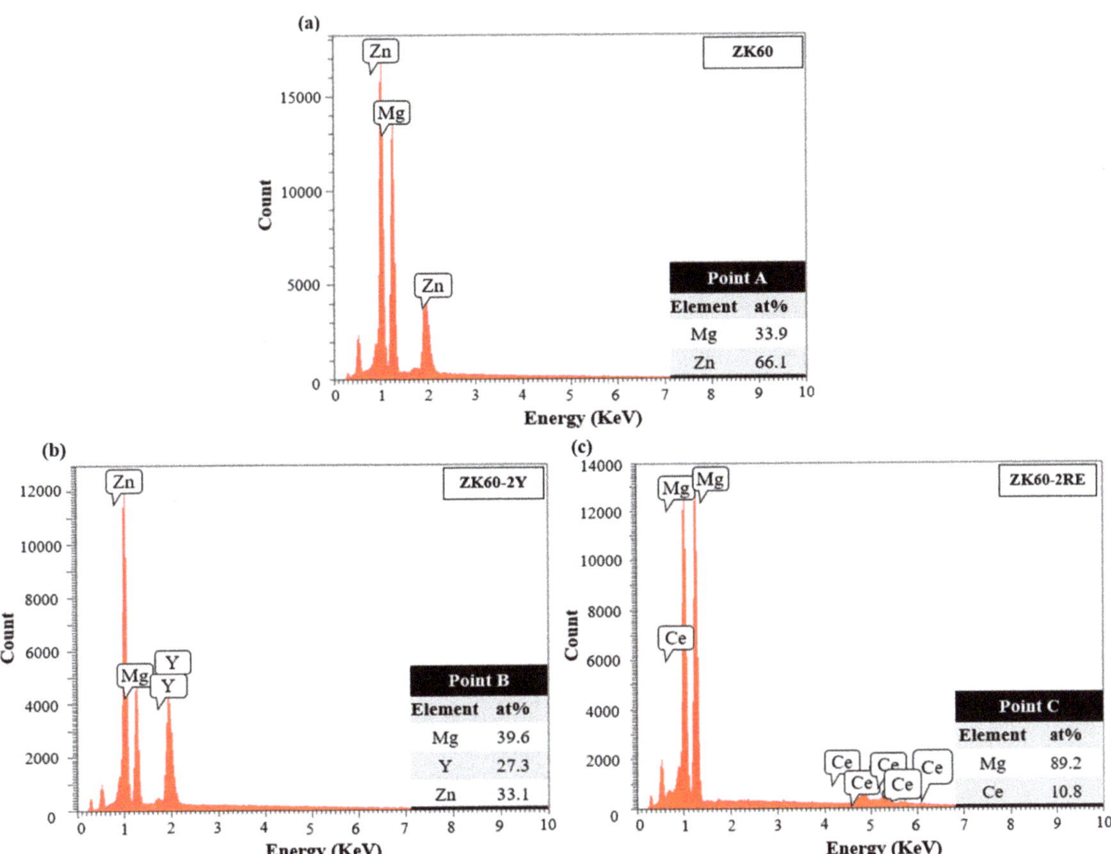

Figure 2. EDS spectra corresponding to (**a**) point A in Figure 1a, (**b**) point B in Figure 1b, and (**c**) point C in Figure 1c.

The structure of the secondary phase particles was studied by XRD. A part of the XRD patterns containing peaks of the precipitates is shown in Figure 3. Using the ICDD PDF-2018 database, the particle structures were identified. It was found that the particles in the ZK60 alloy were Mg_4Zn_7 (PDF card no.: 01-071-9625) with a monoclinic structure. The magnitude of the edge vectors of the cell (i.e., the lattice constants) are a = 2.596 nm, b = 0.524 nm, and c = 1.428 nm. The angles between these edge vectors are $\alpha = 90°$, $\beta = 102.5°$, and $\gamma = 90°$. In the ZK60 alloy, the precipitates were identified as $Mg_3Y_2Zn_3$ (PDF card no.: 00-036-1275) with an fcc structure (the lattice constant is 0.6833 nm). In the ZK60–2RE alloy, the secondary phase particles were identified as $Mg_{41}Ce_5$ (PDF card no.: 01-071-7012) with a tetragonal crystal structure. The lattice constants of this structure are a = 1.478 nm and c = 1.043 nm. It is worth noting that there is reasonable agreement between the chemical compositions determined by EDS (see the former paragraph) and the nominal compositions of the precipitate structures identified by XRD. The slight difference could have been caused by the fact that it was not possible to exclude a fraction of the EDS signal that came from the material beneath the particles. In addition, the actual compositions of the precipitates may have slightly differed from the nominal chemical compositions given in the XRD cards without changing the structure; therefore, these small differences could not be observed by XRD. Similarly to SEM, XRD also suggested a higher amount of precipitates in the ZK60–2Y and ZK60–2RE alloys compared to the ZK60 alloy (see Figure 3).

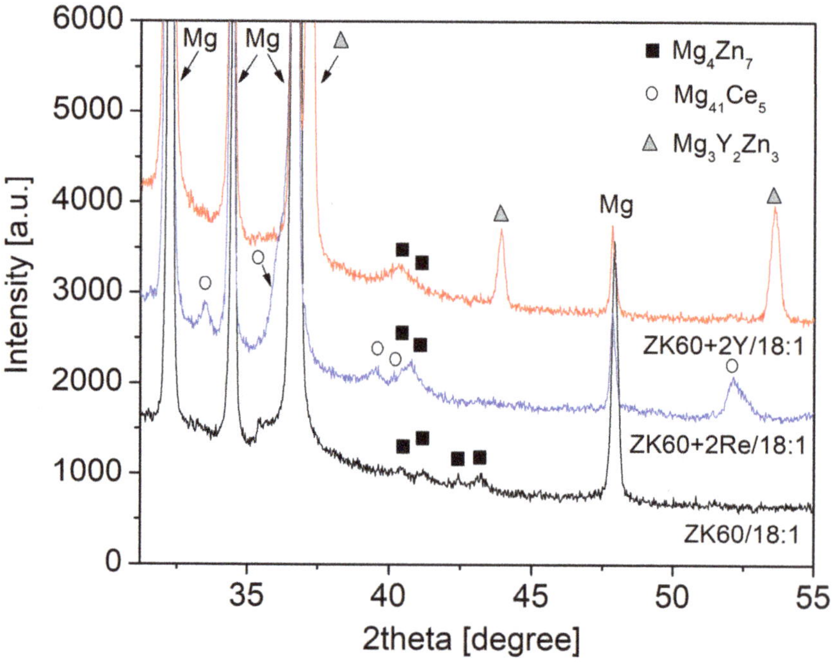

Figure 3. A part of the XRD patterns for the studied alloys.

The grain orientation maps and misorientation angle distribution histograms obtained by EBSD for the three alloys are shown in Figure 4. The analysis of the EBSD images revealed that the grain size of the ZK60 alloy was reduced from 6.5 µm to 2.1 µm and 2.8 µm when Y and Ce-rich RE elements were added to the base alloy, respectively (Figure 4a–c). According to the misorientation angle distribution histograms (see Figure 4d–f), the fraction of HAGBs ($\theta > 15°$) was more than 90% for all three alloys. In addition, the presence of a sharp peak at 30°indicated the occurrence of continuous DRX (CDRX) [25]. Thus, it is suggested that the fine grains in the extruded samples formed due to recrystallization. On the other hand, some large elongated grains were present in the ZK60–2Y and ZK60–2RE alloys (see Figure 4g,h). These grains can be attributed to the segregation of large atoms, such as Ce, La, Nd, and Pr or Y, due to dislocations and grain boundaries which hindered recrystallization during extrusion and increased the required strain to complete the DRX process [18]. On the other hand, coarse secondary phase particles (usually larger than 1 µm) could promote PSN as a DRX mechanism in the ZK60–2Y and ZK60–2RE alloys [26]. It is obvious from Figure 5a that the predominant texture component for all alloys was fiber-like, in which the c-axis of the HCP crystals were aligned perpendicular to ED, and the $(10\bar{1}0)$ and $(2\bar{1}\bar{1}0)$ planes were parallel to the surface analyzed by EBSD. This type of texture is common in extruded Mg alloys and limits their formability due to the limited activation of easy glide systems [2]. It is evident in Figure 5a–c that the texture intensity increased with the addition of Y and Ce-rich RE elements. This type of texture change can be explained by the higher volume fractions of un-DRXed grains (see Figure 4a–c).

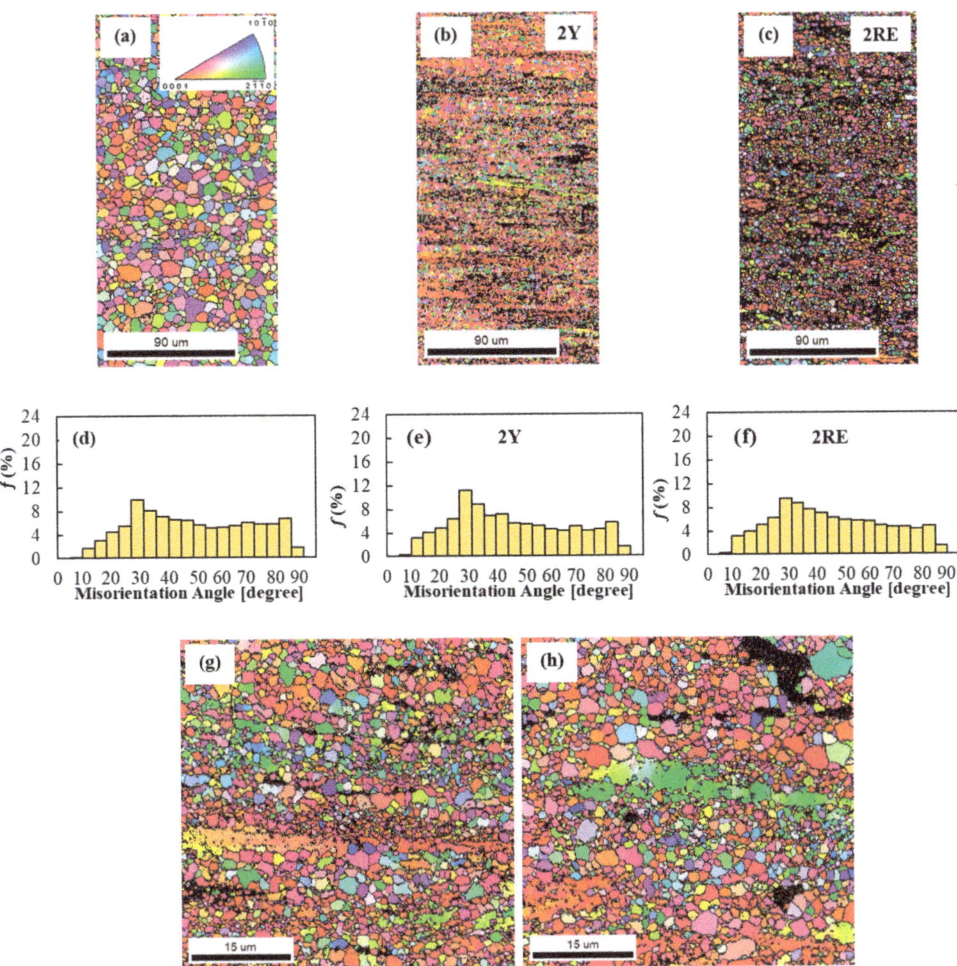

Figure 4. (**a–c**) EBSD grain orientation maps (the extrusion direction is horizontal), and (**d–f**) misorientation angle distribution histograms for ZK60, ZK60–2Y, and ZK60–2RE alloys, and higher magnification EBSD maps for (**g**) ZK60–2Y, and (**h**) ZK60–2RE alloys.

The average crystallite size and the dislocation density were determined by the XLPA method. As an example, Figure 5d demonstrates the CMWP fitting on the X-ray diffraction pattern obtained for the ZK60–2Y alloy. Only a part of the fitted pattern is shown in the figure. Table 1 shows that the crystallite size decreased with the addition of RE alloying elements. This trend is in line with the grain size reduction observed by EBSD. On the other hand, the crystallite size obtained by XLPA was much smaller than the grain size determined by EBSD. This phenomenon is well-known in plastically deformed metallic materials and is caused by the hierarchical nature of the deformed microstructures [22]. Indeed, during deformation, dislocations are formed and arranged into low-angle grain boundaries and/or dipolar walls in order to reduce the energy of the dislocation structure. The volumes separated by these dislocation boundaries exhibit low misorientations which break the coherency of the scattered X-rays; therefore, XLPA detected these volumes as individual crystallites which were considerably smaller than the grain size.

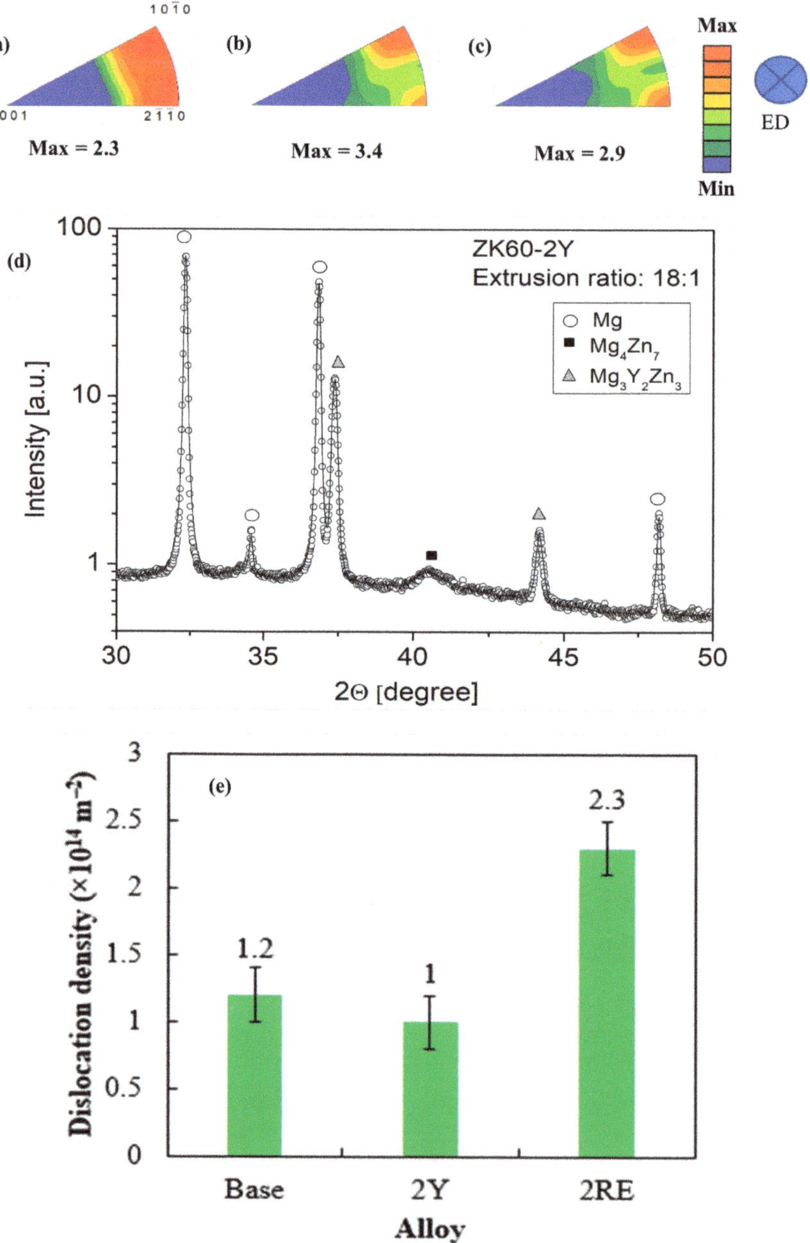

Figure 5. Inverse pole figures for (**a**) ZK60, (**b**) ZK60–2Y, and (**c**) ZK60–2RE alloys, and (**d**) CMWP fitting for the extruded ZK60–2Y alloy. The open circle and the solid line represent the measured data and the fitted XRD pattern, respectively. The intensity is in a logarithmic scale. (**e**) The dislocation density for the three studied alloys determined by CMWP fitting.

XLPA revealed that in the extruded ZK60 alloy the dislocation density was about 1.2×10^{14} m^{-2}, which did not change remarkably due to the addition of 2% Y (see Figure 5e). On the other hand, the dislocation density increased to 2.3×10^{14} m^{-2} in the ZK60–2RE alloy. The higher dislocation density for sample ZK60–2RE can be attributed to the alloying effect since the large RE atoms could hinder the dislocation annihilation during the extrusion either by dissolving into the Mg matrix or by forming precipitates. It is surprising that the addition of 2% Y did not yield an increased dislocation density in the extruded sample. This effect can be understood if we assume that most yttrium was concentrated in the precipitates. Since a major fraction of these particles formed at the grain boundaries, as suggested by SEM (see Figure 1), they had no significant effect on the dislocation multiplication and annihilation occurring in the grain interiors. Therefore, the yttrium addition did not increase the dislocation density compared to the base ZK60 alloy. For sample ZK60–2RE, the relative fraction of RE atoms in the precipitates was much lower than that for specimen ZK60–2Y. Namely, the fractions of RE atoms in $Mg_{41}Ce_5$ and $Mg_3Y_2Zn_3$ particles were 11% and 25%, respectively. On the other hand, the volume fractions of precipitates in the two alloys were similar (see Table 1). Therefore, most probably, the ZK60–2RE alloy contained more solute RE atoms in the grain interiors than the ZK60-2Y alloy did, which possibly contributed to the higher dislocation density after extrusion. In addition, the average size of the Ce, La, Nd, and Pr atoms present in the ZK60-2RE sample was about 5% larger than that of the Y added to the ZK60-2Y alloy. Therefore, the pinning effect of the former elements should have been higher on dislocations compared to that of the latter one. This effect coud have also contributed to the higher dislocation density in sample ZK60-2RE.

3.2. Influence of RE Element Addition on the Mechanical Properties of the Extruded ZK60 Alloy

Figure 6a shows the tensile engineering stress–strain curves for all three alloys. The ultimate tensile strength (UTS) values of the ZK60, ZK60–2Y, and ZK60–2RE alloys were 299, 348, and 337 MPa, while the elongation to failure values were about 30, 21, and 25%, respectively. The formation of the new precipitates and their larger volume fraction due to the addition of Y or Ce-rich elements increased the strength of the two new alloys compared to the ZK60 counterpart. Similar results have been reported for other Mg alloys containing RE elements. For example, according to the research carried out by Sabbaghian et al., the UTS of an extruded Mg–4Zn alloy was enhanced from 301 to 336 MPa after the addition of 1 wt% Gd due to grain boundary hardening, particle strengthening, and texture hardening [27]. Moreover, another study revealed that the UTS of an extruded ZK60 alloy increased from 336 to 378 MPa when 2 wt% Yb was added [15]. In the present case, the addition of Y and Ce-rich RE alloying elements had different strengthening effects. The volume fractions of the secondary phase particles in the ZK60-2Y and the ZK60-2RE alloys were comparable; therefore, their hardening effects could be expected to be similar. On the other hand, the grain size hardening should have been higher for the ZK60–2Y alloy due to its smaller grain size (see Table 1), in accordance with the Hall–Petch relationship [28]. In addition, the strengthening contribution of the crystallographic texture was also higher for sample ZK60-2Y due to the enhanced texture intensity (Figure 5). At the same time, the dislocation density was much lower for ZK60-2Y alloy compared to ZK60-2RE which yielded a reduced dislocation strengthening contribution in the former material. However, the lower dislocation hardening was overwhelmed by the higher grain size and texture strengthening effects, resulting in a higher strength for the ZK60-2Y alloy than for the ZK60-2RE material.

Regarding the ductility, the reduced elongation to failure values for ZK60–2Y and ZK60–2RE alloys compared to the initial alloy was basically caused by the reduced grain size, since for smaller grains, the saturation of the dislocation density usually occurs earlier [22,29]. In addition, an increased amount of precipitates at grain boundaries can cause easier crack nucleation and propagation since these particles may act as stress concentration sites under loading [25,27]. The significance of the weakening effect of precipitates on grain boundary strength was more pronounced in the present ZK60–2Y and ZK60–2RE samples, in which, due to the small grain size, the applied loads were higher than that for the ZK60 alloy. Consequently, the base ZK60 alloy exhibited a greater elongation to failure than the new alloys with RE additives did.

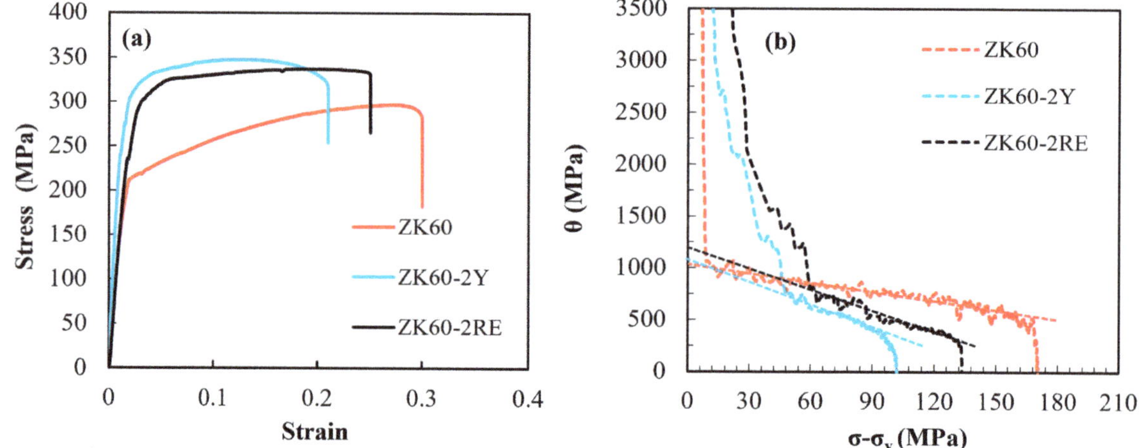

Figure 6. (**a**) Tensile engineering stress–strain curves and (**b**) work hardening curves. The lines fitted on the hardening curve in stage III were used for the determination of the values of θ_0^{III} and σ_s for the three studied alloys.

For investigating the WH behavior, the WH rate (θ) was determined according to the relationship:

$$\theta = \frac{d\sigma}{d\varepsilon},\tag{1}$$

where σ is the true stress and ε is the true strain. Figure 6b shows the θ versus $\sigma - \sigma_y$ curve for the three studied alloys where σ_y is the yield strength. The hardening capacity was characterized by the following equation [30]:

$$H_c = \frac{\sigma_{UTS} - \sigma_y}{\sigma_y}.\tag{2}$$

In Figure 6b, the approximately linear part of the curves is related to stage III which is usually described with the help of the Voce equation, expressed as:

$$\theta = \theta_0^{III}\left(1 - \frac{\sigma}{\sigma_s}\right),\tag{3}$$

where σ_s is the saturation stress. The values of θ_0^{III} and σ_s were obtained from the slope and the intercept of the straight line fitted to the stage III segment of the θ versus $\sigma - \sigma_y$ curve in Figure 6b. For the comparison of the WH behavior of the three alloys, the WH parameters are listed in Table 2. With the addition of 2% RE elements, the θ_0^{III} of the ZK60 alloy increased from about 1016 MPa to the range between 1090 and 1194 MPa, which was due to the stronger initial texture. The hardening capacity (H_c) for the ZK60 alloy was 0.81, which decreased to 0.34 and 0.47 for the ZK60–2Y and ZK60–2RE alloys, respectively. The change in σ_s was similar to H_c, and with the addition of RE, the value of σ_s reduced. Grain refinement in the ZK60–2RE alloy and especially in the ZK60–2Y alloy reduced the WH parameters. A similar result was obtained in a former study in which it was also suggested that the increase in the Mn content from 0 to 1.88 wt% in an extruded Mg–1Sn alloy decreased the H_c, which was attributed to grain refinement [31]. In addition, the uniform plastic deformation stage in the tensile curve was fitted by a power-law constitutive equation:

$$\sigma = K\varepsilon^n,\tag{4}$$

where K is the strength coefficient, and n is the WH exponent. A higher value of n indicates a more pronounced WH. It can be seen in Table 2, that n is influenced by the addition of RE, and its value decreased from 0.3 to 0.13 and 0.19 for the ZK60–2Y and ZK60–2RE alloys, respectively. The reduction of n can be related to the decrease in ductility of RE-containing alloys, as shown in Figure 6a. The lower n value for the RE-containing alloys suggests an early saturation of hardening with the increasing strain which is a typical feature of metallic materials with reduced grain size. Indeed, the addition of RE elements resulted in a significant decrease in grain size as shown in Table 1. Additionally, with the addition of RE, the probability of activation of nonbasal slip systems increased due to the higher stress level in the samples during straining (see Figure 6a) [32]. All of these factors caused the reduction in the WH parameters of the RE-containing alloys compared to the ZK60 alloy. For the ZK60–2Y alloy, which had the smallest grain size, the values of the WH parameters are the lowest among the three studied samples.

Table 2. Extrapolated WH limit in stage III (θ_0^{III}), corresponding saturation stress (σs), hardening capacity (Hc), and work hardening exponent (n) for the three studied alloys.

Alloy	θ_0^{III} (MPa)	σ_s (MPa)	H_c	n
ZK60	1016	170	0.81	0.30
ZK60–2Y	1090	102	0.34	0.13
ZK60–2RE	1194	134	0.47	0.19

4. Conclusions

The influence of the addition of 2 wt% Y and Ce-rich RE elements on the microstructure and mechanical behavior of an extruded ZK60 alloy was investigated. The following conclusions were drawn from the experimental results:

(1) Adding RE elements to the ZK60 alloy caused grain refinement, the formation of new precipitates, and an enhanced volume fraction of precipitates. The grain size reduction from 6.5 μm to 2.1 and 2.8 μm due to the addition of Y and Ce-rich RE elements, respectively, was attributed to the nucleation of new grains due to CDRX and PSN phenomena, and the pinning effects of second phase particles that hindered grain growth.

(2) In the RE-containing alloys, the texture intensity increased, and the highest texture intensity was achieved in the ZK60–2Y alloy. The higher texture intensity was due to the higher volume fractions of un-DRXed grains in the materials alloyed with Y and Ce-rich RE elements. The higher dislocation density in the ZK60-2RE alloy compared to the ZK60-2Y material can be explained by the expected higher concentration of solute atoms and their larger atomic radii compared to sample ZK60-2Y.

(3) The addition of RE elements increased the strength due to the solute and precipitate strengthening, and the grain size hardening. The ZK60–2Y and ZK60–2RE alloys exhibited lower ductility than the base alloy did due to the smaller grain size and the weakening effect of secondary phase particles on the grain boundary strength. These precipitates acted as stress concentration sites under loading, resulting in crack nucleation and/or easier crack propagation along the grain boundaries. The cracking was also facilitated by the higher stress level caused by the small grain size. The ZK60–2Y alloy exhibited the lowest work hardening rate due to the smallest grain size.

Author Contributions: Conceptualization, M.S.; methodology, S.N. and M.S.; validation, M.S. and J.G.; investigation, S.N., A.S., M.S., P.N., K.F. and J.G.; data curation, S.N., A.S. and M.S.; writing—original draft preparation, S.N. and J.G.; writing—review and editing, M.S. and K.F.; visualization, S.N., M.S. and K.F. All authors have read and agreed to the published version of the manuscript.

Funding: This research was financially supported by the International Visegrad Fund (project V4-Japan Joint Research Program, Ref. JP3936) and the National Research, Development and Innovation Office (Contract No.: 2019-2.1.7-ERA-NET-2021-00030).

Institutional Review Board Statement: Not applicable.

Informed Consent Statement: Not applicable.

Data Availability Statement: The raw/processed data required to reproduce these findings cannot be shared at this time as the data also form part of an ongoing study.

Conflicts of Interest: The authors declare no conflict of interest.

References

1. Tsakiris, V.; Tardei, C.; Clicinschi, F.M. Biodegradable Mg Alloys for Orthopedic Implants—A Review. *J. Magnes. Alloys* **2021**, *9*, 1884–1905. [CrossRef]
2. Sabbaghian, M.; Mahmudi, R.; Shin, K.S. Microstructure, Texture, Mechanical Properties and Biodegradability of Extruded Mg–4Zn-xMn Alloys. *Mater. Sci. Eng. A* **2020**, *792*, 139828. [CrossRef]
3. Yu, J.; Zhang, Z.; Wang, Q.; Yin, X.; Cui, J.; Qi, H. Dynamic Recrystallization Behavior of Magnesium Alloys with LPSO during Hot Deformation. *J. Alloys Compd.* **2017**, *704*, 382–389. [CrossRef]
4. Dong, S.; Jiang, Y.; Dong, J.; Wang, F.; Ding, W. Cyclic Deformation and Fatigue of Extruded ZK60 Magnesium Alloy with Aging Effects. *Mater. Sci. Eng. A* **2014**, *615*, 262–272. [CrossRef]
5. Gu, D.; Peng, J.; Wang, J.; Pan, F. Effect of Mn Modification on Microstructure and Mechanical Properties of Magnesium Alloy with Low Gd Content. *Met. Mater. Int.* **2021**, *27*, 1483–1492. [CrossRef]
6. Banijamali, S.M.; Palizdar, Y.; Najafi, S.; Sheikhani, A.; Soltan Ali Nezhad, M.; Valizadeh Moghaddam, P.; Torkamani, H. Effect of Ce Addition on the Tribological Behavior of ZK60 Mg-Alloy. *Met. Mater. Int.* **2021**, *27*, 2732–2742. [CrossRef]
7. Pourbahari, B.; Mirzadeh, H.; Emamy, M.; Roumina, R. Enhanced Ductility of a Fine-grained Mg–Gd–Al–Zn Magnesium Alloy by Hot Extrusion. *Adv. Eng. Mater.* **2018**, *20*, 1701171. [CrossRef]
8. Fakhar, N.; Sabbaghian, M. A Good Combination of Ductility, Strength, and Corrosion Resistance of Fine-Grained ZK60 Magnesium Alloy Produced by Repeated Upsetting Process for Biodegradable Applications. *J. Alloys Compd.* **2021**, *862*, 158334. [CrossRef]
9. Chen, Y.; Zhu, Z.; Zhou, J. Study on the Strengthening Mechanism of Rare Earth Yttrium on Magnesium Alloys. *Mater. Sci. Eng. A* **2022**, *850*, 143513. [CrossRef]
10. Barezban, M.H.; Roumina, R.; Mirzadeh, H.; Mahmudi, R. Effect of Gd on Dynamic Recrystallization Behavior of Magnesium During Hot Compression. *Met. Mater. Int.* **2021**, *27*, 843–850. [CrossRef]
11. Wang, L.; Zhang, Z.; Zhang, H.; Wang, H.; Shin, K.S. The Dynamic Recrystallization and Mechanical Property Responses during Hot Screw Rolling on Pre-Aged ZM61 Magnesium Alloys. *Mater. Sci. Eng. A* **2020**, *798*, 140126. [CrossRef]
12. Wu, J.; Jin, L.; Dong, J.; Wang, F.; Dong, S. The Texture and Its Optimization in Magnesium Alloy. *J. Mater. Sci. Technol.* **2020**, *42*, 175–189. [CrossRef]
13. Yu, H.; Liu, Y.; Liu, Y.; Wang, D.; Xu, Y.; Jiang, B.; Cheng, W.; Huang, L.; Tang, W.; Yu, W. Enhanced Strength-Ductility Synergy in Mg-0.5 Wt%Ce Alloy by Hot Extrusion. *Met. Mater. Int.* **2022**, 1–8. [CrossRef]
14. Li, R.; Fu, G.; Xu, Z.; Su, Y.; Hao, Y. Effect of Dynamically Recrystallized Grains on Rare Earth Texture in Magnesium Alloy Extruded at High Temperature. *Adv. Eng. Mater.* **2018**, *20*, 1700818. [CrossRef]
15. Li, L.; Wang, Y.; Zhang, C.; Wang, T.; Lv, H. Effects of Yb Concentration on Recrystallization, Texture and Tensile Properties of Extruded ZK60 Magnesium Alloys. *Mater. Sci. Eng. A* **2020**, *788*, 139609. [CrossRef]
16. Wang, W.; Chen, W.; Jung, J.; Cui, C.; Li, P.; Yang, J.; Zhang, W.; Xiong, R.; Kim, H.S. Asymmetry Evolutions in Microstructure and Strain Hardening Behavior between Tension and Compression for AZ31 Magnesium Alloy. *Mater. Sci. Eng. A* **2022**, *844*, 143168. [CrossRef]
17. Zhang, D.; Zhang, D.; Zhang, Y.; Chen, S.; Xu, T.; Meng, J. Analysis of Strain Hardening Behavior in a Ductile Mg–Yb Based Alloy. *Mater. Sci. Eng. A* **2021**, *819*, 141462. [CrossRef]
18. Najafi, S.; Sabbaghian, M.; Sheikhani, A.; Nagy, P.; Fekete, K.; Gubicza, J. Effect of Addition of Rare Earth Elements on the Microstructure, Texture, and Mechanical Properties of Extruded ZK60 Alloy. *Met. Mater. Int.* **2022**. [CrossRef]
19. Zhou, B.; Zhu, T.; Jia, H.; Ma, Z.; Hao, T.; Wang, J.; Zeng, X. Revealing the Weak Work-Hardening Behavior in Aged Mg–RE Alloys: A Synchrotron Radiation Diffraction Study. *J. Alloys Compd.* **2023**, *947*, 169705. [CrossRef]
20. Shi, Q.; Wang, C.; Deng, K.; Fan, Y.; Nie, K.; Liang, W. Work Hardening and Softening Behaviors of Mg-Zn-Gd-Ca Alloy Regulated by Bimodal Microstructure. *J. Alloys Compd.* **2023**, *938*, 168606. [CrossRef]
21. Hou, M.; Deng, K.; Wang, C.; Nie, K.; Shi, Q. The Work Hardening and Softening Behaviors of Mg–6Zn-1Gd-0.12Y Alloy Influenced by the VR/VD Ratio. *Mater. Sci. Eng. A* **2022**, *856*, 143970. [CrossRef]
22. Gubicza, J. *X-ray Line Profile Analysis in Materials Science*; IGI Global: Hershey, PA, USA, 2014; ISBN 1466658533.
23. Ribárik, G.; Gubicza, J.; Ungár, T. Correlation between Strength and Microstructure of Ball-Milled Al–Mg Alloys Determined by X-Ray Diffraction. *Mater. Sci. Eng. A* **2004**, *387–389*, 343–347. [CrossRef]
24. Nagy, P.; Kaszás, B.; Csabai, I.; Hegedűs, Z.; Michler, J.; Pethö, L.; Gubicza, J. Machine Learning-Based Characterization of the Nanostructure in a Combinatorial Co-Cr-Fe-Ni Compositionally Complex Alloy Film. *Nanomaterials* **2022**, *12*, 4407. [CrossRef]

25. Sabbaghian, M.; Fakhar, N.; Nagy, P.; Fekete, K.; Gubicza, J. Investigation of Shear and Tensile Mechanical Properties of ZK60 Mg Alloy Sheet Processed by Rolling and Sheet Extrusion. *Mater. Sci. Eng. A* **2021**, *828*, 142098. [CrossRef]

26. Pang, H.; Li, Q.; Chen, X.; Chen, P.; Li, X.; Tan, J. Dynamic Recrystallization Mechanism and Precipitation Behavior of Mg-6Gd-3Y-3Sm-0.5Zr Alloy During Hot Compression. *Met. Mater. Int.* **2023**, *29*, 390–401. [CrossRef]

27. Sabbaghian, M.; Mahmudi, R.; Shin, K.S. Microstructural Evolution, Mechanical Properties, and Biodegradability of a Gd-Containing Mg-Zn Alloy. *Metall. Mater. Trans. A* **2021**, *52*, 1269–1281. [CrossRef]

28. Lv, S.; Meng, F.; Lu, X.; Yang, Q.; Qiu, X.; Duan, Q.; Meng, J. Influence of Nd Addition on Microstructures and Mechanical Properties of a Hot-Extruded Mg−6.0Zn−0.5Zr (Wt.%) Alloy. *J. Alloys Compd.* **2019**, *806*, 1166–1179. [CrossRef]

29. Gubicza, J. *Defect Structure and Properties of Nanomaterials*; Woodhead Publishing: Sawston, UK, 2017; ISBN 0081019181.

30. Liu, T.; Pan, F.; Zhang, X. Effect of Sc Addition on the Work-Hardening Behavior of ZK60 Magnesium Alloy. *Mater. Des.* **2013**, *43*, 572–577. [CrossRef]

31. Liao, H.; Kim, J.; Liu, T.; Tang, A.; She, J.; Peng, P.; Pan, F. Effects of Mn Addition on the Microstructures, Mechanical Properties and Work-Hardening of Mg-1Sn Alloy. *Mater. Sci. Eng. A* **2019**, *754*, 778–785. [CrossRef]

32. Zhang, D.; Zhang, D.; Bu, F.; Li, X.; Li, B.; Yan, T.; Guan, K.; Yang, Q.; Liu, X.; Meng, J. Excellent Ductility and Strong Work Hardening Effect of As-Cast Mg-Zn-Zr-Yb Alloy at Room Temperature. *J. Alloys Compd.* **2017**, *728*, 404–412. [CrossRef]

Communication

Study of Coarse Particle Types, Structure and Crystallographic Orientation Relationships with Matrix in Cu-Cr-Zr-Ni-Si Alloy

Chengdong Xia [1], Chengyuan Ni [1], Yong Pang [1,*], Yanlin Jia [2], Shaohui Deng [1] and Wenhui Zheng [1]

[1] College of Mechanical Engineering, Quzhou University, Quzhou 324000, China; xcd309@163.com (C.X.)
[2] School of Materials Science and Engineering, Central South University, Changsha 410083, China
* Correspondence: thgink@126.com

Abstract: Coarse particles in Cu-0.39Cr-0.24Zr-0.12Ni-0.027Si alloy were studied with scanning electron microscopy and transmission electron microscopy. Three types of coarse particles were determined: a needle-like Cu_5Zr intermetallic phase, a nearly spherical $Cr_{9.1}Si_{0.9}$ intermetallic phase and (Cu, Cr, Zr, Ni, Si)-rich lath complex particles. The crystallographic orientation relationships of the needle-like and nearly spherical coarse particles were also determined. The reasons for formation and the role of the coarse phases in Cu-Cr-Zr alloys are discussed, and some suggestions are proposed to control the coarse phases in the alloys.

Keywords: Cu-Cr-Zr alloy; coarse particle; crystal structure; orientation relationship

1. Introduction

High-strength and high-electrical-conductivity copper alloys have been extensively used in railway contact wires [1,2], integrated circuit lead frame [3], heat exchangers [4] and other electrical and electronic engineering contexts [5]. These kinds of alloys have mainly included Cu-Fe systems, Cu-Ni systems, Cu-Cr systems and Cu-Mg systems [1–7], and Cu-Cr system alloys are potentially the most representative of copper alloys with the best balance of mechanical and electrical properties, especially Cu-Cr-Zr alloys [8–12]. In the past years, many investigations have focused on Cu-Cr-Zr alloys to improve physical and mechanical properties such as strength, conductivity, ductility and thermal stability with alloying, plastic deformation, rotary forging, friction stir processing and equal channel angular pressing [8–12], and the authors of this paper also carried out some studies in this field [13–16]. These previous studies have greatly promoted the developments and applications of Cu-Cr-Zr alloys in various fields.

It has been established unambiguously that Cu-Cr-Zr alloys are mainly strengthened by nanosized precipitates, and the characteristics of the nanosized precipitates such as structure, morphology, crystallographic orientation relationships and precipitation sequence were widely studied in the previous works [17–22]. It was reported that the microstructure had a bimodal particle distribution in Cu-Cr-Zr system alloys, while it has not been reported for the binary Cu-Cr alloy [23,24]. Therefore, some researchers studied the coarse particles in Cu-Cr-Zr system alloys [25–29]. Suzuki et al. [25] considered that the dispersed precipitates located on the {111} plane of Cu were Cu_3Zr phase. Tang et al. [26] reported that intermetallic precipitates on the grain boundary in the Cu-Cr-Zr-Mg alloy were Cu_4Zr. Huang et al. [27] concluded that the coarse intermetallic particles were $Cu_{51}Zr_{14}$ with an hcp structure using transmission electron microscope (TEM) and energy dispersive spectroscopy (EDS) analysis. Theoretical studies conducted by Ge [28] and Cui [29] showed that the $Cu_{51}Zr_{14}$, $Cu_{10}Zr_7$, $CuZr_2$, $CuZr$ phase should exist in Cu-Zr metallic glass. Although the previous researchers confirmed that Cr particles and (Cu, Zr) particles were present in Cu-Cr-Zr system alloys by means of scanning electron microscope (SEM) and TEM investigations, there has been no accurate experimental evidence and unanimous agreement on

Citation: Xia, C.; Ni, C.; Pang, Y.; Jia, Y.; Deng, S.; Zheng, W. Study of Coarse Particle Types, Structure and Crystallographic Orientation Relationships with Matrix in Cu-Cr-Zr-Ni-Si Alloy. *Crystals* **2023**, *13*, 518. https://doi.org/10.3390/cryst13030518

Academic Editors: Andrea Di Schino and Claudio Testani

Received: 27 February 2023
Revised: 14 March 2023
Accepted: 15 March 2023
Published: 17 March 2023

the types and structure of coarse particles or their crystallographic orientation relationships with the matrix so far.

The present study focuses on the coarse particle types, structure and crystallographic orientation relationship with matrix by means of SEM and TEM combined with EDS analysis, electron diffraction and central dark field imaging technique. The purpose is to clarify the controversies about coarse particles in Cu-Cr-Zr alloys and propose some suggestions for designing and processing of the system alloys.

2. Materials and Methods

Material with a composition of Cu-0.39Cr-0.24Zr-0.12Ni-0.027Si (wt.%) was melted using electrolytic copper, pure chromium, magnesium, silicon and copper-13 wt.% zirconium master alloy in a vacuum-induction melting furnace, and then cast in an iron mold with a size of $35 \times 120 \times 180$ mm. The ingot was planed on both sides to remove surface defects and homogenized at 920 °C for 5 h and then hot rolled from 30 mm to 5 mm in thickness, followed by quickly quenching into cold water. Samples were cut and then solution-treated at 960 °C for 1 h in an air atmosphere muffle furnace.

Microstructure was characterized using an FEI Sirion 200 scanning electron microscope equipped with EDS. SEM image observations and EDS analyses of coarse particles were operated at a voltage of 15 kV. TEM specimens were mechanically thinned to 0.1 mm and punched into discs of 3 mm in diameter, and then thinned by jet polishing in a 30 vol.% nitric acid and 70 vol.% methanol solution at about -30 °C. Microstructure observations, electron diffraction and energy dispersive analyses were carried out using a JEM 2100F transmission electron microscope with an accelerating voltage of 200 kV. Jade software was used to determine the crystal structure and lattice parameters, and CaRIne software was applied to construct the cells of the matrix and precipitated phases and simulate the crystal diffraction pattern under the specific zone axes.

3. Results

3.1. SEM Images and EDS Analysis Results of Coarse Particles in Cast Alloy

Figure 1 shows the SEM images and the results of the EDS analysis of coarse particles in cast Cu-Cr-Zr-Ni-Si alloy. It can be seen from the figure that there are a large number of dispersed micron particles both in the grain and at the grain boundary in the as-cast alloy. In the enlarged SEM image, near-spherical and short rod-like particles with a size of about 0.5–3 μm can be observed, as indicated by the arrows in Figure 1b. The corresponding EDS analysis showed that these micron particles mainly contain Si, Cr, Ni and Cu elements.

3.2. SEM Images and EDS Analysis Results of Coarse Particles in Solution-Treated Alloy

Figure 2 shows the SEM images and EDS analysis results of Cu-Cr-Zr-Ni-Si alloy treated by solution. As can be seen from Figure 2a, after high-temperature solution treatment, the large number of micron particles which occurred in the as-cast alloy disappear, and only some coarse particles are distributed both in the grain and at the grain boundary, as shown in Figure 2a,b. This indicates that most of the as-cast micron particles were dissolved into the copper matrix during the high-temperature solution treatment. Two different characteristic particles can be observed in the backscattering electron image, as shown in Figure 2b. Bright white lath particle "A" with a length of about 3 μm and a width of about 1.5 μm was located inside the grains. The results of the EDS analysis showed that the particle is composed of elements of Si, Zr, Cr and Cu. The cross-sectional of dark black particle "B" exhibits a nearly circular shape, and mainly contains Cu, Cr and a small amount of Si (about 1.77 at. %). Since the contrast in a backscattering electron image relates to the atomic number (a conspicuous Z-contrast for Fe_xSn_{1-x}/MgO can be seen due to the great difference in atomic numbers of Fe, Sn, Mg and O [30]), the contrast of particle A mainly contained Zr element (atomic number is 40) and is obviously brighter than particle B, which mainly contains Cr element (atomic number is 24).

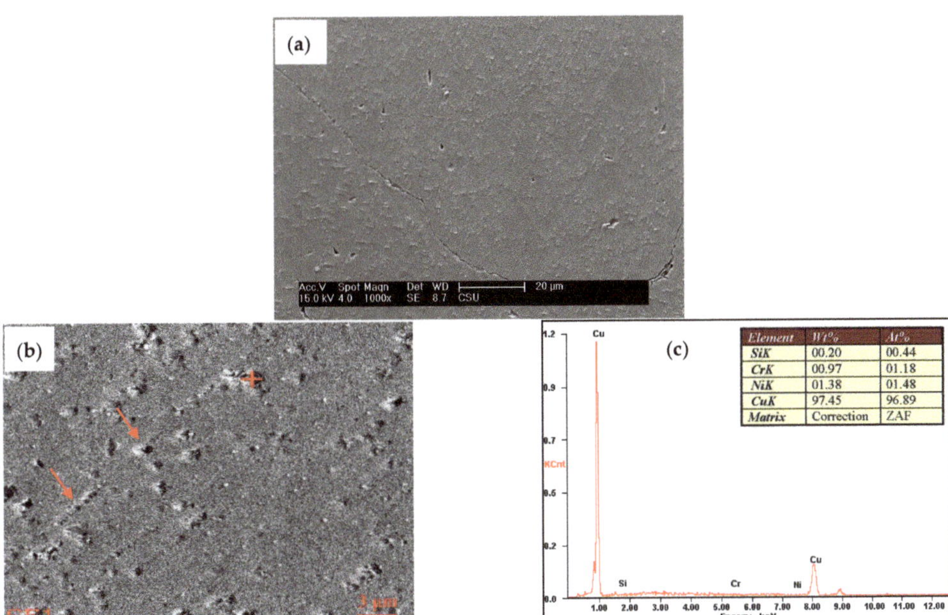

Figure 1. SEM images and EDS analysis results of coarse particles in cast Cu-Cr-Zr-Ni-Si alloy. (**a**) secondary electron image; (**b**) magnified secondary electron image; (**c**) EDS analysis results of selected particle in (**b**).

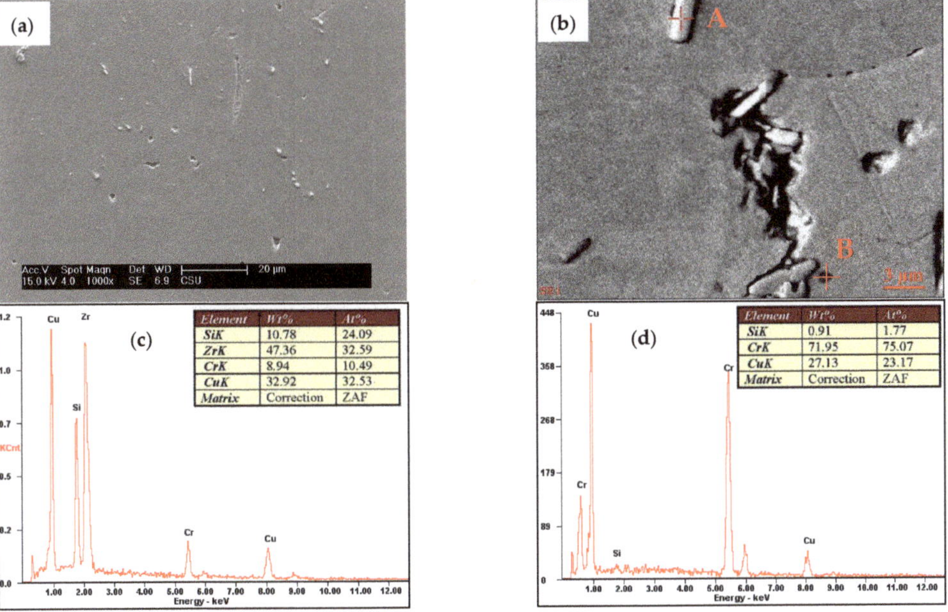

Figure 2. SEM image and EDS analysis results of coarse particles in solution-treated Cu-Cr-Zr-Ni-Si alloy. (**a**) secondary electron image in grain; (**b**) backscattering electron image at the grain boundary; (**c**) EDS analysis results of particle "A"; (**d**) EDS analysis results of particle "B".

3.3. TEM Photographs and EDS Analysis Results of Coarse Particles in Solution-Treated Alloy

Figure 3 presents TEM photographs and EDS analysis results of coarse particles in the solution-treated Cu-Cr-Zr-Ni-Si alloy. It can be seen from Figure 3a that needle-like particles with a length of 100–200 nm (marked by red circle) grow from the grain boundary. The EDS analysis results indicate that the particles are a compound containing Cu and Zr elements. Figure 3c presents a nearly spherical phase with a size of about 200 nm in diameter, and it is proved to be a kind of (Si, Cr, Cu)-rich compound. In addition, a lath phase with a size of 300 nm × 100 nm, which mainly contains Si, Cr, Ni, Zr and Cu elements, can be observed inside the grains in Figure 3e. Similar results had been obtained from other samples which contain coarse particles with the same morphology. It can be inferred that the TEM and EDS analysis results on morphology and the contained elements of Figure 3 are consistent with those of the SEM and EDS analysis in Figure 2.

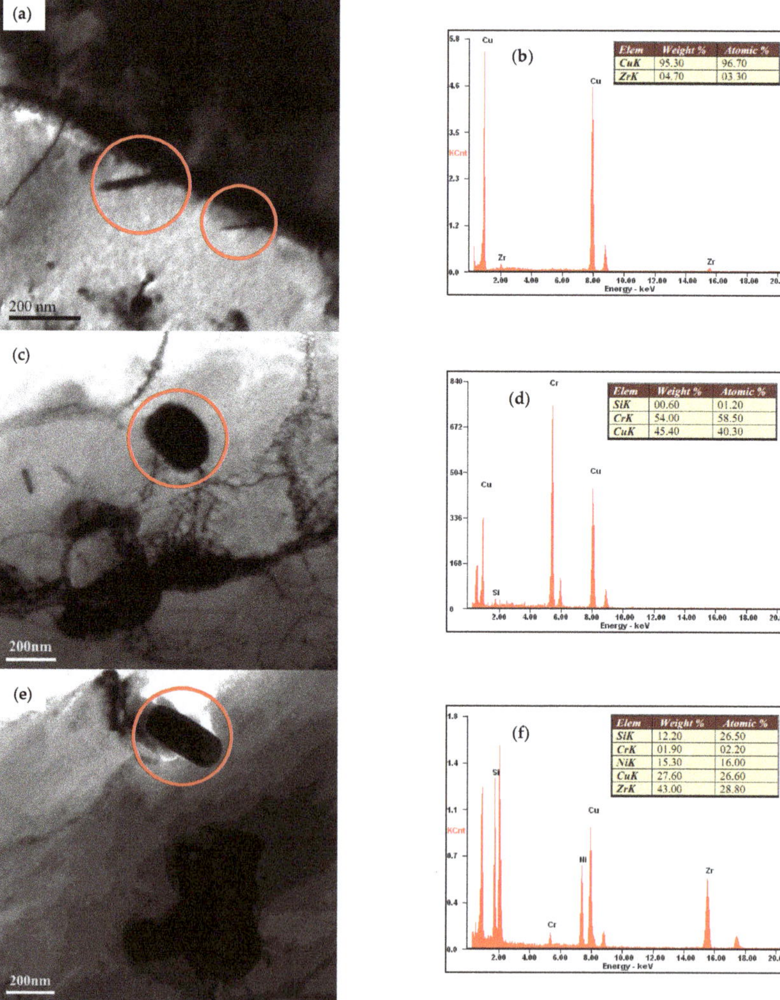

Figure 3. TEM images and EDS analysis results of coarse particles in solution-treated Cu-Cr-Zr-Ni-Si alloy. (**a**,**b**) needle-like particle and EDS analysis results; (**c**,**d**) nearly spherical particle and EDS analysis results; (**e**,**f**) lath-like particle and EDS analysis results.

3.4. TEM Images and Electron Diffraction Results of Coarse Particles in Solution-Treated Alloy

Figure 4 displays TEM images and the electron diffraction pattern of a coarse particle in the Cu-Cr-Zr-Ni-Si alloy. The bright field image and dark field image clearly show a needle-like particle that has a length of 820 nm and width of about 100 nm in Figure 4a, with the corresponding electron diffraction pattern shown in Figure 4c. Two sets of patterns can be distinguished from the electron diffraction pattern: one set of diffraction spots has higher intensity and smaller spacing between crystal planes, while the other set of diffraction spots shows slightly lower intensity and larger crystal face spacing. Angles between the adjacent diffracted spots and the transmitted spot of the two patterns were both measured to be about 54° and 36°, and the distance ratios were both determined to be about 1.1. The results fully comply with the diffraction pattern characteristic of f.c.c crystals under the <011> zone axis. Therefore, it can be inferred that the two sets of diffraction patterns are both induced by f.c.c crystals under <011> zone axes diffraction. The patterns were indexed as shown in Figure 4d. According to the spots distance, the lattice parameters of the two crystals were calculated as 0.3612 nm and 0.687 nm, respectively, and the two phases were identified with reference to the PDF card as follows: one is the Cu matrix and

the other is the Cu_5Zr intermetallic phase, which belongs to $F\overline{4}3\,m$ (216) space group and has an f.c.c structure with a lattice parameter of 0.687 nm. The red-circled spot shown in Figure 4c was selected for an operation of central dark field, and the central dark field image is presented in Figure 4b. It can be further confirmed that the set of patterns is induced by the f.c.c structure Cu_5Zr intermetallic phase.

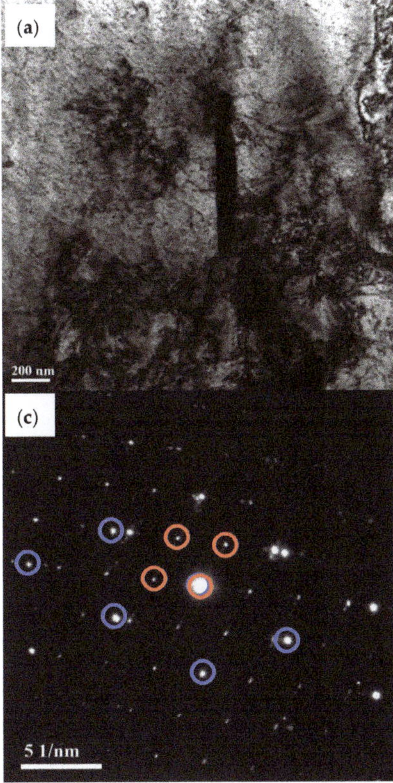

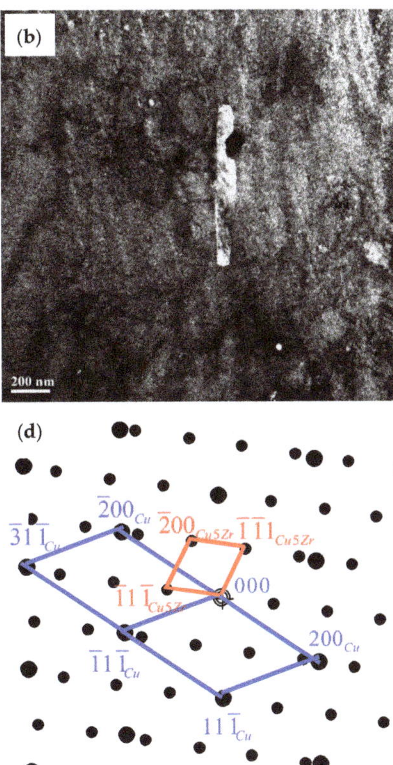

Figure 4. TEM images and electron diffraction pattern of needle-like particle in the Cu-Cr-Zr-Ni-Si alloy (○● Cu; ○● Cu_5Zr). (**a**) bright filed image; (**b**) dark filed image; (**c**) electron diffraction pattern; (**d**) indexing result of (**c**).

According to the electron diffraction pattern and TEM bright field image, it can be ascertained that the Cu_5Zr phase grows along the <111> direction of the Cu matrix. The orientation relationship between the Cu_5Zr phase and the matrix can also be determined as follows: $[011]_{Cu}//[011]_{Cu5Zr}$, $(3\bar{1}\bar{1})_{Cu}//(\bar{1}1\bar{1})_{Cu5Zr}$.

Another nearly spherical particle with a size of approximately 40~120 nm in diameter can be observed in Figure 4. A bright field image and a dark field image of the particle are shown in Figure 5a,b, and the corresponding electron diffraction pattern is presented in Figure 5c. It can be inferred that the extra diffraction is caused by the <001> zone axis of a b.c.c structure phase, and the lattice constant of the phase is 0.288 nm, calculated by the diffraction pattern. Based on the energy spectrum analysis results, which indicate a nearly spherical phase containing Cr and Si elements, it can be concluded that the extra diffraction is induced by a $Cr_{9.1}Si_{0.9}$ intermetallic phase. The indexing result is shown in Figure 5d. According to the indexing result, a Nishiyama–Wassermann (N–W) orientation relationship can be found between the $Cr_{9.1}Si_{0.9}$ phase and Cu matrix, which is that: $[011]_{Cu}//[001]_{Cr9.1Si0.9}$, $(\bar{1}\bar{1}1)_{Cu}//(1\bar{1}0)_{Cr9.1Si0.9}$, $(4\bar{2}\bar{2})_{Cu}//(\bar{1}\bar{1}0)_{Cr9.1Si0.9}//(\bar{2}\bar{2}0)_{Cr9.1Si0.9}$.

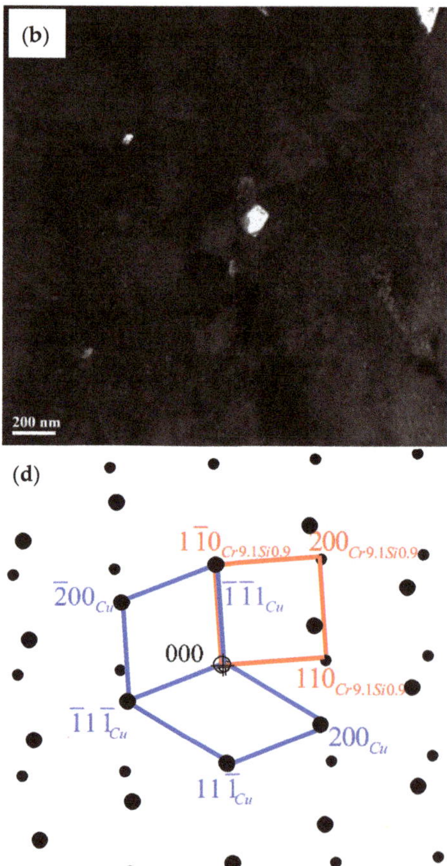

Figure 5. TEM images and electron diffraction pattern of nearly spherical phase in the Cu-Cr-Zr-Ni-Si alloy (○● Cu; ○● $Cr_{9.1}Si_{0.9}$). (**a**) bright filed image; (**b**) dark filed image; (**c**) electron diffraction pattern; (**d**) indexing result of (**c**).

4. Discussion

4.1. The Determination of Coarse Particles in Cu-Cr-Zr System Alloys

Many investigations on the coarse particle types and their roles in Cu-Cr-Zr system alloys have been carried out by means of phase diagram theory, SEM, TEM and EDS, and great progress has been made. Batra et al. [19] investigated coarse particles with a size of about 0.2 μm × 0.4 μm in the Cu-Cr-Zr alloy by electron diffraction and concluded that the particles were b.c.c structure Cr particles. Club-shaped, hexagonal-shaped and spherical-shaped coarse particles were observed in Cu-0.43Cr-0.17Zr- 0.05Mg-0.05RE alloy by Mu et al. [31], and EDS analysis showed that the coarse particles were Cr phase and (Cu, Zr) compound. Though the coarse Cr phase has been confirmed by most researchers, the type and structure of the (Cu, Zr) compounds is still uncertain. Kawakatsu et al. [32] studied the phase equilibrium at the copper corner of the isothermal diagram in Cu-Cr-Zr alloy and suggested that Zr existed in the form of Cu_3Zr in the alloy. Tang et al. [26] considered the coarse particles as a Cu_4Zr intermetallic phase in Cu-Cr-Zr-Mg alloy, but other's work on Cu-Cr-Zr alloy implied that $Cu_{51}Zr_{14}$ should exist [27]. Theoretical studies [33] and SEM experimental studies [34] of Cu-Cr-Zr alloys showed that Cr phase and Cu_5Zr intermetallic phase should be present. These results accord well with those of Holzwarth [23], Correia [35] and Sun [36], who investigated the coarse particles in Cu-Cr-Zr alloy with SEM and EDS analysis.

Due to the spatial resolution limitation of electron microscopes in the mode of EDS (e.g., X-ray beam effectively analyzed by SEM-EDS can only be focused to a few hundred nanometers), the previous EDS studies on the coarse phase with a size of a few microns or a few hundred nanometers in Cu-Cr-Zr alloys could only show a qualitative result, and these failed to precise determine the coarse particle types, structure and orientation relationship with the matrix. However, electron beams can be focused to a few nanometers in TEM electron diffraction mode, which greatly improves the accuracy and resolution of phase analysis.

The present work studied the presence of coarse particles in Cu-Cr-Zr-Ni-Si alloy as well as its chemical composition using SEM, TEM and corresponding EDS analysis, and established relations between the types and morphology of coarse particles. Three types of coarse particles were found in the Cu-Cr-Zr-Ni-Si alloy: needle-like (Cu, Zr) particles with a large aspect ratio, nearly spherical (Cu, Cr, Si) particles and lath (Cu, Cr, Zr, Ni, Si)-rich complex particles with a small aspect ratio. It can be inferred that the morphology of coarse particles varies with the composition.

Furthermore, the types, structure and crystallographic orientation relationships of the coarse particles were precisely examined using TEM electron diffraction and central dark field imaging techniques. Needle-like (Cu, Zr) particles were identified to be an f.c.c structure Cu_5Zr intermetallic phase, which belongs to space group $F\overline{4}3\,m$ (216) with a lattice parameter of 0.687 nm, having $[011]_{Cu}//[011]_{Cu5Zr}$, $(3\overline{1}\overline{1})_{Cu}//(\overline{1}\overline{1}1)_{Cu5Zr}$ orientation relationship with matrix. The Cu_5Zr phase grows approximately along the <111> direction of the copper matrix. As far as the authors know, this is the first time to show the selected area electron diffraction of Cu_5Zr intermetallic phase with a zone axis of [011] and determine the orientation relationship with the matrix. The nearly spherical (Cu, Cr, Si) particles were identified as $Cr_{9.1}Si_{0.9}$ intermetallic phases with a b.c.c structure, which had a typical N–W orientation relationship with the matrix. The relationship is the same as the results reported by Luo [37], Dahmen [38] and Hall [39] in Cu-Cr alloys. From the bright field images and dark field images in Figures 4 and 5, it can also be found that a distinct interface exists between the Cu_5Zr phase and $Cr_{9.1}Si_{0.9}$ phase. In addition, SEM and TEM results showed that the lath-like complex particles were composed of Si, Cr, Ni, Zr and Cu elements; however, these structure types and the orientation relationship need to be further studied.

4.2. The Formation and Control of Coarse Particles in Cu-Cr-Zr System Alloys

The maximum solubility of the chromium and zirconium in equilibrium is 0.7 wt.% and 0.15 wt.% [40], respectively, and the solubility of chromium and zirconium in copper is less than 0.4 wt.% Cr and 0.15 wt.% Zr at 960 °C [41]. Since the composition of Cu-Cr-Zr-Ni-Si alloy is close to or above the solubility limit at the solution temperature, the coarse particles are almost formed during the solidification process. As can be seen from the results of this work, high-temperature solution treatment (960 °C for 1 h) failed to eliminate these coarse particles. The result that the dissolution of second particles was very limited by prolonging the high-temperature solution treatment can be confirmed by Appello [41]. Huang [27] suggested that the Cr-rich phase first solidified molten melt, and Zr-rich phase solidified at the interface of the Cr-rich phase and melt. In other words, the Cr-rich phase became as the solidified core of the Zr-rich phase. Spaic et al. [42] revealed that the Cr-rich phase was the eutectic reaction product formed in the solidification process, which explained the resistance against solution treatment.

It has been reported [43,44] that the second phase particles ranging from 1 nm to 100 nm have a significant strengthening effect on the mechanical properties of the material, so the coarse particles do not contribute to the mechanical strength of the Cu-Cr-Zr-Ni-Si alloy. Holzwarth [45] studied the mechanical properties of the Cu-Cr-Zr alloy under different heat treatment conditions and revealed that the differences in the density and spacing of the coarse particles had little effect on the mechanical properties. Furthermore, due to the difference in the deformation ability of coarse particles and the matrix, it is easy to generate stress concentration at the interface of coarse particles and matrix leading to the occurrence of fatigue and fractures in the hot-working or cold-working process. In addition, the formed coarse particles deplete the amount of solute elements dissolved in the matrix, resulting in the reduction in the precipitation strengthening effect. Therefore, controlling the amount and size of the coarse particles is essential for the precipitation-strengthened Cu-Cr-Zr system alloys. Effective measures can be taken as follows:

(1) Reducing the content of alloying elements to ensure complete dissolution at the corresponding solution temperature according to the phase diagrams, e.g., the content of chromium element in the Cu-Cr-Zr alloys should be less than 0.3 wt.%, and zirconium should be less than 0.12 wt.% when the alloys are solution-treated at 920 °C.

(2) Using master alloys instead of pure metals alloying elements for melting, which could accelerate alloying elements with high melting-point to melt and dissolve into the matrix.

(3) Using modified treatment technology to refine precipitate particles formed in the solidification process.

(4) Using severe plastic deformation in the hot-working or cold-working process, which could partly fragment the coarse particles and redistribute the alloying elements, ultimately relieving the negative effects of coarse particles on the performance of Cu-Cr-Zr alloys.

5. Conclusions

(1) Three types of coarse particles were detected in the Cu-Cr-Zr-Ni-Si alloy and the relations between types, composition and morphology of coarse particles were also established.

(2) Needle-like (Cu, Zr) particles were accurately determined to be an f.c.c structure Cu_5Zr intermetallic phase, and the orientation relationship between the Cu_5Zr phase and Cu matrix was determined as $[011]_{Cu}//[011]_{Cu5Zr}$, $(3\bar{1}\bar{1})_{Cu}//(\bar{1}1\bar{1})_{Cu5Zr}$.

(3) Nearly spherical particles were identified as a $Cr_{9.1}Si_{0.9}$ intermetallic phase with a b.c.c structure, having a typical N–W orientation relationship with the Cu matrix. Another coarse particle with a lath shape was confirmed as a kind of (Cu, Cr, Zr, Ni, Si)-rich complex compound.

Author Contributions: Conceptualization, C.X. and Y.P.; methodology, Y.P. and Y.J.; validation, S.D. and W.Z.; formal analysis, C.X., C.N. and Y.P.; investigation, S.D. and W.Z.; resources, Y.J.; data curation, S.D. and W.Z.; writing—original draft preparation, C.X.; writing—review and editing, C.N., Y.P. and Y.J.; visualization, S.D. and W.Z.; supervision, Y.J.; project administration, C.X. and C.N.; funding acquisition, C.X. and C.N. All authors have read and agreed to the published version of the manuscript.

Funding: This research was funded by the Joint Funds of the Zhejiang Provincial Natural Science Foundation of China (Grant No. LZY23E010001 and LZY22E010002), Zhejiang College Students Innovation and Entrepreneurship Training Program (2022R435A004).

Institutional Review Board Statement: Not applicable.

Informed Consent Statement: Not applicable.

Data Availability Statement: The data used to support the findings of this study are available from the corresponding author upon request.

Conflicts of Interest: The authors declare no conflict of interest.

References

1. Chen, J.; Xiao, X.; Yuan, D.; Guo, C.; Huang, H.; Yang, B. Microstructure and properties of Cu-Cr-Zr alloy with columnar crystal structure processed by upward continuous casting. *J. Alloys Compd.* **2021**, *889*, 161700. [CrossRef]
2. Mao, Q.; Zhang, Y.; Guo, Y.; Zhao, Y. Enhanced electrical conductivity and mechanical properties in thermally stable fine-grained copper wire. *Commun. Mater.* **2021**, *2*, 46. [CrossRef]
3. Zhou, Y.; Zheng, C.; Chen, J.; Chen, A.; Jia, L.; Xie, H.; Lu, Z. Characterizations on Precipitations in the Cu-Rich Corner of Cu-Ni-Al Ternary Phase Diagram. *Crystals* **2023**, *13*, 274. [CrossRef]
4. Li, S.; Fang, M.; Xiao, Z.; Meng, X.; Lei, Q.; Jia, Y. Effect of Cr addition on corrosion behavior of cupronickel alloy in 3.5 wt% NaCl solution. *J. Mater. Res. Technol.* **2023**, *22*, 2222–2238. [CrossRef]
5. Ding, Y.; Xiao, Z.; Fang, M.; Gong, S.; Dai, J. Microstructure and mechanical properties of multi-scale α-Fe reinforced Cu–Fe composite produced by vacuum suction casting. *Mater. Sci. Eng. A* **2023**, *864*, 144603. [CrossRef]
6. Ma, M.; Zhang, X.; Li, Z.; Xiao, Z.; Jiang, H.; Xia, Z.; Huang, H. Effect of Equal Channel Angular Pressing on Microstructure and Mechanical Properties of a Cu-Mg Alloy. *Crystals* **2020**, *10*, 426. [CrossRef]
7. Guo, T.; Tai, X.; Wei, S.; Wang, J.; Jia, Z.; Ding, Y. Microstructure and Properties of Bulk Ultrafine-Grained Cu1.5Cr0.1Si Alloy through ECAP by Route C and Aging Treatment. *Crystals* **2020**, *10*, 207. [CrossRef]
8. Zhang, Y.; Volinsky, A.A.; Hai, T.T.; Chai, Z.; Liu, P.; Tian, B.; Liu, Y. Aging behavior and precipitates analysis of the Cu-Cr-Zr-Ce alloy. *Mater. Sci. Eng. A* **2016**, *650*, 248–253. [CrossRef]
9. Li, J.; Ding, H.; Li, B.; Wang, L. Microstructure evolution and properties of a Cu–Cr–Zr alloy with high strength and high conductivity. *Mater. Sci. Eng. A* **2021**, *819*, 141464. [CrossRef]
10. Huang, A.H.; Wang, Y.F.; Wang, M.S.; Song, L.Y.; Zhu, Y.T. Optimizing the strength, ductility and electrical conductivity of a Cu-Cr-Zr alloy by rotary swaging and aging treatment. *Mater. Sci. Eng. A* **2019**, *746*, 211–216. [CrossRef]
11. Wang, Y.D.; Liu, M.; Yu, B.H.; Wu, L.H.; Xue, P.; Ni, D.R.; Ma, Z.Y. Enhanced combination of mechanical properties and electrical conductivity of a hard state Cu-Cr-Zr alloy via one-step friction stir processing. *J. Mater. Process. Technol.* **2021**, *288*, 116880. [CrossRef]
12. Sousa, T.G.; Moura, I.A.D.B.; Filho, F.D.C.G.; Monteiro, S.N.; Brandão, L.P. Combining severe plastic deformation and precipitation to enhance mechanical strength and electrical conductivity of Cu–0.65Cr–0.08Zr alloy. *J. Mater. Res. Technol.* **2020**, *9*, 5953–5961. [CrossRef]
13. Xia, C.; Zhang, W.; Kang, Z.; Jia, Y.; Wu, Y.; Zhang, R.; Xu, G.; Wang, M. High strength and high electrical conductivity Cu-Cr system alloys manufactured by hot rolling-quenching process and thermomechanical treatments. *Mater. Sci. Eng. A* **2012**, *538*, 295–301. [CrossRef]
14. Xia, C.; Wang, M.; Zhang, W.; Kang, Z.; Jia, Y.; Zhang, R.; Yu, H. Microstructure and properties of a hot rolled-quenched Cu-Cr-Zr-Mg-Si alloy. *J. Mater. Eng. Perform.* **2012**, *21*, 1800–1805. [CrossRef]
15. Xia, C.; Jia, Y.; Zhang, W.; Zhang, K.; Dong, Q.Y.; Xu, G.; Wang, M. Study of deformation and aging behaviors of a hot rolled–quenched Cu–Cr–Zr–Mg–Si alloy during thermomechanical treatments. *Mater. Des.* **2012**, *39*, 404–409. [CrossRef]
16. Pang, Y.; Xia, C.; Wang, M.; Li, Z.; Xiao, Z.; Wei, H.; Chen, C. Effects of Zr and (Ni, Si) additions on properties and microstructure of Cu–Cr alloy. *J. Alloys Compd.* **2014**, *582*, 786–792. [CrossRef]
17. Cheng, J.Y.; Shen, B.; Yu, F.X. Precipitation in a Cu–Cr–Zr–Mg alloy during aging. *Mater. Charact.* **2013**, *81*, 68–75. [CrossRef]
18. Peng, L.; Xie, H.; Huang, G.; Xu, G.; Yin, X.; Feng, X.; Mi, X.; Yang, Z. The phase transformation and strengthening of a Cu-0.71wt% Cr alloy. *J. Alloys Compd.* **2017**, *708*, 1096–1102. [CrossRef]
19. Batra, I.S.; Dey, G.K.; Kulkarni, U.D.; Banerjee, S. Precipitation in a Cu–Cr–Zr alloy. *Mater. Sci. Eng. A* **2002**, *356*, 32–36. [CrossRef]

20. Fujii, T.; Nakazawa, H.; Kato, M.; Dahmen, U. Crystallography and morphology of nanosized Cr particles in Cu-0.2%Cr alloy. *Acta Mater.* **2000**, *48*, 1033–1045.

21. Chbihi, A.; Sauvage, X.; Blavette, D. Atomic scale investigation of Cr precipitation in copper. *Acta Mater.* **2012**, *60*, 4575–4585. [CrossRef]

22. Xia, C.; Pang, Y.; Jia, Y.; Ni, C.; Sheng, X.; Wang, S.; Jiang, X.; Zhou, Z. Orientation relationships between precipitates and matrix and their crystallographic transformation in a Cu–Cr–Zr alloy. *Mater. Sci. Eng. A* **2022**, *850*, 143576. [CrossRef]

23. Holzwarth, U.; Stamm, H. The precipitation behaviour of ITER-grade Cu–Cr–Zr alloy after simulating the thermal cycle of hot isostatic pressing. *J. Nucl. Mater.* **2000**, *279*, 31–45. [CrossRef]

24. Morris, M.A.; Leboeuf, M.; Morris, D.G. Recrystallization mechanisms in a Cu-Cr-Zr alloy with a bimodal distribution of particles. *Mater. Sci. Eng. A* **1994**, *188*, 255–265. [CrossRef]

25. Suzuki, H.; Kanno, M.; Kawakatsu, I. Strength of Cu–Cr–Zr alloy relating to the aged structures. *J. Jpn. Inst. Met.* **1969**, *33*, 628–633. [CrossRef]

26. Tang, N.Y.; Taplin, D.M.R.; Dunlop, G.L. Precipitation and aging in high-conductivity Cu–Cr alloys with additions of zirconium and magnesium. *Mater. Sci. Technol.* **1985**, *1*, 270–275. [CrossRef]

27. Huang, F.; Ma, J.; Ning, H.; Geng, Z. Analysis of phases in a Cu–Cr–Zr alloy. *Scr. Mater.* **2003**, *48*, 97–102.

28. Ge, L.; Hui, X.; Wang, E.R.; Chen, G.L.; Arroyave, R.; Liu, Z.K. Prediction of the glass forming ability in Cu-Zr binary and Cu-Zr-Ti ternary alloys. *Intermetallics* **2008**, *16*, 27–33. [CrossRef]

29. Cui, X.; Zu, F.Q.; Wang, Z.Z.; Huang, Z.Y.; Li, X.Y.; Wang, L.F. Study of the reversible intermetallic phase: B2-type CuZr. *Intermetallics* **2013**, *36*, 21–24. [CrossRef]

30. Myagkov, V.G.; Zhigalov, V.S.; Bykova, L.E.; Solovyov, L.A.; Matsynin, A.A.; Balashov, Y.; Nemtsev, I.V.; Shabanov, A.V.; Bondarenko, G.N. Solid-state synthesis, dewetting, and magnetic and structural characterization of interfacial Fe_xSn_{1-x} layers in Sn/Fe(001) thin films. *J. Mater. Res.* **2021**, *36*, 3121–3133. [CrossRef]

31. Mu, S.G.; Guo, F.A.; Tang, Y.Q.; Cao, X.M.; Tang, M.T. Study on microstructure and properties of aged Cu–Cr–Zr–Mg–RE alloy. *Mater Sci Eng A* **2008**, *475*, 235–240. [CrossRef]

32. Kawakatsu, I.; Suzuki, H.; Kitano, H. Properties of high zirconium Cu-Zr-Cr alloys and their isothermal diagram of the copper corner. *J. Jpn. Inst. Met.* **1967**, *31*, 1253–1257. [CrossRef]

33. Zeng, K.J.; Hamalainen, M. A theoretical study of the phase equilibria in the Cu-Cr-Zr system. *J. Alloys Compd.* **1995**, *220*, 53–61. [CrossRef]

34. Zeng, K.J.; Hamalainen, M.; Lilius, K. Phase relationships in Cu-rich corner of the Cu-Cr-Zr phase diagram. *Scr. Metall. Mater.* **1995**, *32*, 2009–2014. [CrossRef]

35. Correia, J.B.; Davies, H.A.; Sellars, C.M. Strengthening in rapidly solidified age hardened Cu-Cr and Cu-Cr-Zr alloys. *Acta Mater.* **1997**, *45*, 177–190. [CrossRef]

36. Sun, Z.; Guo, J.; Song, X.; Zhu, Y.; Li, Y. Effects of Zr addition on the liquid phase separation and the microstructures of Cu–Cr ribbons with 18–22at.% Cr. *J. Alloys Compd.* **2008**, *455*, 243–248. [CrossRef]

37. Luo, C.P.; Dahmen, U.; Westmacott, K.H. Morphology and crystallography of Cr precipitates in a Cu-0.33 wt% Cr alloy. *Acta Metall. Mater.* **1994**, *42*, 1923–1932. [CrossRef]

38. Dahmen, U. Orientation relationships in precipitation systems. *Acta Metall.* **1982**, *30*, 63–73. [CrossRef]

39. Hall, M.G.; Aaronson, H.I.; Kinsma, K.R. The structure of nearly coherent fcc: Bcc boundaries in a Cu-Cr alloy. *Surf. Sci.* **1972**, *31*, 257–274. [CrossRef]

40. Chakrabarti, D.J.; Laughlin, D.E. The Cr-Cu (chromium-copper) system. *Bull. Alloy Phase Diagr.* **1984**, *5*, 59–68. [CrossRef]

41. Appello, M.; Fenici, P. Solution hot treatment of a CuCrZr alloys. *Mater. Sci. Eng. A* **1988**, *102*, 69–75. [CrossRef]

42. Spaic, S.; Krizman, A.; Marinkovic, V. Structure and properties of low-alloy, hardenable copper alloys. *Metall* **1985**, *39*, 43–48.

43. Martin, J.W. *Micromechanisms in Particle-Hardened Alloys*, 1st ed.; Cambridge University Press: Cambridge, UK, 1980.

44. Lee, S.; Matsunaga, H.; Sauvage, X.; Horita, Z. Strengthening of Cu-Ni-Si alloy using high-pressure torsion and aging. *Mater. Charact.* **2014**, *90*, 62–70. [CrossRef]

45. Holzwarth, U.; Pisoni, M.; Scholz, R.; Stamm, H.; Volcan, A. On the recovery of the physical and mechanical properties of a CuCrZr alloy subjected to heat treatments simulating the thermal cycle of hot isostatic pressing. *J. Nucl. Mater.* **2000**, *279*, 19–30. [CrossRef]

Article

Combined Effect of Substrate Temperature and Sputtering Power on Phase Evolution and Mechanical Properties of Ta Hard Coatings

Cuicui Liu [1,2], Jian Peng [2], Zhigang Xu [3], Qiang Shen [2] and Chuanbin Wang [1,2,*]

1   Chaozhou Branch of Chemistry and Chemical Engineering Guangdong Laboratory, Chaozhou 521000, China
2   State Key Laboratory of Advanced Technology for Materials Synthesis and Processing, Wuhan University of Technology, Wuhan 430070, China
3   Hubei Key Laboratory of Advanced Technology for Automotive Components, Wuhan University of Technology, Wuhan 430070, China
*   Correspondence: chuanbinwang@whut.edu.cn

Abstract: Ta hard coatings were prepared on PCrNi1MoA steel substrates by direct current magnetron sputtering, and their growth and phase evolution could be controlled by adjusting the substrate temperature (T_{sub}) and sputtering power (P_{spu}) at various conditions (T_{sub} = 200–400 °C, P_{spu} = 100–175 W). The combined effect of T_{sub} and P_{spu} on the crystalline phase, surface morphology, and mechanical properties of the coatings was investigated. It was found that higher P_{spu} was required in order to obtain α-Ta coatings when the coatings are deposited at lower T_{sub}, and vice versa, because the deposition energy (controlled by T_{sub} and P_{spu} simultaneously) within a certain range was necessary. At the optimum T_{sub} with the corresponding P_{spu} of 200 °C-175 W, 300 °C-150 W, and 400 °C-100 W, respectively, the single-phased and homogeneous α-Ta coatings were obtained. Moreover, the α-Ta coating deposited at T_{sub}-P_{spu} of 400 °C-100 W showed a denser surface and a finer grain, and as a result exhibited higher hardness (9 GPa), better toughness, and larger adhesion (18.46 N).

Keywords: α-Ta coatings; substrate temperature; sputtering power; combined effect; direct current magnetron sputtering

Citation: Liu, C.; Peng, J.; Xu, Z.; Shen, Q.; Wang, C. Combined Effect of Substrate Temperature and Sputtering Power on Phase Evolution and Mechanical Properties of Ta Hard Coatings. *Metals* **2023**, *13*, 583. https://doi.org/10.3390/met13030583

Academic Editors: Andrea Di Schino and Claudio Testani

Received: 23 February 2023
Revised: 10 March 2023
Accepted: 11 March 2023
Published: 13 March 2023

1. Introduction

Protective coatings with excellent properties, such as high strength, toughness, good wear, and ablation resistance, are essential materials to ensure long-term, stable, and safe service for key artifacts applied in extreme service environments. The widely used Cr plating has a high melting point and large hardness to mitigate ablation and wear at high temperatures. However, the Cr plating is brittle and often cracks, leading to limited protection of artifacts [1–3]. In this case, a series of new ablation-resistant materials, including alloys and ceramics [4,5], have been developed. Among them, Ta coating is considered to be an alternative material to Cr plating because of its higher melting point, more excellent anti-ablation ability, and better toughness [6,7].

It should be noted that Ta coatings display two crystalline phases with entirely distinct characteristics: the stable α-Ta (Im3m space group, a = 3.304 Å) with a body-centered cubic crystal lattice structure, which has a high melting temperature (2996 °C), moderate hardness (8–12 GPa), and good toughness, and the metastable β-Ta (P42$_1$m space group, a = 5.313 Å, c = 10.194 Å) with a tetragonal crystal lattice structure, which is harder (18–20 GPa) but brittle and thermally unstable over 700 °C [8–10]. Magnetron sputtering is commonly used to prepare Ta coatings. Unfortunately, β-Ta always nucleates priorly using this method, which is unsuitable for protective materials. Therefore, many studies have attempted to regulate the phase formation of Ta coatings by adjusting the sputtering conditions, including substrate temperature [11], sputtering power [12], pressure [13], and post treatments [14], etc. Nevertheless, investigations on the phase formation in Ta coatings have yielded some conflicting results. For example, some studies [15] reported that a higher substrate temperature exceeding 365–375 °C was needed to promote α-Ta formation against β-Ta formation. On the contrary, recent experiments showed that pure α-Ta coatings could be fabricated even at a lower temperature of 200 °C [8]. The deposition energy is believed to be the most important factor determining the structural evolution of the coatings. An "energy window" was thus proposed, where α-Ta coatings could be formed between a specific energy range, while other energies higher or lower than this range would be beneficial to β-Ta coatings [16]. For magnetron sputtering, the substrate temperature [11] and sputtering power [12] are typically regarded as two important process variables concerned with the deposition energy, which can affect the surface activity of the substrate and the mobility of deposited Ta particles at the same time. The effects of substrate temperature [11] or sputtering power [12] on the preparation of Ta coatings have been investigated separately; however, their combined effect on the growth and phase evolution of Ta hard coatings has not yet been reported.

In this work, we conducted a comprehensive investigation on the structural evolution of Ta coatings using direct current magnetron sputtering by adjusting the substrate temperature (T_{sub}) and sputtering power (P_{spu}) simultaneously. The combined effect of T_{sub} and P_{spu} is discussed and the relationship between phase structure, surface morphology, mechanical properties, and the deposition parameters are established, to obtain single-phased α-Ta coatings with a fine structure and good mechanical properties.

2. Experimental Procedure

2.1. Preparation

2.1.1. Substrate Pretreatment

PCrNi1MoA steels ($10 \times 10 \times 1$ mm^3 in size) were used as substrates. The substrates were mechanically ground using sandpaper from P240 up to P2000 and then polished by diamond suspension from 3 μm to 1 μm. The polished substrates were ultrasonically cleaned in acetone for 15 min and then dried with cold air. The chemical compositions of the alloy steel are shown in Table 1.

Table 1. Chemical composition of the alloy steel.

Element	C	Si	Mn	P	S	Cr
wt.%	0.361	0.252	0.413	0.011	0.001	1.516
Element	**Ni**	**V**	**Cu**	**Mo**	**Fe**	
wt.%	1.507	0.092	0.071	0.304	Bal	

2.1.2. Deposition Device

This experiment adopts DC magnetron sputtering (DCMS) technology and the equipment is single-target magnetron sputtering instrument (JCZK350A, Juzhi Vacuum Equipment Company, China), which mainly includes a deposition chamber, vacuum system, cooling system, gas supply system, heating system, control system, etc. Figure 1 shows a schematic diagram of the magnetron sputtering.

2.1.3. Deposition Conditions

The cleaned substrates were installed on a substrate holder and loaded into the chamber. A Ta target (purity: 99.95%) was used with dimensions of Φ50 mm × H4 mm. The substrate-to-target distance was 70 mm. A chamber pressure of below 6.0×10^{-4} Pa was reached before coating depositions. In deposition processes, Ar pressure and the deposition time were fixed at 0.5 Pa and 1 h, respectively. The substrate temperature (T_{sub}) was from 200 °C to 400 °C and the sputtering power (P_{spu}) was from 100 W to 175 W. The deposition parameters are listed in Table 2.

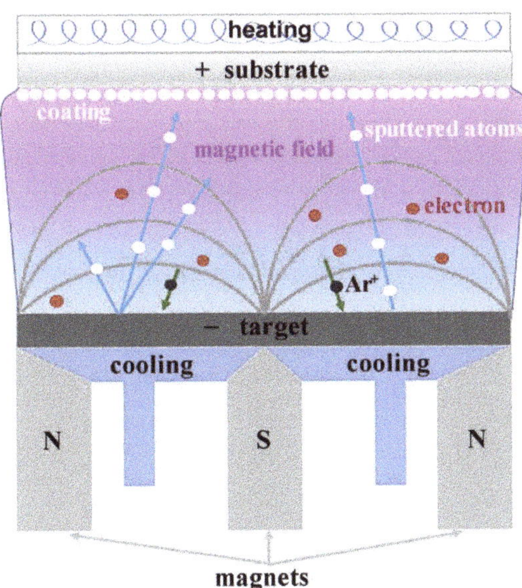

Figure 1. Schematic diagram of magnetron sputtering.

Table 2. Deposition parameters for Ta coatings.

Deposition Conditions	Parameters
Target	99.95% Ta (Φ50 mm × H4 mm)
Substrate	PCrNi1MoA steel (10 mm × 10 mm × 1 mm)
Distance between target and substrate	70 mm
Ar pressure	0.5 Pa
Deposition time	1 h
Substrate temperature (T_{sub})	200–400 °C
Sputtering power (P_{spu})	100–175 W

2.2. Characterization

The crystalline phase of coatings was identified by X-ray diffraction (XRD, Empyrean, PANalytical, the Netherlands) with Cu-K$_\alpha$ radiation ($\lambda = 0.154$ nm) over a 2θ range of 20°–90°. The surface and cross-section morphology were characterized using a field emission scanning electron microscope (FESEM, Quanta-250, FEI Company, Hillsboro, OR, USA) at an accelerating voltage of 20 kV. The three-dimensional profiler (ST400, Nanovea company, Irvine, CA, USA) equipped with confocal optical microprobe was used to collect the microscopic morphology, roughness, and flatness in a specific area of the sample surface. The collection range was 1 mm × 1 mm and the collection accuracy was 0.02 μm.

The hardness and elastic modulus of Ta coatings were measured by a nano-indentation tester (TI-980, Bruker, Billerica, MA, USA) at a loading force of 5 mN. Five measures were recorded for each sample to ensure reliable results and all indentation depths were less than 10% of the coating thickness to avoid the effect of the steel substrate.

The qualitative and quantitative adhesive properties of the samples were analyzed in 2 ways. Firstly, using Rockwell hardness tester (HRD-150, Bangyi Precision meter company, China), the loading force of the Rockwell hardness meter was set to 150 kgf, an indentation was pressed on the sample surface, and we observed the indentation under the optical microscope to evaluate qualitatively the adhesion. Secondly, we quantified the adhesion by the scratch tester (MST, Anton Paar, Austria) with a Rockwell C diamond indenter (200 μm of radius). The scratch length, loading rate, and maximum loading force were 5 mm, 25 N/min, and 30 N, respectively. The larger the measured critical load, the better the coating adhesive performance of the sample.

3. Results and Discussion

3.1. Crystalline Phase

Figure 2 shows XRD patterns of Ta coatings deposited on steel substrates at T_{sub} from 200 °C to 400 °C as a function of power, respectively. The relevant texture coefficient (TC) and full width at half maximum (FWHM) are illustrated in the inset.

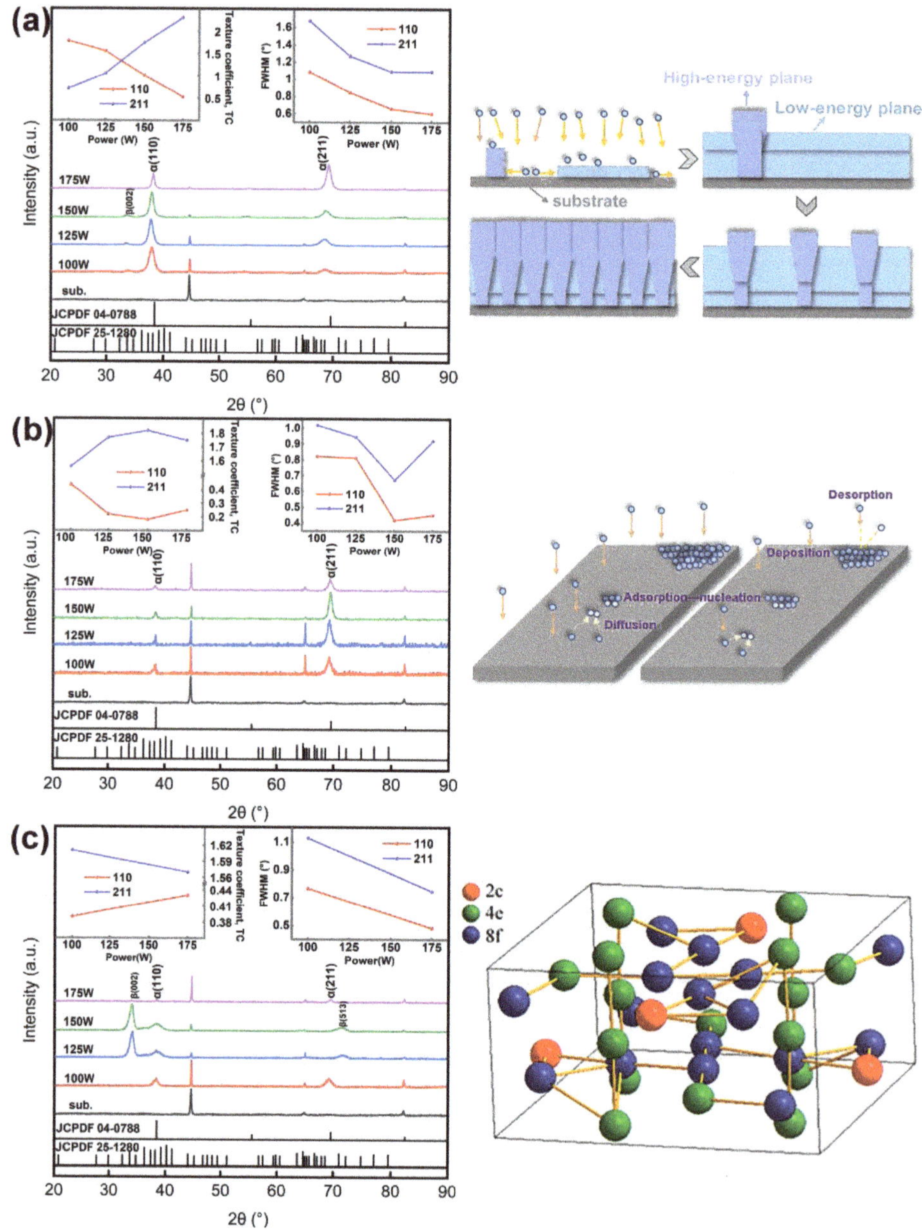

Figure 2. XRD patterns of Ta coatings at different T_{sub} and P_{spu}. (**a**) 200 °C, (**b**) 300 °C, and (**c**) 400 °C.

At T_{sub} = 200 °C (Figure 2a), both α-Ta (JCPDF 04-0788) and β-Ta (JCPDF 25-1280) are detected in the coatings at P_{spu} =100 W, 125 W, and 150 W. When P_{spu} is further increased to 175 W, all diffraction peaks can be indexed to α-Ta, indicating that single-phased α-Ta coatings can be obtained by adjusting P_{spu}. To more accurately characterize the change of the orientation degree, we introduce the texture coefficient $TC_{(hkl)}$, which can be evaluated using the following formula [17,18].

$$TC_{(hkl)} = \frac{I_{(hkl)}/I_{0(hkl)}}{1/N\left[\sum I_{(hkl)}/I_{0(hkl)}\right]}$$

where, $I_{(hkl)}$ and $I_{0(hkl)}$ are the measured intensity of the (hkl) plane and standard intensity in the JCPDF card, respectively, and N is the number of reflections. The larger the $TC_{(hkl)}$, the higher the orientation degree of the (hkl) plane. The fluctuation of $TC_{(110)}$ and $TC_{(211)}$ of α-Ta as a function of P_{spu} is shown in the top-left corner of Figure 2a. It is discovered that the $TC_{(110)}$ value decreases while the $TC_{(211)}$ value grows with increasing P_{spu} from 100 W to 175 W. As is shown in Figure 2a, the different crystallization processes of coatings with different thicknesses will affect the crystallization orientation. In the early growth stage, the plane with low surface energy occupies a large coverage area. When the island growth ends, the plane with high surface energy and a high longitudinal growth rate becomes higher and subsequently expands into the surrounding area, contributing to the compression of the plane with low surface energy. In the middle and late growth stages, the proportion of the plane with high surface energy eventually exceeds the plane with low surface energy. Meanwhile, as can be seen in the top-right corner of Figure 2a, the FWHM of (110) and (211) decrease with increasing power, indicating better crystallinity at the higher power [19]. In addition, the FWHM is related to both crystallite size and the stress of the crystallites. These effects can be separated using the Williamson–Hall method [20]. With the increase in sputtering power, the ionization rate of Ar will lead to the increase in deposition rate and more Ta atoms will accumulate near the nucleation point, resulting in an increase in the grain size of the Ta coating. As can be seen from Figure 2a, the diffraction peaks of coatings are getting closer to the standard peak lines with increasing power, meaning reduced stress. Therefore, the increased crystallite size and reduced stress together promote the decrease in FWHM.

At T_{sub} = 300 °C (Figure 2b), the diffraction peaks are all α-Ta at any powers. However, different degrees of wrinkling and peeling from substrates macroscopically were observed at the P_{spu} of 100 W, 125 W, and 175 W. The highest $TC_{(211)}$ and the lowest $TC_{(110)}$ were obtained at 150 W. Furthermore, the FWHM at 150 W is lower than that of other powers. As is shown in Figure 2b, at the relatively low power, there is not enough energy to diffuse particles reaching the substrate due to the low kinetic energy, leading to poor adhesion and incomplete crystallization. With the increase in the sputtering power, the crystallization of the coating becomes better. However, when the power is too high, the particles deposited on the substrate cannot diffuse enough and are covered by the newly incident particles because of the fast deposition rate, which is not conducive to the crystallization of the coating. Indeed, high nucleation energy may lead to excessive stress, cracks, and other defects inside the coating.

At T_{sub} = 400 °C (Figure 2c), α-Ta was detected as the pure phase at P_{spu} = 100 W. When the P_{spu} was elevated from 125 W to 175 W, the main diffraction peaks were indexed from β-Ta to the mixed phase of α-Ta and β-Ta. The β-Ta phase further grew with crystal planes oriented along (002), which is the densest plane [21], as is shown in Figure 2c. Myers et al. [15] showed that the phase of Ta coating was correlated with its thickness. This conclusion was based on the increase in the substrate temperature with the deposition time, resulting in higher atomic mobility on the coating surface. Gregory Abadias et al. [22] showed that the deposition process parameters did not affect the stability of the crystal phase of α-Ta or β-Ta, indicating that the initial nucleation stage and the energy barrier of the stable α-Ta or β-Ta nucleation were decisive for the phase.

Based on the experimental results in Figure 2, Figure 3 summarizes the evolution of the phase composition of Ta coatings prepared under different T_{sub} and P_{spu} conditions. It can be seen that α-Ta is formed under specific temperature and power conditions. During the deposition process of coatings, the effect of temperature and power on the coating structure can be attributed to the surface activity of the substrate and the energy of the Ta particles reaching the substrate. Accordingly, we speculate that the α-Ta formation may be attributed to the deposition energy which needs to be maintained within a certain range. On the grounds of this finding, we have reviewed some of the literature results. Hua Ren et al. [23] reported that the coatings deposited at intermediate bias voltages had pure α-Ta while they showed either mixed α-Ta with β-Ta or pure β-Ta at higher or lower bias voltages. Kazuhide Ino et al. [24] found that the coatings exhibited β-Ta if the bombarding energy was greater than 25 eV. It was confirmed that an appropriate energy input to the deposition of Ta atoms is required to synthesize α-Ta and excessive high energy may be unnecessary or even deleterious to α-Ta formation.

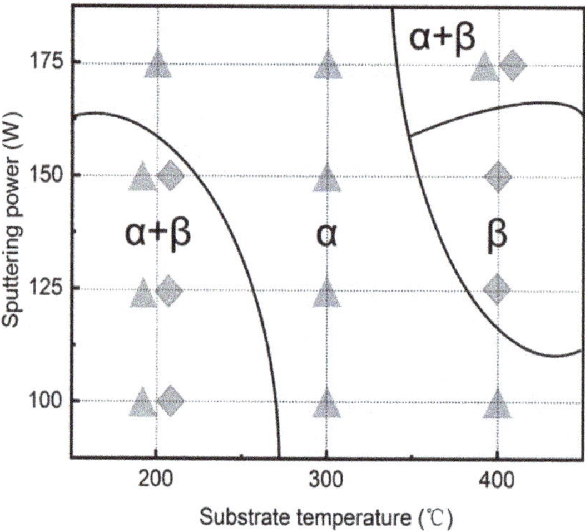

Figure 3. Evolution of phase composition at different T_{sub} and P_{spu}, the α and β were indicated by ▲ and ◆, respectively.

Therefore, T_{sub} and P_{spu} have a combined effect on growth and structural evolution of Ta coatings. They may interact with each other to keep a specific deposition energy. For the single-phased and homogeneous α-Ta coatings, the deposition conditions are 200 °C-175 W, 300 °C-150 W, and 400 °C-100 W.

3.2. Surface Morphology

Figure 4 shows the surface morphology and grain distribution histogram of the Ta coatings deposited at T_{sub}-P_{spu} of 200 °C-175 W (Figure 4a,d), 300 °C-150 W (Figure 4b,e), and 400 °C-100 W (Figure 4c,f), respectively. The size distribution histograms were obtained by measuring more than 100 grains from the FESEM images.

As can be seen from the photographs, the surface morphology of single-phased α-Ta coatings deposited in these different conditions are all characterized by triangular pyramid-shaped particles with clear grain gaps, consistent with results published by M. Grosser et al. [25]. According to the Thornton structural zone model, the deposition pattern of Ta coating is the Z1 structure. This indicates that the diffusion distance of the particles is very small, and the initial nucleus tends to capture the orientation of the atoms on the coating surface at $T/T_m < 0.1$. While the substrate temperature increases from 200 °C–400 °C ($0.1 < T/T_m < 0.3$), the diffusion kinetic energy of surface adsorbed atoms increases, leading to increasing mobility of adsorbed atoms. The initially adsorbed atoms rapidly self-diffuse towards the equilibrium position, eventually trapped by larger grains to form the pyramid-shaped structure. The grain sizes of Ta coatings prepared at

200 °C-175 W and 300 °C-150 W range from 120 nm to 240 nm, while the grains of the sample prepared at 400 °C-100 W are smaller, within 60–120 nm, the surface of which is also denser than that in the other two conditions.

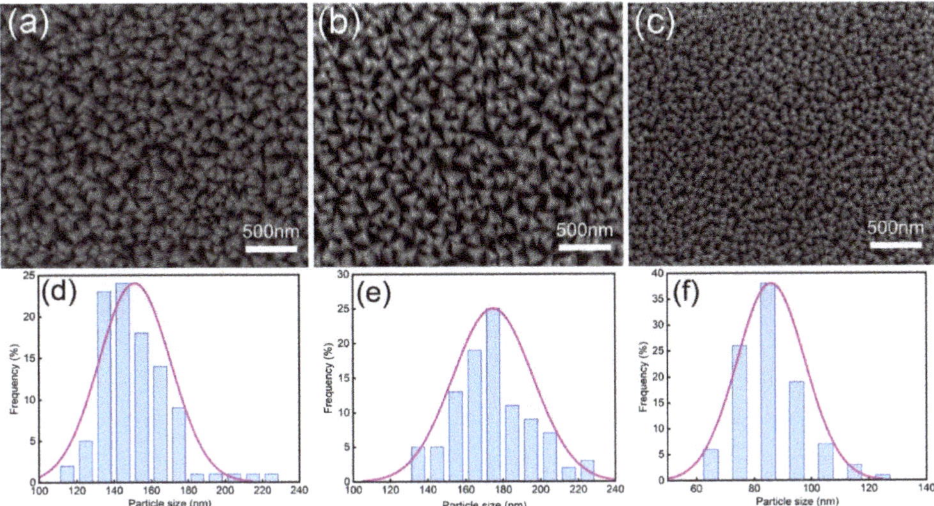

Figure 4. FESEM images of surface morphology and grain size of the Ta coatings. (**a,d**) 200 °C-175 W, (**b,e**) 300 °C-150 W, and (**c,f**) 400 °C-100 W.

Figure 5 shows the 3D profiles of Ta coatings in different deposition conditions. Compared with the microstructural images obtained through SEM observation, the 3D profile contains a larger observation interval which is more suitable for observing the overall morphology distribution of millimeter-level samples. The surface height data of the Ta coating under different conditions were obtained according to the 3D profiles, as is listed in Table 3. Root mean square heights (Sq) of Ta coatings at 200 °C-175 W, 300 °C-150 W, and 400 °C-100 W were 0.0402 µm, 0.0184 µm, and 0.0132 µm, respectively. Arithmetic mean heights (Sa) of Ta coatings at 200 °C-175 W, 300 °C-150 W, and 400 °C-100 W were 0.0319 µm, 0.0145 µm, and 0.0103 µm, respectively. The results indicate that the surface of the Ta coating at 400 °C-100 W is flatter, while the roughness of the Ta coating at 200 °C-175 W is higher. According to the histogram results of the grain distribution in Figure 4, the coating grain size distributions at 400 °C-100 W and 300 °C-150 W have a concentrated grain size distribution, while the coatings at 200 °C-175 W are quite distributed between 130 nm and 170 nm, resulting in higher roughness.

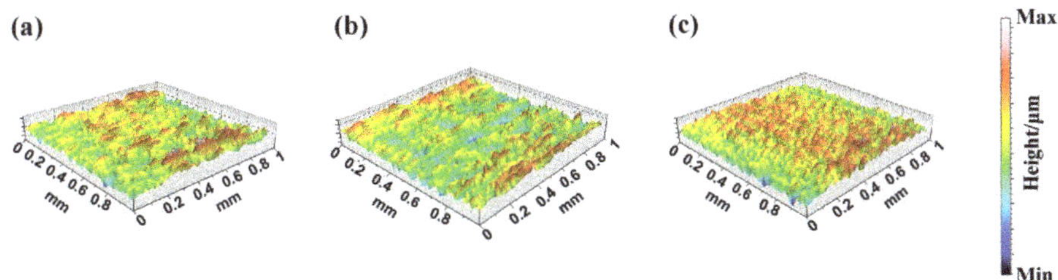

Figure 5. The 3D profile of the Ta coatings. (**a**) 200 °C-175 W, (**b**) 300 °C-150 W, and (**c**) 400 °C-100 W.

Table 3. Height parameters of the Ta coatings in different conditions.

Height Parameters	200 °C-175 W	300 °C-150 W	400 °C-100 W	Header Unabbreviated Form
Sq/μm	0.0402	0.0184	0.0132	Root mean square height
Sp/μm	0.141	0.0677	0.0482	Maximum peak height
Sv/μm	0.143	0.071	0.0678	Maximum pit height
Sz/μm	0.284	0.139	0.116	Maximum height
Sa/μm	0.0319	0.0145	0.0103	Arithmetic mean height

3.3. Cross-Sectional Morphology

Figure 6 shows the cross-sectional morphology of the Ta coatings deposited at T_{sub}-P_{spu} of 200 °C-175 W, 300 °C-150 W, and 400 °C-100 W.

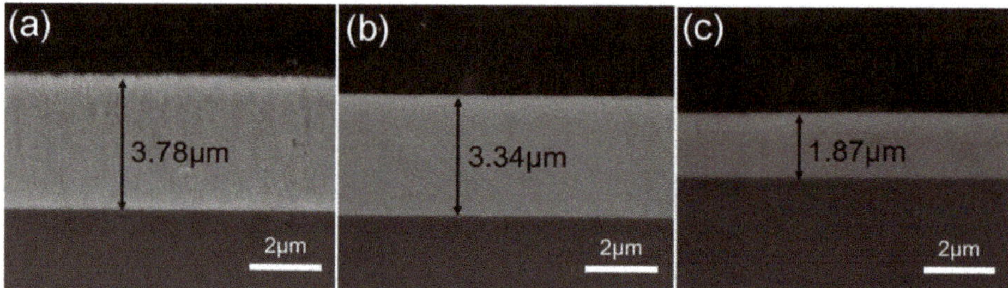

Figure 6. FESEM images of cross-sections of Ta coatings. (**a**) 200 °C-175 W, (**b**) 300 °C-150 W, and (**c**) 400 °C-100 W.

The coatings exhibit a typical columnar crystal growth pattern mainly affected by the shadowing effect. The anisotropy of columnar crystal growth leads to a large competitive advantage for those crystals growing perpendicular to the interface, and columnar crystals in other directions are submerged as the coating thickens. In addition, the thicknesses of Ta coatings are approximately 3.78 μm, 3.34 μm, and 1.87 μm, respectively. When the sputtering power increases, the sputtered Ta atoms from the target increase, resulting in a high sputtering rate. However, according to the theory proposed by Kim et al. [26], a lower deposition rate enables atoms to move through the crystals across larger regions, and atoms are unlikely to be covered by subsequent adatom flux before being dampened in a particular location, potentially improving the mechanical properties of materials.

3.4. Hardness and Modulus

The corresponding hardness and elastic modulus were measured using an indentation tester for these coatings. It can be seen from Figure 7a that hardness and elastic modulus both reach the maximum value under the sputtering condition of 400 °C-100 W, 9 GPa, and 157 GPa, respectively. In general, the high hardness of coatings is usually attributed to their dense structure and small grain size. As shown in Figure 4, the Ta coating at 400 °C-100 W exhibits the densest structure with a minimal grain gap. On the other hand, this is concerned with the grain size that has an effect on mechanical properties of coatings due to the Hall–Petch (H-P) effect [27]. On the basis of the dislocation theory, grain boundaries are obstacles in the movement of dislocations, and refining grains can produce more grain boundaries. If the grain boundary structure does not change, a larger external force is required to generate dislocation plugging, thereby strengthening the material.

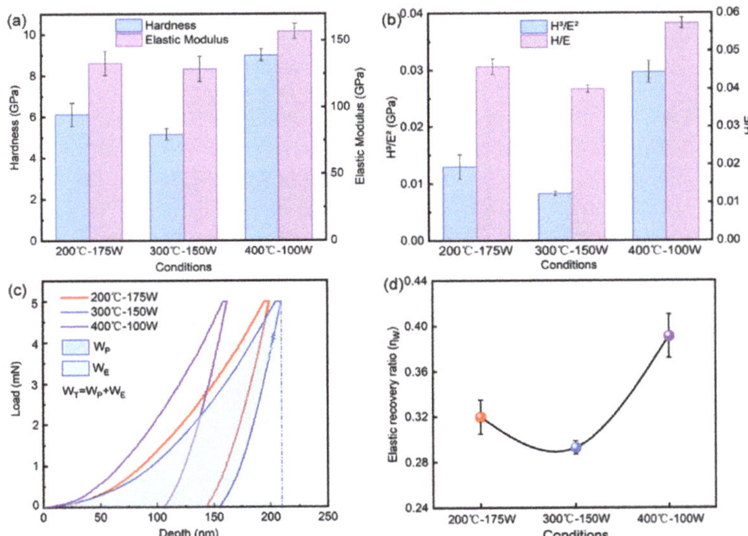

Figure 7. (**a**) H and E, (**b**) H/E and H^3/E^2, (**c**) load–depth curves, and (**d**) elastic recovery ratio of Ta coatings.

Furthermore, the enhanced hardness H is not a single important property of the hard coating. For many applications, H/E (elastic strain failure strength) [28] and H^3/E^2 (plastic deformation strength) [17] calculated by hardness and modulus are more important than their extremely high hardness. H/E and H^3/E^2 are important parameters of coating materials in the field of wear, which are positively related to the toughness of coatings and play a vital role in the crack resistance process [29]. Figure 7b shows the change in H/E and H^3/E^2 calculated from hardness (H) and elastic modulus (E). It can be seen that H/E and H^3/E^2 show similar variation trends, and both reach their highest at 400 °C-100 W, which are 0.057 and 0.030 GPa, respectively. It also demonstrates that the Ta coating at 400 °C-100 W has better crack resistance and toughness, with abilities to endure mechanical damage in abrasive wear applications and withstand impact energy from deformation to fracture [28].

The load–depth curves of Ta coatings at various sputtering conditions are illustrated in Figure 7c. The coatings were complied with an elastoplastic deformation according to the curves. In the nanoindentation test, some of the total energy (W_T) was recovered as reversible elastic energy (W_E), while some was lost as irreversible plastic deformation energy (W_P). The elastic recovery ratio (η_w), which can be defined as $\eta_w = W_E/W_T$, is a measure to express the response of a material to indentation [30,31]. Take the coating prepared at 300 °C-150 W as an example; W_E and W_T are given by the areas under the loading and unloading curves, respectively, as shown in Figure 7c. The coating prepared at 400 °C-100 W has a higher η_w value of 0.39 in Figure 7d, indicating better resilience.

Figure 8 shows the comparison of the H/E and H^3/E^2 of α-Ta coatings prepared in this work with other literatures. In previous studies, the H/E and H^3/E^2 varied within a range of 0.0275–0.055 and 0.0058–0.031 GPa, respectively. Compared with other studies, the α-Ta coating at 400 °C-100 W in our work has higher H/E and comparable H^3/E^2 in contrast to G. Abadias et al. [21], which can be attributed to its high densification and homogeneous structure.

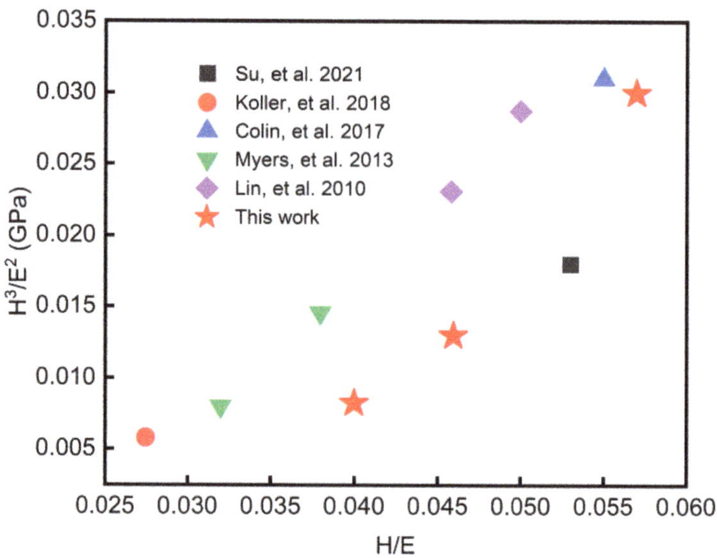

Figure 8. Comparison of the H/E and H^3/E^2 of α-Ta coatings with other literatures [15,21,32–34].

3.5. Adhesion

From actual applications of Ta coatings, the adhesion and integrity of the coating-substrate interface under projected loads are key performance concerns and the basis for improving coating performance and extended applications [11].

The adhesive performance was evaluated according to the reference standard. The standard diagram is shown in Figure 9, where grades one to four are acceptable failure and five to six are unacceptable failure. Figure 10 shows the results of the testing of α-Ta coatings under different conditions. The coatings prepared at 200 °C-175 W and 300 °C-150 W showed cracks and delamination, and the coating prepared at 300 °C-150 W peeled off from the substrate more seriously. Therefore, the coating ratings at 200 °C-175 W and 300 °C-150 W were HF3 and HF4, respectively. The coating prepared at 400 °C-100 W showed good adhesion without obvious delamination around the indentation. Compared with Figure 9, the overall coating rating was HF1. The better adhesion of the coating under 400 °C-100 W may be related to the dense structure and higher hardness.

According to the above analysis with respect to the morphology, hardness, and indentation test, it is determined that 400 °C-100 W is the optimal condition for forming Ta coatings. However, the coating indentation morphology can only qualitatively represent the adhesion of the coating to a certain extent. In this work, a scratch tester was used to quantify the adhesive properties of the Ta coating at 400 °C-100 W. The sudden variation of the acoustic emission and friction force is the deciding factor of every scratch mechanism.

The scratch morphologies of the Ta coating at 400 °C-100 W are shown in Figure 11a. As the applied load increases, the scratch width of the Ta coating surface also gradually increases, and the scratch area is clearly visible. In the scratch test, L_{c1} is the corresponding load when cracks begin to appear and L_{c2} is the minimum load for continuous peeling between the coating and the substrate, indicating the failure of the adhesion of the coating-substrate system. Usually, L_{c2} is used to determine the failure of the coating [35]. From the enlarged image of L_{c1} in Figure 11b, the appearance of small cracks can be seen, and the enlarged image of L_{c2} in Figure 11c demonstrates that the coating begins to peel off from the substrate at this time, exposing the color of the substrate. According to the micro scratch curves in Figure 11d, it can be seen that $L_{c1} = 7.38$ N and $L_c2 = 18.46$ N. The larger the L_{c1}, the stronger the cracking resistance of the coating. The longer the distance between L_{c1} and L_{c2}, the better the coating's resistance to crack propagation [36]. In previous work,

Su et al. [32] found that mono-DC Ta coatings reached optimum adhesion compared with mono-PP and multi-PD coatings. The first fracture of mono-DC coating occurred when the load reached ~320 mN because the main diffraction peaks were β-Ta. Furthermore, Ay Ching Hee et al. [37], using DCMS technology, observed that the initiation of cracking in Ta coatings was at 2.5 N and the second critical load was at 18.6 N, with the coating separating from the substrate.

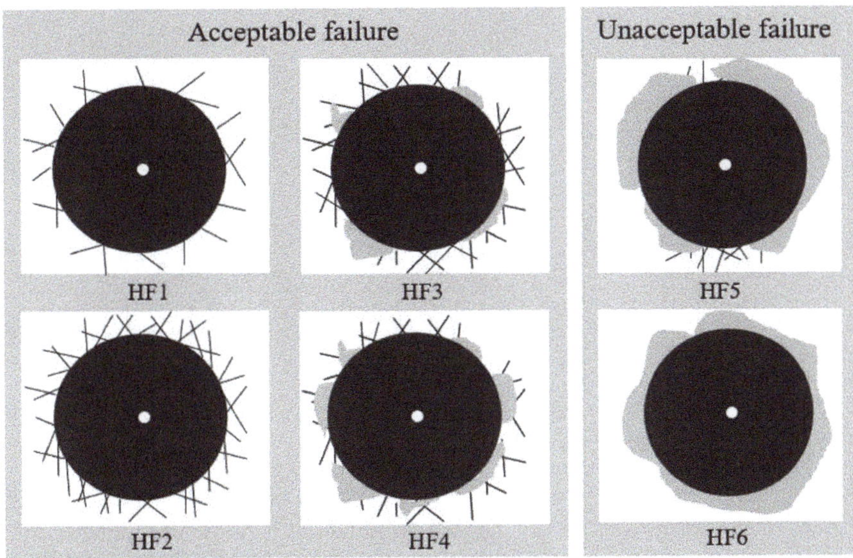

Figure 9. Reference standard for the evaluation of adhesion.

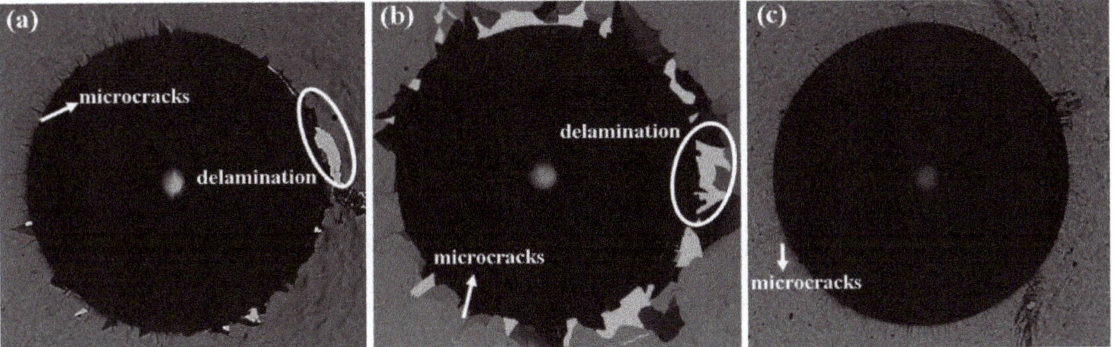

Figure 10. HRC indentation test of Ta coatings. (**a**) 200 °C-175 W, (**b**) 300 °C-150 W, and (**c**) 400 °C-100 W.

Compared with the studies in other literatures [32,37], the Ta coating prepared in this work has good adhesion as well as strong resistance to cracking.

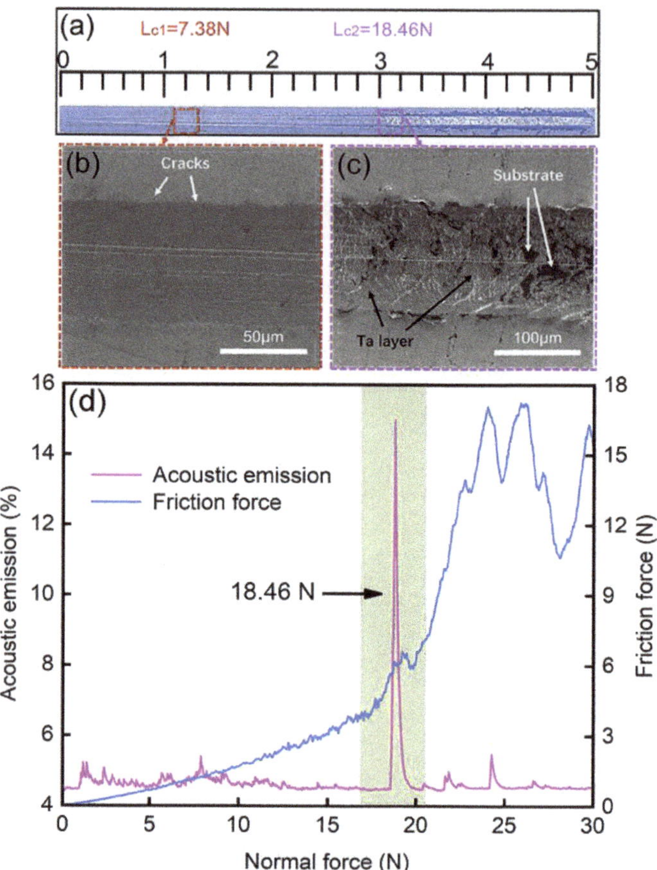

Figure 11. Scratch results of Ta coating prepared at 400 °C-100 W (**a**) OM of the scratch, (**b**,**c**) SEM morphology of the selected areas in Figure 4a, and (**d**) the micro scratch curve.

4. Conclusions

In this study, Ta hard coatings were prepared on PCrNi1MoA steel substrates by direct current magnetron sputtering at various substrate temperatures (T_{sub} = 200–400 °C) and sputtering powers (P_{spu} = 100–175 W). The basic conclusions are as follows:

The growth and phase evolution were observed to be controlled by T_{sub} and P_{spu} simultaneously. Higher P_{spu} was required in order to obtain α-Ta coatings when the coatings are deposited at lower T_{sub}, and vice versa. Therefore, T_{sub} and P_{spu} interacted with each other and had a combined effect on coating growth. At a T_{sub}-P_{spu} at 200 °C-175 W, 300 °C-150 W, and 400 °C-100 W, single-phased and homogeneous α-Ta coatings were obtained.

The α-Ta coating deposited at T_{sub}-P_{spu} at 400 °C-100 W showed a relatively denser structure, finer grain (60–120 nm, and flatter surface, with the root mean square heights (Sq) of 0.0132 μm.

The α-Ta coating deposited at T_{sub}-P_{spu} at 400 °C-100 W exhibited a higher hardness (9 GPa), H/E (0.057), and H^3/E^2 (0.030 GPa), as well as an adhesion of 18.46 N.

Author Contributions: Conceptualization, C.L.; methodology, C.L.; software, C.L.; validation, C.L.; formal analysis, C.L.; investigation, C.L.; resources, J.P., Z.X. and Q.S.; data curation, C.L.; writing—original draft preparation, C.L.; writing—review and editing, J.P. and Z.X.; visualization,

C.L.; supervision, C.W.; project administration, C.W. and Q.S.; funding acquisition, C.W. All authors have read and agreed to the published version of the manuscript.

Funding: This research was funded by Guangdong Major Project of Basic and Applied Basic Research (2021B0301030001), National Key R&D Program of China (2021YFB3802300), and Self-innovation Research Funding Project of Hanjiang Laboratory (HJL202012A001, HJL202012A002, HJL202012A003). And the APC was funded by Guangdong Major Project of Basic and Applied Basic Research (2021B0301030001).

Data Availability Statement: The data that support the findings of this study are all own results of the authors.

Conflicts of Interest: The authors declare no conflict of interest.

References

1. Chen, X.; Yan, Q.; Ma, Q. Influence of the laser pre-quenched substrate on an electroplated chromium coating/steel substrate. *Appl. Surf. Sci.* **2017**, *405*, 273–279. [CrossRef]
2. Sarraf, S.H.; Soltanieh, M.; Aghajani, A. Repairing the cracks network of hard chromium electroplated layers using plasma nitriding technique. *Vacuum* **2016**, *127*, 1–9. [CrossRef]
3. Podgornik, B.; Massler, O.; Kafexhiu, F.; Sedlacek, M. Crack density and tribological performance of hard-chrome coatings. *Tribol. Int.* **2018**, *121*, 333–340. [CrossRef]
4. Peng, X.; Xia, C.; Dai, X.; Wu, A.; Dong, L.; Li, D.; Tao, Y. Ablation behavior of NiCrAlY coating on titanium alloy muzzle brake. *Surf. Coat. Technol.* **2013**, *232*, 690–694. [CrossRef]
5. Wang, D.; Lin, S.S.; Liu, L.Y.; Xue, Y.N.; Yang, H.Z.; Bai, J.L.; Zhou, K.S. Effect of Bias Voltage on Microstructure and Erosion Resistance of CrAlN Coatings Deposited by Arc Ion Plating. *Rare Met. Mater. Eng.* **2020**, *49*, 2583–2590.
6. Matson, D.W.; McClanahan, E.D.; Lee, S.L.; Windover, D. Properties of thick sputtered Ta used for protective gun tube coatings. *Surf. Coat. Technol.* **2001**, *146*, 344–350. [CrossRef]
7. Liu, L.L.; Xu, J.; Lu, X.; Munroe, P.; Xie, Z.H. Electrochemical Corrosion Behavior of Nanocrystalline beta-Ta Coating for Biomedical Applications. *ACS Biomater. Sci. Eng.* **2016**, *2*, 579–594. [CrossRef]
8. Niu, Y.; Chen, M.; Wang, J.; Yang, L.; Guo, C.; Zhu, S.; Wang, F. Preparation and thermal shock performance of thick α-Ta coatings by direct current magnetron sputtering (DCMS). *Surf. Coat. Technol.* **2017**, *321*, 19–25. [CrossRef]
9. Traving, M.; Zienert, I.; Zschech, E.; Schindler, G.; Steinhögl, W.; Engelhardt, M. Phase analysis of TaN/Ta barrier layers in sub-micrometer trench structures for Cu interconnects. *Appl. Surf. Sci.* **2005**, *252*, 11–17. [CrossRef]
10. Zhou, Y.M.; Xie, Z.; Xiao, H.N.; Hu, P.F.; He, J. Effects of deposition parameters on tantalum films deposited by direct current magnetron sputtering. *J. Vac. Sci. Technol. A Vac. Surf. Film.* **2009**, *27*, 109–113. [CrossRef]
11. Gladczuk, L.; Patel, A.; Singh Paur, C.; Sosnowski, M. Tantalum films for protective coatings of steel. *Thin Solid Film.* **2004**, *467*, 150–157. [CrossRef]
12. Al-masha'al, A.; Bunting, A.; Cheung, R. Evaluation of residual stress in sputtered tantalum thin-film. *Appl. Surf. Sci.* **2016**, *371*, 571–575. [CrossRef]
13. Latif, R.; Jaafar, M.F.; Aziz, M.F.; Zain, A.R.M.; Yunas, J.; Majlis, B.Y. Influence of tantalum's crystal phase growth on the microstructural, electrical and mechanical properties of sputter-deposited tantalum thin film layer. *Int. J. Refract. Met. Hard Mater.* **2020**, *92*, 105314. [CrossRef]
14. Navid, A.A.; Hodge, A.M. Controllable residual stresses in sputtered nanostructured alpha-tantalum. *Scr. Mater.* **2010**, *63*, 867–870. [CrossRef]
15. Myers, S.; Lin, J.; Souza, R.M.; Sproul, W.D.; Moore, J.J. The β to α phase transition of tantalum coatings deposited by modulated pulsed power magnetron sputtering. *Surf. Coat. Technol.* **2013**, *214*, 38–45. [CrossRef]
16. Alami, J.; Eklund, P.; Andersson, J.; Lattemann, M.; Wallin, E.; Bohlmark, J.; Persson, P.; Helmersson, U. Phase tailoring of Ta thin films by highly ionized pulsed magnetron sputtering. *Thin Solid Film.* **2007**, *515*, 3434–3438. [CrossRef]
17. Elmkhah, H.; Zhang, T.F.; Abdollah-zadeh, A.; Kim, K.H.; Mahboubi, F. Surface characteristics for the TiAlN coatings deposited by high power impulse magnetron sputtering technique at the different bias voltages. *J. Alloys Compd.* **2016**, *688*, 820–827. [CrossRef]
18. Peng, B.; Xu, Y.X.; Du, J.W.; Chen, L.; Kim, K.H.; Wang, Q. Influence of preliminary metal-ion etching on the topography and mechanical behavior of TiAlN coatings on cemented carbides. *Surf. Coat. Technol.* **2022**, *432*, 128040. [CrossRef]
19. Ferreira, F.; Sousa, C.; Cavaleiro, A.; Anders, A.; Oliveira, J. Phase tailoring of tantalum thin films deposited in deep oscillation magnetron sputtering mode. *Surf. Coat. Technol.* **2017**, *314*, 97–104. [CrossRef]
20. Augustin, A.; Udupa, K.R.; Udaya, B.K. Crystallite size measurement and micro-strain analysis of electrodeposited copper thin film using Williamson-Hall method. *AIP Conf. Proc.* **2016**, *1728*, 020492.
21. Colin, J.J.; Abadias, G.; Michel, A.; Jaouen, C. On the origin of the metastable β-Ta phase stabilization in tantalum sputtered thin films. *Acta Mater.* **2017**, *126*, 481–493. [CrossRef]
22. Abadias, G.; Colin, J.J.; Tingaud, D.; Djemia; Belliard, L.; Tromas, C. Elastic properties of α- and β-tantalum thin films. *Thin Solid Film. Vol.* **2019**, *688*, 137403. [CrossRef]

23. Ren, H.; Sosnowski, M. Tantalum thin films deposited by ion assisted magnetron sputtering. *Thin Solid Film.* **2008**, *516*, 1898–1905. [CrossRef]
24. Ino, K.; Shinohara, T.; Ushiki, T.; Ohmi, T. Ion energy, ion flux, and ion species effects on crystallographic and electrical properties of sputter-deposited Ta thin films. *J. Vac. Sci. Technol. A Vac. Surf. Film.* **1997**, *15*, 2627–2635. [CrossRef]
25. Grosser, M.; Schmid, U. The impact of sputter conditions on the microstructure and on the resistivity of tantalum thin films. *Thin Solid Film.* **2009**, *517*, 4493–4496. [CrossRef]
26. Kim, J.H.; Holloway, P.H. Textured growth of cubic gallium nitride thin films on Si (100) substrates by sputter deposition. *J. Vac. Sci. Technol. A Vac. Surf. Film.* **2004**, *22*, 1591–1595. [CrossRef]
27. Jimenez, M.J.M.; Antunes, V.; Cucatti, S.; Riul, A.; Zagonel, L.F.; Figueroa, C.A.; Wisnivesky, D.; Alvarez, F. Physical and micro-nano-structure properties of chromium nitride coating deposited by RF sputtering using dynamic glancing angle deposition. *Surf. Coat. Technol.* **2019**, *372*, 268–277. [CrossRef]
28. Huang, B.; Liu, L.-T.; Du, H.-M.; Chen, Q.; Liang, D.-D.; Zhang, E.-G.; Zhou, Q. Effect of nitrogen flow rate on the microstructure, mechanical and tribological properties of CrAlTiN coatings prepared by arc ion plating. *Vacuum* **2022**, *204*, 111336. [CrossRef]
29. Leyland, A.; Matthews, A. Design criteria for wear-resistant nanostructured and glassy-metal coatings. *Surf. Coat. Technol.* **2004**, *177–178*, 317–324. [CrossRef]
30. Jha, K.K.; Suksawang, N.; Lahiri, D.; Agarwal, A. Evaluating initial unloading stiffness from elastic work-of-indentation measured in a nanoindentation experiment. *J. Mater. Res.* **2013**, *28*, 789–797. [CrossRef]
31. Xia, Y.; Xu, Z.; Peng, J.; Shen, Q.; Wang, C. In-situ formation, structural transformation and mechanical properties Cr—N coatings prepared by MPCVD. *Surf. Coat. Technol.* **2022**, *44*, 128522. [CrossRef]
32. Su, Y.; Huang, W.; Zhang, T.; Shi, C.; Hu, R.; Wang, Z.; Cai, L. Tribological properties and microstructure of monolayer and multilayer Ta coatings prepared by magnetron sputtering. *Vacuum* **2021**, *189*, 110250. [CrossRef]
33. Koller, C.M.; Marihart, H.; Bolvardi, H.; Kolozsvári, S.; Mayrhofer, P.H. Structure, phase evolution, and mechanical properties of DC, pulsed DC, and high power impulse magnetron sputtered Ta–N films. *Surf. Coat. Technol.* **2018**, *347*, 304–312. [CrossRef]
34. Lin, J.; Moore, J.J.; Sproul, W.D.; Lee, S.L.; Wang, J. Effect of Negative Substrate Bias on the Structure and Properties of Ta Coatings Deposited Using Modulated Pulse Power Magnetron Sputtering. *IEEE Trans. Plasma Sci.* **2010**, *38*, 3071–3078. [CrossRef]
35. Shi, X.W.; Li, X.R.; Yao, N.; Wang, X.C.; Song, K.L.; Zhang, S. A Study on Magnetic Filter Controlling TiN Films Prepared by Arc Ion Plating. *Appl. Mech. Mater.* **2011**, *117–119*, 1071–1075. [CrossRef]
36. Zhang, S.; Sun, D.; Fu, Y.; Du, H. Toughening of hard nanostructural thin films: A critical review. *Surf. Coat. Technol.* **2005**, *198*, 2–8. [CrossRef]
37. Hee, A.C.; Jamali, S.S.; Bendavid, A.; Martin, P.J.; Kong, C.; Zhao, Y. Corrosion behaviour and adhesion properties of sputtered tantalum coating on Ti6Al4V substrate. *Surf. Coat. Technol.* **2016**, *307*, 666–675. [CrossRef]

Article

Archaeometallurgical Analysis of the Provincial Silver Coinage of Judah: More on the Chaîne Opératoire of the Minting Process

Maayan Cohen [1,2], Dana Ashkenazi [3,*], Haim Gitler [4] and Oren Tal [1]

1   Department of Archaeology and Ancient Near Eastern Cultures, Tel Aviv University, Ramat Aviv, Tel Aviv 6997801, Israel
2   Leon Recanati Institute for Maritime Studies, University of Haifa, Haifa 3498838, Israel
3   School of Mechanical Engineering, Tel Aviv University, Ramat Aviv, Tel Aviv 6997801, Israel
4   Israel Museum, Derech Rupin 11, Jerusalem 9171002, Israel
*   Correspondence: danaa@tauex.tau.ac.il or dana@eng.tau.ac.il

Abstract: Silver coins were the first coins to be manufactured by mass production in the southern Levant. An assemblage of tiny provincial silver coins of the local (Judahite standard) and (Attic) *obol*-based denominations from the Persian and Hellenistic period Yehud and dated to the second half of the fourth century BCE were analyzed to determine their material composition. Of the 50 silver coins, 32 are defined as Type 5 (Athena/Owl) of the Persian period Yehud series (ca. 350–333 BCE); 9 are Type 16 (Persian king wearing a jagged crown/Falcon in flight) (ca. 350–333); 3 are Type 24 series (Portrait/Falcon) of the Macedonian period (ca. 333–306 BCE); and 6 are Type 31 (Portrait/Falcon) (ca. 306–302/1 BCE). The coins underwent visual testing, multi-focal light microscope observation, XRF analysis, and SEM-EDS analysis. The metallurgical findings revealed that all the coins from the Type 5, 16, 24, and 31 series are made of high-purity silver with a small percentage of copper. Based on these results, it is suggested that each series was manufactured using a controlled composition of silver–copper alloy. The findings present novel information about the material culture of the southern Levant during the Late Persian period and Macedonian period, as expressed through the production and use of these silver coins.

Keywords: archaeometallurgy; Judah; non-destructive testing analysis; materials characterization; Palestine; silver coins; southern Levant

Citation: Cohen, M.; Ashkenazi, D.; Gitler, H.; Tal, O. Archaeometallurgical Analysis of the Provincial Silver Coinage of Judah: More on the Chaîne Opératoire of the Minting Process. *Materials* **2023**, *16*, 2200. https://doi.org/10.3390/ma16062200

Academic Editors: Andrea Di Schino and Claudio Testani

Received: 13 February 2023
Revised: 6 March 2023
Accepted: 7 March 2023
Published: 9 March 2023

1. Introduction

Analysis of the chemical composition of ancient coins can reveal their alloy content and enable us to determine, among other things, whether certain issues were produced using a controlled and relatively homogenous metallurgical composition. Moreover, the elemental composition of a coin, when compared to a validated dataset, may indicate where that coin was produced [1].

The study of the Provincial silver coinage of Judah was recently revised by Gitler et al. (2023) [2], who presented a typological corpus of the 44 recorded Yehud coin types, as well as a die study of the coins dated to the Late Persian, Macedonian, and Early Hellenistic periods. Here, we present an archaeometallurgical study of some of the more common coin types of the Yehud minting authority (Types 5, 16, 24, and 31) (Figure 1).

Type 5 of the Yehud coinage series (Figure 2) features a helmeted head of Athena in profile, turned to the right on the obverse, while the reverse depicts an owl with the body turned to the right and its head facing front. An olive spray or lily appears in the upper left field, while in the right field, the legend YHD (an abbreviation of the name of the province: Judah) is written in Paleo-Hebrew and Aramaic. These Athenian-styled *gerah* denomination-based issues (1/20 of a local Judahite sheqel; mean weight of 0.48 g based on a sample of 150 coins) were minted in the province of Judah during the Late Persian period between 350 and 333 BCE (see [2]: Chapter III).

With Type 5, the Yehud coinage began a specialization in small denominations in contrast to the larger denominations that had been previously minted for this province (Types 1–4 in the Yehud corpus). Our initial research question was: can we determine with sufficient certainty whether our die-linked issues were not only produced according to a specific standard of metallurgical composition, but also from the same metal batch throughout the minting of the series?

Our study then took this question a step further and sought to determine whether the die-linked issues of Type 5 (Figure 3), which all feature the same obverse (O1) but with five different reverses (R1, R2, R3, R4, and R5), were produced from different metal batches or from the same specific batch that had been used to produce the flans for the entire Type 5 group which is connected to Obverse 1 (O1/R1, O1/R2, O1/R3, O1/R4, and O1/R5 are shown in Figures 2–4). Basically, we sought answers that relate to the chaîne opératoire of the earliest small denomination coins minted in the province of Judah.

Similarly, to other Yehud small denomination coins, Type 5 was struck from dies that were damaged during the striking process but continued to be used (progressive degradation of O1, see [2]: Figure II.1, p. 22). The die damage could have been caused by excessive wear, breaks, or errors. In this peculiar chaîne opératoire, the artisans who struck the coins apparently operated independently of local die engravers and thus were not able to replace the damaged dies or to recut the motifs in order to repair them. Consequently, the damaged obverse dies continued to be used and to produce coins whose obverse motifs are barely discernable [3].

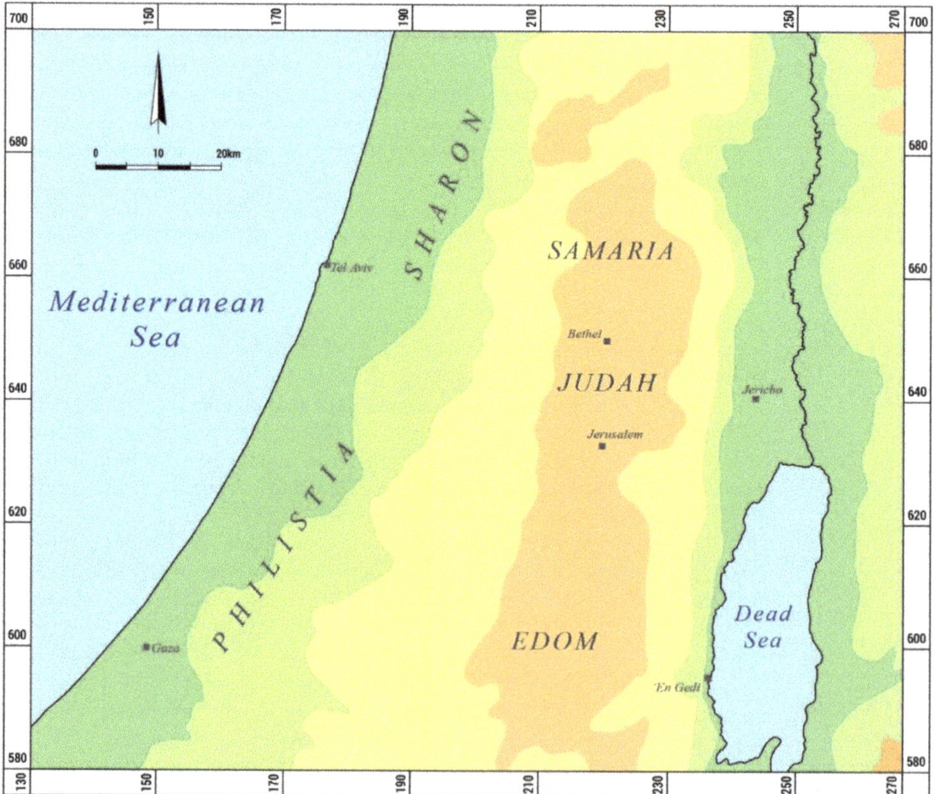

Figure 1. Location map of the research area, showing the fourth-century BCE regional division of southern Palestine.

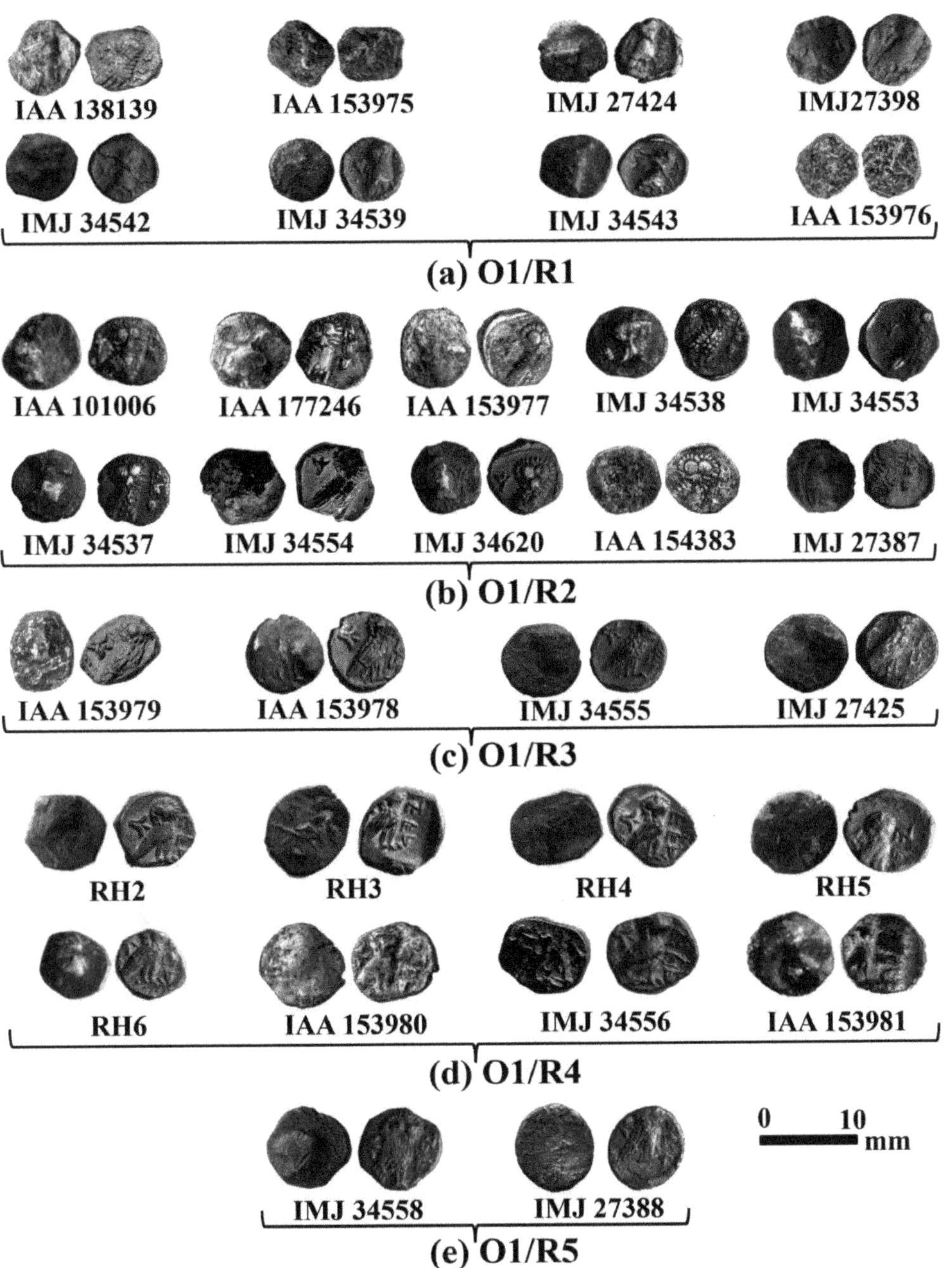

Figure 2. Yehud *gerah* coins—Type 5 of the Yehud series (Athena/Owl) corpus with die combinations O1 (obverse) and R1 (reverse), R2, R3, R4, and R5 ((**a–e**), respectively) with the YHD legend.

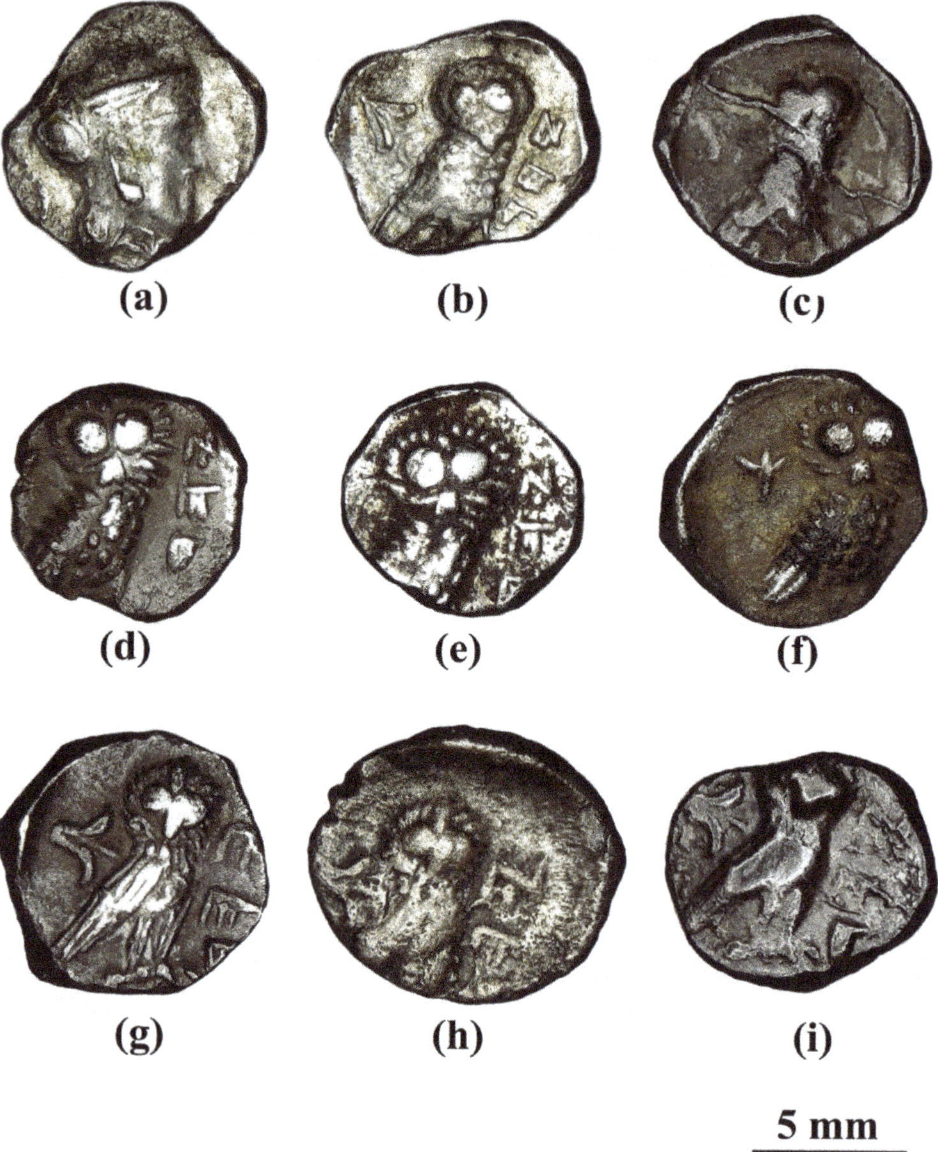

5 mm

Figure 3. Multi-focal light microscope (LM) enlarged observation of the Yehud *gerah* coins Type 5 die-linked connection, with the same obverse (O1) but different reverses (R1, R2, and R4): (**a**) O1/R1: IAA 138139 (Khirbat Qeiyafa excavations), obverse, showing the helmeted head of Athena; (**b**) O1/R1: IAA 138139, reverse, showing a lily, symbol of Jerusalem (left), owl (center), and Aramaic inscription of Judea (YHD, right); (**c**) O1/R1: IMJ 34542, reverse; (**d**) O1/R2: IMJ 34537, reverse; (**e**) O1/R2: IAA 154383 (Khirbat Qeiyafa); (**f**) O1/R2: IMJ 34538, reverse; (**g**) O1/R4: Ramallah area hoard no. 1, IMJ 2006.53.26139 (RH2), reverse; (**h**) O1/R4: Ramallah area hoard no. 5, IMJ 2006.53.26142 (RH5), reverse; and (**i**) O1/R4: IMJ 34556, reverse.

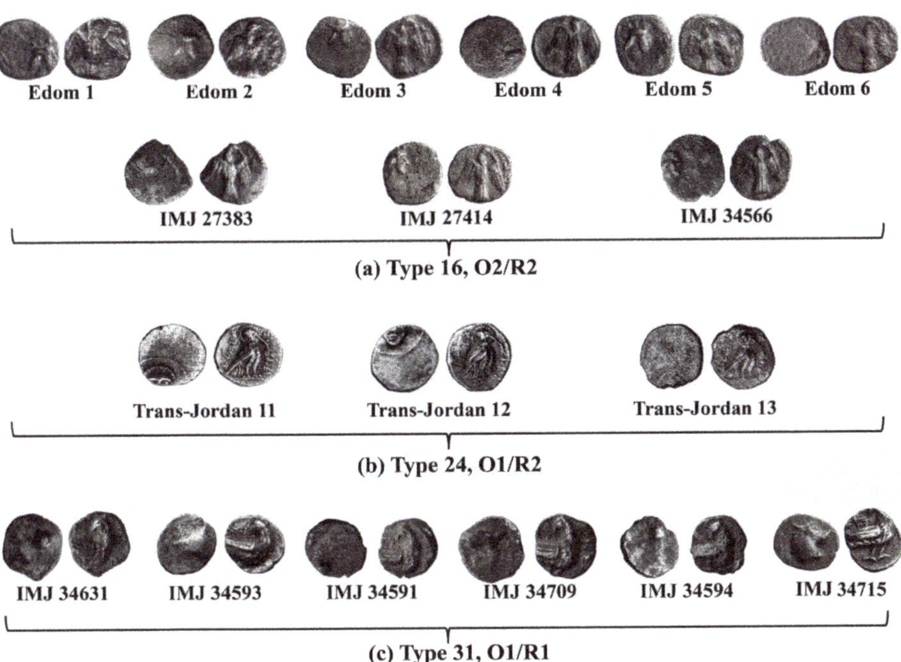

Figure 4. Yehud coins: (**a**) Type 16 O2/R2, (**b**) Type 24 O1/R2, and (**c**) Type 31 O1/R1 of the Yehud series.

A group of 50 specimens was chosen for this study: 32 coins of Type 5 (Figure 3), 9 coins of Type 16 O2/R2 (Figure 4a), 3 coins of Type 24 O1/R2 (Figure 4b), and 6 coins of Type 31 O1/R1 (Figure 4c). The assemblage includes coins found at controlled archaeological excavations and recorded by the Israel Antiquities Authorities (IAA); coins belonging to the Ramallah area hoard, 2006, which belong to the Israel Museum's collection [4]; and coins from the Israel Museum's collection (IMJ).

Because these coins derive from several different sources, we were able to determine whether coins struck from the same pair of dies and found together in the alleged Ramallah area hoard 2006 were produced from the same metal batch or from a different one than that of the coins struck from the same pair of dies and which were either found in controlled archaeological excavations or came from the antiquities market. We assume die-linked coins that appear together in a hoard were probably minted contemporaneously in the striking chain. Hence, our assumption is that their flans were probably prepared from the same metal batch. We also sought to determine the logical striking order of the analyzed Type 5 coins either recovered during controlled archaeological excavations or from the antiquities market based on the deterioration of the obverse and reverse motifs, although such determination is not always precise. We also cannot know how many other coins might have been struck between any two ordered coins in our assemblage during the minting chain process. Our analysis, based on the metallurgical results, may nonetheless help to determine the minting order of a series.

2. Technological Background to Research

The study of ancient silver coins includes aspects associated with minting skills, technological abilities, and political and economic considerations of a particular period, and can assist in understanding the connection between political and economic events [5–7]. Moreover, it can assist in understanding the economic choices of the minting authorities during times of crisis according to conflicts, as well as during times of reduced supply of metal [8]. For example, throughout history, during periods of high inflation, the silver content of coins is lower than the standard. Hence, the concentration of silver during a specific period can be used as an indicator for the estimation of economic, trade, and social conditions [9].

During antiquity, silver was typically produced from silver-rich galena lead sulfide (PbS) ore containing approximately 1–2 wt% Ag [10]. The cupellation method was a multistage process engaging three separate hearths. The first hearth was used for enriching smelted impure lead that contained silver (bullion). Wood fuel was used in order to remelt the lead bullion at high temperature. The lead was oxidized to litharge (PbO), with a melting temperature of 880 °C, inside a bellows-powered tuyères. Additional bullion was added until an appropriate amount of silver-enriched lead was acquired. Next, the silver-enriched lead was moved to a second hearth and oxidized again; however, at this stage, the litharge was removed by sinking iron rods into the hearth to create coated litharge cones on top of the rods. The rods were then removed, the litharge cones were thrown away, and the rods were dipped again, finally leaving silver globules inside the hearth. In the third hearth, several globules were melted and refined to obtain silver ingots and the remaining lead oxide was absorbed in the porous container wall (cupel). The cupellation method is a very effective process for producing silver metal with more than 95% purity [11].

A partial cupellation process resulted in silver alloy with lead presence. Hence, low quantities or absence of lead in the silver alloy points towards a successful silver refining process [12–14]. In addition, high content of lead transforms the silver alloy into a brittle material after a long burial period. Therefore, a good state of preservation of ancient silver objects is often connected to the absence of lead in the alloy.

Pure silver is a shiny white–grey metal with aesthetic appearance; it is a very soft metal that has excellent ductility and malleability [15,16]. Copper was a main alloying element in ancient silver coins. Ancient silver objects, including coins, are commonly available as silver–copper alloys with various ratios of silver and copper [17]. The addition of more than 3 wt% Cu to silver was frequently made to improve the mechanical properties of silver objects and also act as a melting-point depressant. Hence, the presence of more than 3 wt% Cu usually suggests that the copper was intentionally added [12,18–20].

Non-invasive and non-destructive testing (NDT) analyses of coins can be rather challenging due to various factors that change the surface metal composition, including long period corrosion processes and the presence of various corrosion products, tarnish and oxide layers, silver enrichment of the surface, cleaning residues, and conservation treatments [19,21,22]. The patina of ancient objects made of silver alloys may contain Ag_2O, Ag_2S, and/or $AgCl$, which can result in uncertainties concerning the obtained elemental data [9]. Therefore, the detected chemical composition obtained from the surface of the coin can be fundamentally different from the chemical composition over the entire volume of the coin [23]. Moreover, even when bulk material composition of ancient silver coins is obtained by destructive testing methods, there may be uncertainties concerning elemental analysis results that assume homogenous elemental distribution, which is often incorrect [21]. Therefore, measuring the composition in several different areas for each coin is recommended [24].

Surface-enrichment can be performed deliberately for technological and/or economic considerations or naturally due to segregation of the metals during casting and cooling stages, and because of corrosion processes [25]. For example, surface-enrichment, making silver–copper coins look like pure silver (silver depletion), was common during the Roman period. When silver–copper coin blanks were cast, they were kept at red heat condition to oxidize the copper on the external surface. Following the copper oxidation, the blanks were dripped into an organic acid bath which removed the copper from the surface, leaving a silver-enriched layer of up to a few hundred microns deep (usually up to a depth of approximately 200 μm). This process was employed by the Romans on silver–copper alloys with a composition of up to 80 wt% Cu, allowing the treated coins to leave the mint looking as if they were made of pure silver [26,27].

Copper located near the external surface of ancient silver coins can be oxidized and form corrosion products after a long burial period in aggressive environments; for example, cuprite (Cu_2O) and tenorite (CuO) can be formed on the coin surface. When these minerals are removed from the surface, an Ag-rich exterior is achieved depleted in Cu [7]. Nevertheless, ancient silver coins have relatively high durability. In order to examine whether there had been a silver-enrichment of the external surface, Ashkenazi et al. (2017) [19] grounded the surface of a fourth-century BCE silver ring from the Nablus Hoard (Ring B) and found that the Cu wt% concentration at the surface of the ring (before grinding) was similar to the Cu wt% concentration of the bulk alloy (after grinding). This led to the conclusion that the SEM-EDS measurements of the well-preserved fourth-century BCE shiny silver metallic areas of the objects accurately represented the bulk concentration of the silver jewelry from the Samaria and Nablus Hoards. X-ray fluorescence (XRF) and inductively coupled plasma with atomic emission spectrometry (ICP-AES) analyses of southern Palestinian Persian period silver coins also supported this conclusion [19].

3. Experimental Methods and Tests

Because of the rareness of the Yehud silver coins, invasive analysis of the objects was not permitted; thus, the coins were studied by using non-destructive testing (NDT) methods:

(a)   Visual testing (VT) inspection of the coins was performed in order to examine their general preservation condition and locate the better preserved areas of each coin.

(b)   A handheld X-ray fluorescence (XRF) Oxford X-MET8000 (Oxford Instruments, Abingdon, UK) was employed to examine the obverse and reverse of each coin of group Type 5 O1/R1, O1/R2, and O1/R4 to determine the elemental compositions of the surface. The XRF instrument was combined with a Silicon Drift Detector equipped with a 45 kV Rh Target X-ray tube. Each measurement was performed for 30 s with a detected spot diameter of 5 mm. Oxygen could not be detected with this XRF tool according to instrumental limitation.

(c)   A multi-focal digital light microscope (LM) (HIROX RH-2000, Hirox, Limonest, France) with high intensity LED lighting (5700 K color temperature) was used to inspect the general preservation of the surface and to detect microscopic discontinuities and defects. The coins were examined with an improved light sensitivity sensor at high resolution HD (1920×1200) with a multi-focus system combining many levels of light intensity and an integrated stepping motor, as well as powerful 3D software.

(d)   Scanning electron microscopy (SEM) with energy dispersive spectroscopy (EDS) examination was performed with environmental SEM in high vacuum mode and an Everhart–Thornley secondary electron (SE) detector. Both secondary electron (SE) and back-scattered electron (BSE) modes were used. The coins' surface composition (Supplementary Materials, Tables S1–S8) was analyzed by EDS using a Si(Li) liquid-cooled Oxford X-ray detector, calibrated with standard samples from the manufacturer, and providing measurements with a first approximation error of 1% [28]. SEM-EDS analysis accesses only a few micrometers beneath the surface of the analyzed metal; however, it provides very precise information on the morphology of the inspected area [29]. Before SEM-EDS examination, the coins' surfaces were cleaned with ethanol and dried. In order to measure the coins' alloy composition, only bright metal regions according to BSE mode were inspected. An example of SEM-EDS spectra of a typical coin's surface is shown in Supplementary Materials, Figure S1. Different scan areas were examined using EDS between $20\ \mu m \times 20\ \mu m$ and $800\ \mu m \times 800\ \mu m$. An average composition was then calculated by analyzing each coin using 6–8 measurements from different parts of each coin (both obverse and reverse sides were examined) to collect statistical data on each coin (reliable elemental distribution, average composition values and standard deviation), and then the results were normalized to 100 weight percentage (wt%). The alloy composition was calculated after omitting the peaks of oxides, corrosion products, and soil elements. In order to examine if the surface analysis represented the bulk alloy composition, seven Yehud silver coins were locally ground in different areas with 240–320 silicon carbide grit papers to reveal their bulk metal composition. The examined coins were IAA 153976 and IMJ 27424 (Type 5 O1/R1), IAA 101006 and IAA 154383 (Type 5 O1/R2), IMJ 27383 (Type 16 O2/R2), Trans-Jordan 11 (Type 24 O1/R2), and IMJ 34591 (Type 31 O1/R1). The composition of these coins was detected in several areas before (Supplementary Materials, Tables S1, S2 and S6–S8) and after locally grinding their surface (Supplementary Materials, Table S9).

4. Results

VT examination of the external surface of the Yehud *gerah* Type 5 O1/R1, O1/R2, O1/R3, O1/R4, and O1/R5, and of the Yehud series Type 16 (O2/R2), Type 24 (O1/R2), and Type 31 (O1/R1) (Figures 2–4) revealed that these coins were well preserved and covered with grey oxide, but also included areas of exposed shiny silver metal. XRF analysis results of the surface of coins Type 5 O1/R1, O1/R2, and O1/R4 revealed high-purity silver with a small percentage (less than 5 wt%) of copper. Other elements were also detected on the surface of the coins, including Si, Au, Pb, Sn, As, Bi, S, Fe, P, Mg, Mn, Ca, Al, and Ti. One coin in the Type 5 group, O1/R1 (IAA 153976), revealed exceptional composition according to XRF analysis with up to 46.7 wt% Ag and 44.5 wt% Cu. Two coins in the Type 5 group, O1/R2 (IAA 154383 and IMJ 27387), also revealed exceptional composition with up to 69.7 wt% Ag and 35.4 wt% Cu (for coin IAA 154383) and with up to 58.4 wt% Ag and 33.1 wt% Cu (for coin IMJ 27387) according to XRF analysis. One coin in group Type 5, O1/R4 (IAA 153981), also revealed exceptional composition according to XRF analysis with up to 76.2 wt% Ag and 12.6 wt% Cu.

4.1. Yehud Gerah Type 5 O1/R1

Multi-focal LM observation of the Type 5 O1/R1 coins revealed well-preserved silver metal depicting a helmeted head of Athena on the obverse and an owl on the reverse (Figure 5a–c). On the right field of the reverse of the coins, the Paleo-Hebrew inscription YHD appears (Figure 5a and Supplementary Materials, Figure S2b).

The bright areas observed in the SEM BSE mode are silver metal regions and the dark areas according to BSE mode are covered with oxides and some corrosion products (Figure 5d). The SEM-EDS analysis results of eight specimens of the Yehud *gerah* Type 5 O1/R1 coin surfaces (obverse and reverse, Figure 5d, and Supplementary Materials, Figure S2) revealed that the coins were composed of silver, though other elements were also detected, including Cu, Sn, O, Si, Cl, Al, Ca, P, and S (Supplementary Materials, Table S1).

The alloy of the coins revealed a composition of 93.2–100 wt% Ag and up to 6.8 wt% Cu (Supplementary Materials, Table S1), where the average value of the coins' alloy composition after omitting the peaks of oxides, corrosion products, and soil elements was 98.5 ± 2.0 wt% Ag and 1.5 ± 2.0 wt% Cu (where 42 different areas of the Type 5 O1/R1 coins' obverse and reverse sides were measured). It seems that a silver content of approximately 97% was the equivalent of "pure silver". This is entirely to be expected, because elemental silver as measured by modern scientific equipment would not have been available in antiquity. The closest refined silver bullion that could have been achieved by traditional smelting and refining processes would also include traces of gold, lead, and bismuth [2] (p. 334, n. 31).

IAA 153976 coin revealed a different alloy composition of 38.5–100 wt% Ag and up to 61.4 wt% Cu (Supplementary Materials, Table S1). Therefore, this coin was not included in the average composition value and standard deviation (SD) calculations of group Yehud *gerah* Type 5 O1/R1.

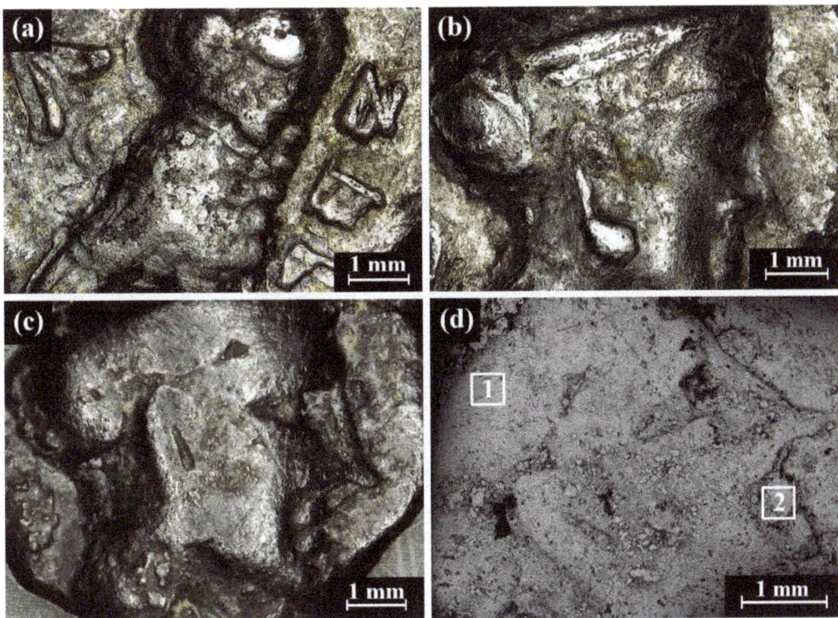

Figure 5. Images of the Yehud *gerah* Type 5 O1/R1 coins: (**a**) IAA 138139 reverse, depicting an owl (multi-focal LM), and (**b–d**) obverses of the coin depicting the head of Athena; (**b**) IAA 138139 (multi-focal LM); (**c**) IMJ 27398 (multi-focal LM); and (**d**) IMJ 27398 (SEM, BSE mode), where the white squares (areas 1 and 2) were examined by EDS analysis (the brighter areas are better preserved metal than the darker areas).

4.2. Yehud Gerah Type 5 O1/R2

Multi-focal LM and SEM observations of the Type 5 O1/R2 coins revealed well-preserved silver metal, where the dark areas are covered with oxides and corrosion products and the bright areas represent shiny silver metal (Supplementary Materials, Figure S3).

The SEM-EDS analysis results of 10 specimens of the Yehud *gerah* Type 5 O1/R2 coin surfaces (obverse and reverse) revealed that the coins were composed of silver, though other elements were also detected, including Cu, O, Si, Cl, Al, Ca, Fe, S, Au, and Pb (Supplementary Materials, Table S2). The bright areas observed in the SEM BSE mode (Figure 6 and Supplementary Materials, Figure S3b) are silver metal regions and the dark areas according to BSE mode are covered with oxides and some corrosion products.

The coins' alloy after omitting the peaks of oxides, corrosion products, and soil elements revealed a composition of 90.2–100 wt% Ag and up to 9.8 wt% Cu, where the average composition value of the alloy of the coins was 96.4 ± 2.5 wt% Ag and 3.6 ± 2.5 wt% Cu (where 53 different areas of the Type 5 O1/R2 coins' obverse and reverse sides were measured).

IAA 154383 revealed a different alloy composition which is between 42.1–96.9 wt% Ag and 3.1–57.9 wt% Cu. IMJ 27387 also revealed a different alloy composition which is between 29.4–98.0 wt% Ag and 2.0–70.6 wt% Cu (Supplementary Materials, Table S2). Therefore, these two coins were not included in the average composition value and SD calculations of group Yehud *gerah* Type 5 O1/R2 coins.

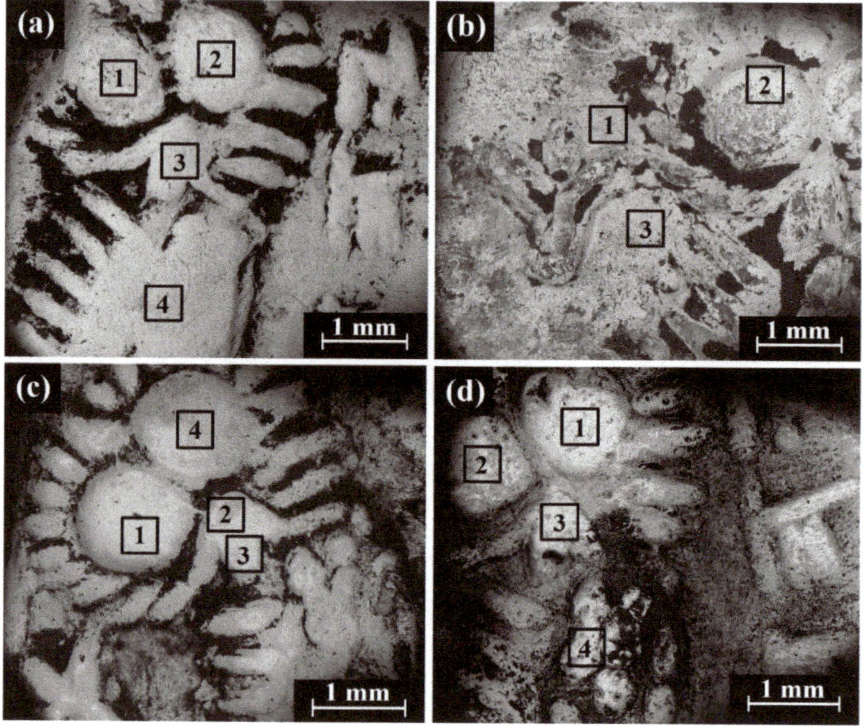

Figure 6. SEM images of the Yehud *gerah* Type 5 O1/R2 coins, with the reverse depicting an owl (BSE mode): (**a**) IAA 177246; (**b**) IAA 153977; (**c**) IMJ 34553; and (**d**) IMJ 34537. The bright areas according to BSE mode (areas 1–4 inside the squares) were examined by SEM-EDS analysis.

4.3. Yehud Gerah Type 5 O1/R3 Coins

SEM observation of the reverse of the Yehud *gerah* Type 5 O1/R3, IAA 153978 coin (Figure 7a,b) shows a lily, the symbol of Jerusalem, on the left and an owl on the right side. Only well-preserved silver alloy areas (bright areas according to BSE mode) were examined by EDS analysis (Figure 7b, areas 1–4). SEM-EDS elemental mapping of the reverse of the IAA 153978 coin (Figure 7c,d) revealed that the bright areas, according to the BSE mode, featured silver metal and the dark grey areas were rich in Cl (Figure 7d and Supplementary Materials, Figure S4), while the elements Cu and O (Supplementary Materials, Figure S4d,e, respectively) were distributed relatively homogeneously.

The SEM-EDS analysis results of four specimens of the Yehud *gerah* Type 5 O1/R3 coins (obverse and reverse surfaces) revealed that the coins were composed of silver, though other elements were also detected, including Cu, O, Si, Cl, Al, Ca, S, and Au (Supplementary Materials, Table S3).

The alloy of the coins after omitting the peaks of oxides, corrosion products, and soil elements revealed a composition of 85.9–100 wt% Ag and up to 14.1 wt% Cu, where the average composition value of the alloy of the coins was 97.6 ± 3.6 wt% Ag and 2.4 ± 3.6 wt% Cu (where 31 different areas of the Type 5 O1/R3 coins' obverse and reverse sides were measured).

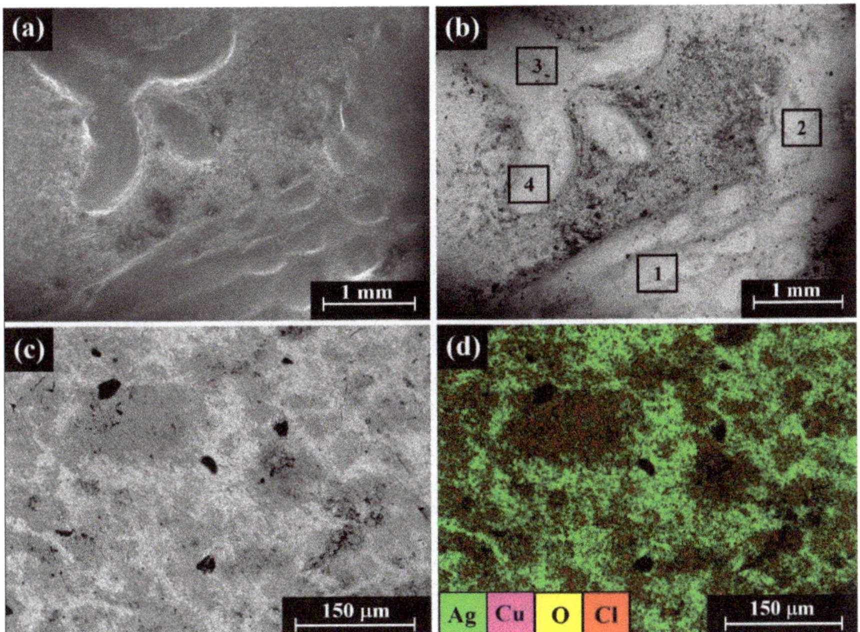

Figure 7. SEM images of the Yehud *gerah* Type 5 O1/R3 coin IAA 153978: (**a**) reverse, images of a lily and an owl, SE mode; (**b**) reverse, BSE mode, where the bright areas 1–4 inside the squares were examined by EDS analysis; (**c**) obverse, BSE mode; and (**d**) elemental mapping, where the green areas are rich in silver and red areas are rich in chlorine.

4.4. Yehud Gerah Type 5 O1/R4

Multi-focal LM observation of the Ramallah area hoard coins (RH2–RH6) revealed relatively smooth and well-preserved surfaces, with the reverse showing an image of a lily on the left field, an owl, and the Paleo-Hebrew and the inscription YHD (Yeh[u]d, right) (Figure 8a). Yet relatively smooth and well-preserved surfaces, as well as scratches and areas covered with oxides and local corrosion products, were observed by SEM (Figure 8b–d).

SEM-EDS surface analysis results of eight specimens of the Yehud *gerah* Type 5 O1/R4, the reverse of RH2–RH6, IAA 153980, and IMJ 34556 revealed that the coins were composed of silver, though other elements were also detected, including Cu, O, Si, Cl, S, Al, and Ca (Supplementary Materials, Table S4).

The alloy of seven Yehud *gerah* Type 5 O1/R4 coins revealed a composition of 95.9–100 wt% Ag and up to 4.1 wt% Cu (Supplementary Materials, Table S4), where the average composition value of the alloy of the coins after omitting the peaks of oxides, corrosion products, and soil elements (O, Si, Cl, S, Al and Ca) was 99.1 ± 1.2 wt% Ag and 0.9 ± 1.2 wt% Cu (where 48 different areas of the Type 5 O1/R4 coins' obverse and reverse sides were measured).

IAA 153981 revealed a different alloy composition of 81.6–98.0 wt% Ag and 2.0–18.4 wt% Cu (Supplementary Materials, Table S4). Thus, this coin was not included in the average composition value and SD calculations of group Yehud *gerah* Type 5 O1/R4.

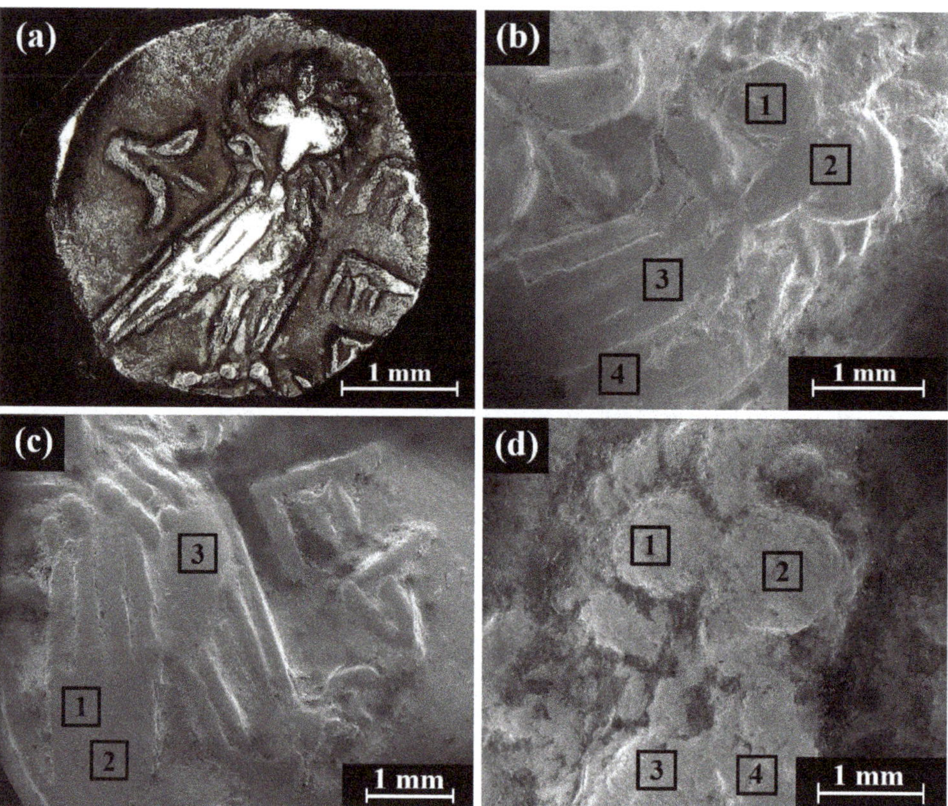

Figure 8. Images of the Yehud *gerah* Type 5 O1/R4, the reverse of the Ramallah area hoard coins (RH): (**a**) general view of coin RH2 (multi-focal LM); (**b**) RH2 (SEM, SE mode); (**c**) RH3 (SEM, SE mode); and (**d**) RH5 (SEM, SE mode). The areas 1–4 inside the squares were examined by SEM-EDS analysis.

4.5. Yehud Gerah Type 5 O1/R5

SEM observations of the Type 5 O1/R5 coins revealed areas of preserved silver metal, with the reverse showing the image of an owl (Figure 9a), where the dark areas according to BSE mode are covered with oxides and corrosion products and the bright areas represent shiny silver metal (Figure 9b). Parallel cracks were observed on the surface of coin IMJ 27388, probably resulting from the striking process and local embrittlement of the silver metal due to local corrosion attack (Figure 9c,d).

The SEM-EDS analysis results of two specimens of the Yehud *gerah* Type 5 O1/R5 coin surfaces (obverse and reverse) revealed that the coins were composed of silver, though other elements were also detected, including Cu, O, Si, Cl, Al, and S (Supplementary Materials, Table S5).

The alloy of the Yehud *gerah* Type 5 O1/R5 coins after omitting the peaks of oxides, corrosion products, and soil elements revealed a composition of 97.6–100 wt% Ag and up to 2.4 wt% Cu, where the average composition value of the alloy of the coins was 99.7 ± 0.8 wt% Ag and 0.3 ± 0.8 wt% Cu (where 14 different areas of the coins' obverse and reverse sides were measured).

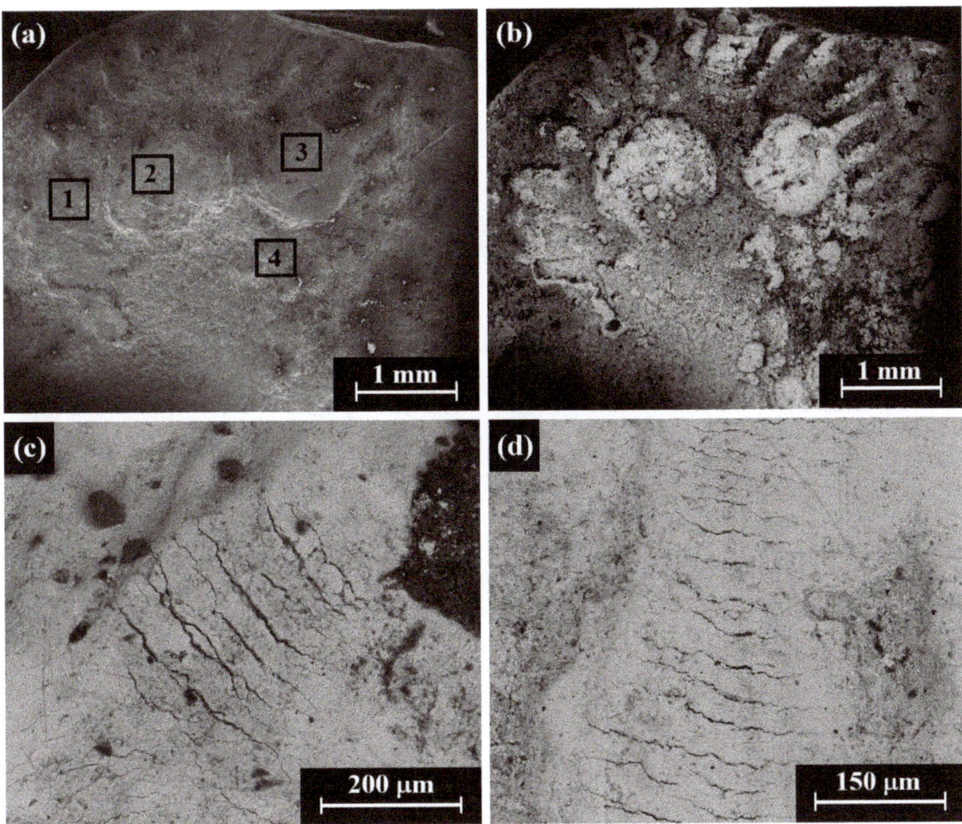

Figure 9. SEM images of the Yehud *gerah* Type 5 O1/R5 reverse side with the image of an owl: (**a**) coin IMJ 34558 (SE mode), where the areas inside the squares 1–4 were examined by EDS analysis; (**b**) coin IMJ 34558 (BSE mode); and (**c**,**d**) coin IMJ 27388 with parallel cracks.

4.6. Yehud Half Gerah Type 16 O2/R2

Multi-focal LM observation of the Edom hoard no. 2 (IMJ 2020.33.2) coin revealed a well-preserved silver metal, with the reverse depicting a falcon in flight (Figure 10a) and the Paleo-Hebrew inscription YHD (Yeh[u]d) on the right field.

The bright areas observed in the SEM BSE mode are silver metal regions and the dark areas according to BSE mode are covered with oxides and some corrosion products (Figure 10d and Supplementary Materials, Figure S6c). The SEM-EDS analysis results of nine specimens of the Yehud half *gerah* Type 16 O2/R2 coins (obverse and reverse, Figure 10b–d and Supplementary Materials, Figure S5) revealed that the surface of the coins were composed of silver, though other elements were also detected, including Cu, O, Si, Cl, Al, Ca, Fe, S, Pb, K, and Mg (Supplementary Materials, Table S6).

The alloy of the Type 16 O2/R2 coins revealed a composition of 98.3–100 wt% Ag and up to 1.7 wt% Cu (Supplementary Materials, Table S6), where the average composition value of the alloy of the coins after omitting the peaks of oxides, corrosion products, and soil elements was 99.9 ± 0.4 wt% Ag and 0.1 ± 0.4 wt% Cu (where 40 different areas of the Type 16 O2/R2 coins' obverse and reverse sides were measured).

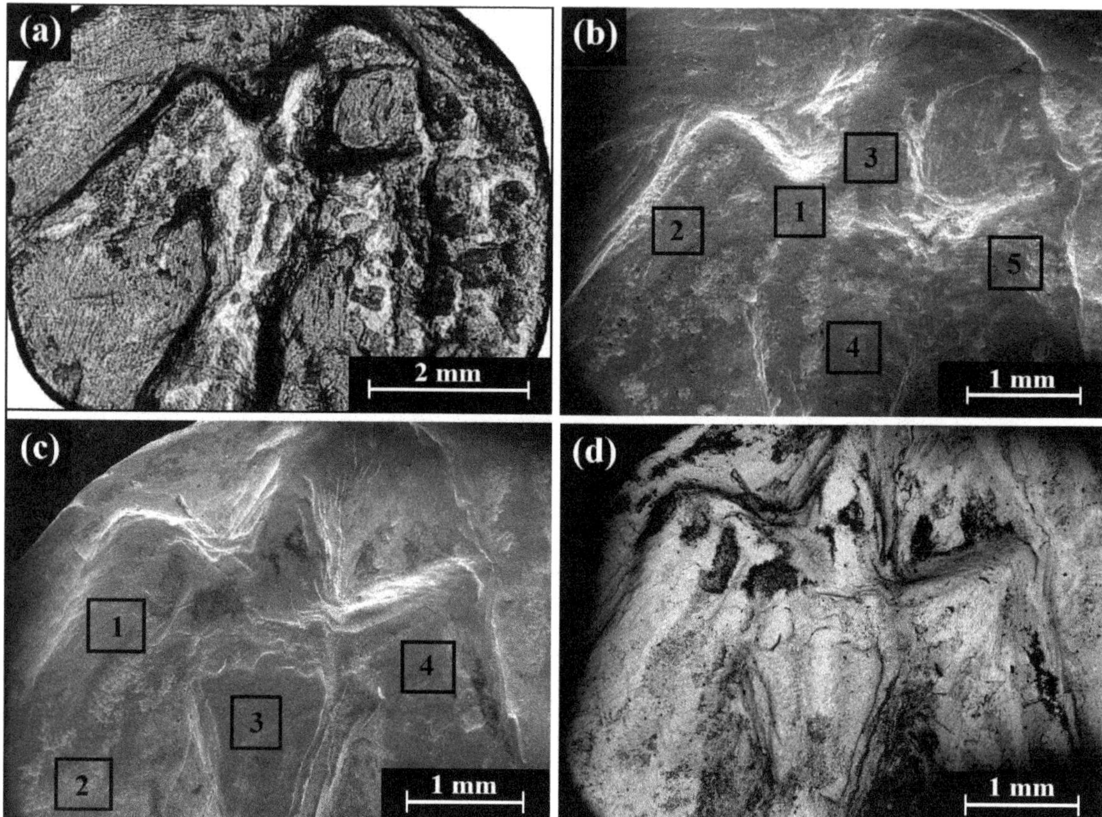

Figure 10. Images of the Edom hoard Type 16 O2/R2 coins, with the reverse side showing a design of a falcon in flight: (**a**) multi-focal LM of coin Edom hoard no. 2 (IMJ 2020.33.2); (**b**) SEM observation of coin Edom hoard no. 2 (SE mode); (**c**) coin Edom hoard no. 3 (SE mode); and (**d**) coin Edom hoard no. 3 (BSE mode). The areas inside the squares 1–5 in (**b**,**c**) were examined by EDS analysis.

4.7. Yehud Attic Standard Quarter Obol Type 24 O1/R2

VT of the Type 24 O1/R2 Yehud quarter *obol* revealed a preserved silver metal. The obverse depicts a facing head and the reverse an owl standing right, head facing. The bright areas observed in the SEM BSE mode are silver metal regions and the dark areas are covered with oxides and some corrosion products (Figure 11 and Supplementary Materials, Figure S6). The SEM-EDS analysis results of three Yehud quarter *obol* Type 24 O1/R2 specimens (obverse and reverse surfaces, Figure 11 and Supplementary Materials, Figure S6 and Table S7) revealed that the coins were composed of silver, though other elements were also detected, including Cu, O, Si, Cl, Al, Ca, and Fe (Supplementary Materials, Table S7).

The alloy of the coins revealed a composition of 100 wt% Ag. No presence of Cu was detected in Type 24 O1/R2 coins (and Supplementary Materials, Table S7, where 20 different areas of the obverse and reverse sides of these coins were measured).

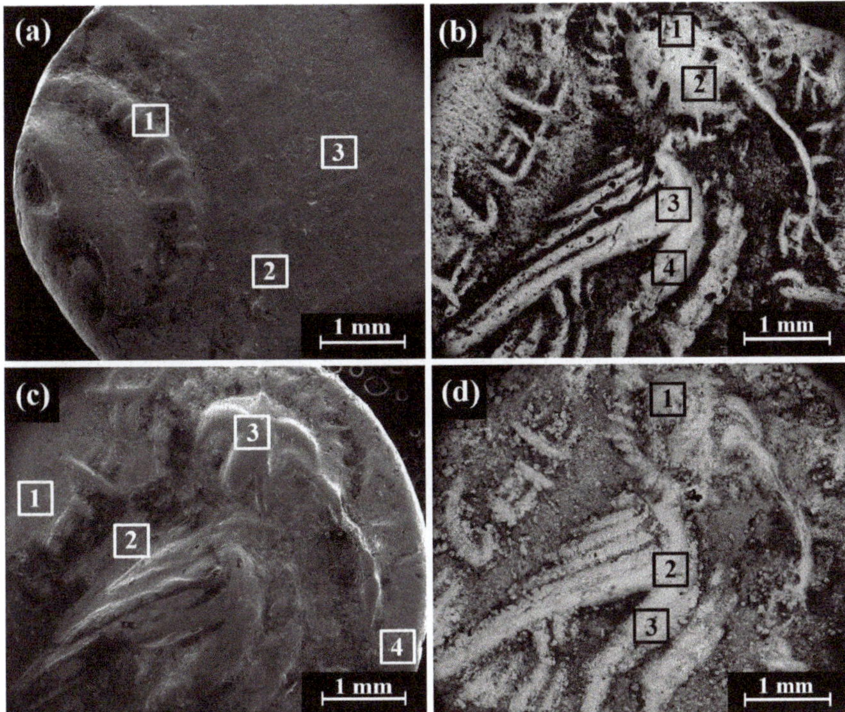

Figure 11. SEM images of the Trans-Jordan hoard coin Type 24 O1/R2: (**a**) coin Trans-Jordan 11 obverse side with a facing head (SE mode); (**b**) coin Trans-Jordan 11 reverse side with and owl (BSE mode); (**c**) coin Trans-Jordan 12 reverse (SE mode); and (**d**) coin Trans-Jordan 12 reverse (BSE mode). The areas 1–4 inside the squares were examined by EDS analysis.

4.8. Yehud Attic Standard Hemiobol Type 31 O1/R1

SEM BSE mode observations of the obverse and reverse of Type 31 O1/R1 coins (head of roaring lion/bird standing, right head reverted) revealed bright areas of well-preserved silver metal, though dark areas covered with oxides and corrosion products were also detected (Figure 12 and Supplementary Materials, Figure S7a).

The SEM-EDS analysis results of the obverse and reverse of the bright surfaces of six specimens of the Yehud *hemiobol* Type 31 O1/R1 coins revealed that they were mostly composed of silver, though other elements were also detected, including Cu, O, Si, Cl, Al, Ca, Fe, S, and Au (Supplementary Materials, Table S8). One exceptionally dark surface area of the reverse of coin IMJ 34593 (according to BSE mode) was also detected by EDS (scanned area of 200 μm × 200 μm), showing a composition of 14.1 wt% Cu, 44.8 wt% O, and 35.0 wt% Si, as well as the presence of 2.7 wt% Al, 1.4 wt% Fe, and less than 1.0 wt% of the elements Mg, S, and K. This measurement was not included in the average composition value and SD calculations, since only well-preserved areas were included in the calculations.

The alloy of Type 31 O1/R1 coins after omitting the peaks of oxides, corrosion products, and soil elements revealed a composition of 91.0–100 wt% Ag and up to 9.0 wt% Cu, where the average composition value of the alloy of the coins was 98.3 ± 3.7 wt% Ag and 1.7 ± 3.7 wt% Cu (where 44 different areas of the obverse and reverse sides of Type 31 O1/R1 coins were measured).

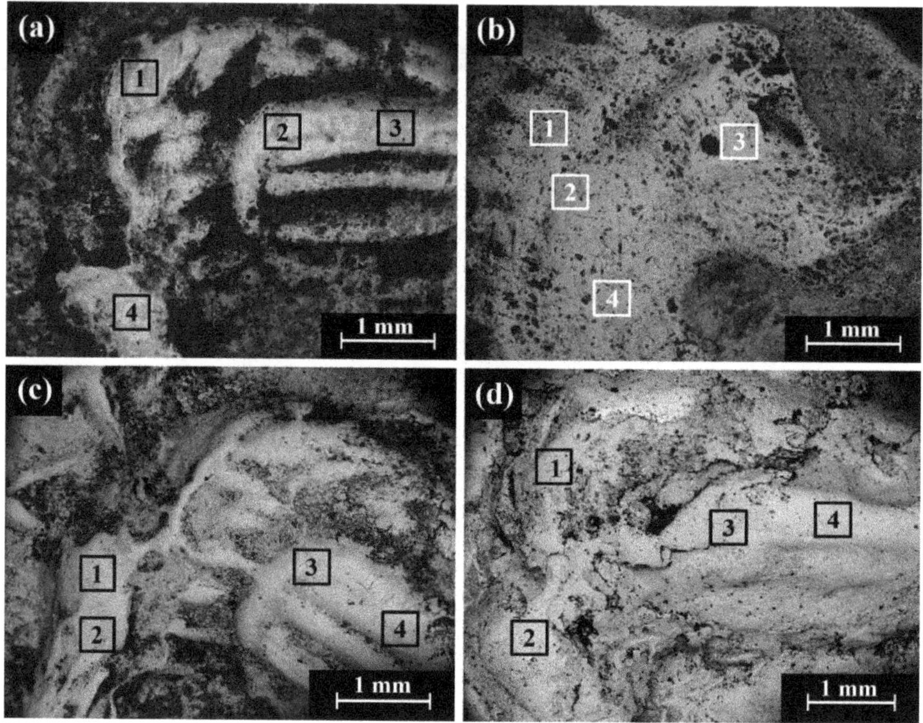

Figure 12. SEM images (BSE mode) of the Yehud *hemiobol* Type 31 O1/R1: (**a**) IMJ 34709 reverse side with an image of a bird standing, right head reverted; (**b**) IMJ 34715 depicts a roaring lion on the obverse; and (**c**) IMJ 34631 and (**d**) IMJ 34594 depict the reverse of these issues. The 1–4 areas inside the squares were examined by EDS analysis.

4.9. Chemical Analysis of the Bulk of the Locally Ground Yehud Coins

The exceptional coins IAA 153976 Type 5 O1/R1 and IAA 154383 Type 5 O1/R2, having high copper content (Supplementary Materials, Tables S1 and S2, respectively), revealed different copper concentration before and after grinding.

The average copper content in the alloy of coin IAA 153976 before grinding was 27.4 ± 20.7 wt% Cu, while the average copper content in the alloy of coin IAA 154383 alloy before grinding was 22.2 ± 18.9 wt% Cu. Yet the average copper content of the ground areas of the alloy of coin IAA 153976 alloy was 32.7 ± 3.7 wt% Cu and of the ground areas of the alloy of coin IAA 154383 was 16.9 ± 6.6 wt% Cu (Supplementary Materials, Table S9). The SD of the exceptional coins was reduced dramatically after grinding the surface.

Good agreement was achieved between the surface and bulk compositions of silver coins with low copper content. For example, the average copper concentration of the alloy of coin IMJ 27424 (O1/R1, Supplementary Materials, Table S1) before grinding was 0.0 ± 0.0 wt% Cu and that of the ground bulk of coin IMJ 27424 was 1.3 ± 1.0 wt% Cu, whereas the average copper content of the alloy of coin IMJ 27383 (Type 16 O2/R2, Supplementary Materials, Table S6) before grinding was 0.2 ± 0.5 wt% Cu and that of the ground bulk of coin IMJ 27383 was 0.0 ± 0.0 wt% Cu (Supplementary Materials, Table S9).

5. Discussion

The current research, as part of an ongoing study on the early indigenous southern Levantine coinages, employed an archaeometallurgical NDT approach in order to analyze die-linked Late Persian period and Macedonian period Yehud silver coins. In the case of the current coins, based on their shiny metallic appearance, it is evident that the coins have been cleaned. Yet such cleaning procedures were not recorded and remain unknown. Nevertheless, even in such circumstances, surface characterization of silver coins can be performed without any additional polishing of the exterior when the coins' surfaces are shiny and the remaining oxides and corrosion products layers are thin enough [19,30,31].

All the examined Yehud coins were shown to be in a good state of preservation, based on VT, multi-focal LM, and SEM SE and BSE observations. Such preservation can be explained by the excellent corrosion resistance of silver, due to the formation of stable Ag_2O on the metal surface [32,33]. Nonetheless, the SEM BSE mode observation combined with EDS analysis revealed the bright areas to constitute exposed silver metal regions and the dark areas as covered with oxides and some corrosion products. The EDS analysis of the surface of the Yehud coins also revealed the presence of the elements O, Si, Cl, Sn, Au, Pb, Al, Ca, Fe, Mg, P, and S (Supplementary Materials, Tables S1–S8). The presence of oxygen and chlorine can arise from both use and storage of the coins and is therefore of little use in shedding light on the metallurgical aspects related to the coins. The absence of lead in most of the examined areas of the coins (lead was only detected by EDS in coin IMJ 34620 (reverse, area 3) of Type 5 O1/R2 and in coin Edom hoard no. 2 (reverse, area 4) of Type 16 O2/R2) indicated that all coins were produced by a successful cupellation refining process. The detection of Cl and S on the surface of the coins was predictable since silver is sensitive to chloride and sulfide ions, leading to the formation of silver chlorides (AgCl) and sulfides (Ag_2S) as the main contamination products [19,33]. The presence of the elements O, Si, Al, Ca, Fe, Mg, and P is related to the existence of corrosion products and soil remains [6,19,20,33], yet the presence of Fe could also be related to natural impurities in the silver ores [9]. In aggressive environments, silver alloy objects often also contain other corrosion products related to base metals, such as cuprite and litharge [33]. No presence of the elements Cr, Ni, and Cd was detected. These elements are typical of modern forgeries [24], indicating the authenticity of all examined silver coins.

The presence of gold in silver alloys was apparently unfamiliar to the ancient minters. Yet ancient silver coins often contain less than 1 wt% Au as an impurity associated with the silver ore. Therefore, the gold content in silver coins can assist to distinguish between silver sources used for coinage [34].

Good agreement was observed between the SEM-EDS analysis results of the surface and the bulk after grinding the surface, and between the SEM-EDS analysis results and the XRF analysis results of Type 5 O1/R1, O1/R2, and O1/R4 coins' surfaces, where with both methods, high-purity silver was detected, with an average copper content of less than 5 wt% Cu, as well as other elements, including, O, Si, Sn, Au, Pb, Al, Ca, Fe, Mg, P, and S. The elements Mn, As, Bi, and Ti were also detected by XRF analysis on the coins' surface but were not detected by EDS analysis, whereas the presence of Cl was detected in the EDS analysis but not by the XRF analysis.

Because of the importance of silver as a representative of the material cultural heritage of different populations throughout history, numerous studies of ancient silver objects, including coins, exist in the literature, studying the chemical composition, microstructure, provenance, manufacturing processes, and the corrosion products of silver items. These studies usually combine NDT and destructive testing methods, for example, LM and SEM observation of metallographic samples, particle-induced X-ray emission (PIXE) analysis, inductively coupled plasma (ICP) analysis, X-ray diffraction (XRD), and X-ray photoelectron spectroscopy (XPS) analysis [19,22,23,27,28]. Some possible NDT methods that can be applied for elemental bulk analysis of silver coins are the neutron activation analysis (NAA) and the prompt-gamma activation analysis (PGAA) [34–39]. Yet these methods have some limitations resulting from the high cross section for neutron capture by silver itself, and therefore the silver data would prevent revealing trace elements that could be useful fingerprint variations in the minting process. There are certain rare-earth metallic elements with large neutron cross sections that could provide interesting data concerning differences in the raw silver source. Another NDT technique that could be used to provide additional information concerning the coins' bulk material composition is the muon induced X-ray emission (MIXE) technique [40].

Ancient silver was commonly produced from silver-rich galena lead sulfide ore using the cupellation process [10]. The presence of Pb and S in some of the measurements is thus probably related to the cupellation process [11,21]. The cupellation technique is very effective for the production of high-purity silver metal, with more than 95 wt% Ag [11], which may explain the high purity of the discussed Yehud silver coins. While the presence of Au and Sn might be related to the ore deposits that were used, the presence of Sn might also be related to the addition of copper to the alloy.

Copper was a main alloying element in ancient silver coins and the addition of copper to silver was often performed to depress the melting point as well as to improve the mechanical properties of the alloy [12,18–20,22]. Since the presence of Sn is quite rare in galena ores, its presence might point to alloying the silver with bronze instead of pure Cu [21]. According to Brocchieri et al. (2020), the detection of high-purity silver coins probably indicates that during the production process of these coins, the mint did not suffer from economic constraints [5]. In addition to the Ag and Cu alloy elements and the corrosion products and soil elements that were detected in all Yehud series coins, other elements were also detected. For example, up to 4.7 wt% Sn was detected in a specific 300 μm × 300 μm scanned area of IAA 153975 (Type 5 O1/R1, Table S1). Up to 1.1 wt% Au was detected in the reverse of coin IMJ 34553 and up to 1.2 wt% Au was detected in the obverse of this coin (Type 5 O1/R2, Table S2). In addition, up to 1.7 wt% Pb was detected in a specific 300 μm × 300 μm scanned area of IMJ 34620 reverse (Type 5 O1/R2, Table S2). This may also support the assumption that each series was manufactured by using a specific composition of silver–copper alloy.

Although SEM-EDS is a valuable NDT tool for surface analysis of ancient silver coins, if the detected objects are covered with a thick oxide layer and contain massive corrosion products, the analysis may not provide a representative characterization of the bulk alloy composition of the object [19,30,41]. Ancient silver objects such as coins sometimes exhibit silver enrichment of the surface and, as a result, a considerable reduction of copper on the surface [33,42]. In silver objects retrieved from a soil environment after a long burial period, a local selective galvanic corrosion attack on the copper-enriched areas may occur, resulting in the diffusion of Cu from the bulk of the Ag object to its surface, causing the formation of a cuprite layer on the external surface of the coin [43,44]. Moreover, under some circumstances, unsuitable conservation methods may cause damage to the ancient coins and affect their surface composition. Nevertheless, when the oxide layer is thin and the shiny metal is exposed beneath, the limitations of SEM-EDS analysis should not prevent the use of this method as a valuable tool for the characterization of ancient silver coins. However, it is essential first to determine whether the composition of the surface layer is representative of the bulk composition of the object [18,19,32,42,45]. In order to do so, seven Yehud silver coins were locally ground in several areas to expose their bulk metal and their composition was examined before and after grinding. Good agreement was achieved between surface and bulk compositions of coins with low copper content (Supplementary Materials, Table S9).

In order to obtain reliable results from the EDS surface analysis, only bright areas with a shiny silver metal appearance were examined here for the calculations of the average value and SD of the alloy composition of each group of coins (Table 1). Our results, after eliminating the peaks of oxides and soil elements, revealed that five of the die combination issues of the Yehud series Type 5 (O1: R1, R2, R3, R4, R5), Type 16 (O2/R2), Type 24 (O1/R2), and Type 31 (O1/R1) are composed of high-purity silver with a small percentage of copper (Table 1).

Table 1. The average alloy composition (Cu wt% content) of the different Yehud Types 5, 16, 24, and 31, the late addition Yehud coins, and the Samaria and Nablus Hoards coins, according to SEM-EDS analysis.

Coin Types	No. of Examined Items	No. of Tests Performed by SEM-EDS	Average Cu (wt%) Content in the Ag-Cu Alloy
Yehud *gerah* Type 5 O1/R1 coins	7	42	1.5 ± 2.0
Yehud *gerah* Type 5 O1/R2 coins	8	53	3.6 ± 2.5
Yehud *gerah* Type 5 O1/R3 coins	4	31	2.4 ± 3.6
Yehud *gerah* Type 5 O1/R4 coins	7	48	0.9 ± 1.2
Yehud *gerah* Type 5 O1/R5 coins	2	14	0.3 ± 0.8
Yehud *gerah* Type 5 coins with exceptional composition	4	28	20.7 ± 19.3
Yehud half *gerah* Type 16 O2/R2 coins	9	40	0.1 ± 0.4
Yehud quarter *obol* Type 24 O1/R2 coins	3	20	0.0 ± 0.0
Yehud *hemiobol* Type 31 O1/R1 coins	6	44	1.7 ± 3.7
Late addition to the Yehud corpus [2]: 270–271, Figure VIII, 84–85	3	22	0.4 ± 0.8
Coins from the Samaria and Nablus Hoards [19]	80	160	4.1 ± 2.4

The copper content in the Type 5 coin alloy is between 0.3 ± 0.8 wt% Cu (for O1/R5) and 3.6 ± 2.5 (for O1/R2). The copper content in the Type 16 coin alloy is 0.1 ± 0.4 wt% Cu (for O2/R2); the copper content in the Type 24 O1/R2 is 0 ± 0 wt% Cu; and the copper content in the Type 31 O1/R1 is 1.7 ± 3.7 wt% Cu (Table 1). Four coins (IAA 153976 of Type 5 O1/R1, IAA 154383 and IMJ 27387 of Type 5 O1/R2, and IAA 153981 of Type 5 O1/R4) were not included in the average composition values and SD calculations of the main group of each coins. The fact that these four coins contain a high amount of copper (20.7 ± 19.3 wt% Cu, Table 1) may indicate that towards the end of the minting process of each series there was a shortage of raw materials and therefore recycled silver was used.

The manufacturing processes of the coins from all the examined groups were similar (casting and striking a blank flan), with some slight differences (between pure silver alloy for the case of Type 24 O1/R2 coins up to 3.6 ± 2.5 wt% Cu for the case of Type 5 O1/R2 coins). Three additional die-connected Yehud coins of the late addition type (a female {?} head to the right with the Aramaic letter yod in the left field on the obverse and an owl standing right, head facing on the reverse) were studied previously (Table 1) [2] and are used here as a reference group with an average composition of 0.4 ± 0.8 wt% Cu, presenting the similarities and differences in terms of metallurgical composition compared with the above-mentioned coin types.

For comparison, the alloy composition of 80 selected Persian period Samarian silver coins from the Samaria and Nablus Hoards, as well as other coins from the Israel Museum collection, was 95.9 ± 2.5 wt% Ag and 4.1 ± 2.5 wt% Cu (SEM-EDS analysis after omitting the peaks of oxides, corrosion products, and soil elements) [19]. Moreover, the average alloy composition of the Persian period jewelry from the Samaria Hoard was 93.4 ± 1.65 wt% Ag and 6.6 ± 1.6 wt% Cu, while that of the jewelry from the Nablus Hoard was 94.9 ± 1.9 wt% Ag and 5.1 ± 1.9 wt% Cu (excluding the joining areas). In order to determine with sufficient certainty whether the current studied die-linked silver coins were produced from the same metal batch throughout the minting of each group, the copper concentration distribution of each series is shown based on the SEM-EDS analysis results after omitting the peaks of oxides, corrosion products, and soil elements (Supplementary Materials, Figure S8), presenting the Cu wt% concentration range vs. the relative no. of measurements (%).

Based on the average values and SD of the alloy composition of each series and the copper concentration distribution of the different groups, each series (including the die-linked issues of Type 5 O1/R1, O1/R2, O1/R3, O1/R4, and O1/R5 subtypes, as well as Type 16 O2/R2, Type 24 O1/R2, and Type 31 O1/R1) was most probably manufactured by using a controlled specific composition of silver–copper alloy.

Based on the current findings, a four-step methodology is suggested for the study of ancient silver coins: (first) a VT should be performed on the objects; (second) the coins should be examined by multi-focal LM in order to establish the areas where shiny silver metal is exposed and whether there are corrosion products on the surface of the coin (the silver alloy composition should be measured only in these shiny areas); (third) initial examination of the coins' surface should be conducted by XRF; and (fourth) the coins should be examined by SEM-EDS analysis in both SE and BSE modes in order to determine the state of preservation of each coin and its alloy composition. Only bright areas observed in the BSE mode should be subjected to EDS analysis in order to determine the alloy composition of each series of coins.

The elaborate iconographic designs on these tiny Persian period and Macedonian period Yehud silver coins demonstrate high artistic and technological abilities (Supplementary Materials, Figures S9–S16). Based on the average alloy composition values and SD of the examined series of the five die combination issues of the Yehud series Type 5 (O1 linked with R1, R2, R3, R4, and R5), each series was struck from a different and controlled specific composition of silver–copper alloy. Approximately 10% of the coins revealed a different alloy composition, however, with a much higher amount of copper and a heterogeneous composition. This implies that at a certain stage of the minting process, a different batch of, possibly recycled, alloy was used rather than the standardized alloy that was recorded for all the other coins of the same die connection.

6. Conclusions

In this study, 50 Late Persian period and Macedonian period silver coins were studied by NDT metallurgical analysis. The results obtained from the SEM-EDS surface analysis of well-preserved areas sufficiently represent the concentrations of the coins' ground bulk. Correspondence was received between the SEM-EDS analysis results and the XRF analysis results. The results show that the coins are made of high purity silver with only small percentages of copper. All the coins were produced by similar techniques of casting flans and striking them by plastic deformation. Based on the average alloy composition values and SD, as well as the copper concentration distribution of the examined series, the five die combination issues of the Yehud series Type 5 O1/R1–O1/R5, Type 16 O2/R2, Type 24 O1/R2, and Type 31 O1/R1 groups, and the late addition Yehud coin type, each series was most likely produced by a different and controlled specific composition of silver–copper alloy. However, four Type 5 exceptional coins (IAA 153976 O1/R1, IAA 154383 O1/R2, IMJ 27387 O1/R2, and IAA 153981 O1/R4) revealed a different alloy composition with a much higher amount of copper and heterogeneity. This implies that at a certain stage of the minting process, a different batch of possibly recycled alloy was used instead of the standardized alloy that was recorded for all other coins of the same die connection. The current four-step methodology revealed novel information concerning the material culture of the southern Levant during the Late Persian period associated with minting production of silver coins. This four-step methodology can be used with additional bulk NDT methods by other researchers to study various ancient silver objects.

Supplementary Materials: The following supporting information can be downloaded at: https://www.mdpi.com/article/10.3390/ma16062200/s1, Figure S1. SEM-EDS analysis showing the spectra of a typical Yehud *gerah* silver coin (Type 5 O1/R3, IAA 153978, reverse); Figure S2. Images of the Yehud *gerah* Type 5 O1/R1 coins: (a) IAA 138139 reverse depicting an owl (SEM, BSE mode); (b) IAA 138139, reverse (multi-focal LM) showing the Paleo-Hebrew inscription YHD (Yeh[u]d); (c) IMJ 27398 obverse depicting a helmeted Athena (SEM, SE mode); (d) IMJ 34543 reverse depicting an owl (SEM, BSE mode); and (e) IMJ 34543 obverse depicting a helmeted Athena (SEM, BSE mode).

The areas inside the squares were examined by EDS analysis; Table S1. SEM-EDS analysis results of the Yehud *gerah* Type 5 O1/R1 coins, where SA represents the scanned area. Only bright areas according to BSE mode of shiny metal were examined by EDS analysis; Figure S3. Images of Yehud *gerah* Type 5 O1/R2: (a) coin IAA 154383, reverse (multi-focal LM), showing the face of an owl; and (b) IAA 154383, reverse (SEM, BSE mode), where the bright areas represent shiny silver metal and the dark areas are covered with oxide and corrosion products. The areas inside the squares were examined by EDS analysis; Table S2. SEM-EDS analysis results of the Yehud *gerah* Type 5 O1/R2 coins, where SA represents the scanned area. Only bright areas according to BSE mode of shiny metal were examined by EDS analysis; Figure S4. Yehud *gerah* Type 5 O1/R3, IAA 153978 (obverse), SEM–EDS elemental mapping: (a) general view of the examined area; (b) the detected elements (Ag, Ca, O, Cl), where the green areas are rich in silver and red areas are rich in chlorine; (c) presence of Ag; (d) presence of Ca; (e) presence of O; and (f) presence of Cl; Table S3. SEM-EDS analysis results of the Yehud *gerah* Type 5 O1/R3 coins, where SA represents the scanned area. Only bright areas according to BSE mode of shiny metal were examined by EDS analysis; Table S4. SEM-EDS analysis results of the Yehud *gerah* Type 5 O1/R4 coins (each scanned area was 300 μm × 300 μm). Only bright areas according to BSE mode of shiny metal were examined by EDS analysis. RH represents Ramallah Hoard (nos. 2–6); Table S5. SEM-EDS analysis results of the Yehud *gerah* Type 5 O1/R5 coins. Only bright areas according to BSE mode of shiny metal were examined by EDS analysis; Figure S5. Yehud half *gerah* Type 16, O2/R2 coin, reverse depicting a falcon in flight: (a) Edom hoard no. 4 (SE mode); (b) Edom hoard no. 5 (SE mode); and (c) Edom hoard no. 6 (BSE mode), where the brighter areas according to the BSE mode are better preserved than the darker areas. The areas inside the squares were examined by EDS analysis; Table S6. SEM-EDS analysis results of the Yehud half *gerah* Type 16 O2/R2 coins. Only bright areas according to BSE mode of shiny metal were examined by EDS analysis; Figure S6. SEM images of the Trans-Jordan hoard Type 24 O1/R2, obverse depicting a facing head: (a) coin Trans-Jordan hoard no. 11 (SE mode); (b) coin Trans-Jordan hoard no. 13 (SE mode); (c,d) coin Trans-Jordan hoard no. 12 (SE mode and BSE mode, respectively). The areas inside the squares were examined by EDS analysis; Table S7. SEM-EDS analysis results of the Yehud quarter *obol* Type 24 O1/R2 coins, where SA represents the scanned area. Only bright areas according to BSE mode of shiny metal were examined by EDS analysis; Figure S7. SEM images of the Yehud *hemiobol* Type 31, O1/R1, reverse depicting a bird standing right, head reverted: (a) IMJ 34593 (BSE mode); and (b) IMJ 34591 (SE mode). The areas inside the squares were examined by EDS analysis; Table S8. SEM-EDS analysis results of the Yehud *hemiobol* Type 31 O1/R1 coins, where SA represents the scanned area. Only bright areas according to BSE mode of shiny metal were examined by EDS analysis; Table S9. SEM-EDS analysis results of the Yehud IAA 153976, IMJ 27424, IAA 101006, IAA 154383, IMJ 27383, Trans-Jordan hoard no. 11, and IMJ 34591 coins after roughly grinding the surface; Figure S8. The copper distribution of each silver coin group based on SEM-EDS chemical analysis results, where the Cu wt% concentration is presented vs. the relative no. of measurements (%): (a) Yehud *gerah* Type 5 O1/R1; (b) Yehud *gerah* Type 5 O1/R2; (c) Yehud *gerah* Type 5 O1/R3; (d) Yehud *gerah* Type 5 O1/R4; (e) Yehud *gerah* Type 5 O1/R5; (f) Yehud half *gerah* Type 16 O2/R2; (g) Yehud quarter *obol* Type 24 O1/R2; and (h) Yehud hemoibol Type 31 O1/R1; Figure S9. Yehud *gerah* coins Type 5 O1/R1 (ca. 350–333 BCE) (Athena/Owl): IAA 138139, IAA 153975, IMJ 27424, IMJ 27398, IMJ 34542, IMJ 34539, IMJ 34543, and IAA 153976; Figure S10. Yehud *gerah* coins Type 5 O1/R2 (ca. 350–333 BCE) (Athena/Owl): IAA 101006, IAA 177246, IAA 153977, IMJ 34538, IMJ 34553, IMJ 34537, IMJ 34554, IMJ 34620, IAA 154383, and IMJ 27387; Figure S11. Yehud *gerah* coins Type 5 O1/R3 (ca. 350–333 BCE) of the Yehud series (Athena/Owl): IAA 153979, IAA 153978, IMJ 34555, and IMJ 27425; Figure S12. Yehud *gerah* coins Type 5 O1/R4 of the Yehud series (Athena/Owl): RH2-RH6, IAA 153980, IMJ 34556, and IAA 153981; Figure S13. Yehud *gerah* coins Type 5 O1/R5 (ca. 350–333 BCE) of the Yehud series (Athena/Owl): IMJ 34558, and IMJ 27388; Figure S14. Yehud coins Type 16 O2/R2 (Persian king wearing a jagged crown/Falcon in flight) (ca. 350–333): Edom hoard nos. 1–6, IMJ 27383, IMJ 27414, and IMJ 34566; Figure S15. Yehud Attic standard quarter *obol*, Type 24 O1/R2 (Facing head/Owl) of the Macedonian period (ca. 320(?)–312 BCE): Trans-Jordan hoard nos. 11, 12, 13; Figure S16. Yehud coins Type 31 O1/R1 (Head of roaring lion/bird standing right, head reverted) (ca. 306–302/1 BCE): IMJ 34631, IMJ 34593, IMJ 34591, IMJ 34709, IMJ 34594, and IMJ 34715.

Author Contributions: Conceptualization, M.C., D.A., H.G. and O.T.; study objectives, M.C., D.A., H.G. and O.T.; research methodology, M.C., D.A., H.G. and O.T.; formal analysis, M.C. and D.A.; discussion of results, M.C., D.A., H.G. and O.T.; conclusions, M.C., D.A., H.G. and O.T.; writing—introduction, H.G. and O.T.; writing—technological background to research, M.C. and D.A.; writing—original draft preparation, D.A.; project administration, H.G. and O.T.; funding acquisition, H.G. and O.T.; supervision, O.T. All authors have read and agreed to the published version of the manuscript.

Funding: This research forms part of an Israel Science Foundation (ISF) project (No. 2883/20); Corpus of Samarian Coinage (ca. Fourth Century BCE); PI: Oren Tal.

Institutional Review Board Statement: Not applicable.

Informed Consent Statement: Not applicable.

Data Availability Statement: The data supporting the reported results can be found in the Supplementary Materials file.

Acknowledgments: The authors are indebted to the ISF and would also like to thank Tomer Reuveni from the Wolfson Applied Materials Research Center at the Tel Aviv University for his SEM technical assistance. We are also grateful to Robert Kool, head of the coin department at the IAA, for providing the controlled archaeological excavation coins for our project.

Conflicts of Interest: The authors declare no conflict of interest. The funders had no role in the design of the study; in the collection, analyses, or interpretation of data; in the writing of the manuscript; or in the decision to publish the results.

References

1. Gitler, H.; Goren, Y.; Konuk, K.; Tal, O.; Peter van Alfen, P.; Weisburd, D. XRF analysis of several groups of electrum coins. In *White Gold, Studies in Early Electrum Coinage*; van Alfen, P., Wartenberg, U., Fischer-Bossert, W., Gitler, H., Konuk, K., Lorber, C.C., Eds.; American Numismatic Society: New York, NY, USA, 2020; pp. 379–422.
2. Gitler, H.; Lorber, C.; Fontanille, J.-P. *The Yehud Coinage: A Study and Die Classification of the Provincial Silver Coinage of Judah*; Israel Numismatic Society: Jerusalem, Israel, 2023.
3. Fontanille, J.-P. Extreme deterioration and damage on Yehud coin dies. *Isr. Numis. Res.* **2008**, *3*, 29–44.
4. Gitler, H. A Hoard of Persian Yehud coins from the environs of Ramallah. *Numisma* **2006**, *250*, 319–324.
5. Brocchieri, J.; Vitale, R.; Sabbarese, C. Characterization of the incuse coins of the Museo Campano in Capua (Southern Italy) by X-ray fluorescence and numismatic analysis. *Nucl. Instrum. Methods Phys. Res. Sect. B Beam Interact. Mater. At.* **2020**, *479*, 93–101. [CrossRef]
6. Fabrizi, L.; Di Turo, F.; Medeghini, L.; Di Fazio, M.; Catalli, F.; De Vito, C. The application of non-destructive techniques for the study of corrosion patinas of ten Roman silver coins: The case of the medieval Grosso Romanino. *Microchem. J.* **2018**, *145*, 419–427. [CrossRef]
7. Hrnjić, M.; Hagen-Peter, G.A.; Birch, T.; Barfod, G.H.; Sindbæk, S.M.; Lesher, C.E. Non-destructive identification of surface enrichment and trace element fractionation in ancient silver coins. *Nucl. Instrum. Methods Phys. Res. Sect. B Beam Interact. Mater. At.* **2020**, *478*, 11–20. [CrossRef]
8. Marussi, G.; Crosera, M.; Prenesti, E.; Cristofori, D.; Callegher, B.; Adami, G. A Multi-Analytical Approach on Silver-Copper Coins of the Roman Empire to Elucidate the Economy of the 3rd Century A.D. *Molecules* **2022**, *27*, 6903. [CrossRef]
9. Zlateva, B.; Lesigyarski, D.; Varbanov, V.; Bonev, V.; Iliev, I. Archaeometrical investigation of Roman silver coins from Bulgaria. *Mediterr. Archaeol. Archaeom.* **2022**, *22*, 23–34.
10. L'Héritier, M.; Baron, S.; Cassayre, L.; Téreygeol, F. Bismuth behavior during ancient processes of silver–lead production. *J. Archaeol. Sci.* **2015**, *57*, 56–68. [CrossRef]
11. Oudbashi, O.; Wanhill, R. Long-term embrittlement of ancient copper and silver alloys. *Heritage* **2021**, *4*, 2287–2319. [CrossRef]
12. Mortazavi, M.; Naghavi, S.; Khanjari, R.; Agha-Aligol, D. Metallurgical study on some Sasanian silver coins in Sistan Museum. *Archaeol. Anthropol. Sci.* **2018**, *10*, 1831–1840. [CrossRef]
13. Nerantzis, N. Pre-industrial iron smelting and silver extraction in North-Eastern Greece: An archaeometallurgical approach. *Archaeometry* **2016**, *58*, 624–641. [CrossRef]
14. Vogl, J.; Paz, B.; Völling, E. On the ore provenance of the Trojan silver artefacts. *Archaeol. Anthropol. Sci.* **2019**, *11*, 3267–3277. [CrossRef]
15. Boev, I. Chemical composition of the silver tetradrachms from the locality Isar Marvinci determined with the application of the SEM-EDS method. *Nat. Resour. Technol.* **2021**, *15*, 43–47.
16. Liu, S.; Li, Z.; Zhou, Y.; Li, R.; Xie, Z.; Liu, C.; Hu, G.; Hu, D. Optimizing the thermal treatment for restoration of brittle archaeological silver artifacts. *Herit. Sci.* **2022**, *10*, 21. [CrossRef]

17. Tayyari, J.; Emami, M.; Agha-Aligol, D. Identification of microstructure and chemical composition of a silver object from Shahrak-e Firouzeh, Nishapur, Iran (~2nd millennium BC). *Surf. Interfaces* **2021**, *25*, 101168. [CrossRef]

18. Alinezhad, Z.; Dehpahlavan, M.; Rashti, M.L.; Oliaiy, P. Elemental analysis of Seleucid's silver coins from Hamadan Museum by PIXE technique. *Radiat. Phys. Chem.* **2019**, *158*, 165–174. [CrossRef]

19. Ashkenazi, D.; Gitler, H.; Stern, A.; Tal, O. Metallurgical investigation on fourth century BCE silver jewellery of two hoards from Samaria. *Sci. Rep.* **2017**, *7*, 40659. [CrossRef]

20. Ashkenazi, D.; Gitler, H.; Stern, A.; Tal, O. Archaeometallurgical characterization and manufacturing technologies of fourth century BCE silver jewelry: The Samaria and Nablus Hoards as test case. *Metallogr. Microstruct. Anal.* **2018**, *7*, 387–413. [CrossRef]

21. Davis, G.; Gore, D.B.; Sheedy, K.A.; Albarède, F. Separating silver sources of Archaic Athenian coinage by comprehensive compositional analyses. *J. Archaeol. Sci.* **2020**, *114*, 105068. [CrossRef]

22. Doménech-Carbó, M.T.; Di Turo, F.; Montoya, N.; Catalli, F.; Doménech-Carbó, A.; De Vito, C. FIB-FESEM and EMPA results on Antoninianus silver coins for manufacturing and corrosion processes. *Sci. Rep.* **2018**, *8*, 10676. [CrossRef]

23. Bakirov, B.A.; Kichanov, S.E.; Khramchenkova, R.K.; Belushkin, A.V.; Kozlenko, D.P.; Sitdikov, A.G. Studies of coins of medieval Volga Bulgaria by neutron diffraction and tomography. *J. Surf. Investig. X-ray Synchrotron Neutron Tech.* **2020**, *14*, 376–381. [CrossRef]

24. Westner, K.J.; Birch, T.; Kemmers, F.; Klein, S.; Höfer, H.E.; Seitz, H.M. Rome's rise to power. Geochemical analysis of silver coinage from the Western Mediterranean (fourth to second centuries BCE). *Archaeometry* **2000**, *62*, 577–592. [CrossRef]

25. Di Fazio, M.; Di Turo, F.; Medeghini, L.; Fabrizi, L.; Catalli, F. New insights on medieval Provisini silver coins by a combination of nondestructive and micro-invasive techniques. *Microchem. J.* **2019**, *144*, 309–318. [CrossRef]

26. Hampshire, B.V.; Butcher, K.; Ishida, K.; Green, G.; Paul, D.M.; Hillier, A.D. Using negative muons as a probe for depth profiling silver Roman coinage. *Heritage* **2019**, *2*, 400–407. [CrossRef]

27. Ortega-Feliu, I.; Moreno-Suárez, A.I.; Gómez-Tubío, B.; Ager, F.J.; Respaldiza, M.A.; García-Dils, S.; Rodríguez-Gutiérrez, O. A comparative study of PIXE and XRF corrected by Gamma-ray Transmission for the non-destructive characterization of a gilded roman railing. *Nucl. Instrum. Methods Phys. Res. Sect. B Beam Interact. Mater. At.* **2010**, *268*, 1920–1923. [CrossRef]

28. Newbury, D.E.; Ritchie, N.W.M. Is scanning electron microscopy/energy dispersive X-ray spectrometry (SEM/EDS) quantitative? *Scanning* **2013**, *35*, 141–168. [CrossRef]

29. Debernardi, P.; Corsi, J.; Borghi, A.; Cossio, R.; Gambino, F.; Ghignone, S.; Scherillo, A.; Re, A.; Giudice, A.L. Some insight into "bronze quadrigati": A multi-analytical approach. *Archaeol. Anthropol. Sci.* **2022**, *14*, 133. [CrossRef]

30. Pitarch, A.; Queralt, I. Energy dispersive X-ray fluorescence analysis of ancient coins: The case of Greek silver drachmae from the Emporion site in Spain. *Nucl. Instrum. Methods Phys. Res. Sect. B* **2010**, *268*, 1682–1685. [CrossRef]

31. Pitarch, A.; Queralt, I.; Alvarez-Perez, A. Analysis of Catalonian silver coins from the Spanish war of independence period (1808–1814) by energy dispersive X-ray fluorescence. *Nucl. Instrum. Methods Phys. Res. Sect. B* **2011**, *269*, 308–312. [CrossRef]

32. Ortega-Feliu, I.; Scrivano, S.; Gómez-Tubío, B.; Ager, F.J.; de la Bandera, M.L.; Respaldiza, M.A.; Navarro, A.D.; San Martín, C. Technical characterization of the necklace of El Carambolo hoard (Camas, Seville, Spain). *Microchem. J.* **2018**, *139*, 401–409. [CrossRef]

33. Pięta, E.; Lekki, J.; del Hoyo-Meléndez, J.M.; Paluszkiewicz, C.; Nowakowski, M.; Matosz, M.; Kwiatek, W.M. Surface characterization of medieval silver coins minted by the early Piasts: FTIR mapping and SEM/EDX studies. *Surf. Interface Anal.* **2018**, *50*, 78–86. [CrossRef]

34. Gordus, A.A. Neutron activation analysis of archaeological artefacts. *Philos. Trans. R. Soc. Lond. A* **1970**, *269*, 165–174.

35. Ashworth, M.J.; Abeles, T.P. Neutron activation analysis and archaeology. *Nature* **1966**, *210*, 9–11. [CrossRef]

36. Bolewski, A.; Matosz, M.; Pohorecki, W.; del Hoyo-Meléndez, J.M. Comparison of neutron activation analysis (NAA) and energy dispersive X-ray fluorescence (XRF) spectrometry for the non-destructive analysis of coins minted under the early Piast dynasty. *Radiat. Phys. Chem.* **2020**, *171*, 108699. [CrossRef]

37. Das, D.D.; Sharma, N.; Chawla, P.A. Neutron activation analysis: An excellent nondestructive analytical technique for trace metal analysis. *Crit. Rev. Anal. Chem.* **2023**, 1–17. [CrossRef]

38. Szentmiklósi, L.; Maróti, B.; Kis, Z. Prompt-gamma activation analysis and neutron imaging of layered metal structures. *Nucl. Instrum. Methods Phys. Res. A* **2021**, *1011*, 165589. [CrossRef]

39. Kasztovszky, Z.; Panczyk, E.; Fedorowicz, W.; Révay, Z.; Sartowska, B. Comparative archaeometrical study of Roman silver coins by prompt gamma activation analysis and SEM-EDX. *J. Radioanal. Nucl. Chem.* **2005**, *265*, 193–199. [CrossRef]

40. Biswas, S.; Gerchow, L.; Luetkens, H.; Prokscha, T.; Antognini, A.; Berger, N.; Cocolios, T.E.; Dressler, R.; Indelicato, P.; Jungmann, K.; et al. Characterization of a continuous muon source for the non-destructive and depth-selective elemental composition analysis by muon induced X- and gamma-rays. *Appl. Sci.* **2022**, *12*, 2541. [CrossRef]

41. Moreno-Suárez, A.I.; Ager, F.J.; Scrivano, S.; Ortega-Feliu, I.; Gómez-Tubío, B.; Respaldiza, M.A. First attempt to obtain the bulk composition of ancient silver–copper coins by using XRF and GRT. *Nucl. Instrum. Methods Phys. Res. Sect. B* **2015**, *358*, 93–97. [CrossRef]

42. Beck, L.; Bosonnet, S.; Réveillon, S.; Eliot, D.; Pilon, F.J.N.I. Silver surface enrichment of silver–copper alloys: A limitation for the analysis of ancient silver coins by surface techniques. *Nucl. Instrum. Methods Phys. Res. Sect. B* **2004**, *226*, 153–162. [CrossRef]

43. Oudbashi, O.; Shekofteh, A. Chemical and microstructural analysis of some Achaemenian silver alloy artefacts from Hamedan, western Iran. *Period. Mineral.* **2015**, *84*, 419–434.

44. Flament, C.; Marchetti, P. Analysis of ancient silver coins. *Methods Phys. Res. Sect. B Beam Interact. Mater. At.* **2004**, *226*, 179–184. [CrossRef]
45. Tate, J. Some problems in analysing museum material by nondestructive surface sensitive techniques. *Methods Phys. Res. Sect. B Beam Interact. Mater. At.* **1986**, *14*, 20–23. [CrossRef]

Article

Abnormal Trend of Ferrite Hardening in a Medium-Si Ferrite-Martensite Dual Phase Steel

Ali Khajesarvi [1], Seyyed Sadegh Ghasemi Banadkouki [1,*], Seyed Abdolkarim Sajjadi [2] and Mahesh C. Somani [3]

[1] Mining Technologies Research Center, Department of Mining and Metallurgical Engineering, Yazd University, Yazd 98195, Iran
[2] Department of Materials Science and Engineering, Ferdowsi University of Mashhad, Mashhad 91775, Iran
[3] Materials and Mechanical Engineering, Centre for Advanced Steels Research, University of Oulu, 90014 Oulun Yliopisto, Finland
* Correspondence: sghasemi@yazd.ac.ir; Tel.: +98-913-3516-649

Abstract: In this paper, the effects of carbon, Si, Cr and Mn partitioning on ferrite hardening were studied in detail using a medium Si low alloy grade of 35CHGSA steel under ferrite-martensite/ferrite-pearlite dual-phase (DP) condition. The experimental results illustrated that an abnormal trend of ferrite hardening had occurred with the progress of ferrite formation. At first, the ferrite microhardness decreased with increasing volume fraction of ferrite, thereby reaching the minimum value for a moderate ferrite formation, and then it surprisingly increased with subsequent increase in ferrite volume fraction. Beside a considerable influence of martensitic phase transformation induced residual compressive stresses within ferrite, these results were further rationalized in respect of the extent of carbon, Si, Cr and Mn partitioning between ferrite and prior austenite (martensite) microphases leading to the solid solution hardening effects of these elements on ferrite.

Keywords: medium Si low alloy steel; step-quenched heat treatment; ferrite-martensite/ferrite-pearlite DP microstructure; alloying element partitioning; ferrite hardening variation

Citation: Khajesarvi, A.; Banadkouki, S.S.G.; Sajjadi, S.A.; Somani, M.C. Abnormal Trend of Ferrite Hardening in a Medium-Si Ferrite-Martensite Dual Phase Steel. *Metals* **2023**, *13*, 542. https://doi.org/10.3390/met13030542

Academic Editor: Andrea Di Schino

Received: 18 January 2023
Revised: 25 February 2023
Accepted: 6 March 2023
Published: 8 March 2023

1. Introduction

The low carbon, low alloy ferrite-martensite dual-phase (DP) steels are characterized by a mixture of soft ferritic matrix phase in conjunction with dispersed hard martensite. Attractive engineering properties are nowadays attained in these steels by the use of innovative heat treatments [1], which allow significant improvement in continuous yielding behavior, resulting in superior strength–ductility combinations in association with rapid strain hardening at early stage of plastic deformation due to the formation of a favorable microstructure [2–6]. These engineering properties are imparted by some of the variable microstructural parameters, such as ferrite and martensite volume fractions, their morphology and distribution, ferrite grain size, and the complex interaction of microphases with each other causing a considerable variation in mechanical behavior of low alloy DP steels [7–10]. Bag et al. [11,12] studied the impact and tensile properties of high martensite containing low alloy, ferrite-martensite DP steels and reported that an equal amount of finely divided ferrite and martensite microphases facilitate an optimum combination of high ductility and strength with good impact toughness. Using a 0.11% C, 1.6% Mn, 0.73% Si steel, Cai et al. [13] measured room temperature tensile properties of various samples with different ferrite and martensite morphologies, but with almost the same martensite volume fraction, derived from different primary microstructures. They reported that relatively higher strengths were attained in the DP samples, which received an intermediate quenching heat treatment, whereby the martensite content after intercritical annealing was very fine and fibrous morphology in the microstructures. Chang et al. [14] studied the effect of ferrite grain size on the tensile properties of a low alloy DP steel and presented that the effect of ferrite grain size on the yield strength was much stronger than on the tensile strength.

A careful literature review of relevant articles regarding the structure-property relationships of low alloy DP steels indicate that the individual hardening behavior of ferrite and martensite microphases have been one of the noteworthy research topics in physical metallurgy of advanced high strength low alloy steels [15–17]. Kumar et al. [8] have reported that the ferrite hardness changed as a function of its volume fraction in a low alloy ferrite-bainite DP steel. They found that about 8% alteration in ferrite microhardness can be related with the ferrite-bainite DP samples consisting of 10 to 50% volume fraction of ferrite in conjunction with the remaining bainite regions. Another experimental work has also been conducted by Kumar et al. [8] showing that the mechanical behavior of low alloy ferrite-martensite DP steels was relevant to the size of ferrite grains. Also, the interaction between ferrite and martensite microconstituents has been believed to introduce unpinned dislocations generated within ferrite, which can affect the ferrite strain hardening in the low alloy ferrite-martensite DP steels [18,19]. Therefore, the ferrite hardening in ferrite–martensite DP steels has been presented to be variable depending on various parameters such as volume fraction, morphology and grain size of ferrite, and of course, the role of carbon and other alloying elements partitioning between ferrite and prior austenite (martensite) microphases are questionable and have been not followed significantly. Accordingly, it has been tried to find out the effect of carbon and other alloying elements partitioning on the ferrite hardening of different ferrite–martensite DP microstructures using a commercial grade of medium Si low alloy 35CHGSA steel by means of microhardness measurements and EDS analyses.

2. Materials and Methods

In the present investigation, a commercial grade of medium silicon low alloy 35CHGSA steel was used with the chemical composition indicated in Table 1. The heat treatment schedules were designed to achieve multiphase microstructures including various volume fractions of ferrite, pearlite, martensite and retained austenite microconstituents. The practical heat treatment processes practical the following sequential stages: (a) normalizing after austenitising at 900 °C for 15 min, (b) re-austenitising at 900 °C for 15 min to get fully homogenous prior austenite grains, (c) step-quenching (SQ) in a salt bath at 720 °C for 1–30 min to achieve different volume fractions of ferrite, and (d) rapid water quenching to transform all the remaining prior austenite to martensite. For each set of ferrite–martensite and ferrite–pearlite DP microstructure, three specimens were heat-treated in order to confirm the reproducibility of DP microstructures with respect to volume fractions of ferrite, pearlite, and martensite. In this way, the related specimens were marked as SQ1, SQ5, SQ15, and SQ30, representing the SQ holding times of 1, 5, 15 and 30 min, respectively, in salt bath maintained at 720 °C prior to water quenching. The applied heat treatment cycles are indicated schematically in Figure 1.

The metallography of heat-treated specimens was carried out on the transverse section surfaces relative to hot rolling direction of as-received strap specimens according to ASTM E 3 standard [20]. The polished specimens were etched with a 2% Nital solution (2 mL HNO_3 and 98 mL C_2H_5OH) [21] to reveal various microstructural features with good contrasting resolution. The ferrite, pearlite, and martensite volume fractions were measured according to the ASTM E562-02 standard [22]. The microstructural observations were carried out using a Olympus-PMG3 light microscope and a TESCAN-MIRA 3-XMU field-emission scanning electron microscope (FE-SEM) operating at an accelerated voltage of 15 kV. Both the spot and line scan energy dispersive X-ray spectroscopy (EDS) analysis techniques were largely applied to identify the carbon, Si, Cr and Mn concentrations in different locations of ferrite. Microhardness tests were conducted within ferrite, pearlite, and martensite microphases using a load of 5 g for ferrite grain and a somewhat higher load of 10 g for pearlite and martensite areas, applied for 20 s duration using a FM700 Future Tech microhardness tester, and the related data were presented as Vickers hardness numbers

(VHNs). In addition, microhardness tests were also carried out at different locations of select ferrite grains, but with a small load of 1 g applied for 20s duration loading time using a Future Tech microhardness tester machine model FM700. Two different locations of each step-quenched heat-treated samples were considered for conducting microhardness tests: (1) the central area of ferrite grain, and (2) the ferrite region close to the ferrite-martensite interfaces. The microhardness data were collected from at least five readings per sample, and the data were reported as Vickers hardness numbers (VHNs).

Table 1. The chemical composition of researched low alloy, medium silicon commercial grade of 35CHGSA steel (in wt%).

C	Si	Mn	Cr	S	P	Mo	Ni	Ti	V	Fe
0.35	1.25	0.89	1.18	0.01	0.01	0.01	0.04	0.03	0.01	Balance

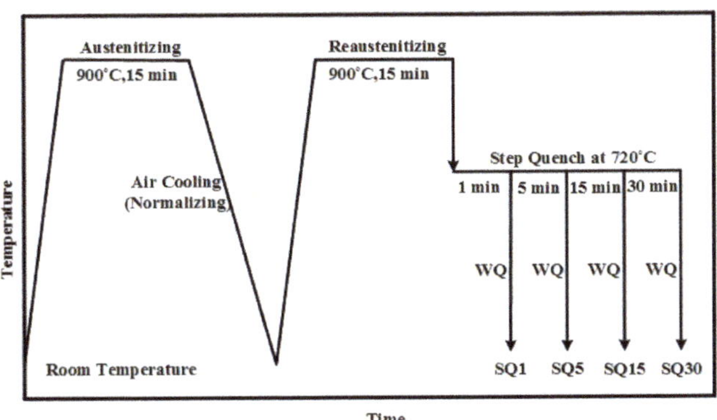

Figure 1. The schematic diagrams indicating heat treatment cycles used to achieve microstructures involving different volume fractions of ferrite, pearlite and martensite microphases. WQ: water quench; SQ: step quench [10].

The X-ray diffraction analysis was also carried out on a AW-DX300 Asenware X-ray diffractometer using copper Kα radiation (1.54184 Å) and angular step size (in 2θ) of 0.05 degrees with a time step of 1 s. The residual stresses were quantified using the Sin$^2\Psi$ analysis technique [23,24]. The following expression was used to calculate the amount of residual stress within various SQ specimens [24]:

$$\frac{d_{\phi\Psi} - d_0}{d_0} = \frac{1+v}{E}\,\sigma_\phi \sin^2\Psi - \frac{v}{E}(\sigma_{11} - \sigma_{22}) \tag{1}$$

where $d_{\phi\Psi}$ is the d-spacing of diffracted planes oriented at an angle Ψ from the surface normal, d_0 is the d-spacing of (hkl) planes at $\Psi = 0$ deg, v is poisson's ratio, E is young's modulus, σ_{11} and σ_{22} are principal stresses parallel to the surface, and σ_ϕ is the stress acting in a chosen direction of surface at an angle ϕ from σ_{11}. The σ_ϕ factor accounts for differences in specimen orientation, and is related to σ_{11} and σ_{22} parameters. The term $(\sigma_{11} - \sigma_{22})$ is constant for a given specimen, and Equation (1) shows that a linear variation between $d_{\phi\Psi}$ and $\sin^2\Psi$ is expected [23]. The term $(d_{\phi\Psi} - d_0)/d_0$ can be plotted against $\sin^2\Psi$ for several Ψ angles for a particular diffracted plane, and, the stress σ_ϕ can be calculated from the slope. When, the deviation of d with $\sin^2\Psi$ data occurs, the residual stress can be determined using additional mathematical and experimental treatments.

3. Results and Discussion

3.1. Microstructural Characterization

Figure 2 represents typical light microstructures along with super imposed electron micrographs (at top right hand side) of SQ specimens heat-treated at 720 °C for various holding times. These micrographs show a significant increase in volume fraction, polygonality, grain size, and continuity of ferrite as a consequence of longer isothermal holding time at 720 °C. For more information, the progress of prior austenite decomposition in various SQ heat-treated specimens was measured in term of microconstituent volume fractions, and the associated results are given in Table 2. A careful microstructural analysis indicated that the microstructures of SQ1, SQ5, and SQ15 heat-threated samples were characterized by just a mixture of ferrite and martensite microphases, introducing ferrite-martensite DP microstructures, while in the case of SQ30 heat-threated samples, a maximum volume fraction of 15% ferrite along with the remaining 85% pearlite microconstituents formed in the microstructures, representing the typical ferrite-pearlite DP microstructures. In other words, with the increase in SQ holding time up to 30 min in the intercritical $\alpha+\gamma$ region, there was enough time available for carbon atoms to diffuse from ferrite into the prior austenite and equilibrate and this enabled enough carbon concentration within the prior austenite regions for subsequent pearlite formation.

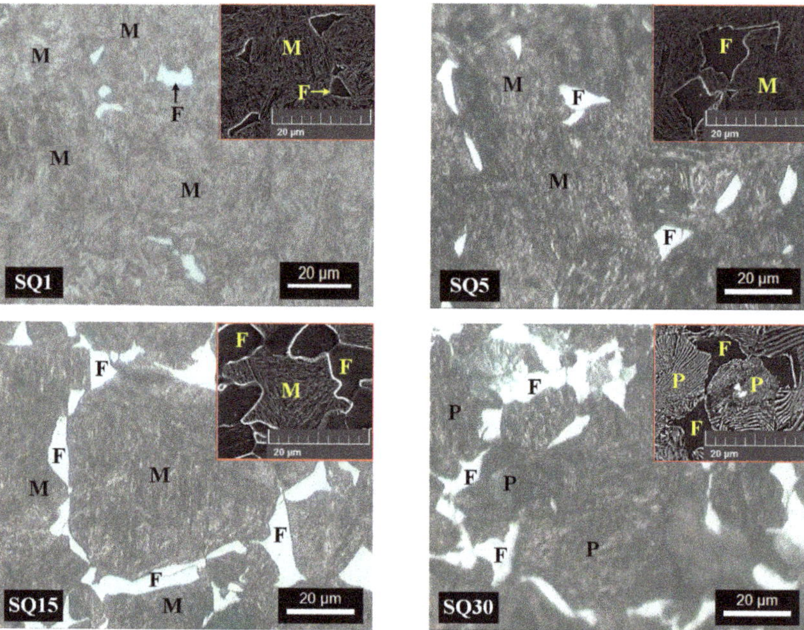

Figure 2. Light and FE-SEM micrographs of various SQ heat-treated specimens depicting the formation of ferrite, pearlite, and martensite microphases in the microstructures. The blocky ferrite grains were mostly surrounded by martensite areas in the SQ1 and SQ5 heat-treated specimens, while continuous grain boundary ferrite was developed in the SQ15 and SQ30 heat-treated samples. Microstructural features were characterized by ferrite-martensite DP microconstituents for SQ1, SQ5, and SQ15 samples, while ferrite-pearlite microphases were formed in the SQ30 heat-treated ones. Ferrite, pearlite, and martensite microconstituents are marked with F, P, and M symbols, respectively.

Table 2. The progress of ferrite and pearlite formations realized at 720 °C for various SQ holding times. The pearlite formation started after 15 min isothermal holding time.

Sample Mark	SQ Holding Time (min)	Ferrite Volume Fraction (%)	Martensite Volume Fraction (%)	Pearlite Volume Fraction (%)
SQ1	1	3	97	-
SQ5	5	6	94	-
SQ15	15	13	87	-
SQ30	30	15	0	85

3.2. Ferrite Hardening Variation

The alteration of ferrite hardening as a function of SQ holding time in conjunction with the changes in ferrite volume fraction, are illustrated in Figure 3. Figure 3a illustrates that the progress of ferrite formation can be characterized with a typical S-shaped curve indicating that the phase transformation of prior austenite to ferrite can be rationalized by a general diffusional nature of nucleation and growth for ferrite formation during SQ holding at 720 °C. The variation of ferrite grain size against SQ holding time is indicated in Figure 3b. The abnormal trend in ferrite hardness as a function of SQ holding time is illustrated in Figure 3c. Accordingly, the ferrite microhardness initially decreased from 352 to 217HV5g with the increase in SQ holding time from 1 to 15 min, beyond which it surprisingly increased to 245HV5g with further increase in SQ holding time to 30 min. It is interesting to emphasize that the minimum ferrite microhardness occurred in the ferrite-martensite DP specimens SQ15 consisting of 13% ferrite volume fraction with remainder phase fraction as martensite, and subsequent increase in ferrite hardness was realized for the SQ30 samples with essentially ferrite-pearlite microstructures. The minimum ferrite hardening response can be fully supported by the verity that the SQ heat treatment at longer holding time can be related to the more ferrite formation, in addition to the occurrence of significantly thermally activated relaxation, which decreases the accumulation of transformational residual stresses causing a lower hardness level in ferrite. In addition to the occurrence of these ferrite softening phenomena during longer holding times at 720 °C, the abnormal higher ferrite hardness realized in SQ30 samples suggests that another ferrite hardening mechanism must be operational, such as solid solution hardening effects caused by diffusion of substitutional alloying elements leading to simultaneous hardening of ferrite during the progress of its formation.

To examine the hardening alteration of ferrite grains more precisely, the microhardness test has been accomplished at different locations within certain ferrite grains using various SQ heat-treated samples. Typical light optical micrograph in Figure 4 shows an example of microhardness method followed, indicating indentation impressions taken from various positions (the central regions of ferrite grains and the ferrite regions adjacent to the ferrite-prior austenite interfaces) of ferrite grains using SQ5 heat-treated samples. It is obvious that the minimum ferrite microhardness has been associated to the central location of ferrite grains, and it increases as the microhardness test location is moved towards ferrite-prior austenite interfaces, thus clarifying that the deformation resistance of ferrite is related to the position of microhardness testing location. Therefore, as the microhardness location is moved from the central location of ferrite grains towards the ferrite area adjacent to the ferrite-prior austenite interfaces, the average ferrite microhardness increased from 122 to 145HV1g for the SQ5 specimens.

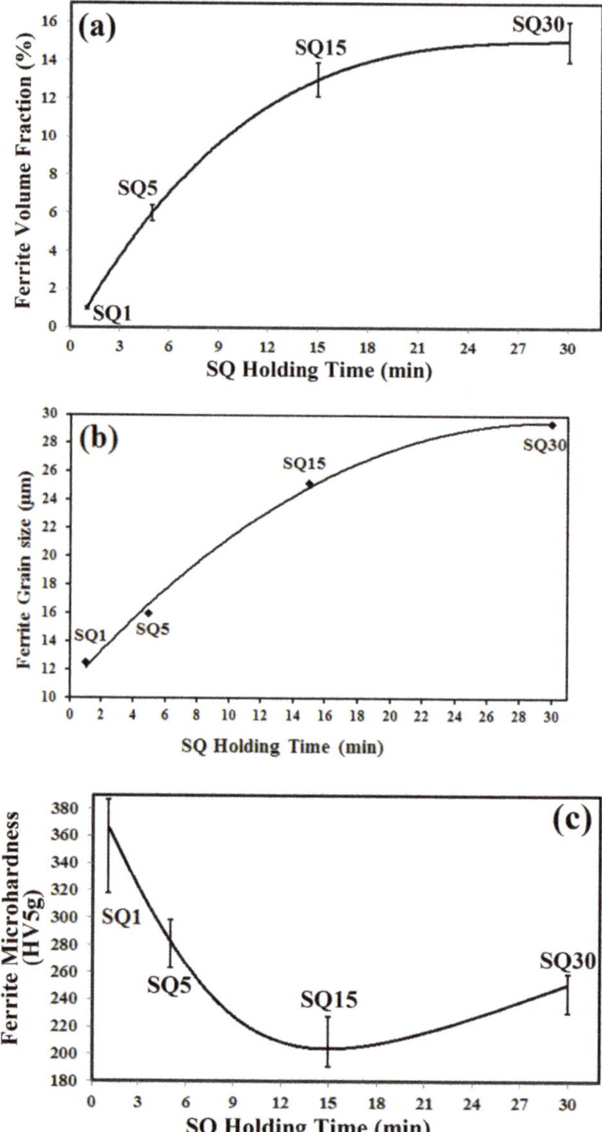

Figure 3. Changes in: (**a**) volume fraction of ferrite as a function of SQ holding time; (**b**) ferrite grain size against SQ holding time; and (**c**) ferrite microhardness versus SQ holding time. An abnormal trend in respect of ferrite hardening occurred during the progress of ferrite formation in the SQ30 heat-treated samples. The ferrite microhardness measurements are related with a fixed loading force of 5 g.

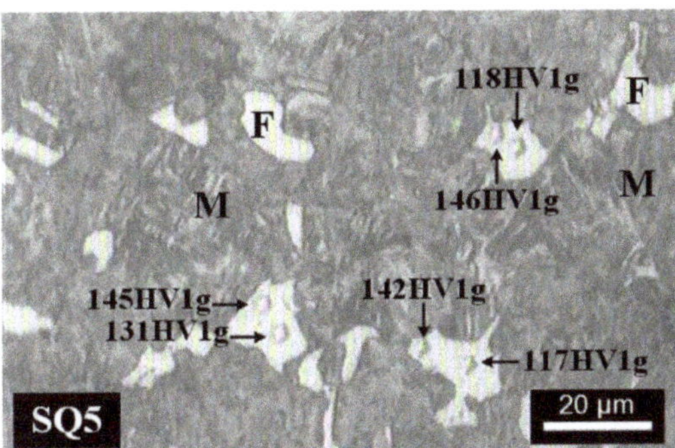

Figure 4. Typical light micrograph with superimposed locations at which ferrite microhardness tests were carried out and the associated ferrite microhardness data showed the alteration of ferrite hardening within the specific ferrite grains recorded on heat-treated SQ5 ferrite-martensite DP specimens. A greater ferrite hardening response was realized in the areas adjacent to the martensite in contrast to the central areas of ferrite grains. The ferrite microhardness data are measured with a fixed loading force of 1 g. F: Ferrite; M: Martensite [25].

3.3. Alloying Element Partitioning between Ferrite and Prior Austenite

3.3.1. Spot EDS Analysis for Alloying Concentration

The EDS analysis was accomplished widely at several positions within central ferrite grains and the central vicinity martensite (prior austenite) regions in order to investigate in detail the alteration of carbon, Si, Cr and Mn partitioning within ferrite and prior austenite regions developed during the progress of ferrite formation at 720 °C using various SQ heat-treated samples. Although, the measurement of carbon concentration within ferrite and prior austenite (martensite) microphases by EDS analysis technique accompanied some overestimation, but this technique can still be used as a comparable study in order to identify the alteration of carbon concentration within ferrite and prior austenite microphases as reported by several investigators [7,26,27]. In this way, the results of spot EDS analysis for carbon, Si, Cr, and Mn concentrations within the central locations of ferrite and martensite microphases are summarized in Tables 3 and 4 for various SQ heat-treated samples. For a better comparison of carbon, Si, Cr and Mn partitioning within ferrite and prior austenite with the progress of ferrite formation at 720 °C, the concentrations of these alloying elements as a function of SQ holding times are shown in Figure 5 for central ferrite and martensite areas. A careful investigation of the results depicted in Table 3 and Figure 5a illustrates that the mean level of carbon concentration within the central region of ferrite grains decreased continuously from 6.32 EDSNs for the short time ferrite-martensite SQ1 samples to 5.87 EDSNs for the long time ferrite-pearlite ones in conjunction with a concomitant increase in carbon concentration of prior austenite from 10.64 to 11.87 EDSNs, respectively. These qualitative results illustrate that a higher level of carbon concentration occurred within the prior austenite regions associated with the pearlite formation (SQ30) as a consequence of greater ferrite areas, in comparison to the prior austenite formation related to the short time treated SQ1 specimens. The Si concentration within the central areas of ferrite grains increased with a gentle slope from 1.23 to 1.51 EDSNs, while this was almost constant for the central location of prior austenite areas (Figure 5b, Table 3). The mean Cr concentration also increased within the central ferrite grain from 0.83 to 1.09 EDSNs with the increase in SQ holding times from 1 to 30 min (Figure 5c, Table 4). However, the amount of mean Cr concentration in the central region of martensite areas did not change for the SQ1, SQ5, and SQ15 samples with ferrite-martensite microstructures (nearly 0.93 EDSNs), while it increased suddenly to 1.25 EDSNs for the SQ30 sample with ferrite-pearlite microstructures. These results indicate that Si and Cr atoms are partitioned from growing ferrite-prior austenite interfaces towards the ferrite phase constituents during the progress of ferrite formation, and Cr atoms, on the other hand, promote pearlite formation at the lateral stages of prior austenite phase transformation in SQ30 specimens heat treated for the prolonged duration. Although, the Mn content of central ferrite grains has been almost constant, the amount of mean Mn concentration increased in the central region of martensite areas from 0.50 to 0.72 EDSNs with the increase in SQ holding time from 1 to 15 min. For longer isothermal holding of 30 min, the amount of mean Mn concentration within the central region of pearlite increased to 0.78 EDSNs, emphasizing that Mn atoms did partition to the prior austenite side during the progress of ferrite formation in the SQ heat-treated specimens (Figure 5d, Table 4).

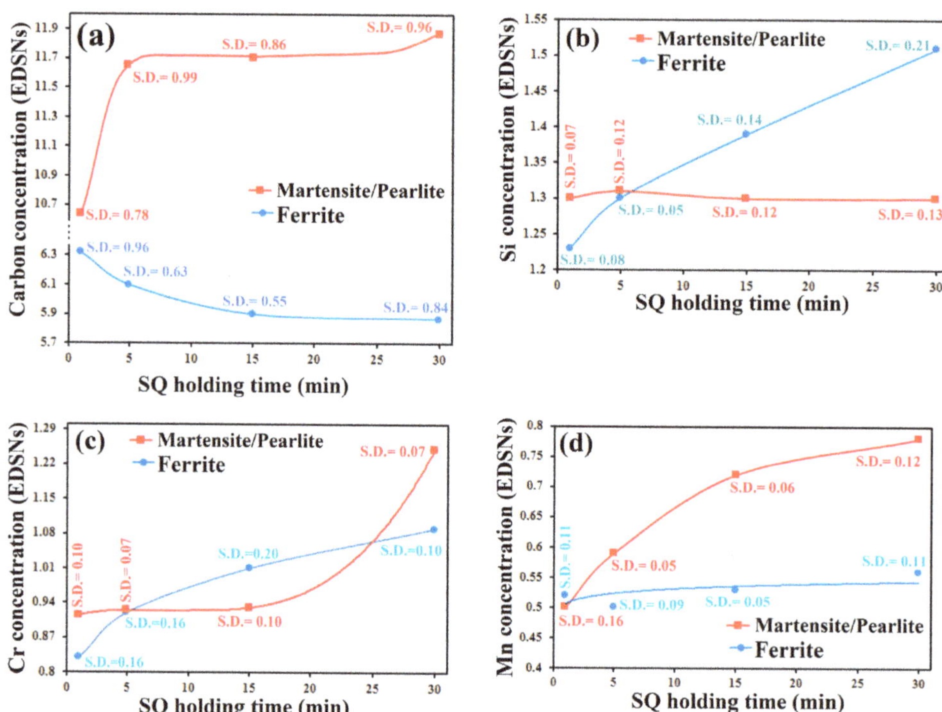

Figure 5. Changes in carbon (**a**), Si (**b**), Cr (**c**), and Mn (**d**) concentrations as a function of SQ holding time for central regions of ferrite and martensite/pearlite areas using various SQ heat-treated samples.

Table 3. The results of mean EDS analysis with the related average (Ave.) and standard deviation (S.D.) for carbon and Si concentrations within the central regions of ferrite grains and martensite areas taken from various SQ heat-treated samples.

Sample Mark	Carbon Content (EDSNs)				Si Content (EDSNs)			
	Ferrite		Martensite		Ferrite		Martensite	
	Ave.	S.D.	Ave.	S.D.	Ave.	S.D.	Ave.	S.D.
SQ1	6.32	0.96	10.64	0.78	1.23	0.08	1.30	0.07
SQ5	6.10	0.63	11.67	0.99	1.30	0.05	1.32	0.12
SQ15	5.90	0.55	11.69	0.86	1.39	0.14	1.30	0.12
			Pearlite				Pearlite	
SQ30	5.87	0.84	11.87	0.96	1.51	0.21	1.30	0.13

Table 4. The results of mean EDS analysis with the related average (Ave.) and standard deviation (S.D.) for Cr and Mn concentrations within the central regions of ferrite grains and martensite areas taken from various SQ heat-treated samples.

Sample Mark	Cr Content (EDSNs)				Mn Content (EDSNs)			
	Ferrite		Martensite		Ferrite		Martensite	
	Ave.	S.D.	Ave.	S.D.	Ave.	S.D.	Ave.	S.D.
SQ1	0.83	0.16	0.92	0.10	0.52	0.11	0.50	0.16
SQ5	0.92	0.16	0.93	0.07	0.50	0.09	0.59	0.05
SQ15	1.01	0.20	0.93	0.10	0.53	0.05	0.72	0.06
			Pearlite				Pearlite	
SQ30	1.09	0.10	1.25	0.07	0.56	0.11	0.78	0.12

3.3.2. Line Scan EDS Analysis for Alloying Concentration

To study the ferrite hardening mechanisms and related possibility of solid solution hardening effects of carbon and other alloying elements, the EDS line scan analyses were accomplished at various positions of ferrite grains as illustrated in Figure 6. The general microstructures of short (SQ5) and long time (SQ30) treated samples are depicted by electron micrographs presented in Figure 6a,b, respectively. The associated results of EDS analyses for C, Si, Cr, and Mn concentrations across the ferrite grains are illustrated in Figure $6a_1$–a_4,b_1–b_4 taken from short (SQ5) and long time (SQ30) treated specimens, respectively. The carbon concentration from the central position of ferrite grains toward the ferrite area adjacent to the ferrite-prior austenite interfaces increased from 5.13 to 7.25 and 3.85 to 10.35 EDSNs, as respective SQ holding time increased from 5 to 30 min (Figure $6a_1$,b_1). These results show that the prior austenite to ferrite phase transformation was related to a greater carbon concentration within ferrite areas formed at pre-existing defect regions of prior austenite grain boundaries and that the progress of ferrite formation was accompanied by further carbon rejection from ferrite to the remaining austenite regions, causing a considerable gradient in the carbon concentration across ferrite grains. The Si concentration from the ferrite region close to the ferrite-prior austenite interfaces, towards the central location of ferrite grains, increased from 1.12 to 1.44 and 1.07 to 1.54 EDSNs (Figure $6a_2$,b_2), and also the Cr concentration increased from 0.78 to 1.00 and 0.75 to 1.11 EDSNs, with respective SQ holding time increasing from 5 to 30 min (Figure $6a_2$,a_3,b_2,b_3). These results illustrate that Si and Cr atoms are distributed from the growing ferrite-prior austenite interfaces towards the ferrite grains. The Mn concentration, on the other hand, increased from 0.44 to 0.73 and 0.51 to 0.78 EDSNs for central locations of ferrite grains towards the ferrite regions adjacent to the ferrite-prior austenite interfaces of SQ5 and SQ30 samples, respectively (Figure $6a_4$,b_4). The higher carbon and Mn concentrations of ferrite areas close to the ferrite-prior austenite areas can be associated to the considerable contribution of solid solution hardening for these ferrite regions.

3.4. Ferrite/Martensite Residual Stress Analysis

Typical XRD analysis has been employed to detect and estimate the microstructural microconstituents and the associated results are shown in Figure 7 for various SQ heat-treated specimens. All the patterns show almost the same BCC diffracted ferrite/martensite planes emphasizing that the peaks corresponding to ferrite or martensite microphases occurred in the same diffracted angles and it is typically difficult to distinguish between martensite and ferrite by XRD analysis because of low tetragonality of martensite developed in this low carbon low alloy steel. Approximately very small peaks of retained austenite (RA) can be also detected for ferrite-martensite DP specimens.

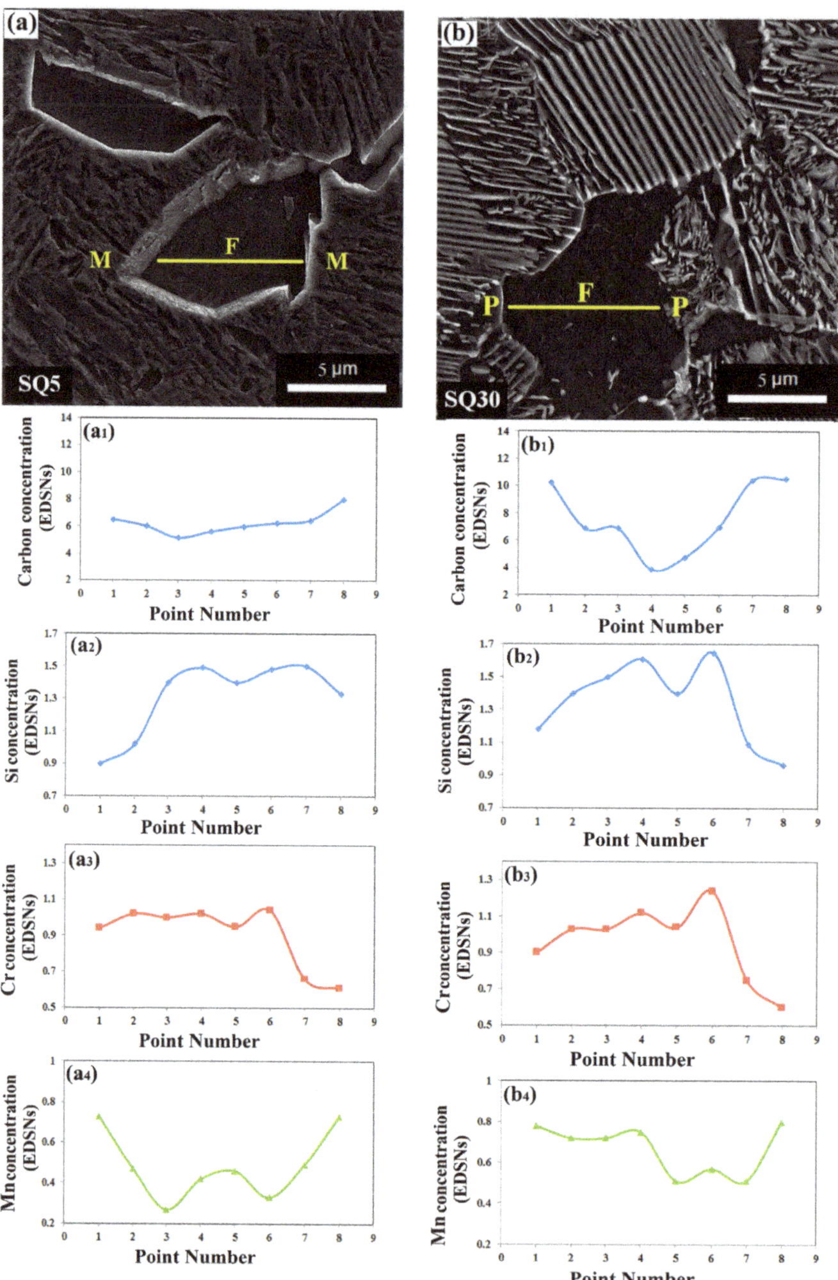

Figure 6. Electron micrographs along with carbon, Si, Cr and Mn EDS line-scan curves taken from short (SQ5) and long time (SQ30) treated samples presented in (**a**,**a₁**–**a₄**,) and (**b**,**b₁**–**b₄**); respectively. (**a**,**b**) represent electron micrographs in conjunction with hypothetical EDS scan lines expanded within ferrite grains followed in turn by (**a₁**–**a₄**); and (**b₁**–**b₄**) indicating the respective qualitative curves for carbon, Si, Cr and Mn concentrations within ferrite grains, respectively. Ferrite grains, pearlite and martensite are labeled as F, P and M symbols, respectively.

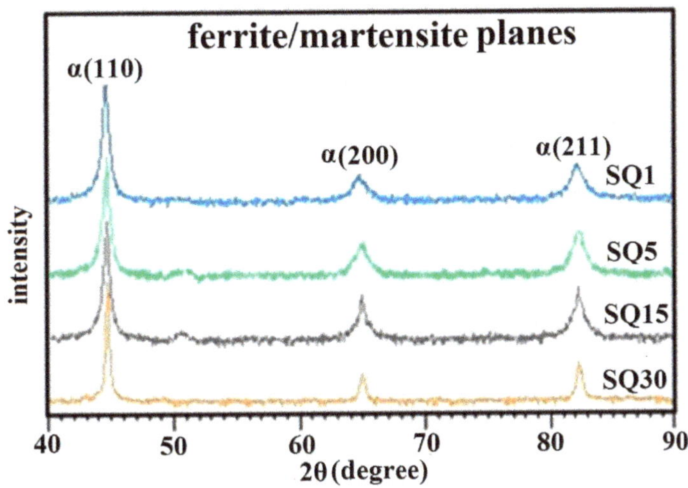

Figure 7. Comparison of XRD patterns of various SQ heat-treated specimens.

Residual stress "Sin$^2\Psi$" analysis was carried out using the various (211) diffracted ferrite/martensite planes at Ψ angles including −30, −20, −10, 0, 15, 30 and 45 deg from the XRD results shown in Figure 7. A ferrite/martensite peak was chosen for the analysis because of the compositional effects, such as solute carbon variation in the prior austenite, which could also cause shifting of ferrite/martensite peak. Figure 8 shows the variation in residual stress values calculated from the Sin$^2\Psi$ analysis results obtained from various SQ heat-treated specimens. For these calculations, E ~ 177 GPa and $v = 0.26$ were used [28]. From the Sin$^2\Psi$ analyses, it has been determined that the d-spacing decreases with increasing Ψ, indicating that ferrite/martensite is under higher residual compressive stress condition in short-time treated SQ specimens. This phenomenon can be to correlated a higher mutual ferrite-martensite interaction, thereby generating a considerable density of geometrically necessary dislocations within lower ferrite containing SQ specimens. On the other hand, with increasing in SQ holding time at 720 °C, the progress of ferrite formation was enhanced, resulting in more recovered ferrite in the microstructures.

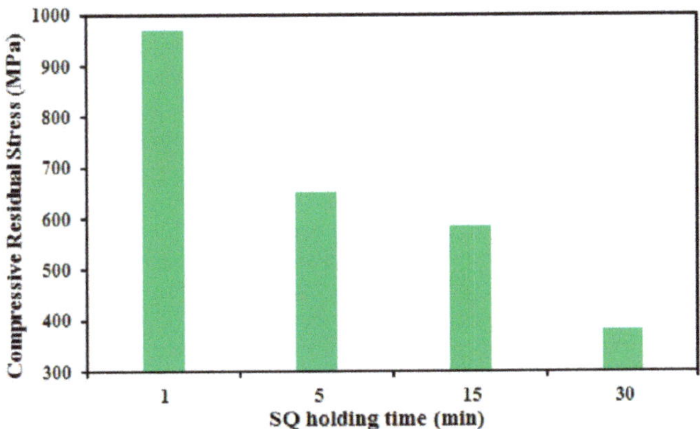

Figure 8. Compressive residual stress values (σ_ϕ) obtained from Sin$^2\Psi$ analyses versus SQ holding time.

3.5. Ferrite Hardening Mechanism

The experimental results indicate that not only the ferrite hardening is quite variable as a function of volume fraction of ferrite, but also it has changed across a specific ferrite grain (from central location towards the ferrite-martensite interfaces) for a particular SQ heat-treated sample (Figures 3 and 4). For ferrite-martensite DP samples containing large fractions of martensite, a significant contribution to ferrite hardening can be associated to the high level of carbon concentration and of course, residual compressive stresses extended within finer ferrite grains (Table 3, Figures 5a and 8). Fine ferrite grains with high carbon concentrations indicate that in addition to the other ferrite hardening mechanisms such as refinement of ferrite crystallite size as well as ferrite-martensite interaction, the solid solution hardening effect of C should be considered to account for the variations in ferrite hardness of short-time treated SQ specimens. Ferrite grains with higher carbon concentrated will have inevitable occurrence of lattice distortion and creation of stress field around solute iron atoms that can be contributed in part to the higher mechanical behavior of ferrite in the ferrite-martensite DP samples containing higher martensite volume fraction. On the other hand, it is obvious that the comprehensive partitioning of carbon will be faster at ferrite-prior austenite interfacial regions with significant density of defects. This is accompanied by the progress of prior austenite to ferrite phase transformation that can be associated to the possibility of greater segregation of carbon atoms to the induced dislocations as well as increased interaction of iron atoms with its stress fields leading to a greater capability to constrain the mobility of geometrically necessary dislocations, which raises the ferrite resistance to deformation through solid solution hardening mechanism [29,30].

The higher ferrite hardening in SQ heat-treated samples containing lower volume fraction of ferrite can be made according to the enhanced interaction of martensite with fine ferrite grains generating higher residual compressive stresses within ferrite (Figure 8). The formation of greater martensite volume fraction in the short-time treated SQ samples means that a smaller ferrite crystallite size is often surrounded by greater numbers of martensitic packets. As a result, the lower fraction of an individual ferrite grain in the short-time treated ferrite-martensite SQ samples experiences a higher localized compressive residual stresses in comparison with the long-time treated ferrite-martensite SQ samples, and hence, the ferrite hardness increases because of the lower mean spacing between dislocations [15,31,32]. This ferrite hardening mechanism seems to be more and more effective in the ferrite-martensite DP specimens containing a higher volume fraction of martensite, beside the occurrence of finer grain boundary ferrite crystals in comparison to the those of DP specimens containing lower volume fractions of martensite. Therefore, it is reasonable to conclude that the short-time treated SQ specimens are characterized by a higher density of dislocations caused by shear and extensive strain generated by the associated martensitic phase transformation, since this path is expected to minimize the accommodation strain energy for the formation of ferrite-martensite interfaces [33,34].

An abnormal trend in ferrite microhardness data occurred in respect of ferrite formation in the case of prolonged-time treated ferrite-pearlite SQ30 samples in comparison to the ferrite-martensite DP microstructures of shorter-time treated SQ samples (Figure 3c). This is interesting to emphasize that an abnormal high ferrite hardness occurred in the SQ30 heat-treated samples containing the maximum level of 15% ferrite with lower carbon concentration in association with the remaining 85% soft pearlite regions. These results indicate that the abnormal ferrite hardening cannot be related to the solid solution hardening effect of carbon and ferrite-martensite interaction, and that this ferrite hardening phenomenon can be associated to a higher redistribution of Si and Cr atoms within ferrite as shown in Figure 9. The Si and Cr concentrations within the central regions of ferrite grains increased by gentle slopes comprising 1.39 to 1.51 and 1.01 to 1.09 EDSNs, respectively, as the SQ holding time increased from 15 to 30 min, while the Si and Cr concentrations were almost constant for the central martensite areas. Therefore, intense solid solution hardening effects of Si and Cr atoms would give rise to the greater hardening of resultant ferrite crystals in the prolonged-time treated SQ30 samples (Figures 3c and 9).

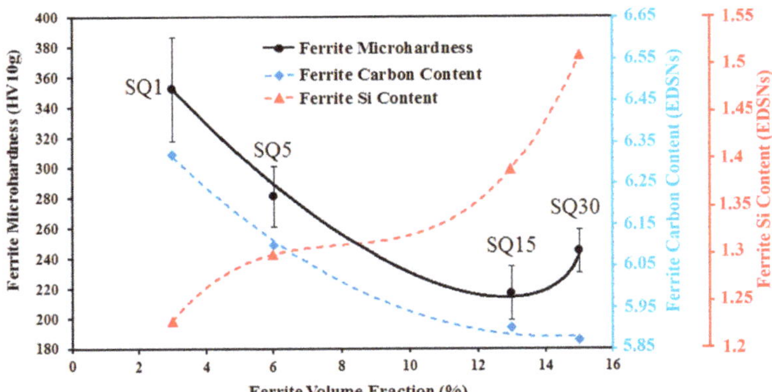

Figure 9. Changes in ferrite microhardness, ferrite carbon content and ferrite Si content as a function of progress of ferrite formation for the various SQ heat treated samples.

In addition to a greater solid solution hardening effect caused by relatively higher carbon concentration in ferrite areas adjacent to the ferrite-prior austenite interfaces, the corresponding higher ferrite hardness can also be in part related to the prior austenite to martensite phase transformation causing localized higher residual compressive stresses within adjacent ferrite regions. This is due to the generation of a high dislocation density within ferrite, following water quenching from 720 °C to room temperature. The accommodation of compressive residual stresses generated during prior austenite to martensite phase transformation can also be related to the generation of a greater level of mobile dislocation in the vicinity of ferrite areas leading to a greater magnitude of ferrite microhardness that increases from the central position toward the interfacial ferrite regions [35–37]. Therefore, besides uneven partitioning of a greater concentration of carbon within ferrite region adjacent to martensite, the mutual ferrite-martensite interaction generating a considerable density of geometrically necessary dislocations within ferrite can be also considered to be responsible in part for the higher hardening of ferrite area close to the ferrite-martensite interfaces.

4. Conclusions

Ferrite hardening alteration was studied in a medium silicon low alloy commercial grade of 35CHGSA steel under ferrite-martensite/ferrite-pearlite microstructures in relation to possibility of carbon, Si, Cr and Mn partitioning between ferrite and prior austenite microphases over a wide range of ferrite volume fractions. The conclusions are as followings:

1. The prior austenite to ferrite phase transformation has proceeded consistently with increase in SQ holding time at 720 °C. By increasing of SQ holding time from 1 to 30 min, the volume fraction of ferrite increased from 3% to a maximum value of 15%, respectively.
2. Both ferrite-martensite and ferrite-pearlite DP microstructures were realized during SQ holding time extending over 30 min at 720 °C. For SQ holding time lower than 15 min, only the ferrite-martensite DP microphases formed in the microstructures, while a mixture of ferrite and pearlite microphase constituents were realized in the prolonged-time treated SQ30 samples.
3. The ferrite hardening is completely variable with volume fraction of ferrite in the SQ samples. At first, the average ferrite microhardness sharply decreased from 352 to 217HV5g with the increase in volume fraction of ferrite from 3 to 13% under ferrite-martensite DP microstructures, respectively. Then, the average ferrite microhardness was abnormally higher from its lowest value of 217HV5g corresponding to the ferrite volume fraction of 13%, to 245HV5g with a marginal increase in ferrite volume

fraction to 15%, but with remaining fraction comprising of pearlite in SQ30 heat-treated samples.

4. A significant alteration in ferrite hardening also occurred within a given ferrite grain of a particular ferrite-martensite DP microstructure. The ferrite microhardness increased from 122 to 145HV1g with increasing distance from the central ferrite areas toward the ferrite-martensite interfaces of coarse ferrite grains realized in the SQ5 samples.

5. In contrast to almost constant Mn content of ferrite, the average Si and Cr concentrations for ferrite grains increased along gentle slopes from 1.23 to 1.51 and 0.83 to 1.09 EDSNs, respectively, with SQ holding time increasing from 1 to 30 min, respectively. The further intense solid solution hardening effects of Si and Cr would give rise to the abnormally greater hardening of resultant ferrite grains in the prolonged-time treated SQ30 samples.

6. The carbon concentration of ferrite is completely variable depending on the progress of ferrite formation. The average ferrite carbon concentration diminished from 6.32 to 5.90EDSNs with raising SQ holding time from 1 to 15 min, respectively. The average carbon concentration has also increased from 5.13 to 7.25EDSNs as the indentation location was moved from the central locations of ferrite grains towards the regions adjacent to the ferrite-prior austenite interfaces of SQ5 samples. The higher carbon concentration can be related to the more solid solution hardening of ferrite.

7. The residual compressive stresses decreased from 971 to 382 MPa with the increase in SQ holding time from 1 to 30 min at 720 °C. The higher residual compressive stresses of short time treated SQ specimens are associated in part to the higher ferrite hardening of large martensite containing DP microstructures.

Author Contributions: Conceptualization, S.S.G.B., S.A.S. and M.C.S.; methodology, S.S.G.B., S.A.S. and M.C.S.; software, A.K.; validation, A.K.; formal analysis, A.K. and S.S.G.B.; investigation, A.K.; resources, A.K., S.S.G.B., S.A.S. and M.C.S.; data curation, A.K. and S.S.G.B.; writing—original draft preparation, A.K. and S.S.G.B.; writing—review and editing, A.K., S.S.G.B. and M.C.S.; visualization, A.K. and M.C.S.; supervision, A.K.; project administration, S.S.G.B.; funding acquisition, S.S.G.B. and M.C.S.; All authors have read and agreed to the published version of the manuscript.

Funding: This research received no external funding.

Acknowledgments: One of the authors, M.C. Somani, would like to express his gratitude to the Academy of Finland for funding this research under the auspices of the Genome of Steel (Profi3) through project #311934.

References

1. Gaggiotti, M.; Albini, L.; Di Nunzio, P.E.; Di Schino, A.; Stornelli, G.; Tiracorrendo, G. Ultrafast Heating Heat Treatment Effect on the Microstructure and Properties of Steels. *Metals* **2022**, *12*, 1313. [CrossRef]
2. Abedini, O.; Behroozi, M.; Marashi, P.; Ranjbarnodeh, E.; Pouranvari, M. Intercritical heat treatment temperature dependence of mechanical properties and corrosion resistance of dual phase steel. *Mater. Res.* **2019**, *22*, 1–10. [CrossRef]
3. Alipour, M.; Torabi, M.A.; Sareban, M.; Lashini, H.; Sadeghi, E.; Fazaeli, A.; Habibi, M.; Hashemi, R. Finite element and experimental method for analyzing the effects of martensite morphologies on the formability of DP steels. *Mech. Based Des. Struct. Mach.* **2020**, *48*, 525–541. [CrossRef]
4. Ko, Y.G.; Lee, C.W.; Namgung, S.; Shin, D.H. Strain hardening behavior of nanostructured dual-phase steel processed by severe plastic deformation. *J. Alloy. Compd.* **2010**, *504*, 452–455. [CrossRef]
5. Mulidrán, P.; Spišák, E.; Tomáš, M.; Majerníková, J.; Bidulská, J.; Bidulský, R. Impact of Blank Holding Force and Friction on Springback and Its Prediction of a Hat-Shaped Part Made of Dual-Phase Steel. *Materials* **2023**, *16*, 811. [CrossRef]
6. Schino, A.D. Analysis of phase transformation in high strength low alloyed steels. *Metalurgija* **2017**, *56*, 349–352.
7. Wang, C.; Shi, J.; Cao, W.; Dong, H. Characterization of microstructure obtained by quenching and partitioning process in low alloy martensitic steel. *Mater. Sci. Eng. A* **2010**, *527*, 3442–3449. [CrossRef]
8. Kumar, A.; Singh, S.B.; Ray, K. Influence of bainite/martensite-content on the tensile properties of low carbon dual-phase steels. *Mater. Sci. Eng. A.* **2008**, *474*, 270–282. [CrossRef]

9. Ashrafi, H.; Shamanian, M.; Emadi, R.; Saeidi, N. Correlation of tensile properties and strain hardening behavior with martensite volume fraction in dual-phase steels. *Trans. Indian Inst. Met.* **2017**, *70*, 1575–1584. [CrossRef]

10. Khajesarvi, A.; Ghasemi Banadkouki, S.S.; Sajjadi, S.A. Effect of tempering heat treatment on mechanical properties of a medium silicon low alloy ferrite–martensite DP steel. *Int. J. Iron Steel Soc. Iran* **2022**, *19*, 20–29.

11. Bag, A.; Ray, K.; Dwarakadasa, E. Influence of martensite content and morphology on the toughness and fatigue behavior of high-martensite dual-phase steels. *Metall. Mater. Trans. A* **2001**, *32*, 2207–2217. [CrossRef]

12. Bag, A.; Ray, K.; Dwarakadasa, E. Influence of martensite content and morphology on tensile and impact properties of high-martensite dual-phase steels. *Metall. Mater. Trans. A* **1999**, *30*, 1193–1202. [CrossRef]

13. Cai, X.-L.; Feng, J.; Owen, W. The dependence of some tensile and fatigue properties of a dual-phase steel on its microstructure. *Metall. Trans. A* **1985**, *16*, 1405–1415. [CrossRef]

14. Peng-Heng, C.; Preban, A. The effect of ferrite grain size and martensite volume fraction on the tensile properties of dual phase steel. *Acta Metall.* **1985**, *33*, 897–903. [CrossRef]

15. Kadkhodapour, J.; Schmauder, S.; Raabe, D.; Ziaei-Rad, S.; Weber, U.; Calcagnotto, M. Experimental and numerical study on geometrically necessary dislocations and non-homogeneous mechanical properties of the ferrite phase in dual phase steels. *Acta Mater.* **2011**, *59*, 4387–4394. [CrossRef]

16. Byun, Y.S.; KIM, I.S.; KIM, S.J. Yielding and Strain Aging Behaviors of an Fe-0.07 C-1.6 Mn Dual-phase Steel. *Trans. Iron Steel Inst. Jpn.* **1984**, *24*, 372–378. [CrossRef]

17. Jiang, Z.; Guan, Z.; Lian, J. The relationship between ductility and material parameters for dual-phase steel. *J. Mater. Sci.* **1993**, *28*, 1814–1818. [CrossRef]

18. Ebrahimian, A.; Banadkouki, S.G. Effect of alloying element partitioning on ferrite hardening in a low alloy ferrite-martensite dual phase steel. *Mater. Sci. Eng. A* **2016**, *677*, 281–289. [CrossRef]

19. Monia, S.; Varshney, A.; Sangal, S.; Kundu, S.; Samanta, S.; Mondal, K. development of highly ductile spheroidized steel from high C (0.61 wt.% C) low-alloy steel. *J. Mater. Eng. Perform.* **2015**, *24*, 4527–4542. [CrossRef]

20. *Astm E3-01*; Standard Guide for Preparation of Metallographic Specimens. Testing, American Society for and Materials, Astm: West Conshohocken, PA, USA, 2009.

21. Vander Voort, G.F. *Metallography and Microstructures*; ASM International Materials Park: Geauga County, OH, USA, 2004; Volume 9.

22. HE, J.; Schoenung, J. *ASTM E–562–02–Standard Test Method for Determining Volume Fraction by Systematic Manual Point*; American Society for Testing and Materials: West Conshohocken, PA, USA, 2005; Volume 3.

23. Cullity, B.D. *Elements of X-Ray Diffraction*; Addison: Wesley, MA, USA, 1978; pp. 127–131.

24. Noyan, I.C.; Cohen, J.B.; Noyan, I.C.; Cohen, J.B. Determination of strain and stress fields by diffraction methods. *Residual Stress Meas. Diffr. Interpret.* **1987**, *1*, 117–163.

25. Khajesarvi, A.; Banadkouki, S.S. Investigation of carbon and silicon partitioning on ferrite hardening in a medium silicon low alloy ferrite-martensite dual-phase steel. *Int. J. Iron Steel Soc. Iran* **2020**, *17*, 25–33.

26. Zarchi, H.R.K.; Khajesarvi, A.; Banadkouki, S.S.G.; Somani, M.C. Microstructural evolution and carbon partitioning in interstitial free weld simulated api 5l x60 steel. *Rev. Adv. Mater. Sci.* **2019**, *58*, 206–217. [CrossRef]

27. Soleimani, M.; Mirzadeh, H.; Dehghanian, C. Processing route effects on the mechanical and corrosion properties of dual phase steel. *Met. Mater. Int.* **2020**, *26*, 882–890. [CrossRef]

28. Mizubayashi, H.; Yumoto, S.; Li, H.; Shimotomai, M. Young's modulus of single phase cementite. *Scr. Mater.* **1999**, *40*, 773–777. [CrossRef]

29. Li, Y.; Li, W.; Min, N.; Liu, H.; Jin, X. Homogeneous elasto-plastic deformation and improved strain compatibility between austenite and ferrite in a co-precipitation hardened medium Mn steel with enhanced hydrogen embrittlement resistance. *Int. J. Plast.* **2020**, *133*, 102805. [CrossRef]

30. Cong, J.; Li, J.; Fan, J.; Misra, R.D.K.; Xu, X.; Wang, X. Effect of austenitic state before ferrite transformation on the mechanical behavior at an elevated temperature for seismic-resistant and fire-resistant constructional steel. *J. Mater. Res. Technol.* **2021**, *13*, 1220–1229. [CrossRef]

31. Calcagnotto, M.; Ponge, D.; Demir, E.; Raabe, D. Orientation gradients and geometrically necessary dislocations in ultrafine grained dual-phase steels studied by 2D and 3D EBSD. *Mater. Sci. Eng. A* **2010**, *527*, 2738–2746. [CrossRef]

32. Sodjit, S.; Uthaisangsuk, V. Microstructure based prediction of strain hardening behavior of dual phase steels. *Mater. Des.* **2012**, *41*, 370–379. [CrossRef]

33. Morooka, S.; Tomota, Y.; Kamiyama, T. Heterogeneous deformation behavior studied by in situ neutron diffraction during tensile deformation for ferrite, martensite and pearlite steels. *ISIJ Int.* **2008**, *48*, 525–530. [CrossRef]

34. Calcagnotto, M.; Adachi, Y.; Ponge, D.; Raabe, D. Deformation and fracture mechanisms in fine-and ultrafine-grained ferrite/martensite dual-phase steels and the effect of aging. *Acta Mater.* **2011**, *59*, 658–670. [CrossRef]

35. Jacques, P.; Furnémont, Q.; Mertens, A.; Delannay, F. On the sources of work hardening in multiphase steels assisted by transformation-induced plasticity. *Philos. Mag. A* **2001**, *81*, 1789–1812. [CrossRef]

36. Erdogan, M.; Priestner, R. Effect of epitaxial ferrite on yielding and plastic flow in dual phase steel in tension and compression. *Mater. Sci. Tech.* **1999**, *15*, 1273–1284. [CrossRef]
37. Ostash, O.; Kulyk, V.; Poznyakov, V.; Haivorons'kyi, O.; Markashova, L.; Vira, V. Fatigue crack growth resistance of welded joints simulating the weld-repaired railway wheels' metal. *Arch. Mater. Sci. Eng.* **2017**, *86*, 49–55. [CrossRef]

Article

Experimental Study on Anchoring Performance of Short-Lapped-Rebar Splices with Pre-Set Holes and Spiral Hoops

Quanwei Liu [1,2,*], Xi Liu [2], Ran Chen [2], Zhengyi Kong [1,*] and Tengfei Xiang [1]

[1] School of Civil Engineering, Anhui University of Technology, Ma'anshan 243032, China
[2] School of Civil Engineering, Chang'an University, Xi'an 710061, China
* Correspondence: lqw@ahut.edu.cn (Q.L.); zsh2007@ahut.edu.cn (Z.K.)

Abstract: The precast concrete structure has the advantage of a short construction period, less labor consumption, and less pollution. The lapped-rebar splice is a type of connection for assembled reinforced concrete shear walls in the precast concrete structure. In this study, the anchoring performance of a short-lapped-rebar splice with a corrugated metal duct and spiral hoops is investigated. A total of 30 specimens were designed considering the influences of the rebar diameter and the lapped length, and the tension testing of the splice was carried out. The results show that the specimens with 0.15 times the suggested length in GB 50010-2010 fail by the fracture of rebar, while the specimens with 0.1 times and 0.05 times the suggested length show the pull-out failure of rebar. The ultimate bond strength of specimens with the suggested length is higher than that of the conventional specimens. The stress of the anchored rebar in the short-lapped-rebar splices is distributed symmetrically along the longitudinal direction. The maximum bond stress of the anchored rebar reaches 35 MPa, which is approximately 1.4 times that in the conventional specimens. A semi-empirical model for predicting the ultimate bond strength of the short-lapped-rebar splice is proposed, and it shows good agreement with tested values; the average error estimated from the proposed model is only 4.49%.

Keywords: short-lapped-rebar splice; spiral hoop; corrugated metal duct; anchoring performance; ultimate bond strength

Citation: Liu, Q.; Liu, X.; Chen, R.; Kong, Z.; Xiang, T. Experimental Study on Anchoring Performance of Short-Lapped-Rebar Splices with Pre-Set Holes and Spiral Hoops. *Metals* **2023**, *13*, 530. https://doi.org/10.3390/met13030530

Academic Editors: Andrea Di Schino and Claudio Testani

Received: 12 February 2023
Revised: 3 March 2023
Accepted: 4 March 2023
Published: 6 March 2023

1. Introduction

Due to the application of quick-assembled rebar technology (i.e., grouted sleeve connections, metal or plastic duct connections, post-tensioned connections, and steel plate connections), the precast concrete structures were developed rapidly in China. As corrugated metal duct connections [1] and lapped connections with the spiral hoops [2] have the advantages of fast assembly, reliable connection, and convenient construction, they are widely utilized in the precast concrete structure.

Previously, corrugated metal duct connections [1] were mainly utilized for connecting the foundation and column in bridge engineering. Raynor et al. [3] first proposed a post-tensioned metal duct connection, and the experiment of metal duct connections under tension testing was performed. The results showed the metal duct connections exhibited high bearing capacity and good ductility. Matsumoto et al. [4] studied the bonding behavior of the corrugated duct connection, and a model for the anchorage length of rebar was suggested. Brenes [5] discussed the major parameters affecting the mechanical behavior of the corrugated duct connection, and a bond-stress-slip model was developed. Steuck et al. [6] evaluated the minimum anchorage length of large-size rebar in the corrugated duct connection. Galvis and Correal [7] proposed a model for the anchorage length of two or three bundles of metal ducts in the corrugated duct connection.

The corrugated duct connection was then developed in precast shear walls or frames, as shown in Figure 1 [8]. Here, a metal duct was first embedded in the formwork of the precast specimen, and the metal duct was closely connected to the embedded rebar with fixed iron wire. Then, the metal duct was bent to the outside of the formwork at the highest location for the later pouring of grout. Finally, the grout material was poured into the metal duct after the anchored rebar was inserted into the metal duct during the in-site assembly. Zhi et al. [8] experimentally investigated the seismic behavior of the corrugated duct connection with the lapped rebar splice in shear walls. Seifi et al. [9] found that the corrugated duct connection with a transverse confinement can effectively limit the development of cracks in seismic loading. In Tazarv and Saiidi [10] and Hofer et al.'s [11] cyclic testing, the good seismic behavior of the precast frames with the corrugated duct connection was exhibited.

Figure 1. Corrugated metal duct connections [8].

As discussed above, there is extensive research focusing on the mechanical behavior of the corrugated duct connection. However, the concrete near the zone of the connection is easily cracked. For this reason, Ma et al. [12] proposed a new type of lapped connection with spiral hoops and steel rod rotated holes, as shown in Figure 2. The stirrups with spiral hoops were embedded in the scope of overlapping reinforcement and were then preset in the formwork. After the pre-hardening of precast concrete, the holes with ribs were formed by pulling out the rotated steel rod. The experiment of the pull-out splice testing was then performed by Ma et al. [12] and Zhang [13], and the results showed this type of lapped connection can effectively connect precast shear walls. From the results of seismic testing, Gu et al. [14,15] pointed out the mechanical behavior of the precast shear walls with the lapped connection was similar to that of shear walls in cast. Based on Jiang et al.'s test [16], the precast shear walls with the lapped connection exhibited better ductility performance than that of shear walls in cast.

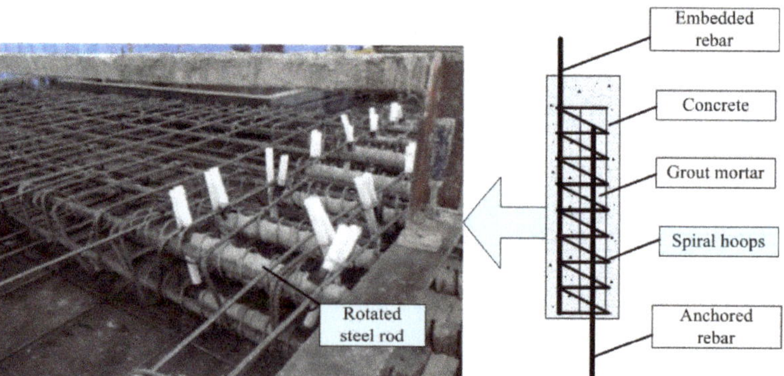

Figure 2. Lapped connection [12].

However, the spiral hoops and the steel rod rotated in the lapped connections are difficult to fix during the precast process, and the rotation of the steel rod may cause the cracking of the concrete. To solve this kind of problem, a new type of connection named the short-lapped-rebar splice is proposed in this study, as shown in Figure 3. Compared with Ma et al.'s method [12], the anchorage length in this new type of connection is shorter. In addition, the spiral hoops are welded on the embedded rebar to be fixed, and the metal duct avoiding the rotation of the steel rod is utilized. To investigate the mechanical behavior (i.e., failure mode, ultimate strength, and strain variation) of the short-lapped-rebar splices, the experiment of the short-lapped-rebar splices with the pre-set holes and spiral hoops was tested in this study. A model for predicting the ultimate bond strength of the short-lapped-rebar splices is developed.

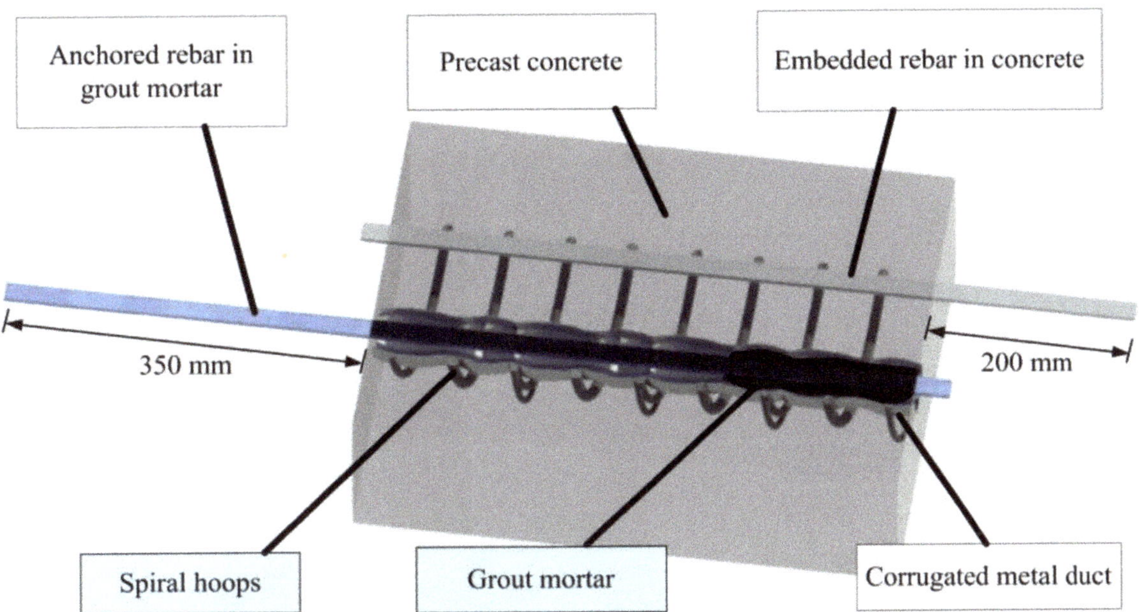

Figure 3. The short-lapped-rebar splices.

2. Experimental Program

A total of 30 specimens for the short-lapped-rebar splices were designed considering the influences of the lapped length and the rebar diameter. The pull-out tests were then carried out to investigate the bond–slip behavior of the short-lapped-rebar splices.

2.1. Design of Specimens

In this study, the dimension for the short-lapped-rebar splices was 200 mm × 200 mm × 300 mm, as shown in Figure 4. A corrugated metal duct with 300 mm in length, 40 mm in diameter, 0.2 mm in wall thickness, and 5 mm in wave height was utilized. The spiral hoops diameter was selected as 8 mm and three different diameters of rebar (i.e., 12 mm, 16 mm and 20 mm) were selected as the anchored rebar or the embedded rebar. The rebar at the anchorage end was 200 mm longer than the splices and the rebar at the loading end was 350 mm longer than the splices considering the length of the loading jack, as shown in Figure 3. The inner diameter and the spacing of the spiral hoops are 75 mm and 50 mm, respectively.

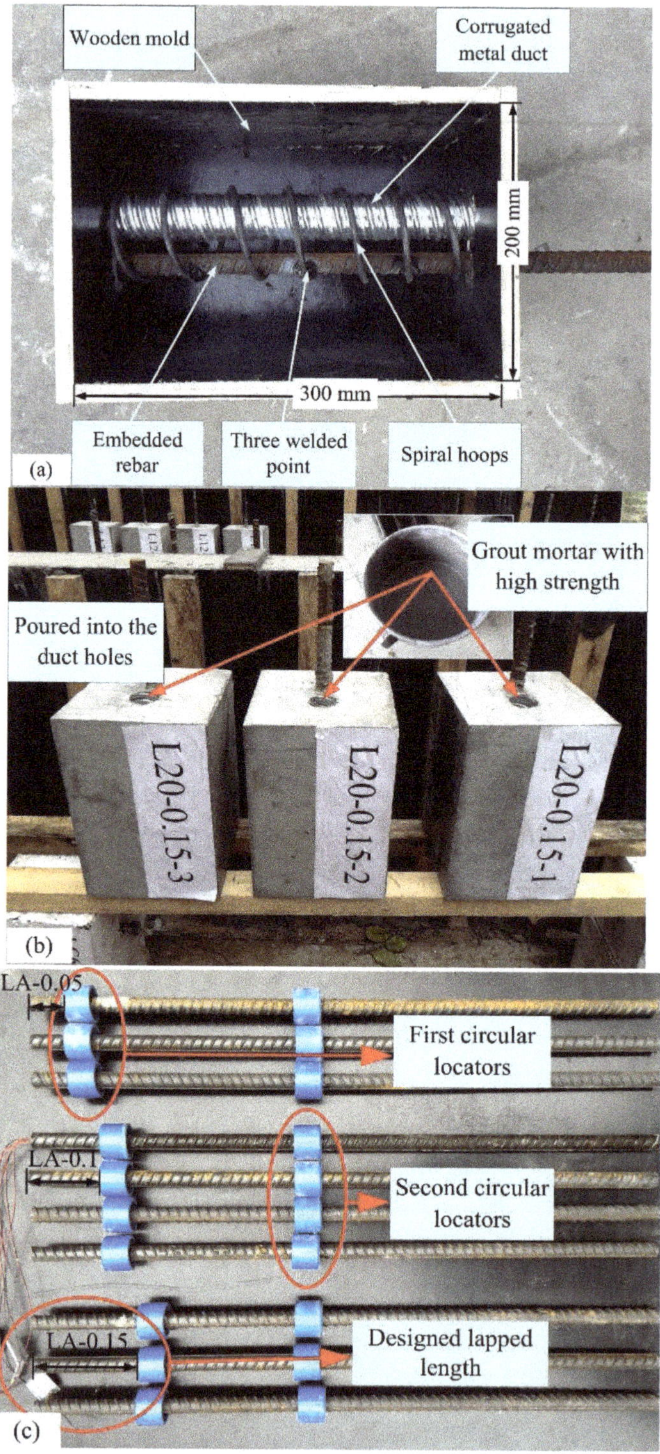

Figure 4. *Cont.*

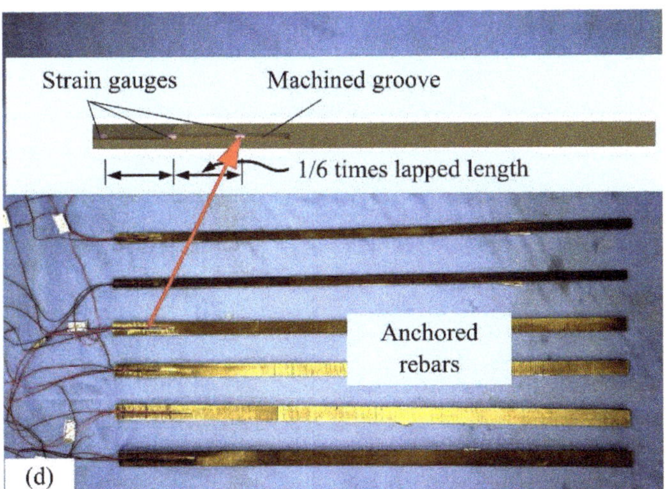

Figure 4. Precast construction process. (**a**) Mold with spiral hoops, rebar, and duct (**b**) Grout pouring (**c**) Anchored rebar with the locators (**d**) Arrangement of strain gauges.

2.2. Material Properties

According to GB17671-1999 [17], the average compression strength of CGM-340 high-strength non-shrinkage grout after 28 days was 84.33 MPa, which is established from the compression testing of three specimens, as shown in Table 1. Similarly, the average compression strength of concrete from compression testing was 59.30 MPa according to GB17671-1999 [17], as shown in Table 2. Tensile testing was also performed to obtain the material property of rebar according to GB1499.2-2007 [18], as shown in Table 3. For example, the average yield strength and the ultimate strength of rebar with the diameter of 12 mm were 430.08 MPa and 574.43 MPa, respectively.

Table 1. Compressive strength of high-strength grouted mortar.

Number	Compressive Strength (MPa)	Average Compressive Strength (MPa)
1	79.69	
2	89.06	84.33
3	84.25	

Table 2. Compressive strength of concrete.

Number	Compressive Strength (MPa)	Average Compressive Strength (MPa)
1	57.28	
2	60.24	59.30
3	60.38	

Table 3. Mechanical properties of rebar.

d/mm	Yield Strength (MPa)	Average Yield Strength (MPa)	Ultimate Strength (MPa)	Average Ultimate Strength (MPa)	Young's Modulus E ($\times 10^5$)
	447.99		598.03		
12	423.09	430.08	570.15	574.43	1.90
	419.16		555.11		
	441.22		593.57		
16	434.96	435.58	590.17	590.29	1.98
	430.56		587.15		

Table 3. *Cont.*

d/mm	Yield Strength (MPa)	Average Yield Strength (MPa)	Ultimate Strength (MPa)	Average Ultimate Strength (MPa)	Young's Modulus E ($\times 10^5$)
20	449.83 451.25 443.79	448.29	594.62 597.45 592.34	594.80	2.07

2.3. Precast Construction Process

The precast construction process of the short-lapped-rebar splices can be divided into five steps: First, a unique wooden mold was made and the metal duct and the embedded rebar were fixed on the mold, as shown in Figure 4a. Then, the spiral hoops were welded on the embedded rebar. The concrete was then poured and the mold was kept for seven days. Figure 4b shows the specimen with the pre-set duct hole after removing the mold. After that, the anchored rebar with two circular plastic locators (Figure 4c) was installed in the duct holes. Note that the first circular locator was used to ensure the designed lapped length, and the second circular locator was used to fix the anchored rebar to the center of the duct. The high-strength grouted mortar was finally poured into the duct holes and cured for 28 days.

2.4. Testing Setup

To obtain the strain variation of rebar, six strain gauges were arranged at each 1/6 of the lapped length, as shown in Figure 4d. Three displacement meters (i.e., SM1, SM2, and SM3) (Jiangsu Donghua Testing Technology Co., Ltd, Taizhou, China) for obtaining the displacements at the loading end, the anchorage end, and the specimen were also arranged, as shown in Figure 5.

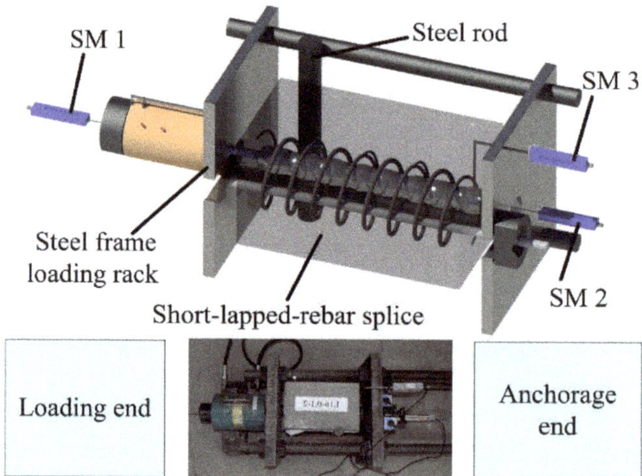

Figure 5. Testing schematic.

The tensile testing of the short-lapped-rebar splices was performed using a steel frame loading rack (Beijing Tianyijiashi International Technology Co., Ltd, Beijing, China) with the maximum bearing capacity of 1000 kN, as shown in Figure 5. The loading rate for the rebar with 12 mm in diameter was 6 kN/min and that for the rebar with the diameter of 16 mm and 20 mm was 10 kN/min. To avoid the eccentricity, two steel rods were added at both sides of the steel frame loading rack.

3. Test results and Discussion

3.1. Failure Modes

As expected, two different failure modes, namely, the fracture of rebar and the pull-out failure of rebar, were found, as shown in Table 4.

Table 4. Experimental results.

Specimen LA-B-C	Failure Mode	Yield Strength (MPa)	Ultimate Strength or Ultimate Bond Strength (MPa)
L12-0.15-1	Rebar fracture	396.41	558.83
L12-0.15-2	Rebar fracture	395.52	571.48
L12-0.15-3	Rebar fracture	398.35	579.45
L16-0.15-1	Rebar fracture	394.85	591.41
L16-0.15-2	Rebar fracture	400.08	590.42
L16-0.15-3	Rebar fracture	413.91	614.45
L20-0.15-1	Rebar fracture	446.50	601.24
L20-0.15-2	Rebar fracture	-	-
L20-0.15-3	Rebar fracture	425.57	602.01
L12-0.10-2	Rebar fracture	396.14	591.21
L12-0.10-4	Rebar fracture	378.27	467.53
L16-0.10-4	Rebar fracture	398.64	521.55
L12-0.10-1	Rebar pull-out	400.12	30.61
L12-0.10-3	Rebar pull-out	399.06	30.3
L16-0.10-1	Rebar pull-out	406.50	31.04
L16-0.10-2	Rebar pull-out	419.09	30.09
L16-0.10-3	Rebar pull-out	403.41	30.64
L20-0.10-1	Rebar pull-out	434.78	28.37
L20-0.10-2	Rebar pull-out	421.91	27.04
L20-0.10-3	Rebar pull-out	422.10	28.12
L20-0.10-4	Rebar pull-out	433.66	27.45
L12-0.05-1	Rebar pull-out	-	32.21
L12-0.05-2	Rebar pull-out	-	31.13
L12-0.05-3	Rebar pull-out	-	36.73
L16-0.05-1	Rebar pull-out	-	41.62
L16-0.05-2	Rebar pull-out	-	39.89
L16-0.05-3	Rebar pull-out	-	37.31
L20-0.05-1	Rebar pull-out	-	37.00
L20-0.05-2	Rebar pull-out	-	35.38
L20-0.05-3	Rebar pull-out	-	37.54

Note: L represents the short-lapped-rebar splices, A represents the diameter of rebar, B represents the ratio of the length of the lapped-rebar splices to the suggested length of splices (550 mm for rebar with 12 mm diameter, 750 mm for rebar with 16 mm diameter, and 940 mm for rebar with 20 mm diameter) in GB 50010-2010 [19], C represents the number of specimens (i.e., 1, 2, 3, and 4) in each group, and 4 represents the specimens with strain gauges in machined grooves, as shown in Figure 4d.

For the specimen with 0.15 times of the suggested length in GB 50010–2010 shown in Table 4 [19] (LA-0.15), the bond strength between grout and rebar was strong enough to resist the tensile loading, and the failure mode for the fracture of rebar was shown, as shown in Figure 6a. Clearly, all anchored rebar fractured at the center of splices. In addition, a wide inclined crack appeared between the anchored rebar and embedded rebar at the anchorage end of the splice, as shown in Figure 6a. This may be attributed to the small eccentricity of tensile loading. For the specimens with 0.10 times and 0.05 times the suggested length in Table 4 (LA-0.10, LA-0.05), the bond strength between grout and rebar was not strong enough to resist the tensile loading, and the failure mode for the rebar pull-out failure was shown, as shown in Figure 6b,c.

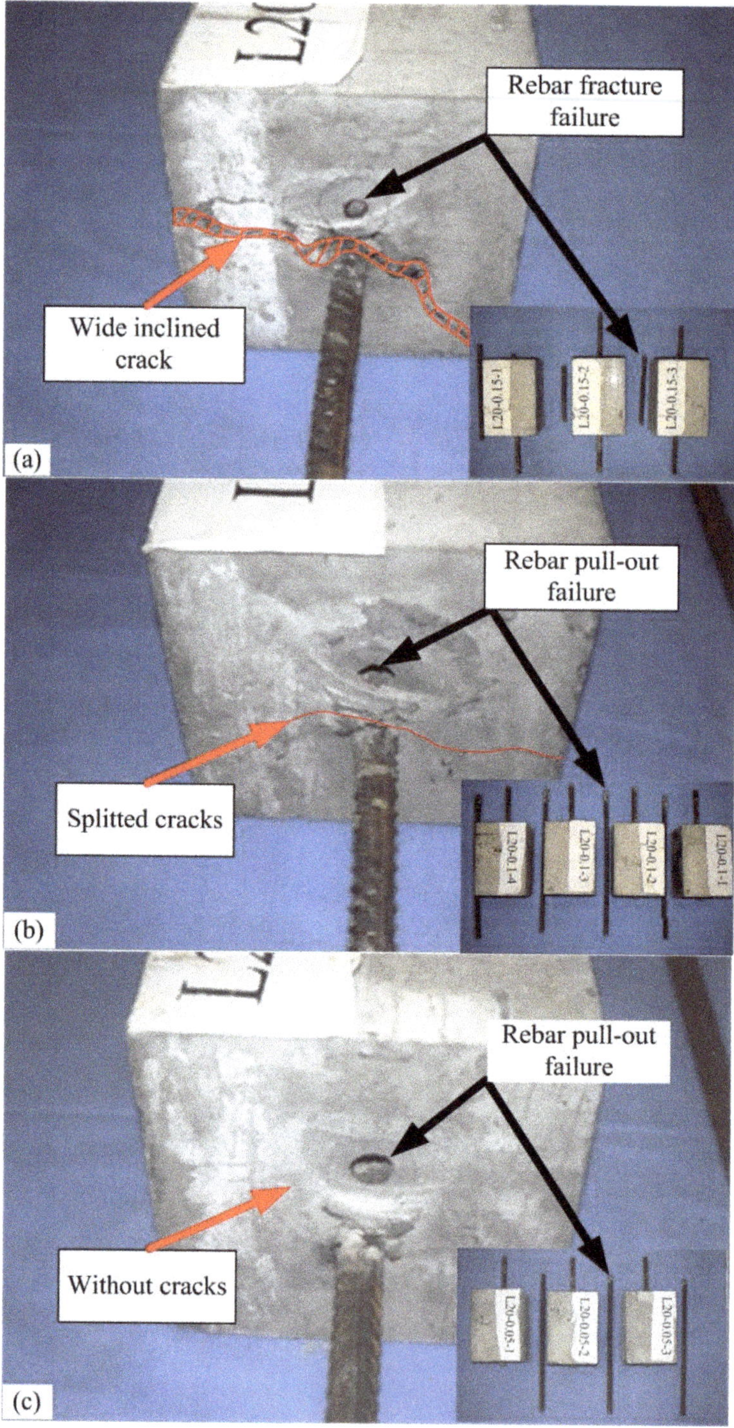

Figure 6. *Cont.*

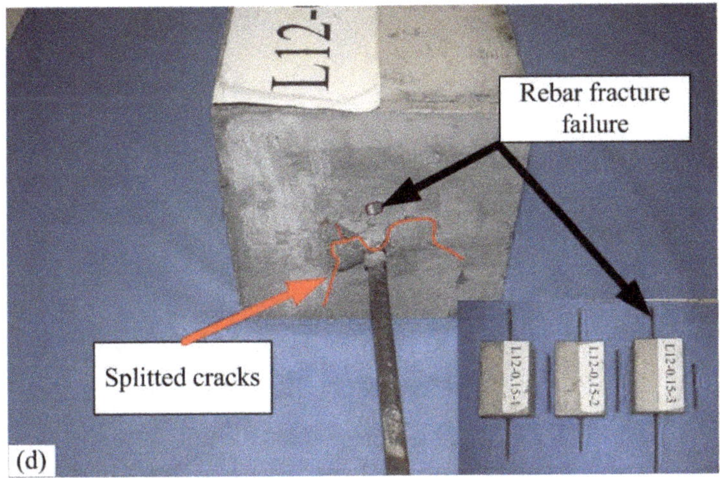

Figure 6. Failure modes of the short-lapped-rebar splices (**a**) Rebar fracture failure (L20-0.15) (**b**) Rebar pull-out failure (L20-0.1) (**c**) Rebar pull-out failure (L20-0.05) (**d**) Rebar fracture failure (L12-0.15).

In contrast, Specimen LA-0.1 or Specimen LA-0.05 shows the pull-out failure of rebar, as shown in Figure 6b,c. Clearly, the shear failure of grout mortar was shown. Note that spitted cracks were also found on the anchored rebar of Specimen LA-0.1.

As shown in Figure 6d, the rebar diameter had little effect on the failure mode in the short-lapped-rebar splices, but the development of cracks was influenced by the rebar diameter. For example, Specimen L12-0.15 shows smaller cracks than that of Specimen L20-0.15. This may be attributed to the decrease in area (diameter) and yielding forces of rebar.

3.2. Ultimate Strength

Figure 7 shows the load–displacement curves of the short-lapped-rebar splices under tensile loading. Clearly, Specimen L20-0.15 (Figure 7c) showed an obvious yielding stage and hardening stage and it had higher strength than Specimen L20-0.10 and L20-0.05. In contrast, the yielding stage was not exhibited for Specimen L20-0.05, and it fractured at an early stage due to the limited bond strength between grout and rebar. Table 4 lists the yield strength and the ultimate strength, including the ultimate bond strength of the short-lapped-rebar splices. Compared with the conventional splices without high strength grout [20] or transverse spiral hoops [21], the ultimate strength of the short-lapped-rebar splices was significantly improved.

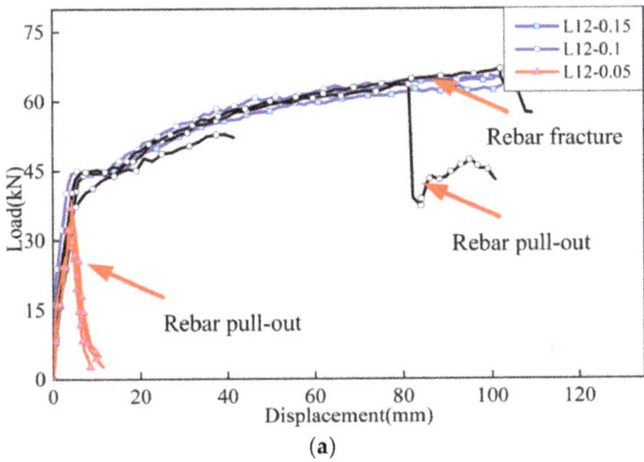

Figure 7. *Cont.*

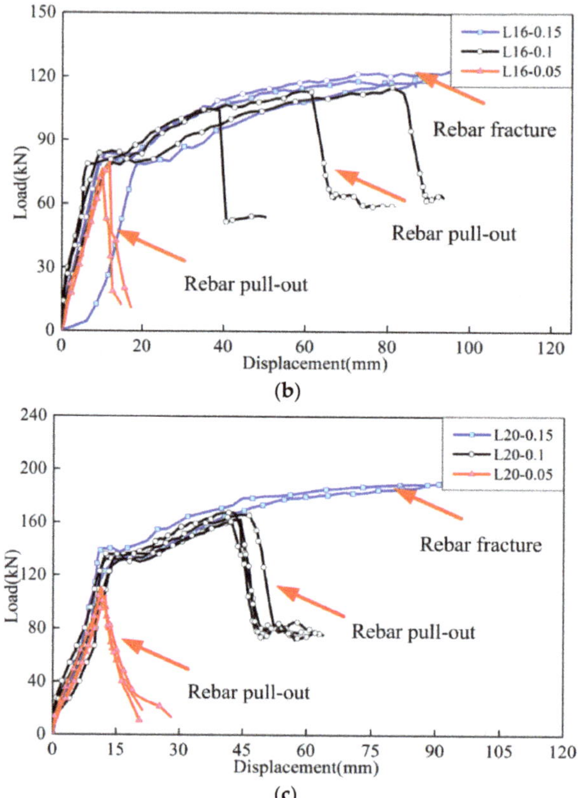

Figure 7. Load–displacement curves. (**a**) L12 (**b**) L16 (**c**) L20.

As listed in Table 4, the effect of the rebar diameter on the ultimate strength of the short-lapped-rebar splices is not significant. For example, the ultimate strength of Specimen L20-0.15-1 is 1.12 times of that of Specimen L12-0.15-1 when the rebar diameter increases from 12 mm to 20 mm.

3.3. Strain Variation of Rebar

Strain variations in the anchored rebar along the longitudinal direction at different loading levels are depicted in Figure 8. Here, x_i is the distance from the loading end to the anchorage end. Clearly, the strain of the anchored rebar increases from the loading end to the anchorage end, as shown in Figure 8. A similar phenomenon is also found in Kang's test [22], as shown in Figure 8.

According to Xu's method [23], the average bond stress τ_i between the grout mortar and the anchored rebar can be established from the strain of the anchored rebar. Figure 9 illustrates the distribution of the average bond stress τ_i along the longitudinal direction at different loading levels. Compared with the stress distribution of the anchored rebar in conventional specimens shown in Xu's test [23], the stress of the anchored rebar in the short-lapped-rebar splices is distributed symmetrically along the longitudinal direction, as shown in Figure 9. In contrast, the stress of the anchored rebar in conventional specimens is relatively high at the anchorage end. In addition, the bond stress of the short-lapped-rebar splices is much higher than that of the conventional specimens shown in Xu's test [23] and GB50010-2010 [19].

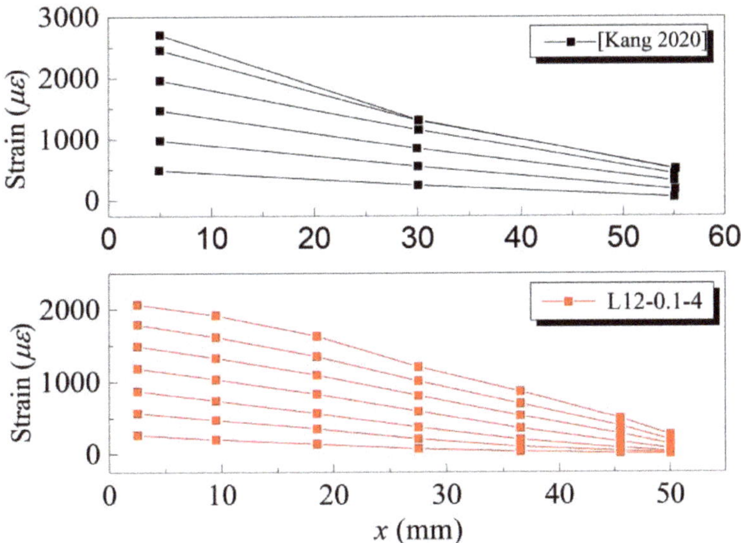

Figure 8. Strain variations of x_i–average strain at different loading levels [22].

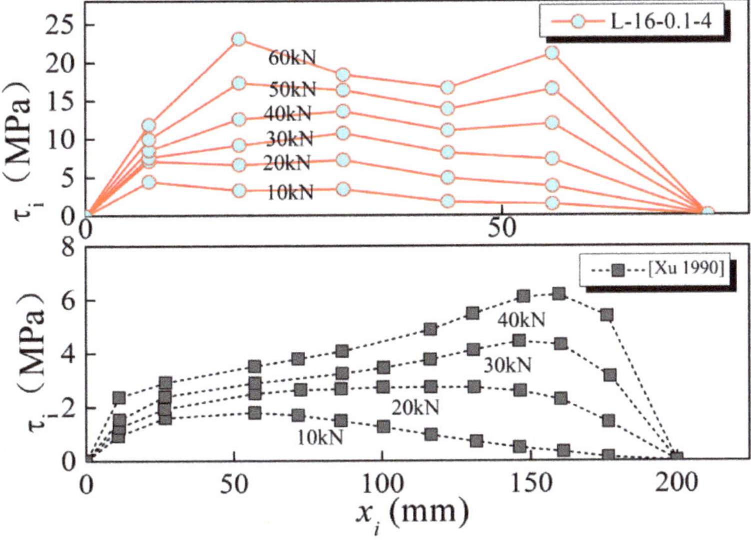

Figure 9. Strain distributions of $x_i\tau_i$ under different loading levels [23].

3.4. Bond Slip Behavior

On the basis of the values of SMs, the slip displacement s between the grout mortar and the anchored rebar can be obtained from Equation (1).

$$s = (S_{SM1} - S_{SM2}) - (S_{SM2} - S_{SM3}) \tag{1}$$

where $SM1$ is the displacement of the anchored rebar in the loading end, $SM2$ is the displacement of the anchored rebar in the anchorage end, and $SM3$ is the displacement of the short-lapped-rebar splices.

The total bond stress between the grout mortar and the anchored rebar is the sum of the bond stress τ_i from the loading end to the anchorage end, as shown below.

$$\tau = \frac{P}{\pi d l_l} \tag{2}$$

where P is the axial load applied to the loading end of the anchored rebar, d is the diameter of the anchored rebar, and l_l is the lapped length of the anchored rebar.

Figure 10 shows the relationships between the total bond stress τ and the slip displacement s. Clearly, the total bond stress increases linearly with the increase in the slip displacement. The maximum bond stress (the ultimate bond strength) reaches 35 MPa, which is approximately 1.4 times that in conventional specimens [23,24].

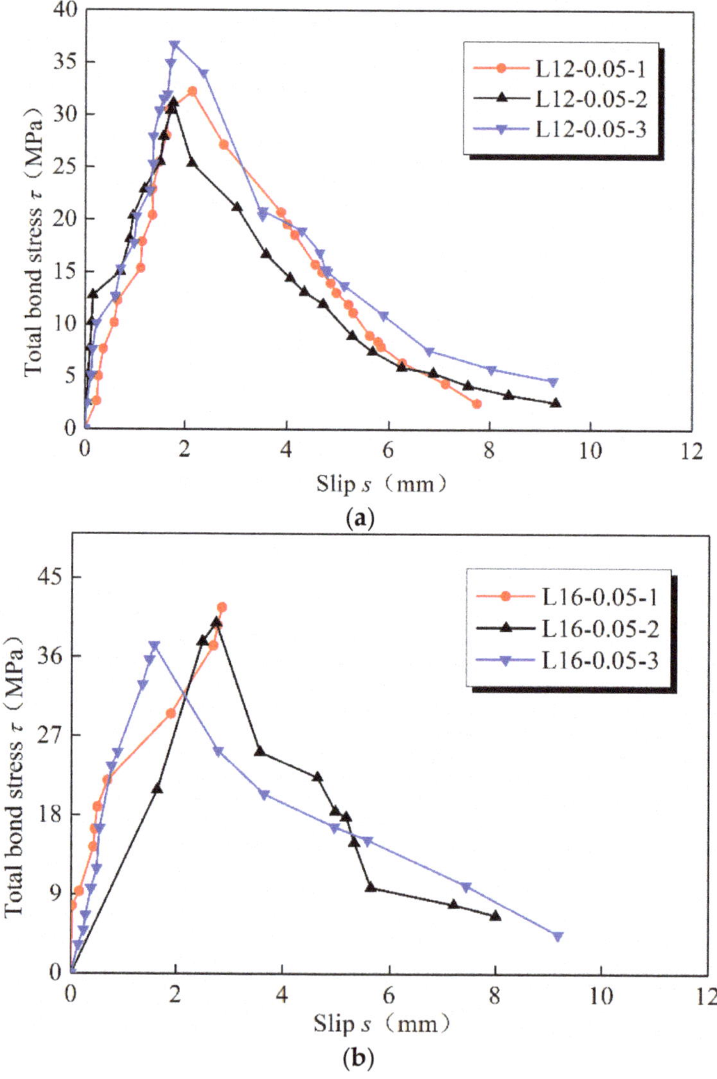

Figure 10. *Cont.*

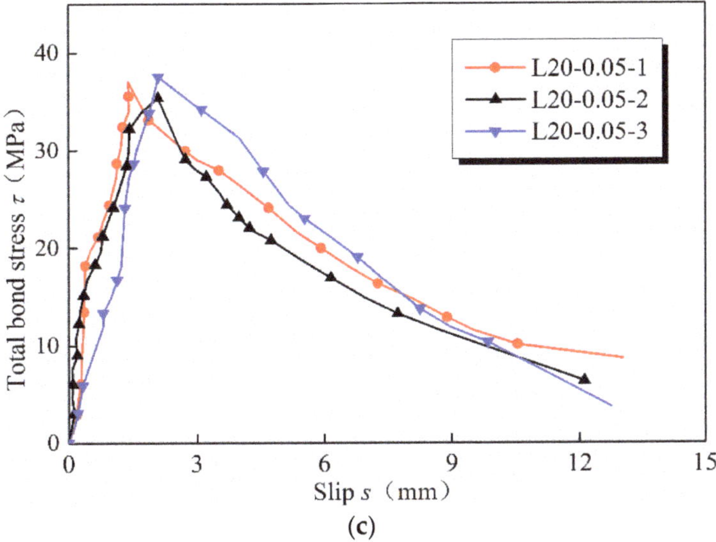

Figure 10. Bond and slip relationships in Specimens. (**a**) L12-0.05 (**b**) L16-0.05 (**c**) L20-0.05.

4. Models for the Ultimate Bond Strength

4.1. Current Models

Previously, variety formulas for the ultimate bond strength of anchoring or lapping rebar were provided in GB 50010-2010 [19], AS-3600 [25], ACI 318-05 [26], and Wu [27], as shown in Equations (3)–(6).

$$\tau_u = (0.82 + 0.9\frac{d}{l_a})(1.6 + 0.7\frac{c}{d} + 20\rho_{sv})f_t \tag{3}$$

$$\tau_u = 0.265(\frac{c}{d} + 0.5)\sqrt{f_u} \tag{4}$$

$$\tau_u = 0.083(1.2 + 3\frac{c}{d} + 50\frac{d}{l_a})\sqrt{f_u} \tag{5}$$

$$\tau_u = (0.36 + 30.81\frac{d}{l_s})(2.48 - 6.2\frac{d}{D} + 46.9\rho_{sv})f_t \tag{6}$$

where τ_u is the ultimate bond strength, d is the rebar diameter, l_a is the anchorage length, l_s is the lapped length, c is the concrete cover thickness, ρ_{sv} is the spiral hoop ratio, f_u is the ultimate compressive strength of concrete or grout mortar, and f_t is the ultimate tensile strength of concrete or grout mortar, which can be established from Equation (7) [28].

$$f_t = 0.26f_u^{2/3} \tag{7}$$

The comparison of the ultimate bond strength between various models and test data is shown in Figure 11. As the high-strength grouted mortar and the spiral hoop effect is not considered in AS-3600 [25] and ACI 318-05 [26], the estimated values from these two specifications are much smaller than the tested values, and the maximum error reaches 20%. Wu's model [27] shows obvious difference with the test data as his model is developed based on the experiment of the long-lapped-rebar splices. GB 50010-2010 [19] is a little underestimated as the maximum error reaches −23%. For predicting the behavior of the short-lapped-rebar splice, it is necessary to develop a more accurate model for the ultimate bond strength.

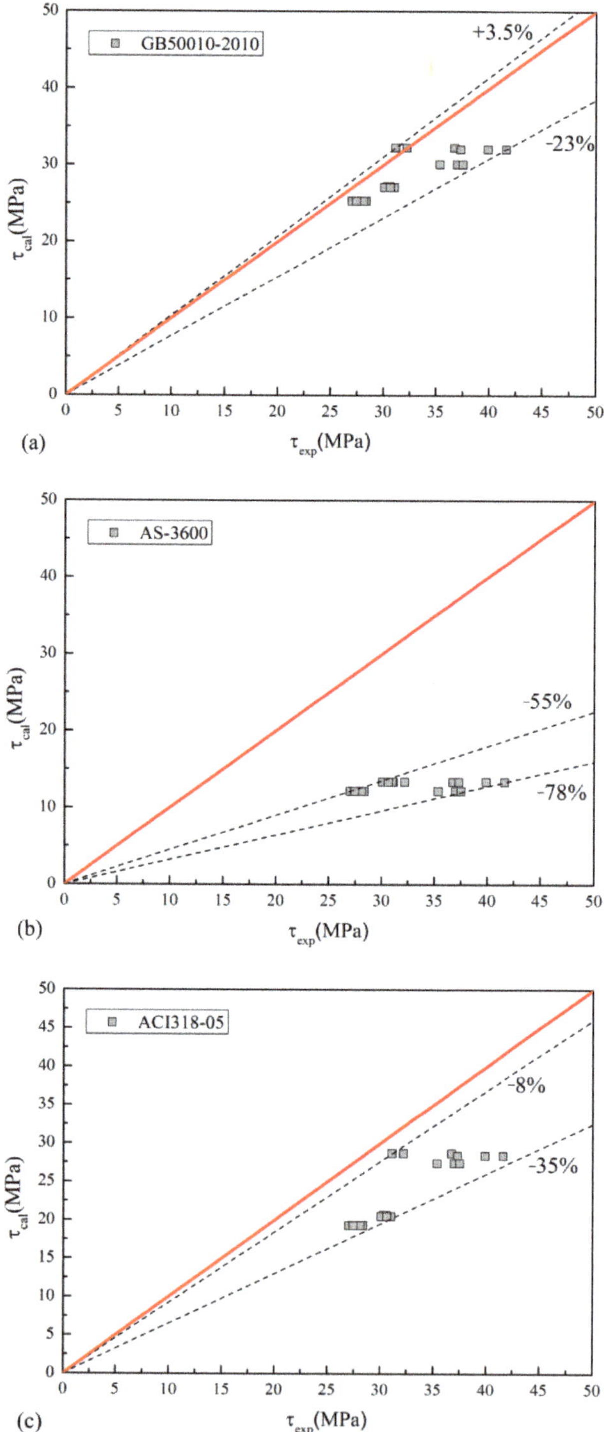

Figure 11. *Cont.*

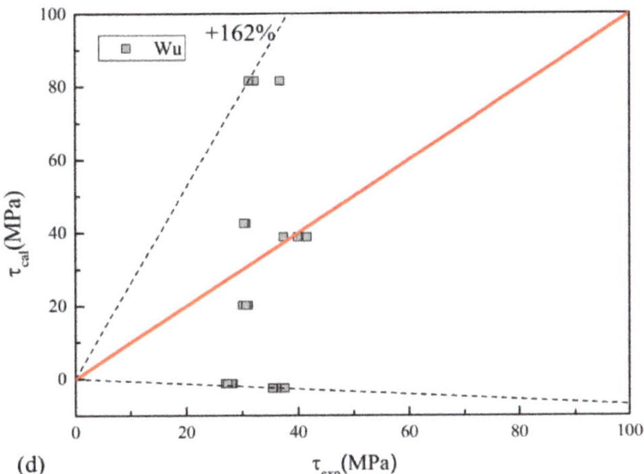

(d)

Figure 11. Comparison of ultimate bond strength between various models and test data. (**a**) GB50010-2010 [19] (**b**) AS-3600 [25] (**c**) ACI 318-05 [26] (**d**) Wu [27].

4.2. A Semi-Empirical Model for the Ultimate Bond Strength

Figures 12–14 illustrate the mechanical mechanism of the short-lapped-rebar splice. Similar to conventional splices, the shear force is carried by the friction force, mechanical interlocking force, and chemical cemented force. However, in conventional splices, cracks easily appear on the concrete near the zone of the connection. In the short-lapped-rebar splice, the concrete is constrained by the spiral hoops, which can effectively limit the development of cracks and increase the ultimate bond strength. As illustrated in Figure 12, the shear force is transferred from the anchored rebar to the embedded rebar through grout, metal duct, and concrete.

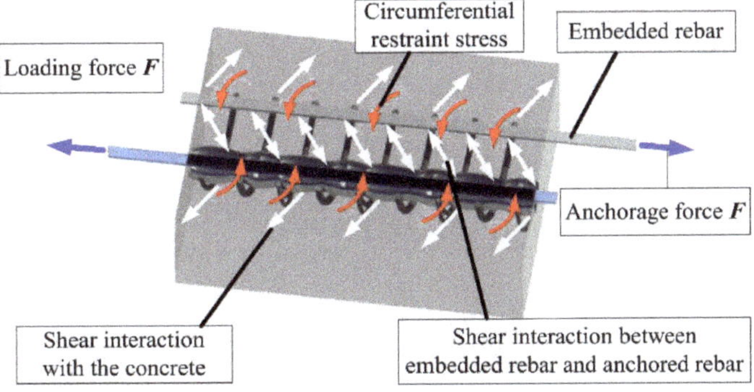

Figure 12. Bond mechanism of the short-lapped-rebar splices.

Figure 13 illustrates the stress distribution of the interaction surface between the anchored rebar and the grout mortar along the longitudinal direction. Under the action of external anchorage force F, the interaction surface between the rebar rib and grout mortar is subjected to extrusion stress p and friction stress μp. On the basis of equivalent conditions, the stress in the horizontal and circumferential direction can be obtained, as shown in Equation (8).

$$\begin{cases} \tau = p \sin \alpha + \mu p \cos \alpha \\ q = p \cos \alpha - \mu p \sin \alpha \end{cases} \tag{8}$$

where τ is the bond strength between the rebar and grout mortar; q is the circumferential compressive stress; p is the extrusion stress between the rebar and the grout mortar; μ is the friction coefficient between concrete and rebar, which is 0.3; and α is the inclined angle between rebar rib and grout mortar, which is 45 degrees [19,29].

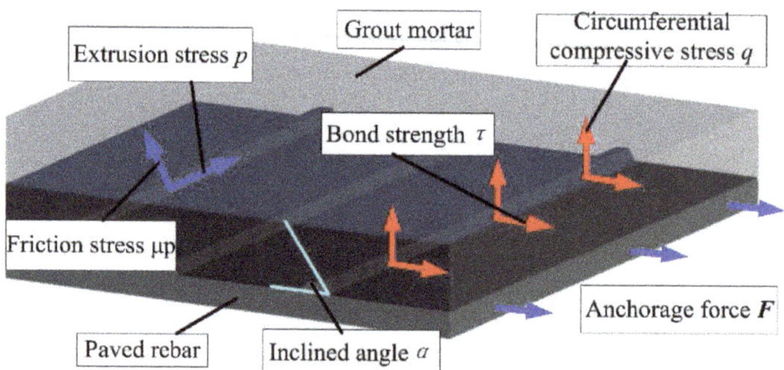

Figure 13. Stress distribution of the interaction surface.

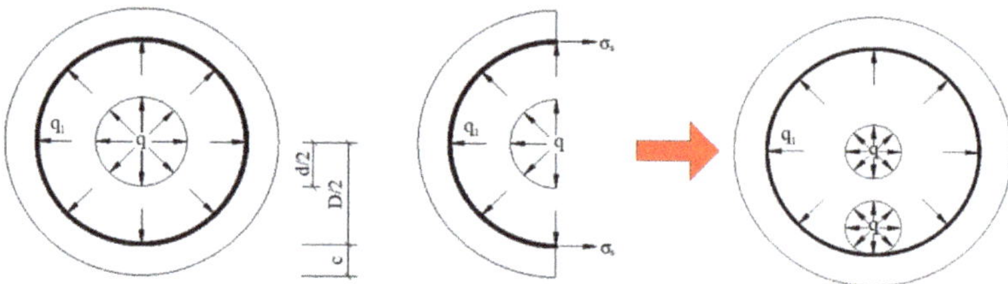

Figure 14. Stress distributed on the cross section.

The stress distributed on the cross section of the short-lapped-rebar splice is shown in Figure 14. Note that the assumption that the stress and cracks only spread to the outer bound of the spiral hoop is made. The constraining effect of the spiral hoop can be considered as a thick-walled cylinder with two rebar subjected to uniformly distributed stress, as shown in Figure 14. Based on the thick-walled cylinder theory [29], the circumferential tensile stress of the grout mortar or concrete at a certain point can be obtained from Equation (9).

$$\begin{cases} q\pi d = \frac{1}{2}q_1\pi D \\ \sigma_\theta = \frac{q_1(D/2)^2}{(c+D/2)^2-(D/2)^2}\left[1 + \frac{(c+D/2)^2}{r^2}\right] \end{cases} \tag{9}$$

where q_1 is the compressive stress of the grout mortar, c is the protective layer of concrete, D is the inner diameter of the spiral hoop, and r is the distance from the certain point to the center of specimen, σ_θ is the circumferential tensile stress at the certain position.

Similar with Xu's method [30], a constraint coefficient, β, is utilized to consider the constraint effect of the spiral hoop, as shown below.

$$\sigma_\theta|_{r=D/2} = \beta f_t \tag{10}$$

where β is the constraint coefficient of the spiral hoop, and it is suggested to be 1.2, according to Xu's test [30].

On the basis of Equations (7)–(10), the ultimate bond strength can be obtained, as shown below.

$$\tau_u = 1.12\frac{D}{d}\frac{(c+D/2)^2-(D/2)^2}{(c+D/2)^2+(D/2)^2}f_t \tag{11}$$

As shown in Figure 15 and Table 5, there is a significant error between theoretical and tested values. The reason for that is that the influences of the lapped length, the rebar diameter, and the protective layer are not fully considered. As a result, another coefficient for considering the effects of the lapped length, the rebar diameter, and the protective layer is proposed, as shown below.

$$\tau'_u = \eta \tau_u \tag{12}$$

where τ_u is the ultimate bond strength, and η is the affecting coefficient considering the effects of the lapped length, the rebar diameter, and the protective layer.

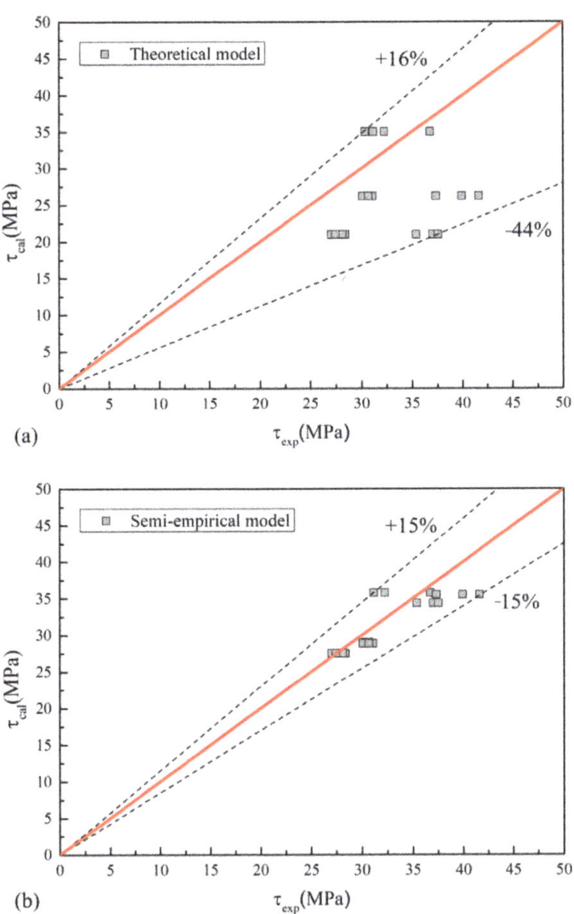

Figure 15. Comparison of the ultimate bond strength between estimated values and test data (**a**) Theoretical model (**b**) Semi-empirical model.

From the results of the test, the affecting coefficient has good relations with the rebar diameter, the ratio of the rebar diameter to the lapped length, and the ratio of the protective layer to the rebar diameter, as shown in Equation (13).

$$\eta = \begin{cases} \left(0.08\frac{d}{l_l} + 0.007\frac{c}{d} + 0.02\right)d & \frac{c}{d} \leq 5.0 \\ \left(0.08\frac{d}{l_l} + 0.055\right)d & \frac{c}{d} > 5.0 \end{cases} \tag{13}$$

where τ_u is the ultimate bond strength, d is the anchored rebar diameter, l_l is the lap length, and c is the concrete cover thickness.

Based on Equations (7)–(13), the ultimate bond strength of the short-lapped-rebar splice can be obtained. The comparison of the ultimate bond strength between the proposed models and the tested values is shown in Figure 15 and Table 5. The average error estimated from the proposed model is only 4.49%, and the maximum error varies from -15% to 15%. Clearly, the proposed model is more accurate than the existing models.

Table 5. Comparison of the ultimate bond strength between theoretical and semi-empirical values and test data.

Specimen	τ_{exp} (MPa)	f_{cu} (MPa)	l_l (mm)	d (mm)	c/d	τ_u (MPa)	τ'_u (MPa)	Error (%)
L12-0.10-1	30.61	84.33	55	12	5	35.00	30.43	0.59
L12-0.10-3	30.3	84.33	55	12	5	35.00	30.43	0.42
L12-0.05-1	32.21	84.33	28	12	5	35.00	37.50	14.10
L12-0.05-2	31.13	84.33	28	12	5	35.00	37.50	16.98
L12-0.05-3	36.73	84.33	28	12	5	35.00	37.50	2.05
L16-0.10-1	31.04	84.33	75	16	5	26.25	30.27	2.55
L16-0.10-2	30.09	84.33	75	16	5	26.25	30.27	0.58
L16-0.10-3	30.64	84.33	75	16	5	26.25	30.27	1.23
L16-0.05-1	41.62	84.33	38	16	5	26.25	37.25	11.74
L16-0.05-2	39.89	84.33	38	16	5	26.25	37.25	7.09
L16-0.05-3	37.31	84.33	38	16	5	26.25	37.25	0.17
L20-0.10-1	28.37	84.33	94	20	4.5	21.00	28.78	1.42
L20-0.10-2	27.04	84.33	94	20	4.5	21.00	28.78	6.04
L20-0.10-3	28.12	84.33	94	20	4.5	21.00	28.78	2.28
L20-0.10-4	27.45	84.33	94	20	4.5	21.00	28.78	4.61
L20-0.05-1	37	84.33	47	20	4.5	21.00	35.93	2.98
L20-0.05-2	35.38	84.33	47	20	4.5	21.00	35.93	1.52
L20-0.05-3	37.54	84.33	47	20	4.5	21.00	35.93	4.48

Note: τ_{exp} represents the values of tested data, τ_u represents the values of theoretical values, τ'_u represents the values of semi-empirical model.

5. Conclusions

In this study, a new type of connection (the short-lapped-rebar splices) in precast concrete structure was developed, and the failure modes, strain distribution, bond slip behavior, and bond strength of the connection were experimentally investigated. A semi-empirical model was proposed to predict the ultimate bond strength of the short-lapped-rebar splices. The following conclusions are made:

(1) Two different types of failure modes for the short-lapped-rebar splices, namely, the fracture of rebar and the pull-out failure of rebar, are found.

(2) The short-lapped-rebar splices have higher ultimate strength or ultimate bond strength than that of the conventional lapped splices.

(3) The stress of the anchored rebar in the short-lapped-rebar splices is distributed symmetrically along the longitudinal direction, and the maximum bond stress is approximately twice that of the conventional specimens.

(4) A semi-empirical model considering the effect of the spiral hoop is developed to predict the ultimate bond strength of the short-lapped-rebar splices, which shows good agreement with test data.

Note that the cyclic behavior of the short-lapped-rebar splices and structural reliability of the short-lapped-rebar splices are not included in this study; the determination of the lapped length of the rebar should be further studied before the design of the short-lapped-rebar splices.

Author Contributions: Conceptualization, Q.L. and X.L.; methodology, Q.L.; validation, Q.L., X.L. and R.C.; formal analysis, Q.L.; investigation, Q.L.; resources, X.L.; data curation, R.C.; writing—original draft preparation, Q.L.; writing—review and editing, Z.K.; visualization, T.X.; supervision, Z.K.; project administration, Q.L.; funding acquisition, Q.L. All authors have read and agreed to the published version of the manuscript.

Funding: This research was funded by (1) [Anhui University of Technology Foundation] grant number [QZ202015]; (2) [National Natural Science Foundation of China] grant number [52078042].

Data Availability Statement: The data presented in this study are available upon request from the corresponding author.

Acknowledgments: All the authors are very grateful to the anonymous reviewers and editors for their careful review and critical comments.

Conflicts of Interest: The authors declare no conflict of interest.

References

1. Tazarv, M.; Shrestha, G.; Saiidi, M.S. State-of-the-art review and design of grouted duct connections for precast bridge columns. *Structures* **2021**, *30*, 895–909. [CrossRef]
2. Zhi, Q.; Kang, L.; Jia, L.; Xiong, J.; Guo, Z. Seismic performance of precast shear walls prestressed via post-tensioned high strength bars placed inside grouted corrugated pipes. *Eng. Struct.* **2021**, *237*, 112153. [CrossRef]
3. Raynor, D.J.; Lehman, D.E.; Stanton, J.F. Bond-Slip Response of Reinforcing Bars Grouted in Ducts. *ACI Struct. J.* **2002**, *99*, 568–576.
4. Matsumoto, E.E.; Waggoner, M.C.; Kreger, M.E.; Vogel, J.; Wolf, L. Development of a precast concrete bent-cap system. *PCI J.* **2008**, *53*, 74–99. [CrossRef]
5. Brenes, F.J. Anchorage of Grouted Vertical Duct Connections for Precast Bent Caps. Ph.D Thesis, University of Texas, Austin, TX, USA, 2005.
6. Steuck, K.P.; Eberhard, M.O.; Stanton, J.F. Anchorage of Large-Diameter Reinforcing Bars in Ducts. *ACI Struct. J.* **2009**, *106*, 506–513. [CrossRef]
7. Galvis, F.A.; Correal, J.F. Anchorage of Bundled Bars Grouted in Ducts. *ACI Struct. J.* **2018**, *115*, 415–424. [CrossRef]
8. Zhi, Q.; Xiong, J.; Yang, W.; Liu, S. Experimental Study on Shear Transfer Performance of Uncracked Monolithic Steel Fiber Concrete. *KSCE J. Civ. Eng.* **2022**, *26*, 5210–5221. [CrossRef]
9. Seifi, P.; Henry, R.S.; Ingham, J.M. In-plane cyclic testing of precast concrete wall panels with grouted metal duct base connections. *Eng. Struct.* **2019**, *184*, 85–98. [CrossRef]
10. Tazarv, M.; Saiidi, M.S. UHPC-filled duct connections for accelerated bridge construction of RC columns in high seismic zones. *Eng. Struct.* **2015**, *99*, 413–422. [CrossRef]
11. Hofer, L.; Zanini, M.A.; Faleschini, F.; Toska, K.; Pellegrino, C. Seismic behavior of precast reinforced concrete column-to-foundation grouted duct connections. *Bull. Earthq. Eng.* **2021**, *19*, 5191–5218. [CrossRef]
12. Ma, C.; Jiang, H.; Wang, Z. Experimental investigation of precast RC interior beam-column-slab joints with grouted spiral-confined lap connection. *Eng. Struct.* **2019**, *196*, 109317. [CrossRef]
13. Zhang, H.S. Experimental Study on Plug-in Filling Hole for Lap-Joint of Steel Bar of PC Concrete Structure. Ph.D. Thesis, Harbin Institute of Technology, Harbin, China, 2009. (In Chinese)
14. Gu, Q.; Wu, R.; Ren, J.; Tan, Y.; Tian, S.; Wen, S. Effect of position of non-contact lap splices on in-plane force transmission performance of horizontal joints in precast concrete double-face superposed shear wall structures. *J. Build. Eng.* **2022**, *51*, 104197. [CrossRef]
15. Gu, Q.; Dong, G.; Ke, Y.; Tian, S.; Wen, S.; Tan, Y.; Gao, X. Seismic behavior of precast double-face superposed shear walls with horizontal joints and lap spliced vertical reinforcement. *Struct. Concr.* **2020**, *21*, 1973–1988. [CrossRef]
16. Jiang, J.; Luo, J.; Xue, W.; Hu, X.; Qin, D. Seismic performance of precast concrete double skin shear walls with different vertical connection types. *Eng. Struct.* **2021**, *245*, 112911. [CrossRef]
17. *GB17671-1999*; Method of Testing Cements-Determination of Strength. Chinese Code: Beijing, China, 1999. (In Chinese)
18. *GB1499 2-2007*; Steel for the Reinforcement of Concrete—Part 2: Hot rolled ribbed bars. Chinese Code: Beijing, China, 2007. (In Chinese)
19. *GB50010-2010*; Code for Design of Concrete Structures (English Version). China Architecture & Building Press: Beijing, China, 2009.
20. Gu, Q.; Dong, G.; Wang, X.; Jiang, H.; Peng, S. Research on pseudo-static cyclic tests of precast concrete shear walls with vertical rebar lapping in grout-filled constrained hole. *Eng. Struct.* **2019**, *189*, 396–410. [CrossRef]
21. Ma, J.; Zhu, J.; Bai, G.; Zheng, W. Experimental study on anchoring performance of steel bar-corrugated pipe grouted connection. *Structures* **2021**, *34*, 1834–1842. [CrossRef]
22. Kang, S.B.; Wang, S.; Long, X.; Wang, D.-D.; Wang, C.-Y. Investigation of dynamic bond-slip behavior of reinforcing bars in concrete. *Constr. Build. Mater.* **2020**, *262*, 120824. [CrossRef]
23. Xu, Y.L. Experimental Study of Anchorage Properties for Deformed Bars in Concrete. Ph.D. Thesis, Tsinghua University, Beijing, China, 1990. (In Chinese)
24. Elsayed, M.; Nehdi, M.L. Experimental and analytical study on grouted duct connections in precast concrete construction. *Mater. Struct.* **2017**, *50*, 198. [CrossRef]
25. *AS3600-2009*; Concrete Structures. Standards Australia: Sydney, Australia, 2009.
26. *ACI 318-05*; Building Code Requirements for Structural Concrete and Commentary. ACI Committee 318. Structural Building Code. American Concrete Institute: Farmington Hills, MI, USA, 2005.
27. Wu, Z. Study on Mechanical Properties of Prefabricated Hole with Embedded Metal Bellows for Constraint Grout-Filled Lap Connection of Steel Bar. Master's Dissertation, Xi'an University of Architecture and Technology, Xi'an, China, 2019. (In Chinese)

28. Guo, Z.H.; Shi, X.D. *Reinforced Concrete Theory and Analyse*; Tsinghua University Press: Beijing, China, 2003. (In Chinese)
29. Wu, T.; Liu, Q.W.; Cheng, R.; Liu, X. Experimental study and stress analysis of mechanical performance of grouted sleeve splice. *Eng. Mech.* **2017**, *34*, 68–75. (In Chinese)
30. Xu, F.; Wang, K.; Wang, S.; Li, W.; Liu, W.; Du, D. Experimental bond behavior of deformed rebars in half-grouted sleeve connections with insufficient grouting defect. *Constr. Build. Mater.* **2018**, *185*, 264–274. [CrossRef]

 *aerospace*

Article

Decision Science Driven Selection of High-Temperature Conventional Ti Alloys for Aeroengines

Ramachandra Canumalla [1,*,†] **and Tanjore V. Jayaraman** [2,*,†]

1   Weldaloy Specialty Forgings, Warren, MI 48089, USA
2   Department of Mechanical Engineering, University of Michigan-Dearborn, Dearborn, MI 48128, USA
*   Correspondence: r.canumalla@weldaloy.com (R.C.); tvjraman@umich.edu (T.V.J.)
†   These authors contributed equally to this work.

Abstract: Near-α Ti alloys find themselves in advanced aeroengines for applications of up to 600 °C, mainly as compressor components owing to their superior combination of ambient- and elevated-temperature mechanical properties and oxidation resistance. We evaluated, ranked, and selected near-α Ti alloys in the current literature for high-temperature applications in aeroengines driven by decision science by integrating multiple attribute decision making (MADM) and principal component analysis (PCA). A combination of 12 MADM methods ranked a list of 105 alloy variants based on the thermomechanical processing (TMP) conditions of 19 distinct near-α Ti alloys. PCA consolidated the ranks from various MADMs and identified top-ranked alloys for the intended applications as: Ti-6.7Al-1.9Sn-3.9Zr-4.6Mo-0.96W-0.23Si, Ti-4.8Al-2.2Sn-4.1Zr-2Mo-1.1Ge, Ti-6.6Al-1.75Sn-4.12Zr-1.91Mo-0.32W-0.1Si, Ti-4.9Al-2.3Sn-4.1Zr-2Mo-0.1Si-0.8Ge, Ti-4.8Al-2.3Sn-4.2Zr-2Mo, Ti-6.5Al-3Sn-4Hf-0.2Nb-0.4Mo-0.4Si-0.1B, Ti-5.8Al-4Sn-3.5Zr-0.7Mo-0.35Si-0.7Nb-0.06C, and Ti-6Al-3.5Sn-4.5Zr-2.0Ta-0.7Nb-0.5Mo-0.4Si. The alloys have the following metallurgical characteristics: bimodal matrix, aluminum equivalent preferably ~8, and nanocrystalline precipitates of Ti_3Al, germanides, or silicides. The analyses, driven by decision science, make metallurgical sense and provide guidelines for developing next-generation commercial near-α Ti alloys. The investigation not only suggests potential replacement or substitute for existing alloys but also provides directions for improvement and development of titanium alloys over the current ones to push out some of the heavier alloys and thus help reduce the engine's weight to gain advantage.

Keywords: near-α Ti alloys; aeroengine applications; multiple attribute decision making

Citation: Canumalla, R.; Jayaraman, T.V. Decision Science Driven Selection of High-Temperature Conventional Ti Alloys for Aeroengines. *Aerospace* **2023**, *10*, 211. https://doi.org/ 10.3390/aerospace10030211

Academic Editor: Sebastian Heimbs

Received: 10 February 2023
Revised: 22 February 2023
Accepted: 22 February 2023
Published: 24 February 2023

1. Introduction and Background

The selection of materials for aeroengine applications to meet the stringent requirements of high specific strength, good creep and fatigue resistance, high fracture toughness, oxidation and corrosion resistance, and so forth, is a challenge. To explore suitable materials for applications, including compressor blades, at temperatures of up to ~500 °C, an effort to select materials using Cambridge Engineering Selector (CES) software was attempted by maximizing several material performance indices, such as resistance to bending, fatigue, specific stiffness, and so on [1]. The analysis revealed titanium (Ti) alloys provide the best performance in temperatures of up to ~500 °C considering the cost and other trade-offs among the other competing alloy systems, viz., low alloy steels, stainless steels, nickel-based superalloys, etc. However, once the selection is zoomed down to Ti alloys, as per the analyses in [1], it is imperative to focus on the choice of apt Ti alloys for applications where strength-efficient structures and corrosion resistance are immanent, including aeroengines [2].

Since the beginning of the historical evolution in 1954, the high-temperature conventional Ti alloys, also known as near-α Ti alloys [3–8], are the choice class among the five different categories of Ti alloys for applications in compressor components in temperatures of up to ~600 °C in aeroengines [9,10]. The most advanced current commercial near-α

Ti alloys are IMI834 and Ti-1100, with the capability for applications up to ~600 °C [11,12]. However, several investigations have shown that low tensile ductility at room temperature is a concern, which is attributed to various reasons, such as the precipitation of silicides, silicides aided by Ti_3Al, Ti_3Al aided by silicides, Ti_3Al, etc. [3]. Therefore, alternative thermomechanical processing (TMP) and stability of the microstructures in service conditions are currently being investigated to mitigate the low tensile ductility at room temperature (that is designer specific) in near-α Ti alloys, which is critical for compressor components in aeroengines. The standard processing condition for the most current commercial alloy (IMI 834), suitable for up to ~600 °C, is typically considered the benchmark. However, generating creep, fatigue, fracture toughness, etc., obtaining data for the intended application/s on every one of those alternatives and variations become time-consuming, tedious, and expensive. Thus, to advance research and perform testing in a limited, faster, less expensive, and more sensible way, it is necessary to sort and select a few alloys, among the several alternative alloys available in the current literature, based on the important and easy to obtain room temperature tensile properties, by adopting decision science driven methods.

Material selection, a holistic approach of selecting an optimal material from a list of materials that is best suited for a given design and application, typically involves compromises between various material properties (mechanical, physical, chemical, etc.), cost, availability, environmental effects, to name a few [13]. The most common approach to material selection is Ashby's material-selection approach—popularly referred to as the materials property chart approach [13–15]. The less common techniques include multiple attribute decision making (MADM) [16–23], cost per unit property method [15,24], Pareto-optimal solutions [15], and artificial intelligence methods (e.g., neural networks) [15,25,26]. MADM refers to making preference decisions over the available alternatives (list of materials) characterized by multiple, usually conflicting attributes (i.e., properties) [22,23]. MADM techniques find applications widely in various industries, including but not limited to logistics, management, manufacturing, and so on [27]. In this paper, we compile, evaluate, sort, and select near-α Ti alloys in the current literature for high-temperature applications in aeroengines, driven by decision science integrating MADM and principal component analysis (PCA). A combination of 12 MADM methods ranks a list of 105 alloy variants based on the TMP conditions of 19 different near-α Ti alloys (the majority are 'research' alloys). PCA, a powerful tool that transforms a multi-dimensional dataset into two dimensions [28–30], consolidates the ranks from various MADMs and identifies the ten top-ranking alloy variants for the intended applications.

2. Methods

Figure 1 presents a flowchart of the decision science driven selection of near-α Ti alloys from the literature for applications in compressor parts in aeroengines. The literature data comprises 105 variants (based on the TMP routes) of 19 distinct near-α Ti alloys. The method consists of three key routines: (i) Literature data (compilation of the near-α Ti alloys), (ii) Ranking (ranking by MADM methods), and (iii) Analyses (rank consolidation by PCA and interpretation).

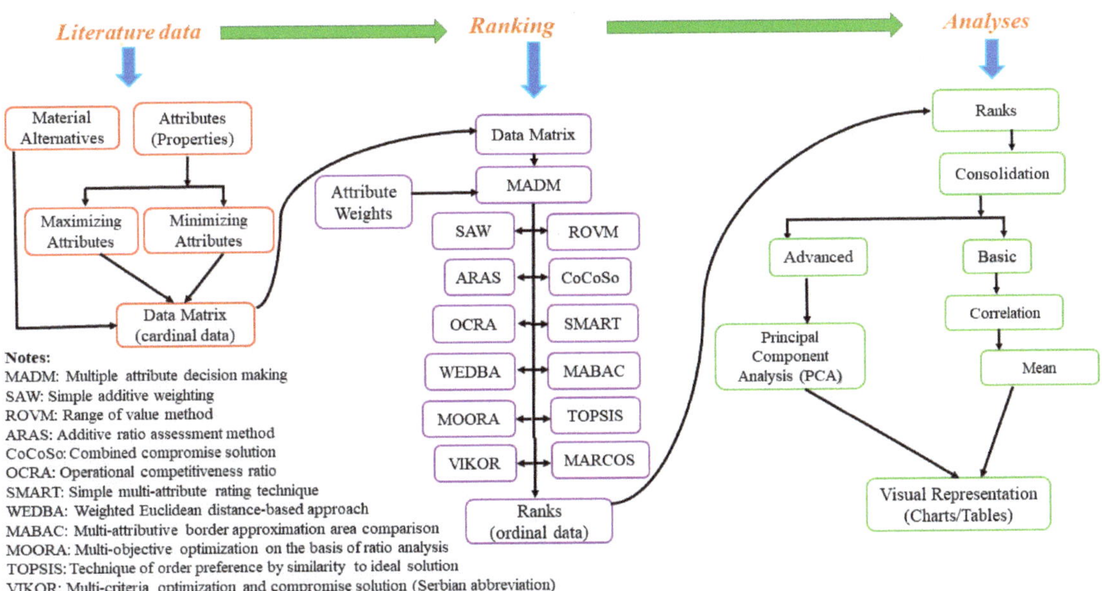

Figure 1. The flowchart of decision science driven analyses of the near-α Ti alloys. It comprises three routines: literature data, ranking, and analyses.

2.1. Literature Data

We compiled a list of near-α Ti alloys (alternatives) and their room-temperature mechanical properties (attributes) from the literature. Table A1 (in Appendix A) presents the alternatives, the near-α Ti alloys, screened for the current study primarily from peer-reviewed journals and conference proceedings [31–52]. The table presents the nominal chemistry, processing conditions, and imminent microstructure. Eleven of the above 19 alloys are 'research' alloys (viz., WJZ-Ti, KIMS, JZ1, JZ2, JL, LD-Ti423, TMC-Ti213, TKT-1, TKT-2, TKT-3, and PC), implying they were fabricated and processed on a laboratory scale (under development) followed by characterization and testing. Eight of the 19 are current commercial alloys (IMI685, IMI829, IMI834, Ti-1100, Ti6242S, TA19, TA29, and Ti60). We identified room temperature % elongation (%EL), yield strength (YS), and ultimate tensile strength (UTS) as the properties (attributes) for the current investigation. For a targeted application, such as the compressor blade, the material needs to satisfy the desired room-temperature attributes (i.e., %EL, YS, and UTS) before examining the other important attributes, namely, the high-temperature properties, including creep resistance, oxidation resistance, and corrosion resistance to optimize the alloy. In the parlance of MADM, all of the identified attributes (%EL, YS, and UTS) are maximizing (or beneficial) attributes, suggesting, for most applications, that the alloys ought to have the following combination: high %EL, high YS, and high UTS. Table A2 (in Appendix A) is the decision matrix comprising the alternatives (near-α Ti alloys) and attributes (properties %EL, YS, and UTS) in the literature [31–52].

2.2. Ranking

We evaluated the decision matrix (Table A2) by several multiple MADM methods. MADM refers to making preference decisions by evaluating and prioritizing alternatives on multiple attributes [22,23]. Distinct components of the MADM are (i) the decision matrix, which comprises the alternatives and the attributes, and (ii) attribute weights: the priorities of attributes are expressed quantitatively according to the MADM theory—they quantify the relative importance of each of the attributes [22,23,53]. The attribute weights are typical of three types [53]: (a) objective weights—based on the decision matrix utilizing mathematical models without considering the decision maker's preferences (e.g., mean weighing, standard deviation method, entropy, etc.), (b) subjective weights, based on the preference derived from the evaluations of the experts (from their previous experience) or designers (constraints of design), or both, and (c) integrated weights, as the name suggests, both objective and subjective weighting are combined to determine the weights. We adopted objective and subjective attribute weights in this investigation.

We evaluated the weights by assigning equal weights (1/3) for each of the attributes based on the understanding of these materials and their intended application. We identified twelve MADM methods to evaluate the data matrix and rank the alloys, including the simple additive weighting (SAW) [22,23,53–55], range of value method (ROVM) [56,57], additive ratio assessment method (ARAS) [58–60], combined compromise solution (CoCoSo) [61–63], operational competitiveness ratio (OCRA) [64–66], simple multi-attribute rating technique (SMART) [22,53,67,68], weighted Euclidean distance-based approach (WEDBA) [23,69,70], multi-attributive border approximation area comparison (MABAC) [71,72], multi-objective optimization on the basis of ratio analysis (MOORA) [73,74], technique of order preference by similarity to ideal solution (TOPSIS) [22,53,75,76], multi-criteria optimization and compromise solution (VIKOR)—the Serbian name is VIse Kriterijumska Optimizacija Kompromisno Resenje—method [77–79], and measurement of alternatives and ranking according to compromise solution (MARCOS) [80,81]. Each MADM approach comprises a unique mathematical aggregation procedure to rank the alternatives. The MADMs identified were diverse. Applying such distinct aggregation procedures is likely to generate a robust set of ranks of the alternatives. The ranks produced by each method, as would be expected, are likely to deviate from one another; nevertheless, the correlation among the various techniques is expected to strengthen the reliability of the results. The modus operandi was soft coded in Microsoft Excel, as formulated in the respective references of MADMs.

2.3. Analyses

The ranks obtained by various MADMs were correlated. We evaluated Spearman's correlation coefficients [82,83] among the ranks obtained from the 12 MADMs. We consolidated the ranks from various MADMs by estimating their mean and by principal component analysis (PCA). PCA, a multivariate technique, reduces the dimensionality of a dataset consisting of several interrelated variables by transforming to a new set of variables termed the principal components (PCs), which are uncorrelated and are ordered so that the first few PCs (typically one or two) retain most of the variation present in the original data [28,29]. The score plot presents a visual representation of the rank evaluation. The analyses were carried out using the commercial software Minitab® 20.

3. Results and Discussions

Figure 2 presents the ranks of the near-α Ti alloys from the literature evaluated by the 12 MADMs. The ranks of the alloys represented as points in the figure by nature are discrete; thin dotted lines for a better visual effect connect the ranks assessed by each of the MADMs. Despite the unique mathematical aggregation procedures in various MADMs, the peaks and troughs of several MADMs somewhat coincide. For example, several MADMs assign similar ranks to WJZ-Ti-2, TKT-2, WJZ-Ti-1, PC-IMDF4, and KIMS-2 (green-shaded). Moreover, the rank assigned by various MADMs to most alloys differs significantly, for instance, as in the alloys designated as Ti-1100-5, IMI834-5 and JZ2-3 (pink shaded). Table 1 presents the Spearman rank (S_ρ) that correlates ranks evaluated by the 12 MADMs. For example, the S_ρ between CoCoSo and ROVM, MABAC and WEDBA, or MARCOS and TOPSIS is >0.95, which indicates strong correlations. On the contrary, S_ρ between ARAS and SMART or TOPSIS and SMART is less than <0.3, which is expected owing to the distinct mathematical aggregation formulation in various MADMs. Out of the 66 combinations of MADM pairs, ~72% have rank correlations equal to or above 0.70, which elicits the robustness and validity of the ranking of the near-α Ti alloys. Therefore, it is imperative to consolidate the ranks obtained from various MADMs. Based on S_ρ among all various combinations of MADMs, it is practical to consolidate the rankings evaluated by the 12 different MADM evaluations. Accordingly, the mean-based (arithmetic mean) rank consolidation of Figure 2 is shown in Figure 3. The ranks of the top ten data points are WJZ-Ti-2, WJZ-Ti-1, TKT-2, TA19-2, TKT-6, TKT-1, TA19-1, KIMS-2, IMI834-2, and PC-IMDF4 in that order.

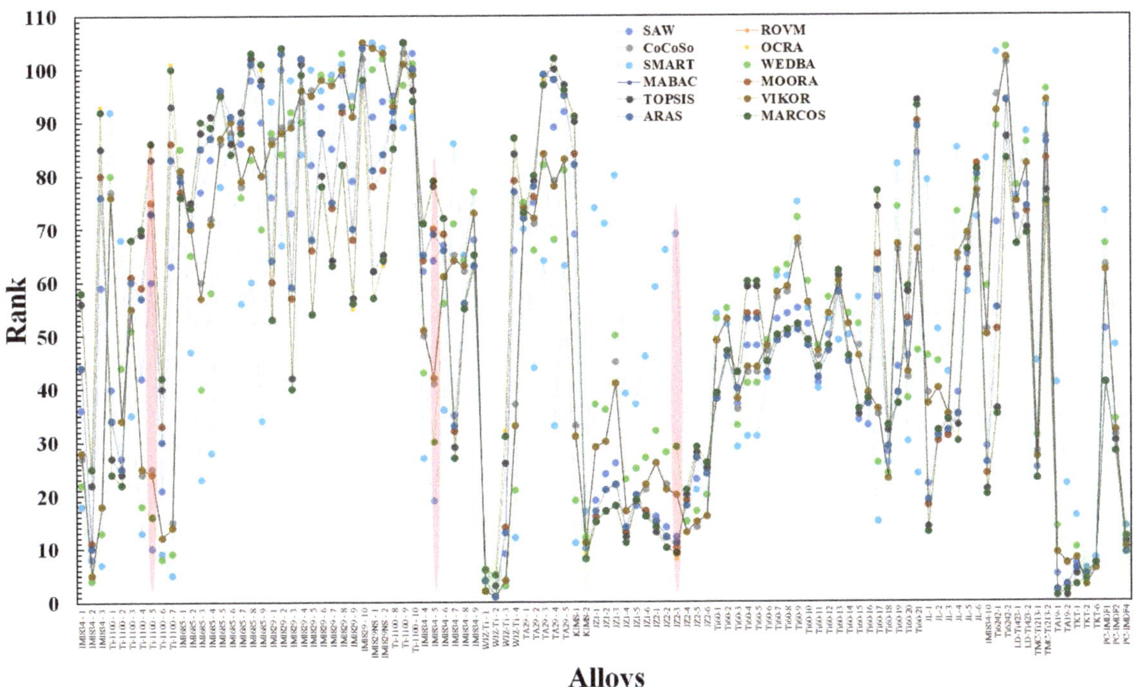

Figure 2. The ranks of the near-α Ti alloys from the literature evaluated by the 12 multiple attribute decision making (MADM) methods. Several MADMs assign relatively similar ranks (green shaded) to WJZ-Ti-2, TKT-2, WJZ-Ti-1, PC-IMDF4, and KIMS-2, while Ti-1100-5, IMI834-5, and JZ2-3 are assigned diverse set of ranks (pink shaded).

Table 1. The Spearman rank (S_ρ) correlation of the near-α Ti alloys ranks from the literature evaluated by the 12 multiple attribute decision-making (MADM) methods.

	SAW	ROVM	CoCoSo	OCRA	SMART	WEDBA	MABAC	MOORA	TOPSIS	VIKOR	ARAS
ROVM	0.902										
CoCoSo	0.903	0.999									
OCRA	0.906	0.661	0.660								
SMART	0.530	0.799	0.800	0.172							
WEDBA	0.826	0.983	0.981	0.544	0.884						
MABAC	0.902	1.000	0.999	0.661	0.799	0.983					
MOORA	0.973	0.794	0.794	0.975	0.357	0.694	0.794				
TOPSIS	0.925	0.698	0.696	0.998	0.216	0.584	0.698	0.984			
VIKOR	0.902	1.000	0.999	0.661	0.799	0.983	1.000	0.794	0.698		
ARAS	0.980	0.813	0.812	0.967	0.383	0.716	0.813	0.999	0.978	0.813	
MARCOS	0.907	0.665	0.663	1.000	0.176	0.548	0.665	0.976	0.998	0.665	0.968

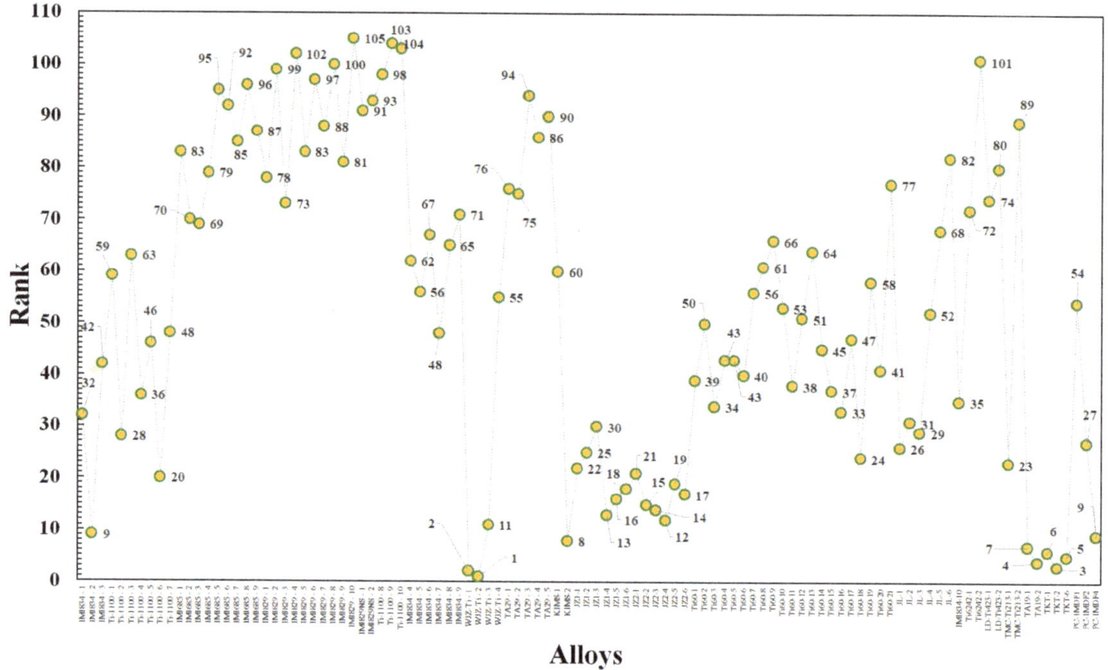

Figure 3. The arithmetic mean-based rank consolidation of the near-α Ti alloys from the literature evaluated by the 12 multiple attribute decision making (MADM) methods. The ranks of the top 10 data points are WJZ-Ti-2, WJZ-Ti-1, TKT-2, TA19-2, TKT-6, TKT-1, TA19-1, KIMS-2, IMI834-2, and PC-IMDF4 in that order.

Figure 4 is the score plot that presents the consolidated rank by PCA, of the near-α Ti alloys. It is the plot of the first two components (*PC1* and *PC2*), post-reduction of the data dimensionality (i.e., ranks from 12 MADMs) into a two-dimensional space. Table 2 presents the eigenvalues (and their proportion) that capture the variation of the distribution of each principal component. The first principal component (*PC1*) captures ~82% of the variation or scatter in the original data, while the second principal (*PC2*) describes ~17% of the variation. Since *PC1* captures nearly 82% of the variation in the initial 12 dimensions (sets of ranks), it approximates the rank of near-α Ti alloys. An imaginary reference line perpendicular to *PC1* traversing from left to right (−6 to 6) indicates the overall ranks of the near-α Ti alloys. The alloy grades WJZ-Ti, TKT-2, TA19, TKT-6, TKT-1, KIMS, IMI834, and PC top the list, followed by JZ1, JZ2, Ti-1100, and so on. The ranks of the top ten data points are WJZ-Ti-2, WJZ-Ti-1, TKT-2, TA19-2, TKT-6, TKT-1, TA19-1, KIMS-2, IMI834-2, and PC-IMDF4 in that order (the data points within the box in Figure 4), while certain variants of WJZ-Ti, JZ1, and JZ2 also appear promising (the data points close to the box). The top-ranked alloys by PCA-based consolidation are strikingly similar to the top-ranked alloys evaluated by mean-based consolidation. Specifically, the PCA-based consolidation refines the IMI834-2 (rank#9) and PC-IMDF4 (rank#9) assigned by mean-based consolidation to rank#9 and #10, respectively. Therefore, it is logical to label the score plot of PCA-based consolidated ranks as a 'rank chart'.

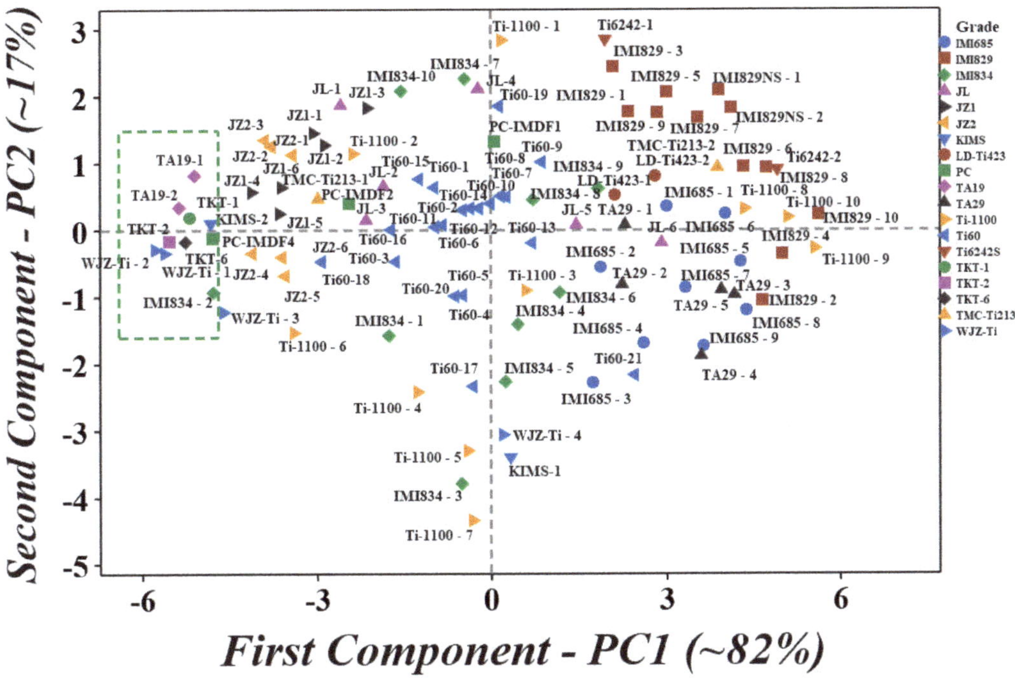

Figure 4. Score plot by principal component analysis (PCA) of the ordinal data, i.e., PCA-based rank consolidation of the near-α Ti alloys evaluated by the 12 MADM methods. The top-ranked alloy variants evaluated by PCA-based consolidation are strikingly similar to the top-ranked alloy variants evaluated by the mean-based consolidation. Specifically, the PCA-based consolidation refines the IMI834-2 (rank#9) and PC-IMDF4 (rank#9) assigned by the mean-based consolidation to rank#9 and #10, respectively.

Table 2. The eigenvalues and their proportion by the principal component analysis (PCA) of the ranks of the near-α Ti alloys from the literature by the 12 multiple attribute decision making (MADM) methods.

	PC1	PC2	PC3	PC4	PC5	PC6	PC7	PC8	PC9	PC10	PC11	PC12
Eigenvalue	9.833	2.044	0.089	0.022	0.005	0.003	0.002	0.001	0.000	0.000	0.000	0.000
Proportion	0.819	0.170	0.007	0.002	0.000	0.000	0.000	0.000	0.000	0.000	0.000	0.000
Cumulative	0.819	0.990	0.997	0.999	1.000	1.000	1.000	1.000	1.000	1.000	1.000	1.000

For deeper insight into the rank chart (Figure 4) of near-α Ti alloys, Figure 5a–d presents the score plots through the lens of various categories. Here, the region of interest (green box) corresponds to the top 10 alloy variants. Key inferences from the figures are as follows: (i) majority (seven out of 10) of the data points in the area of interest have aluminum equivalent to 8 (Figure 5a), (ii) all of the data points in the region of interest have a bimodal matrix, i.e., primary α + transformed β (Figure 5b), (iii) among the top ten data points, five (WJZ-Ti-1, TKT-2, TKT-1, TA19-2, and TA19-1) have no precipitates; one of them, WJZ-Ti-2, has precipitates Ti_3Al in αp-1 (inside primary α); one of them (TKT-6) has germanide precipitates; one has silicide precipitates (KIMS-2—Hf in silicide and no Zr); and two (IMI834-2 and PC-IMDF4) have no information regarding the precipitates (Figure 5c) based on the chemistry, thermomechanical processing and the thermal treatments, these two variants would highly likely have Ti_3Al and silicides; and lastly (iv) among the top 10 data points, five have no precipitates, four of them have nanocrystalline precipitates, and one has no information about any precipitate (Figure 5d). These analyses suggest guidelines for developing next-generation commercial near-α Ti alloys. The alloy design strategy for near-α Ti alloys for high-temperature applications with a combination of high *YS*, high *UTS*, and high *%EL* has two distinct options: (i) a combination of the aluminum equivalent to 8 and a bimodal matrix (primary α + transformed β) with no precipitates, (ii) a combination of the aluminum equivalent to 8, bimodal matrix, and nanocrystalline Ti₃Al or germanide or silicide (no Zr, but Hf, as the silicides containing Hf, do not reduce ductility, however, Hf provides solid solution strengthening [3]) precipitates in α.

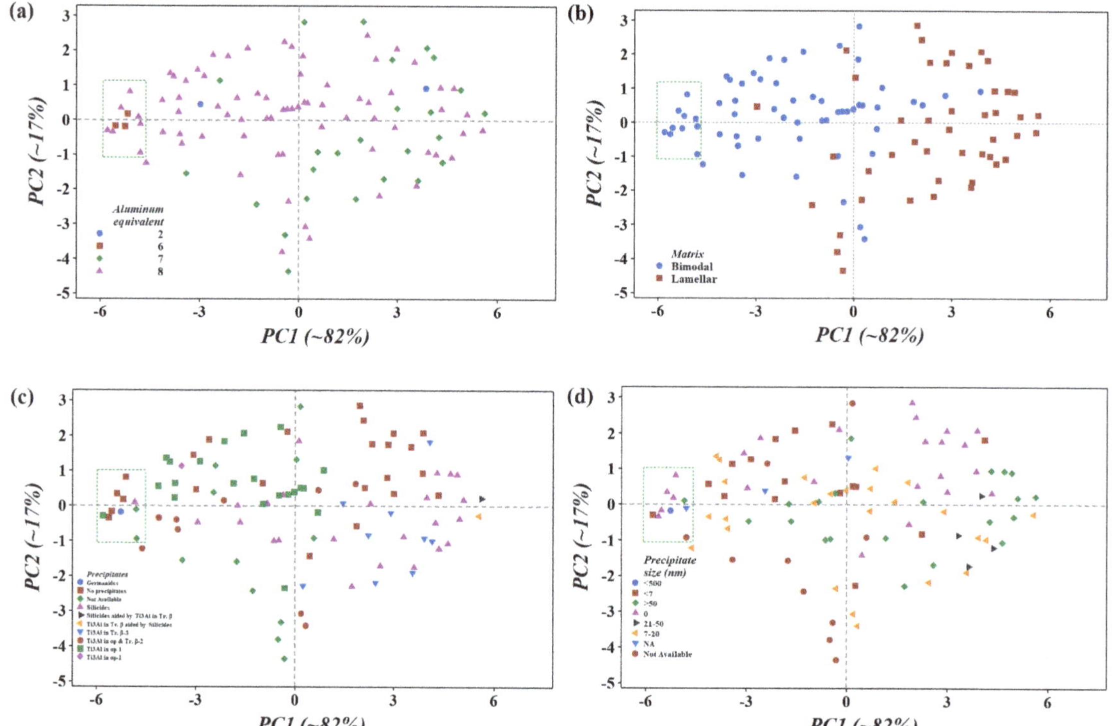

Figure 5. Score plots by the principal component analysis (PCA) of the ordinal data, i.e., PCA-based rank consolidation of the near-α Ti alloys evaluated by the 12 MADM methods through the lens of, i.e., categorized based on (**a**) aluminum equivalent, (**b**) matrix, (**c**) precipitates, and (**d**) precipitate size. The region of interest (green box) shows the top-ranked ten alloy variants.

In this investigation, we compile, evaluate, sort, and select near-α Ti alloys in the current literature for high-temperature applications in aeroengines, driven by decision science, by integrating MADM and principal component analysis (PCA). The evaluation provided valuable insight into potential existing materials ('research alloys') to focus on further research and development for commercialization. Among the top-ranked ten alloy variants (WJZ-Ti-2, WJZ-Ti-1, TKT-2, TA19-2, TKT-6, TKT-1, TA19-1, KIMS-2, IMI834-2, and PC-IMDF4), seven variants belong to the six 'research grade' alloys (WJZ-Ti, TKT-2, TKT-6, TKT-1, KIMS, and PC), while the data point IMI834-2 is a variant of an existing commercial alloy IMI834. Thus, all of these alloys appear to be strong contenders for large-scale development and testing. Additionally, in the future, newly discovered novel high-temperature Ti alloys (conventional and high-entropy alloys) can be included in the list and evaluated to assess their relative position in the rank chart and infer their potential to replace existing materials. In the near future, we plan to expand the decision science driven material selection by including several other relevant mechanical properties as they become available. Lastly, this effort (i) validates the decision science driven MADM coupled with PCA for sorting, ranking, and material selection, (ii) weeds out the alloys that need not be pursued further with time-consuming experimental studies to generate data on additional attributes that are required for use for the intended application/s, and (iii) provide directions for advancing alloys that are under development or suggest some critical improvements for possible newer alloys by providing metallurgical perspectives. Developing a methodology that applies decision science principles to compile and sort a relatively large literature data, select or identify top-ranked alloys, unearth metallurgical patterns, and recommend guidelines for developing next-generation commercial near-α Ti alloys for aeroengines is the novelty of the investigation.

4. Summary and Conclusions

We compiled, evaluated, ranked, and selected near-α Ti alloys in the current literature for high-temperature applications in aeroengines, driven by decision science by integrating MADM and principal component analysis (PCA). A combination of 12 MADM methods ranked a list of 105 alloy variants based on the thermomechanical processing (TMP) conditions of 19 different near-α Ti alloys. PCA consolidated the ranks from various MADMs and identified ten top-ranked alloy variants for the intended application/s. The ten top-ranked alloy variants are WJZ-Ti-2, WJZ-Ti-1, TKT-2, TA19-2, TKT-6, TKT-1, TA19-1, KIMS-2, IMI834-2, and PC-IMDF4 in that order and they correspond to the following eight alloys: Ti-6.7Al-1.9Sn-3.9Zr-4.6Mo-0.96W-0.23Si, Ti-4.8Al-2.2Sn-4.1Zr-2Mo-1.1Ge, Ti-6.6Al-1.75Sn-4.12Zr-1.91Mo-0.32W-0.1Si, Ti-4.9Al-2.3Sn-4.1Zr-2Mo-0.1Si-0.8Ge, Ti-4.8Al-2.3Sn-4.2Zr-2Mo, Ti-6.5Al-3Sn-4Hf-0.2Nb-0.4Mo-0.4Si-0.1B, Ti-5.8Al-4Sn-3.5Zr-0.7Mo-0.35Si-0.7Nb-0.06C, and Ti-6Al-3.5Sn-4.5Zr-2.0Ta-0.7Nb-0.5Mo-0.4Si. The top-ranked alloys evaluated by PCA-based consolidation are strikingly similar to the top-ranked alloys evaluated by mean-based consolidation. The top-ranked alloys suggest the following metallurgical characteristics: bimodal matrix (primary α + transformed β), aluminum equivalent preferably up to 8, and nanocrystalline precipitates of Ti_3Al, germanides, or silicides. The analyses driven by decision science made metallurgical sense. It provides guidelines for developing next-generation commercial near-α Ti alloys. The alloy design strategy for near-α Ti alloys for high-temperature applications with a combination of high *YS*, high *UTS*, and high *%EL* has two distinct options: (i) a combination of the aluminum equivalent to 8 and a bimodal matrix with no precipitates, or (ii) a combination of the aluminum equivalent to 8, bimodal matrix, and nanocrystalline Ti_3Al or germanide or silicide (not Zr, but Hf, as the silicides containing Hf do not reduce ductility, to the contrary, Hf provides solid solution strengthening) precipitates in α. Thus, novel alloys could be developed based on these directions for the future. A similar analysis could include data from newer exotic experimental materials, such as composites, for compressor parts.

Author Contributions: Conceptualization, T.V.J. and R.C.; Methodology, T.V.J.; Software, T.V.J.; Validation, R.C. and T.V.J.; Formal Analysis, T.V.J. and R.C.; Investigation, T.V.J. and R.C.; Data Curation, R.C.; Writing—Original Draft Preparation, T.V.J.; Writing—Review & Editing, T.V.J. and R.C.; Visualization, T.V.J.; Supervision, R.C. and T.V.J.; Project Administration, R.C. and T.V.J.; Funding Acquisition, R.C. and T.V.J. All authors have read and agreed to the published version of the manuscript.

Funding: Weldaloy Specialty Forgings (Research and Development Account# 8860.00) Institute of Advanced Vehicle Systems (grant# 052349), University of Michigan-Dearborn.

Institutional Review Board Statement: Not applicable.

Informed Consent Statement: No applicable.

Data Availability Statement: The data used in the analyses is presented in Tables A1 and A2.

Acknowledgments: The author R. Canumalla thanks the Weldaloy Specialty Forgings management for all their support; the author T. V. Jayaraman, thanks the Department of Mechanical Engineering, College of Engineering and Computer Science, University of Michigan-Dearborn for all their support.

Conflicts of Interest: The authors declare no conflict of interest.

Appendix A

Table A1. The alternatives, a list of 105 variants of 19 distinct near-α Ti alloys identified for the investigation, chemistry (nominal composition), fabrication and processing conditions, and microstructure; alloy designation is the unique identifier assigned to the variants.

Sl#	Alloy	Chemistry (Nominal)	Processing Step 1	Processing Step 2	Microstructure Description	Alloy Designation	Ref.
1	IMI834	Ti-5.8Al-4Sn-3.5Zr-0.7Mo-0.35Si-0.7Nb-0.06C	834-($\alpha + \beta$)ST1025 °C OQ	700 °C	Micro 1-Bimodal-αp (15 vol.%/15–20 µm) & Tr.β	IMI834-1	
2	IMI834	Ti-5.8Al-4Sn-3.5Zr-0.7Mo-0.35Si-0.7Nb-0.06C	834-TMT-($\alpha + \beta$)ST1000WQ	600 °C-4 h	Micro 2-Bimodal-higher amount of αp than Micro1	IMI834-2	
3	IMI834	Ti-5.8Al-4Sn-3.5Zr-0.7Mo-0.35Si-0.7Nb-0.06C	834-TMT-βST1080 °C WQ	600 °C-4 h	Micro 3-Lamellar-Tr.β	IMI834-3	
4	Ti-1100	Ti-5.8Al-2.7Sn-4Zr-0.4Mo-0.45Si	Ti-1100 °C Forged at 980 °C AQ	Unaged	Micro A-Bimodal-αp (15 vol.%/15–20 µm) & Tr.β	Ti-1100-1	
5	Ti-1100	Ti-5.8Al-2.7Sn-4Zr-0.4Mo-0.45Si	Ti-1100 °C ($\alpha + \beta$) ST940 °C WQ	600 °C-4 h	Micro B-Bimodal and finer than Micro A	Ti-1100-2	[31]
6	Ti-1100	Ti-5.8Al-2.7Sn-4Zr-0.4Mo-0.45Si	Ti-1100 °C ($\alpha + \beta$) ST980 °C WQ	600 °C-4 h	Micro C-Bimodal coarse compared to Micro B but comparable to Micro-A	Ti-1100-3	
7	Ti-1100	Ti-5.8Al-2.7Sn-4Zr-0.4Mo-0.45Si	Ti-1100 °C-βST1020 °C WQ	600 °C-4 h	Micro D-Lamellar-Prior β grain size 200 µm	Ti-1100-4	
8	Ti-1100	Ti-5.8Al-2.7Sn-4Zr-0.4Mo-0.45Si	Ti-1100 °C-βST1060WQ	600 °C-4 h	Micro E-Lamellar-Prior β grain size 500 to 600 µm	Ti-1100-5	
9	Ti-1100	Ti-5.8Al-2.7Sn-4Zr-0.4Mo-0.45Si	Ti-1100 °C-TMT-($\alpha + \beta$) ST1000 °C WQ	600 °C-4 h	Micro F-Bimodal-finer compared to Micro C	Ti-1100-6	
10	Ti-1100	Ti-5.8Al-2.7Sn-4Zr-0.4Mo-0.45Si	Ti-1100 °C-TMT-βST1060 °C WQ	600 °C-4 h	Micro G-Lamellar-finer compared to Micro E	Ti-1100-7	
11	IMI685	Ti-6.18Al-5.27Zr-0.5Mo-0.28Si	685-βST1050 °C-WQ	Unaged	Lamellar α'—No precipitates	IMI685-1	
12	IMI685	Ti-6.18Al-5.27Zr-0.5Mo-0.28Si	685-βST1050 °C-WQ	550 °C-24 h	Lamellar α'—No precipitates	IMI685-2	
13	IMI685	Ti-6.18Al-5.27Zr-0.5Mo-0.28Si	685-βST1050 °C-WQ	650 °C-24 h	Lamellar-Silicides S1 & S2—NO Ti$_3$Al	IMI685-3	[32]
14	IMI685	Ti-6.18Al-5.27Zr-0.5Mo-0.28Si	685-βST1050 °C-WQ	700 °C-24 h	Lamellar-Silicides S2—NO Ti$_3$Al	IMI685-4	
15	IMI685	Ti-6.18Al-5.27Zr-0.5Mo-0.28Si	685-βST1050 °C-WQ	800 °C-24 h	Lamellar~0.1µm Silicides S2—NO Ti$_3$Al	IMI685-5	
16	IMI685	Ti-6.18Al-5.27Zr-0.5Mo-0.28Si	685-βST1050 °C-WQ	700 °C-24 h	Lamellar-finer Silicides S2/41.2 nm—NO Ti$_3$Al	IMI685-6	
17	IMI685	Ti-6.18Al-5.27Zr-0.5Mo-0.28Si	685-βST1050 °CWQ6CR	700 °C-24 h	Lamellar-finer Silicides S2/38.6 nm—NO Ti$_3$Al	IMI685-7	
18	IMI685	Ti-6.18Al-5.27Zr-0.5Mo-0.28Si	685-βST1050 °CWQ12CR	700 °C-24 h	Lamellar-finer Silicides S2/33.4 nm—NO Ti$_3$Al	IMI685-8	[33]
19	IMI685	Ti-6.18Al-5.27Zr-0.5Mo-0.28Si	685-βST1050 °CWQ15CR	700 °C-24 h	Lamellar-finer Silicides S2/28.5 nm-NO Ti$_3$Al	IMI685-9	

Table A1. *Cont.*

Sl#	Alloy	Chemistry (Nominal)	Processing Step 1	Processing Step 2	Microstructure Description	Alloy Designation	Ref.
20	IMI829	Ti-6.1Al-3.3Sn-3.2Zr-1Nb-0.5Mo-0.32Si	829-βST1050 °C-WQ	Unaged	Lamellar α'—No precipitates	IMI829-1	
21	IMI829	Ti-6.1Al-3.3Sn-3.2Zr-1Nb-0.5Mo-0.32Si	829-βST1050 °C-WQ	625 °C-24 h	Lamellar—Silicides S2 only-No Ti$_3$Al	IMI829-2	
22	IMI829	Ti-6.1Al-3.3Sn-3.2Zr-1Nb-0.5Mo-0.32Si	829-βST1050 °C-OQ	Unaged	Lamellar α'—No precipitates	IMI829-3	
23	IMI829	Ti-6.1Al-3.3Sn-3.2Zr-1Nb-0.5Mo-0.32Si	829-βST1050 °C-OQ	625 °C-24 h	Lamellar-Silicides S2 only-No Ti$_3$Al	IMI829-4	
24	IMI829	Ti-6.1Al-3.3Sn-3.2Zr-1Nb-0.5Mo-0.32Si	829-βST1050 °C-AC	Unaged	Lamellar/Widmanstatten-No precipitates-No Silicides or Ti$_3$Al	IMI829-5	[34]
25	IMI829	Ti-6.1Al-3.3Sn-3.2Zr-1Nb-0.5Mo-0.32Si	829-βST1050 °C-AC	625 °C-24 h	Lamellar/Widmanstatten-Silicides S2 only-No Ti$_3$Al	IMI829-6	
26	IMI829	Ti-6.1Al-3.3Sn-3.2Zr-1Nb-0.5Mo-0.32Si	829-βST1050 °C-FC	Unaged	Aligned alpha/Lamellar—No precipitates	IMI829-7	
27	IMI829	Ti-6.1Al-3.3Sn-3.2Zr-1Nb-0.5Mo-0.32Si	829-βST1050 °C-FC	625 °C-24 h	Aligned alpha/LamellarS2—Ti$_3$Al	IMI829-8	
28	IMI829	Ti-5.54Al-3.48Sn-2.95Zr-0.97Nb-0.34Mo-0.28Si	829-βST1050 °C-AC	Unaged	Lamellar- No pecipitates	IMI829-9	
29	IMI829	Ti-5.54Al-3.48Sn-2.95Zr-0.97Nb-0.34Mo-0.28Si	829-βST1050 °C-AC	625 °C-2 h-AC-575 °C-1000 h-AC	Lamellar—Ti$_3$Al (5 nm)	IMI829-10	[35]
30	IMI829	Ti-5.51Al-3.48Sn-3.04Zr-0.99Nb-0.33Mo < 0.02Si	829NS-βST1050 °C-AC	Unaged	Lamellar-No precipitates	IMI829NS-1	
31	IMI829	Ti-5.51Al-3.48Sn-3.04Zr-0.99Nb-0.33Mo < 0.02Si	829NS-βST1050 °C-AC	625 °C-2 h-AC-575 °C-1000 h-AC	Lamellar—Ti$_3$Al (5 nm)	IMI829NS-2	
32	Ti-1100	Ti-6Al-2.8Sn-4Zr-0.4Mo-0.45Si	Ti1100-βST1093 °C-AC	Unaged (593C-8 h-AC)	Lamellar-No precipitates	Ti-1100-8	
33	Ti-1100	Ti-6Al-2.8Sn-4Zr-0.4Mo-0.45Si	Ti1100-βST1093 °C-AC	Overaged (Unaged + 593C-180 K min-AC)	Lamellar-13 nm Ti$_3$Al in Tr β and 175 × 35 nm Silicides	Ti-1100-9	[36]
34	Ti-1100	Ti-6Al-2.8Sn-4Zr-0.4Mo-0.45Si	Ti1100-βST1093 °C-AC	PAHT (Unaged + 593C-60 K min + 750C-4 h-AC)	Lamellar-only Silicides (~100 nm)-NO Ti$_3$Al	Ti-1100-10	
35	IMI834	Ti-5.07Al-3.08Sn-3.45Zr-0.31Mo-0.2Si-0.66Nb-0.04C	834-βST1080 °C-cooled to (α + β)1010 °C-1 h-WQ	Unaged	Lamellar-No precipitates	IMI834-4	
36	IMI834	Ti-5.07Al-3.08Sn-3.45Zr-0.31Mo-0.2Si-0.66Nb-0.04C	834-βST1080 °C-cooled to (α + β)1010 °C-1 h-WQ	700 °C-2 h-AC	Lamellar—Ti$_3$Al (5 nm) in Tr. β and Silicides	IMI834-5	[37]
37	IMI834	Ti-5.07Al-3.08Sn-3.45Zr-0.31Mo-0.2Si-0.66Nb-0.04C	834-βST1080 °C-cooled to (α + β)1010 °C-1 h-WQ	825 °C-2 h-WQ	Lamellar—100 to 175 nm Silicides (Ti$_3$Al dissolved at 825 °C)	IMI834-6	

Table A1. *Cont.*

Sl#	Alloy	Chemistry (Nominal)	Processing Step 1	Processing Step 2	Microstructure Description	Alloy Designation	Ref.
38	IMI834	Ti-5.78Al-4.54Sn-4.05Zr-0.70Nb-0.52Mo-0.44Si-0.055C	834-($\alpha + \beta$)ST1020 °C-2 h-OQ (12-15%αp)	600 °C-4 h	Bimodal-Ti$_3$Al in only αp	IMI834-7	
39	IMI834	Ti-5.78Al-4.54Sn-4.05Zr-0.70Nb-0.52Mo-0.44Si-0.055C	834-($\alpha + \beta$)ST1020 °C-2 h-OQ- (12-15%αp)	650 °C-4 h	Bimodal-Ti$_3$Al in αp & Tr. β and Silicides S2	IMI834-8	[38]
40	IMI834	Ti-5.78Al-4.54Sn-4.05Zr-0.70Nb-0.52Mo-0.44Si-0.055C	834-($\alpha + \beta$)ST1020 °C-2 h-OQ- (12-15%αp)	700 °C-4 h	Bimodal-Ti$_3$Al in αp & Tr. β and Silicides S2	IMI834-9	
41	WJZ-Ti	Ti-6.7Al-1.9Sn-3.9Zr-4.6Mo-0.96W-0.23Si	834-($\alpha + \beta$)ST940 °C-2 h-AC	Unaged	Bimodal-No precipitates	WJZ-Ti-1	
42	WJZ-Ti	Ti-6.7Al-1.9Sn-3.9Zr-4.6Mo-0.96W-0.23Si	834-($\alpha + \beta$)ST940 °C-1 h-AC	600 °C-2 h	Bimodal-6 nm Ti$_3$Al in αp	WJZ-Ti-2	[39]
43	WJZ-Ti	Ti-6.7Al-1.9Sn-3.9Zr-4.6Mo-0.96W-0.23Si	834-($\alpha + \beta$)ST940 °C-1 h-AC	750 °C-2 h	Bimodal-7 nm Ti$_3$Al in αp & Tr.β	WJZ-Ti-3	
44	WJZ-Ti	Ti-6.7Al-1.9Sn-3.9Zr-4.6Mo-0.96W-0.23Si	834-($\alpha + \beta$)ST940 °C-1 h-AC	750 °C-12 h	Bimodal-15 nm Ti$_3$Al in αp & Tr.β	WJZ-Ti-4	
45	TA29	Ti-5.8Al-4Sn-4Zr-0.7Nb-1.5Ta-0.4Si-0.06C	TA29-βST (at > 1050 °C)	750 °C-2 h	Lamellar—~100 nm Silicides -small number at IPB	TA29-1	
46	TA29	Ti-5.8Al-4Sn-4Zr-0.7Nb-1.5Ta-0.4Si-0.06C	TA29-βST (at > 1050 °C) + 750 °C-2 h	650 °C-8 h	Lamellar—~100 nm Silicides at IPB & Ti$_3$Al (<5 nm)	TA29-2	
77	TA29	Ti-5.8Al-4Sn-4Zr-0.7Nb-1.5Ta-0.4Si-0.06C	TA29-βST (at > 1050 °C) + 750 °C-2 h	650 °C-100 h	Lamellar—~100 nm-Silicides at IPB and some inside &Ti$_3$Al in Tr. β (~8 nm)	TA29-3	[40]
48	TA29	Ti-5.8Al-4Sn-4Zr-0.7Nb-1.5Ta-0.4Si-0.06C	TA29-βST (at > 1050 °C) + 750 °C-2 h	650 °C-500 h	Lamellar—~100 nm Silicides at IPB and inside-IPB & Ti$_3$Al in Tr. β (26 nm L x13 nm thk.)	TA29-4	
49	TA29	Ti-5.8Al-4Sn-4Zr-0.7Nb-1.5Ta-0.4Si-0.06C	TA29-βST (at > 1050 °C) + 750 °C-2 h	650 °C-1000 h	Lamellar—~100 nm Silicides at IPB and inside & Ti$_3$Al in Tr. β (~20 nm dia.)	TA29-5	
50	KIMS	Ti-6.5Al-3Sn-4Hf-0.2Nb-0.4Mo-0.4Si-0.1B	KIMS-($\alpha + \beta$)ST-1 h-WQ	650 °C-5 h	Bimodal—Ti$_3$Al αp & Tr.β uniformly and Silicides-~80 nm	KIMS-1	[41]
51	KIMS	Ti-6.5Al-3Sn-4Hf-0.2Nb-0.4Mo-0.4Si-0.1B	KIMS-($\alpha + \beta$)ST-1 h-WQ	700 °C-2 h-AC	Bimodal—150 nm Silicides-IPB	KIMS-2	
52	JZ1	Ti-5.6Al-4.8Sn-2Zr-1Mo-0.35Si	JZ1-($\alpha + \beta$)ST-1005 °C-2 h-AC	700 °C-2 h-AC	Bimodal—No precipitates	JZ1-1	
53	JZ1	Ti-5.6Al-4.8Sn-2Zr-1Mo-0.35Si	JZ1-($\alpha + \beta$)ST-1005 °C-2 h-AC	700 °C-5 h-AC	Bimodal—Ti$_3$Al in αp + Silicides (~100 nm)	JZ1-2	
54	JZ1	Ti-5.6Al-4.8Sn-2Zr-1Mo-0.35Si	JZ1-($\alpha + \beta$)ST-1005 °C-2 h-AC	700 °C-15 h-AC	Bimodal—Ti$_3$Al in αp + Silicides (~100 nm)	JZ1-3	
55	JZ1	Ti-5.6Al-4.8Sn-2Zr-1Mo-0.35Si	JZ1-($\alpha + \beta$)ST-1005 °C-2 h-AC	700 °C-2 h-AC-600 °C-100 h	Bimodal—Ti$_3$Al in αp + Silicides	JZ1-4	
56	JZ1	Ti-5.6Al-4.8Sn-2Zr-1Mo-0.35Si	JZ1-($\alpha + \beta$)ST-1005 °C-2 h-AC	700 °C-5h-AC-600 °C-100 h	Bimodal—Ti$_3$Al in αp + Silicides (~100 nm)	JZ1-5	
57	JZ1	Ti-5.6Al-4.8Sn-2Zr-1Mo-0.35Si	JZ1-($\alpha + \beta$)ST-1005 °C-2 h-AC	700 °C-15h-AC-600 °C-100 h	Bimodal—Ti$_3$Al in αp + Silicides (~100 nm)	JZ1-6	[42]

Table A1. *Cont.*

Sl#	Alloy	Chemistry (Nominal)	Processing Step 1	Processing Step 2	Microstructure Description	Alloy Designation	Ref.
58	JZ2	Ti-6Al-4.8Sn-2Zr-1Mo-0.35Si	JZ2-($\alpha + \beta$)ST-1015 °C-2 h-AC	760 °C-2 h-AC	Bimodal—Ti$_3$Al in αp + Silicides	JZ2-1	
59	JZ2	Ti-6Al-4.8Sn-2Zr-1Mo-0.35Si	JZ2-($\alpha + \beta$)ST-1015 °C-2 h-AC	760 °C-5h-AC	Bimodal—Ti$_3$Al in αp + Silicides (~100 nm)	JZ2-2	
60	JZ2	Ti-6Al-4.8Sn-2Zr-1Mo-0.35Si	JZ2-($\alpha + \beta$)ST-1015 °C-2 h-AC	760 °C-10 h-AC	Bimodal—Ti$_3$Al in αp + Silicides	JZ2-3	
61	JZ2	Ti-6Al-4.8Sn-2Zr-1Mo-0.35Si	JZ2-($\alpha + \beta$)ST-1015 °C-2 h-AC	760 °C-2 h-AC	Bimodal—Ti$_3$Al in αp & Tr.β + Silicides	JZ2-4	
62	JZ2	Ti-6Al-4.8Sn-2Zr-1Mo-0.35Si	JZ2-($\alpha + \beta$)ST-1015 °C-2 h-AC	760 °C-5h-AC-600 °C-100 h	Bimodal—Ti$_3$Al in αp & Tr.β + Silicides (~100 nm)	JZ2-5	
63	JZ2	Ti-6Al-4.8Sn-2Zr-1Mo-0.35Si	JZ2-($\alpha + \beta$)ST-1015 °C-2 h-AC	760 °C-10 h-AC-600 °C-100 h	Bimodal—Ti$_3$Al in αp & Tr.β + Silicides	JZ2-6	
64	Ti60	Ti-5.8Al-4Sn-3.5Zr-0.4Mo-0.4Nb-1Ta-0.4Si-0.06C	TA60-($\alpha + \beta$)ST1010 °C-2 h-OQ	Unaged	Bimodal—No precipitates	Ti60-1	
65	Ti60	Ti-5.8Al-4Sn-3.5Zr-0.4Mo-0.4Nb-1Ta-0.4Si-0.06C	TA60-($\alpha + \beta$)ST1010 °C-2 h-OQ	650 °C-2 h-AC	Bimodal-No Ti$_3$Al+ small number of Silicides 100 nm-Stage 1	Ti60-2	
66	Ti60	Ti-5.8Al-4Sn-3.5Zr-0.4Mo-0.4Nb-1Ta-0.4Si-0.06C	TA60-($\alpha + \beta$)ST1010 °C-2 h-OQ	650 °C-4 h-AC	Bimodal-No Ti$_3$Al+ small number of Silicides-100 nm-Stage 1	Ti60-3	
67	Ti60	Ti-5.8Al-4Sn-3.5Zr-0.4Mo-0.4Nb-1Ta-0.4Si-0.06C	TA60-($\alpha + \beta$)ST1010 °C-2 h-OQ	650 °C-8 h-AC	Bimodal-No Ti$_3$Al + Silicides—100 nm-Stage 1	Ti60-4	
68	Ti60	Ti-5.8Al-4Sn-3.5Zr-0.4Mo-0.4Nb-1Ta-0.4Si-0.06C	TA60-($\alpha + \beta$)ST1010 °C-2 h-OQ	650 °C-16 h-AC	Bimodal-No Ti$_3$Al + Silicides—100 nm—Stage 1	Ti60-5	
69	Ti60	Ti-5.8Al-4Sn-3.5Zr-0.4Mo-0.4Nb-1Ta-0.4Si-0.06C	TA60-($\alpha + \beta$)ST1010 °C-2 h-OQ	700 °C-2 h-AC	Bimodal-No Ti$_3$Al + Silicides 100 nm Stage 1	Ti60-6	
70	Ti60	Ti-5.8Al-4Sn-3.5Zr-0.4Mo-0.4Nb-1Ta-0.4Si-0.06C	TA60-($\alpha + \beta$)ST1010 °C-2 h-OQ	700 °C-4 h-AC	Bimodal-Ti$_3$Al in αp + Silicides	Ti60-7	
71	Ti60	Ti-5.8Al-4Sn-3.5Zr-0.4Mo-0.4Nb-1Ta-0.4Si-0.06C	TA60-($\alpha + \beta$)ST1010 °C-2 h-OQ	700 °C-8 h-AC	Bimodal-Ti$_3$Al in αp + Silicides	Ti60-8	[43]
72	Ti60	Ti-5.8Al-4Sn-3.5Zr-0.4Mo-0.4Nb-1Ta-0.4Si-0.06C	TA60-($\alpha + \beta$)ST1010 °C-2 h-OQ	700 °C-16 h-AC	Bimodal-Ti$_3$Al in αp + Silicides	Ti60-9	
73	Ti60	Ti-5.8Al-4Sn-3.5Zr-0.4Mo-0.4Nb-1Ta-0.4Si-0.06C	TA60-($\alpha + \beta$)ST1010 °C-2 h-OQ	700 °C-24 h-AC	Bimodal-Ti$_3$Al in αp + Silicides	Ti60-10	
74	Ti60	Ti-5.8Al-4Sn-3.5Zr-0.4Mo-0.4Nb-1Ta-0.4Si-0.06C	TA60-($\alpha + \beta$)ST1010 °C-2 h-OQ	700 °C-48 h-AC	Bimodal-Ti$_3$Al in αp + Silicides	Ti60-11	
75	Ti60	Ti-5.8Al-4Sn-3.5Zr-0.4Mo-0.4Nb-1Ta-0.4Si-0.06C	TA60-($\alpha + \beta$)ST1010 °C-2 h-OQ	750 °C-2 h-AC	Bimodal-Silicides-Ti$_3$Al in αp	Ti60-12	
76	Ti60	Ti-5.8Al-4Sn-3.5Zr-0.4Mo-0.4Nb-1Ta-0.4Si-0.06C	TA60-($\alpha + \beta$)ST1010 °C-2 h-OQ	750 °C-4 h-AC	Bimodal-Ti$_3$Al in αp + Silicides	Ti60-13	

Table A1. *Cont.*

Sl#	Alloy	Chemistry (Nominal)	Processing Step 1	Processing Step 2	Microstructure Description	Alloy Designation	Ref.
77	Ti60	Ti-5.8Al-4Sn-3.5Zr-0.4Mo-0.4Nb-1Ta-0.4Si-0.06C	TA60-(α + β)ST1010 °C-2 h-OQ	750 °C-8 h-AC	Bimodal-Ti$_3$Al in αp + Silicides	Ti60-14	
78	Ti60	Ti-5.8Al-4Sn-3.5Zr-0.4Mo-0.4Nb-1Ta-0.4Si-0.06C	TA60-(α + β)ST1010 °C-2 h-OQ	750 °C-16 h-AC	Bimodal-Ti$_3$Al in αp + Silicides—100 nm	Ti60-15	
79	Ti60	Ti-5.8Al-4Sn-3.5Zr-0.4Mo-0.4Nb-1Ta-0.4Si-0.06C	TA60-1035C-near β forged-(α + β)ST1010 °C-2 h-AC	700 °C-2 h-AC	Bimodal-No Ti$_3$Al + Silicides-~100 nm	Ti60-16	
80	Ti60	Ti-5.8Al-4Sn-3.5Zr-0.4Mo-0.4Nb-1Ta-0.4Si-0.06C	TA60-1035C-near β forged-(α + β) ST1010 °C-2 h-AC	700 °C-2 h-AC-600C-100 h-AC	Bimodal-Ti$_3$Al in αp + Silicides ~200 nm	Ti60-17	
81	Ti60	Ti-5.8Al-4Sn-3.5Zr-0.4Mo-0.4Nb-1Ta-0.4Si-0.06C	TA60-1035C-near β forged-(α + β)ST1010 °C-2 h-AC	700 °C-2 h-A-700 °C-100 h-AC	Bimodal-only Silicides possibly (dissolution of Ti$_3$Al)	Ti60-18	[44]
82	Ti60	Ti-5.8Al-4Sn-3.5Zr-0.4Mo-0.4Nb-1Ta-0.4Si-0.06C	TA60-1035C-near β forged-(α + β)ST1010 °C-2 h-AC	700 °C-2 h-AC-750 °C-100 h-AC	Bimodal-only Silicides possibly (dissolution of Ti$_3$Al)	Ti60-19	
83	Ti60	Ti-5.8Al-4Sn-3.5Zr-0.4Mo-0.4Nb-1Ta-0.4Si-0.06C	TA60-1070C- β forged-(α + β)ST1010 °C-2 h-AC	700 °C-2 h-AC	Lamellar + No Ti$_3$Al + Silicides ~100 nm	Ti60-20	
84	Ti60	Ti-5.8Al-4Sn-3.5Zr-0.4Mo-0.4Nb-1Ta-0.4Si-0.06C	TA60-1070C-β forged-(α + β)ST1010 °C-2 h-AC	700 °C-2 h-AC-600 °C-100 h-AC	Lamellar-Ti$_3$Al in Tr. β + Silicides ~100 nm	Ti60-21	
85	JL	Ti-5.6Al-4.3Sn-3Zr-1Mo-0.8Nd-0.34Si	JL-(α + β) forged-(α + β)ST1010 °C-2 h-AC	Unaged	Bimodal—No precipitates (αp is ~10% vf.;~150 nm)	JL-1	
86	JL	Ti-5.6Al-4.3Sn-3Zr-1Mo-0.8Nd-0.34Si	JL-(α + β) forged-(α + β)ST1010 °C-2 h-AC	700 °C-2 h-AC	Bimodal—Ti$_3$Al in αp + Silicides ~125 nm	JL-2	
87	JL	Ti-5.6Al-4.3Sn-3Zr-1Mo-0.8Nd-0.34Si	JL-(α + β) forged-(α + β)ST1010 °C-2 h-AC	700 °C-12 h-AC	Bimodal—Ti$_3$Al in αp & Tr. β + Silicides ~125 nm	JL-3	[45]
88	JL	Ti-5.6Al-4.3Sn-3Zr-1Mo-0.8Nd-0.34Si	JL-(α + β) forged-βST1030 °C-2 h-AC	Unaged	Lamellar—No precipitates (colony size; ~350 nm)	JL-4	
89	JL	Ti-5.6Al-4.3Sn-3Zr-1Mo-0.8Nd-0.34Si	JL-(α + β) forged-βST1030 °C-2 h-AC	700 °C-2 h-AC	Lamellar—Ti$_3$Al in Tr. β + Silicides ~125 nm	JL-5	
90	JL	Ti-5.6Al-4.3Sn-3Zr-1Mo-0.8Nd-0.34Si	JL-(α + β) forged-βST1030 °C-2 h-AC	700 °C-12 h-AC	Lamellar—Ti$_3$Al in Tr. β + Silicides ~125 nm	JL-6	
91	IMI834	Ti-5.8Al-4Sn-3.5Zr-0.7Mo-0.35Si-0.7Nb-0.06C	834-(α + β) ST1020 °C-2 h-AC	700 °C-2 h-AC	Bimodal—Ti$_3$Al in αp + Silicides ~125 nm	IMI834-10	[46]

Table A1. *Cont.*

Sl#	Alloy	Chemistry (Nominal)	Processing Step 1	Processing Step 2	Microstructure Description	Alloy Designation	Ref.
92	Ti6242S	Ti-5.7Al-1.9Sn-3.7Zr-1.9Mo-0.09Si	Ti6242S-($\alpha + \beta$) Hot rolled βST1052 °C-1 h-CC	Unaged	Lamellar-No precipitates	Ti6242-1	[47]
93	Ti6242S	Ti-5.7Al-1.9Sn-3.7Zr-1.9Mo-0.09Si	Ti6242S-($\alpha + \beta$) Hot rolled β ST1052 °C-1 h-CC	760 °C-600 h-AC	Lamellar-Silicides (>100 nm)	Ti6242-2	
94	LD-Ti423	Ti-8Al-1Cr-1V-0.5Fe-0.1Si	VIM-Open die forged at (β)1100 °C-Hot rolled at ($\alpha + \beta$)1000 °C-50%Reduction (Rolled plate) or HR	NA	Heterogenous microstructure/similar to bimodal (major equiaxed α_p 11 µm and small amount of acicular αs and β phases in the interstices of equiaxed α)	LD-Ti423-1	[48]
95	LD-Ti423	Ti-8Al-1Cr-1V-0.5Fe-0.1Si	VIM-Open die forged at (β)1100 °C-Hot rolled at ($\alpha + \beta$)1000 °C-50%Reduction (Rolled plate)	HR +($\alpha + \beta$) ST at 1000 °C-1 h-AC-Aged at 560 °C-4 h-AC (STA)	Heterogenous microstructure/similar to bimodal (major equiaxed α_p 20µm and small amount of acicular αs and β phases in the interstices of equiaxed α)	LD-Ti423-2	
96	TMC-Ti213	Ti-2Al-1.3V	Thermomechanical consolidation (TMC) of TiH$_2$ and Ti6Al4V at a mass ratio of 2:1, and extruded to produce Ti-2Al-1.3V at 16 to 1 ratio at around 1200 °C	TMC-Vac Anneal 700 °C-6 h-FC	lamellar α/β (lamellae of 0.9µm thk and ave. lamellar colony size of 15.4 µm)	TMC-Ti213-1	[49]
97	TMC-Ti213	Ti-2Al-1.3V	Thermomechanical consolidation (TMC) of TiH$_2$ and Ti6Al4V at a mass ratio of 2:1, and extruded to produce Ti-2Al-1.3V at 16 to 1 ratio at around 1200 °C	TMC-Vac Anneal 700 °C-6 h-FC-980 °C-1 h-AC	equiaxed α grains and α/β lamellar structured domains (βt or β transformed structure)	TMC-Ti213-2	
98	TA19	Ti-6.6Al-1.75Sn-4.12Zr-1.91Mo-0.32W-0.1Si	α-β field rolled plate, tempered in the α-β field	970 °C-1 h-AC (for equiaxed structure (EM))	equiaxed α grains (42%Vol.; 12 µm dia) and α/β lamellar (αs 10 µm length × 400 nm width); g.b α 420 nm width	TA19-1	[50]
99	TA19	Ti-6.6Al-1.75Sn-4.12Zr-1.91Mo-0.32W-0.1Si	α-β field rolled plate, tempered in the α-β field	1015 °C-35s-cooled at 20 °C/s (for semi equiaxed structure (S-EM))	Semi-equiaxed α grains (41%Vol.; 13 µm dia) and α/β lamellar (αs 11 µm length × 410 nm width); g.b α 850 nm width	TA19-2	

Table A1. *Cont.*

Sl#	Alloy	Chemistry (Nominal)	Processing Step 1	Processing Step 2	Microstructure Description	Alloy Designation	Ref.
100	TKT-1	Ti-4.8Al-2.3Sn-4.2Zr-2Mo	Double melted; β forged qt 1100 °C-groove rolled to 50% reduction at $(\alpha + \beta)$ 960 °C to rods of 14mm dia	Annealed at 950 °C-1 h-AC-aged at 590 °C-8 h-AC	Bimodal microstructure comprising primary equiaxed α and α-β lamellar (transformed β) structures	TKT-1	
101	TKT-2	Ti-4.8Al-2.2Sn-4.1Zr-2Mo-1.1Ge	Double melted; β forged qt 1100 °C-groove rolled to 50% reduction at $(\alpha + \beta)$ 960 °C to rods of 14mm dia	Annealed at 950 °C-1 h-AC-aged at 590 °C-8 h-AC	Bimodal microstructure comprising primary equiaxed α and α-β lamellar (transformed β) structures	TKT-2	[51]
102	TKT-6	Ti-4.9Al-2.3Sn-4.1Zr-2.1Mo-0.1Si-0.8Ge	Double melted; β forged qt 1100 °C-groove rolled to 50% reduction at $(\alpha + \beta)$ 960 °C to rods of 14mm dia	Annealed at 950 °C-1 h-AC-aged at 590 °C-8 h-AC	Bimodal microstructure comprising primary equiaxed α and α-β lamellar (transformed β) structures with $(TiZr)_6(SiGe)_3$ precipitates	TKT-6	
103	PC	Ti-6Al-3.5Sn-4.5Zr-2.0Ta-0.7Nb-0.5Mo-0.4Si	Induction skull melted ingot 80mm dia × 120mm length and then 1100 °C β upset forged to a total height reduction of 75% (reduction ratio of 4 to 1)	ONE-Isothermal Multidirectional Forging (IMDF) at 1020 °C-annealed at 650 °C-6 h	Mainly α laths (lamellar mainly—mean size 2.13 µm) with small amount of equiaxed α	PC-IMDF1	
104	PC	Ti-6Al-3.5Sn-4.5Zr-2.0Ta-0.7Nb-0.5Mo-0.4Si	Induction skull melted ingot 80 mm dia × 120 mm length and then 1100C β upset forged to a total height reduction of 75% (reduction ratio of 4 to 1)	TWO-IMDFs at 1020 °C-annealed at 650 °C-6 h	prior β boundaries start to disappear and α laths transform to spheroidal α grains (equiaxed α—mean size 1.85 µm)	PC-IMDF2	[52]
105	PC	Ti-6Al-3.5Sn-4.5Zr-2.0Ta-0.7Nb-0.5Mo-0.4Si	Induction skull melted ingot 80 mm dia × 120 mm length and then 1100 °C β upset forged to a total height reduction of 75% (reduction ratio of 4 to 1)	FOUR-IMDFs at 1020 °C-annealed at 650 °C-6 h	large number of spheroidal α grains transformed from lamellae	PC-IMDF4	

Note: ST: solution treated; h: hours; AC: air cooled; OQ: oil quenched; WQ: water quenched; α_p: primary alpha; Tr. β: transformed beta; bimodal matrix: primary α + transformed β; and lamellar matrix: completely transformed β.

Table A2. Database of 105 variants of 19 distinct near-α Ti alloys identified from the literature, which form the decision matrix comprising the alternatives and their attributes.

Alloy Desig.	Grade	Al.eq.	Matrix	Precipitates	Size	YS	UTS	%El	Ref.
IMI834-1	IMI834	8	Bimodal	NA	NA	1040	1125	9	
IMI834-2	IMI834	8	Bimodal	NA	NA	1200	1255	13	
IMI834-3	IMI834	8	Lamellar	NA	NA	1110	1220	4	
Ti-1100-1	Ti-1100	7	Bimodal	NA	NA	900	965	13.5	
Ti-1100-2	Ti-1100	7	Bimodal	NA	NA	965	1050	14	[31]
Ti-1100-3	Ti-1100	7	Bimodal	NA	NA	995	1090	8	
Ti-1100-4	Ti-1100	7	Lamellar	NA	NA	1050	1160	7.5	
Ti-1100-5	Ti-1100	7	Lamellar	NA	NA	1080	1190	5	
Ti-1100-6	Ti-1100	7	Bimodal	NA	NA	1100	1200	10	
Ti-1100-7	Ti-1100	7	Lamellar	NA	NA	1150	1250	2.5	
IMI685-1	IMI685	7	Lamellar	NP	0	919	1058	7.2	
IMI685-2	IMI685	7	Lamellar	NP	0	966	1090	7.3	
IMI685-3	IMI685	7	Lamellar	Silicides	>50	1020	1132	4.7	[32]
IMI685-4	IMI685	7	Lamellar	Silicides	>50	1005	1102	4.8	
IMI685-5	IMI685	7	Lamellar	Silicides	>50	954	1038	3.75	
IMI685-6	IMI685	7	Lamellar	Silicides	21–50	917.5	1021	5.55	
IMI685-7	IMI685	7	Lamellar	Silicides	21–50	980	1064	4.9	[33]
IMI685-8	IMI685	7	Lamellar	Silicides	21–50	978	1060	2.5	
IMI685-9	IMI685	7	Lamellar	Silicides	21–50	1025.5	1067	2.65	
IMI829-1	IMI829	8	Lamellar	NP	0	886	970	10	
IMI829-2	IMI829	8	Lamellar	Silicides	>50	1005	1023	2	
IMI829-3	IMI829	8	Lamellar	NP	0	863	951	11	
IMI829-4	IMI829	8	Lamellar	Silicides	>50	960	1005	3	[34]
IMI829-5	IMI829	8	Lamellar	NP	0	867	942	10	
IMI829-6	IMI829	8	Lamellar	Silicides	>50	858	975	7	
IMI829-7	IMI829	8	Lamellar	NP	0	866	946	9	
IMI829-8	IMI829	8	Lamellar	Silicides	>50	851	953	6.5	
IMI829-9	IMI829	7	Lamellar	NP	0	861	977.5	9.6	
IMI829-10	IMI829	7	Lamellar	Silicides aided by Ti_3Al in Tr.β	>50	881.5	953	3.1	[35]
IMI829NS-1	IMI829	7	Lamellar	NP	0	800	904	9.4	
IMI829NS-2	IMI829	7	Lamellar	Ti_3Al in Tr.β-3	<7	819.5	891.5	9.05	
Ti-1100-8	Ti-1100	8	Lamellar	NP	0	915	1000	5.5	
Ti-1100-9	Ti-1100	8	Lamellar	Ti_3Al in Tr.β aided by Silicides	7–20	955	982	0.18	[36]
Ti-1100-10	Ti-1100	8	Lamellar	Silicides	>50	895	980	4.15	
IMI834-4	IMI834	7	Lamellar	NP	0	987	1128	7.5	
IMI834-5	IMI834	7	Lamellar	Ti_3Al in Tr.β-3	<7	1028	1134	6.5	[37]
IMI834-6	IMI834	7	Lamellar	Silicides	>50	980	1098	7.5	

Table A2. *Cont.*

Alloy Desig.	Grade	Al.eq.	Matrix	Precipitates	Size	YS	UTS	%El	Ref.
IMI834-7	IMI834	8	Bimodal	Ti_3Al in αp-1	<7	905	1037	13	
IMI834-8	IMI834	8	Bimodal	Ti_3Al in αp & Tr.β-2	7–20	953	1075	9.5	[38]
IMI834-9	IMI834	8	Bimodal	Ti_3Al in αp & Tr.β-2	7–20	933	1060	8.7	
WJZ-Ti-1	WJZ-Ti	8	Bimodal	NP	0	1195	1300	18	
WJZ-Ti-2	WJZ-Ti	8	Bimodal	Ti_3Al in αp-1	<7	1325	1450	18	
WJZ-Ti-3	WJZ-Ti	8	Bimodal	Ti_3Al in αp & Tr.β-2	7–20	1230	1250	12	[39]
WJZ-Ti-4	WJZ-Ti	8	Bimodal	Ti_3Al in αp & Tr.β-2	7–20	1100	1120	5	
TA29-1	TA29	8	Lamellar	Silicides	>50	995	1062	7.5	
TA29-2	TA29	8	Lamellar	Ti_3Al in Tr.β-3	<7	990	1075	6.5	
TA29-3	TA29	8	Lamellar	Ti_3Al in Tr.β-3	7–20	975	1060	3	[40]
TA29-4	TA29	8	Lamellar	Ti_3Al in Tr.β-3	7–20	1018	1085	2.5	
TA29-5	TA29	8	Lamellar	Ti_3Al in Tr.β-3	7–20	975	1060	3.5	
KIMS-1	KIMS	8	Bimodal	Ti_3Al in αp & Tr.β-2	7–20	1113.8	1133.9	4.07	[41]
KIMS-2	KIMS	8	Bimodal	Silicides	>50	1032.6	1124.6	16.94	
JZ1-1	JZ1	8	Bimodal	NP	0	965	1030	15.5	
JZ1-2	JZ1	8	Bimodal	Ti_3Al in αp-1	<7	965	1040	15	
JZ1-3	JZ1	8	Bimodal	Ti_3Al in αp-1	<7	940	1030	15	
JZ1-4	JZ1	8	Bimodal	Ti_3Al in αp-1	<7	1000	1060	16	
JZ1-5	JZ1	8	Bimodal	Ti_3Al in αp-1	<7	1000	1070	14.5	
JZ1-6	JZ1	8	Bimodal	Ti_3Al in αp-1	7-20	990	1060	15	
JZ2-1	JZ2	8	Bimodal	Ti_3Al in αp-1	<7	965	1060	15.5	
JZ2-2	JZ2	8	Bimodal	Ti_3Al in αp-1	7–20	960	1060	16.5	[42]
JZ2-3	JZ2	8	Bimodal	Ti_3Al in αp-1	7–20	960	1050	17	
JZ2-4	JZ2	8	Bimodal	Ti_3Al in αp & Tr.β-2	7–20	1040	1110	14	
JZ2-5	JZ2	8	Bimodal	Ti_3Al in αp & Tr.β-2	7–20	1040	1110	12.5	
JZ2-6	JZ2	8	Bimodal	Ti_3Al in αp & Tr.β-2	7–20	1030	1100	13	
Ti60-1	Ti60	8	Bimodal	NP	0	960	1080	11	
Ti60-2	Ti60	8	Bimodal	Silicides	>50	962	1082	10	
Ti60-3	Ti60	8	Bimodal	Silicides	>50	1000	1100	10	
Ti60-4	Ti60	8	Bimodal	Silicides	>50	1000	1100	9	
Ti60-5	Ti60	8	Bimodal	Silicides	>50	1000	1100	9	
Ti60-6	Ti60	8	Bimodal	Silicides	>50	982	1082	10	
Ti60-7	Ti60	8	Bimodal	Ti_3Al in αp-1	<7	970	1060	10	[43]
Ti60-8	Ti60	8	Bimodal	Ti_3Al in αp-1	<7	960	1070	10	
Ti60-9	Ti60	8	Bimodal	Ti_3Al in αp-1	7–20	940	1060	10	

Table A2. *Cont.*

Alloy Desig.	Grade	Al.eq.	Matrix	Precipitates	Size	YS	UTS	%El	Ref.
Ti60-10	Ti60	8	Bimodal	Ti$_3$Al in αp-1	7–20	960	1080	10	
Ti60-11	Ti60	8	Bimodal	Ti$_3$Al in αp-1	7–20	980	1085	10	
Ti60-12	Ti60	8	Bimodal	Ti$_3$Al in αp-1	<7	960	1082	10	
Ti60-13	Ti60	8	Bimodal	Ti$_3$Al in αp-1	7–20	970	1080	9	
Ti60-14	Ti60	8	Bimodal	Ti$_3$Al in αp-1	7–20	965	1080	10	
Ti60-15	Ti60	8	Bimodal	Ti$_3$Al in αp-1	7–20	955	1077	11.5	
Ti60-16	Ti60	8	Bimodal	Silicides	>50	1010	1060	11	
Ti60-17	Ti60	8	Bimodal	Ti$_3$Al in αp-1	7–20	1050	1120	7	
Ti60-18	Ti60	8	Bimodal	Silicides	>50	1021	1090	12	[44]
Ti60-19	Ti60	8	Bimodal	Silicides	>50	950	1018	11.5	
Ti60-20	Ti60	8	Lamellar	Silicides	>50	1020	1080	9	
Ti60-21	Ti60	8	Lamellar	Ti$_3$Al in Tr.β-3	7–20	1040	1100	4	
JL-1	JL	8	Bimodal	NP	0	944	1029	15.6	
JL-2	JL	8	Bimodal	Ti$_3$Al in αp-1	<7	985	1057	12.2	
JL-3	JL	8	Bimodal	Ti$_3$Al in αp & Tr.β-2	<7	994	1066	12	[45]
JL-4	JL	8	Lamellar	NP	0	933	1020	12.4	
JL-5	JL	8	Lamellar	Ti$_3$Al in Tr.β-3	7–20	981	1052	8.4	
JL-6	JL	8	Lamellar	Ti$_3$Al in Tr.β-3	7–20	970	1042	6.5	
IMI834-10	IMI834	8	Bimodal	Ti$_3$Al in αp-1	<7	945	1012	14.5	[46]
Ti6242-1	Ti6242S	7	Lamellar	NP	0	837	946	12	[47]
Ti6242-2	Ti6242S	7	Lamellar	Silicides	>50	875	925	5.8	
LD-Ti423-1	LD-Ti423	8	Bimodal	NP	0	948	1046	8.3	[48]
LD-Ti423-2	LD-Ti423	8	Bimodal	NP	0	923	1013	8	
TMC-Ti213-1	TMC-Ti213	2	Lamellar	NP	0	996	1059	13.9	[49]
TMC-Ti213-2	TMC-Ti213	2	Bimodal	NP	0	876	994	7.4	
TA19-1	TA19	8	Bimodal	NP	0	984	1067	23.9	[50]
TA19-2	TA19	8	Bimodal	NP	0	1022	1113	22.8	
TKT-1	TKT-1	6	Bimodal	NP	0	980	1175	20	
TKT-2	TKT-2	6	Bimodal	NP	0	1075	1285	19	[51]
TKT-6	TKT-6	6	Bimodal	Germanides	<500	1060	1255	18	
PC-IMDF1	PC	8	Lamellar	NA	NA	982	1020	10.6	
PC-IMDF2	PC	8	Bimodal	NA	NA	1004	1043	12.7	[52]
PC-IMDF4	PC	8	Bimodal	NA	NA	1072	1118	15.6	

Note: NA: not available; NP: no precipitates.

References

1. Ikpe, E.; Ikechukwu, O.; Ebunilo, P.O.; Ikpe, E. Material selection for high pressure (HP) compressor blade for an aircraft engine. *Int. J. Adv. Mater. Res.* **2016**, *2*, 59–65.
2. Foltz, J.; Gram, M. Introduction to titanium and its alloys. In *ASM Handbook, Heat Treating of Nonferrous Alloys, Volume 4E*; Totten, G.E., MacKenzie, D.S., Eds.; ASM International: Materials Park, OH, USA, 2016.

3. Canumalla, R. On the low tensile ductility at room temperature in high temperature titanium alloys. *SCIREA J. Metall. Eng.* **2020**, *4*, 16–51.
4. Williams, J.C.; Boyer, R.R. Opportunities and issues in the application of titanium alloys for aerospace components. *Metals* **2020**, *10*, 705. [CrossRef]
5. Rao, M.N. Materials for gas turbines—An overview. In *Advances in Gas Turbine Technology*; Intechopen: Rijeka, Croatia, 2011; pp. 293–314.
6. Gogia, K. High-temperature titanium alloys. *Def. Sci. J.* **2005**, *55*, 149–173. [CrossRef]
7. Singh, K.; Ramachandra, C. Characterization of silicides in high temperature titanium alloys. *J. Mater. Sci.* **1997**, *32*, 229–234. [CrossRef]
8. Ramachandra, C.; Singh, V.; Rao, P.R. On silicides in high temperature Ti alloys. *Def. Sci. J.* **1986**, *36*, 207–220. [CrossRef]
9. Zhang, T.; Liu, C.-T. Design of titanium alloys by additive manufacturing: A critical review. *Adv. Powder Mater.* **2022**, *1*, 100014. [CrossRef]
10. Tshephe, S.T.; Akinwamide, S.O.; Olevsky, E.; Olubambi, P.A. Additive manufacturing of titanium-based alloys—A review of methods, properties, challenges, and prospects. *Heliyon* **2022**, *8*, e09041. [CrossRef]
11. *TIMETAL 834 Datasheet*; Titanium Metals Corporation: Warrensville Heights, OH, USA, 2000.
12. *TIMETAL 1100 Datasheet*; Titanium Metals Corporation: Warrensville Heights, OH, USA, 2008.
13. Ashby, M.F. *Materials and the Environment Eco-Informed Material Choice*, 1st ed.; Elsevier: Amsterdam, The Netherlands, 2009.
14. Ashby, M.F. *Materials Selection in Mechanical Design*, 5th ed.; Elsevier: Amsterdam, The Netherlands, 2017.
15. Jahan, A.; Ismail, M.Y.; Sapuan, S.M.; Mustapha, F. Materials screening and choosing methods—A review. *Mater. Des.* **2010**, *31*, 696–705. [CrossRef]
16. Anojkumar, L.; Ilangkumaran, M.; Sasirekha, V. Comparative analysis of MCDM methods for pipe material selectionin sugar industry. *Expert Syst. Appl.* **2014**, *41*, 2964–2980. [CrossRef]
17. Zhou, C.; Yin, G.F.; Hu, X.B. Multi-objective optimization of material selection for sustainable products: Artificial neural networks and genetic algorithm approach. *Mater. Des.* **2009**, *30*, 1209–1215. [CrossRef]
18. Rao, R.V.; Davim, J.P. A decision-making framework model for material selection using a combined multiple attribute decision-making method. *Int. J. Adv. Manuf. Technol.* **2008**, *35*, 751–760. [CrossRef]
19. Gupta, N. Material selection for thin-film solar cells using multiple attribute decision making approach. *Mater. Des.* **2011**, *32*, 1667–1671. [CrossRef]
20. Shanian, A.; Savadogo, O. A material selection model based on the concept of multiple attribute decision making. *Mater. Des.* **2006**, *27*, 329–337. [CrossRef]
21. Chauhan, A.; Vaish, R. Magnetic material selection using multiple attribute decision making approach. *Mater. Des.* **2012**, *36*, 1–5. [CrossRef]
22. Tzeng, G.H.; Huang, J.J. *Multiple Attribute Decision Making Methods and Applications*; CRC Press: Boca Raton, FL, USA, 2011.
23. Rao, R.V. *Decision Making in the Manufacturing Environment, Using Graph Theory and Fuzzy Multiple Attribute Decision Making Methods*; Springer: Berlin/Heidelberg, Germany, 2007.
24. Farag, M.M. *Materials and Process Selection in Engineering*; Elsevier Science and Technology: Amsterdam, The Netherlands, 1979.
25. Cherian, R.P.; Smith, L.N.; Midha, P.S. A neural network approach for selection for powder metallurgy materials and process parameters. *Artif. Intell. Eng.* **2000**, *14*, 39–44. [CrossRef]
26. Perez-Benitez, J.A.; Padovese, L.R. Feature Selection and Neural Network for analysis of microstructural changesin magnetic materials. *Expert Syst. Appl.* **2011**, *38*, 10547–10553. [CrossRef]
27. Zavadskas, K.; Turskis, Z.; Kildiene, S. State of art surveys of overviews on MCDM/MADM methods. *Technol. Econ. Dev. Econ.* **2014**, *20*, 165–179. [CrossRef]
28. Rajan, K. Materials Informatics. *Mater. Today* **2005**, *8*, 38–45. [CrossRef]
29. Cadima, J.; Jolliffe, I.T. Principal Component Analysis: A review and recent developments. *Phil. Trans. R. Soc. A* **2016**, *374*, 20150202.
30. George, L.; Hrubiak, R.; Rajan, K.; Saxena, S.K. Principal component analysis on properties of binary and ternary hydrides and a comparison of metal versus metal hydride properties. *J. Alloys Compd.* **2009**, *478*, 731–735. [CrossRef]
31. Peters, M.; Lee, Y.T.; Grundhoff, K.J.; Schurmann, H.; Welsch, G. Influence of processing on microstructure and mechanical properties of Ti-1100 and IMI 834. In Proceedings of the Minerals and Metals Society Fall Meeting, Detroit, MI, USA, 7–11 October 1991; pp. 533–548.
32. Ramachandra, C.; Singh, V. Effect of silicides on tensile properties and fracture of alloy Ti-6Al-5Zr-0.5Mo-0.25Si from 300 to 823 K. *J. Mater. Sci.* **1988**, *23*, 835–841. [CrossRef]
33. Ramachandra, C.; Singh, V. Effect of thermomechanical treatments on size and distribution of silicides and tensile properties of alloy Ti-6AI-5Zr-0.5Mo-0.25Si. *Metall. Trans. A* **1988**, *19*, 389–391. [CrossRef]
34. Sridhar, G.; Sarma, D.S. Structure and properties of a near-alpha titanium alloy after beta solution treatment and aging at 625 °C. *Metall. Trans. A* **1988**, *19*, 3025–3033. [CrossRef]
35. Woodfield, P.; Postans, P.J.; Loretto, M.; Smallman, R. The effect of long-term high temperature exposure on the structure and properties of the titanium alloy Ti-5331S. *Acta Metall.* **1988**, *36*, 507–515. [CrossRef]

36. Madsen, A.; Ghonem, H. Separating the effects of T3Al and silicide precipitates on the tensile and crack growth behavior at room temperature and 593 °C in a near-alpha titanium alloy. *J. Mater. Eng. Perform.* **1995**, *4*, 301–307. [CrossRef]
37. Srinadh, K.V.S.; Nidhi, S.; Singh, V. Role of Ti3Al/Silicides on tensile properties of Timetal 834 at various temperatures. *Bull. Mater. Sci.* **2007**, *30*, 595–600. [CrossRef]
38. Cope, M.T.; Hill, M.J. The influence of aging temperature on the mechanical properties of IMI 834. In Proceedings of the 6th World Conference on Titanium, Cannes, France, 6–9 June 1988.
39. Zhang, W.J.; Song, X.Y.; Hui, S.X.; Ye, W.J.; Yu, Y.; Li, Y.F. α2 phase precipitation behavior and tensile properties at room temperature and 650 °C in an ($α + β$) titanium alloy. *Rare Met.* **2019**, *40*, 3261–3268. [CrossRef]
40. Li, J.; Cai, J.; Xu, Y.; Xiao, W.; Huang, X.; Ma, C. Influences of thermal exposure on the microstructural evolution and subsequent mechanical properties of a near-$α$ high temperature titanium alloy. *Mater. Sci. Eng. A* **2020**, *774*, 138934. [CrossRef]
41. Narayana, P.; Kim, S.W.; Hong, J.K.; Reddy, N.; Yeom, J.T. Tensile properties of a newly developed high-temperature titanium alloy at room temperature and 650 °C. *Mater. Sci. Eng.* **2018**, *718*, 287–291. [CrossRef]
42. Zhang, J.; Peng, N.; Wang, Q.; Wang, X. A new aging treatment way for near $α$ high temperature titanium alloys. *J. Mater. Sci. Technol.* **2009**, *25*, 454–458.
43. Jia, W.; Zeng, W.; Yu, H. Effect of aging on the tensile properties and microstructures of a near-alpha titanium alloy. *Mater. Des.* **2014**, *58*, 108–115. [CrossRef]
44. Jia, W.; Zeng, W.; Liu, J.; Zhou, Y.; Wang, Q. Influence of thermal exposure on the tensile properties and microstructures of Ti60 titanium alloy. *Mater. Sci. Eng. A* **2011**, *2011*, 511–518. [CrossRef]
45. Liu, J.; Li, S.; Li, D.; Yang, R. Effect of aging on fatigue-crack growth behavior of a high temperature titanium alloy. *Mater. Trans.* **2004**, *5*, 1577–1585. [CrossRef]
46. Singh, N.; Gauthama; Singh, V. Low cycle fatigue behavior of Ti alloy IMI 834 at room temperature. *Mater. Sci. Eng. A* **2002**, *325*, 324–332. [CrossRef]
47. Allison, J.E.; Cho, W.; Jones, J.W.; Donlon, W.T.; Lasecki, J.V. The influence of elevated temperature on the mechanical behavior of alpha/beta titanium alloys. In Proceedings of the Sixth World Conference on Titanium, Cannes, France, 6–9 June 1988.
48. Zhu, S.; Zhu, C.; Luo, D.; Zhang, X.; Zhou, K. Development of a Low-Density and High-Strength Titanium Alloy. *Metals* **2023**, *13*, 251. [CrossRef]
49. Zhang, Y.; Wu, X.; Zhang, B.; Zhang, Y.; Zhang, D. Microstructure and mechanical properties of a novel PM Ti–2Al–1.3V alloy in different heat treatment conditions. *Heat Treat. Surf. Eng.* **2022**, *4*, 43–53. [CrossRef]
50. Luo, M.; Lin, T.; Zhou, L.; Li, W.; Liang, Y.; Han, M.; Liang, Y. Deformation Behavior and Tensile Properties of the Semi-Equiaxed Microstructure in Near Alpha Titanium Alloy. *Materials* **2021**, *14*, 3380. [CrossRef]
51. Kitashima, T.; Suresh, K.S.; Yamba-Mitarai, Y. Effect of germanium and silicon additions on the mechanical properties of a near-$α$ titanium alloy. *Mater. Sci. Eng. A* **2014**, *597*, 212–218. [CrossRef]
52. Zhang, J.; Guo, C.X.; Zhang, S.Z.; Feng, H.; Chen, C.Y.; Zhang, H.Z.; Cao, P. Microstructural manipulation and improved mechanical properties of a near $α$ titanium alloy. *Mater. Sci. Eng. A* **2020**, *771*, 138569. [CrossRef]
53. Jahan, A.; Edwards, K.L.; Bahraminasab, M. *Multi-Criteria Decision Analysis—For Supporting the Selection of Engineering Materials in Product Design*, 2nd ed.; Elsevier: Amsterdam, The Netherlands, 2016.
54. Memariani, A.; Amini, A.; Alinezhad, A. Sensitivity analysis of simple additive weighting method (SAW): The results of change in the weight of one attribute on the final ranking alternatives. *J. Ind. Eng.* **2009**, *4*, 13–18.
55. Afshari, A.; Mojaahed, M.; Yusuff, R.M. Simple additive weighting approach to personal selection problem. *Int. J. Innov. Mgt. Technol.* **2010**, *1*, 511–515.
56. Madic, M.; Radovanovic, M.; Manic, M. Application of the ROV method of selection of cutting fluids. *Decis. Sci. Lett.* **2016**, *5*, 245–254. [CrossRef]
57. Jha, G.K.; Chatterjee, P.; Chatterjee, R.; Chakraborty, S. Suppliers selection in a manufacturing environment using a range of value method. *J. Mech. Eng.* **2013**, *3*, 16–22. [CrossRef]
58. Zavadskas, K.; Turskis, Z.; Vilutiene, T. Multiple criteria analysis of foundation installment alternatives by applying additive ratio assessment (ARAS) method. *Arch. Civil Mech. Eng.* **2010**, *10*, 123–141. [CrossRef]
59. Zavadskas, K.; Turskis, Z. A new additive ratio assessment (ARAS) method in multicriterial decision making. *Technol. Econ. Dev. Econ.* **2010**, *16*, 159–172. [CrossRef]
60. Stanujic, S.; Jovanovic, R. Measuring a quality of faculty website using ARAS method. *Contemp. Issues Bus. Manag. Educ.* **2012**, *2012*, 545–554.
61. Yazdabi, M.; Zavadskas, E.K.; Turskis, Z. A combined compromise solution (CoCoSo) method for multi-criteria decision-making problems. *Manag. Desicion* **2019**, *57*, 2501–2519.
62. Qiyas, M.; Naeem, M.; Khan, S.; Abdullah, S.; Botmart, T.; Shah, T. Decision support system based on CoCoSo method with the picture fuzzy information. *J. Math.* **2022**, *2022*, 1476233. [CrossRef]
63. Mefgouda, B.; Idoudi, H. A novel MADM based netwrok interface selection approach with rank reversal avoidance in HWNs. In Proceedings of the 2022 IEEE Wireless Communications and Networking Conference, Austin, TX, USA, 10–13 April 2022.
64. Chatterjee, P.; Chakraborty, S. Material selection using preferential ranking methods. *Mater. Des.* **2012**, *35*, 384–393. [CrossRef]

65. Erdogan, S.; Aydin, S.; Balki, M.K.; Sayin, C. Operational evaluation of thermal barrier coated diesel engine fueled with biodiesel/diesel blend by using MCDM method base on engine performance, emission and combustion characteristics. *Renew. Energy* **2020**, *151*, 698–706. [CrossRef]
66. Darji, V.P.; Rao, R.V. Intelligent multi-criteria decision making methods for material selection in sugar industry. *Procedia Mater. Sci.* **2014**, *5*, 2585–2594. [CrossRef]
67. Siregar, D.; Arisandi, D.; Usman, A.; Irwan, D.; Rahim, R. Research of simple multi-attribute rating technique for decision support. *J. Phys. Conf. Ser.* **2017**, *930*, 012015. [CrossRef]
68. Patel, M.R.; Vashi, M.P.; Bhatt, B.V. SMART-Multi-criteria decision-making technique for use in planning activities. In Proceedings of the New Horizons in Civil Engineering, Surat, India, 25–26 March 2017.
69. Rao, R.V.; Singh, D. Evaluating flexible manufacturing systems using euclidean distance-based integrated approach. *Int. J. Risk Manag.* **2012**, *3*, 32–53. [CrossRef]
70. Rao, R.V.; Singh, D. Weighted Euclidean distance-based approach as a multiple attribute decision making method for plant or facility layout design selection. *Int. J. Ind. Eng. Comput.* **2012**, *3*, 365–382.
71. Alinezhad, A.; Khalili, J. *New Methods and Applications of Multiple Attribute Decision Making (MADM)*; Springer Nature: Berlin/Heidelberg, Germany, 2019; p. 193.
72. Tesic, D.; Radovanovic, M.; Bozanic, D.; Pamucar, D.; Milic, A.; Puska, A. Modification of the DIBR and MABAC methods by applying rough numbers and its application in making decisions. *Information* **2022**, *13*, 353. [CrossRef]
73. Chakraborty, S. Applications of the MOORA method for decision making in manufacturing environment. *Int. J. Adv. Manuf. Technol.* **2011**, *4*, 1155–1166. [CrossRef]
74. Dominguez, L.P.; Mojica, K.Y.S.; Pabon, L.C.O.; Diaz, M.C.C. Application of the MOORA method for the evaluation of the industrial maintenance system. *J. Phys. Conf. Ser.* **2018**, *1126*, 012018. [CrossRef]
75. Triantaphyllou, E.; Shu, B.; Sanchez, S.N.; Ray, T. Multi-criteria decision making: An operations research approach. In *Encyclopedia of Electrical and Electronics Engineering*; Webster, H.G., Ed.; John Wiley & Sons: New York, NY, USA, 1998; Volume 15, p. 175.
76. Deng, H.; Yeh, C.H.; Willis, R.J. Inter-company comparison using modified TOPSIS with objective weights. *Comput. Oper. Res.* **2000**, *27*, 963–973. [CrossRef]
77. Gul, M.; Celik, E.; Aydin, N.; Gumus, A.T.; Guneri, A.F. A state of the art literature review of VIKOR and its fuzzy extensions on applications. *Appl. Soft Comput.* **2016**, *46*, 60–89. [CrossRef]
78. Sayadi, M.M.; Heydari, M.; Shahanaghi, K. Extension of VIKOR method for decision making problem with interval numbers. *Appl. Math. Model.* **2009**, *33*, 2257–2262. [CrossRef]
79. Mohanty, P.P.; Mahapatra, S.S. A compromise solution by VIKOR method for ergonomically designed product with optimal set of design characteristics. *Procedia Mater. Sci.* **2014**, *6*, 633–640. [CrossRef]
80. Trung, D. Multi-criteria deciison making under MARCOS method and weighting methods: Applied to milling, grinding, and turning processes. *Manuf. Rev.* **2022**, *9*, 3.
81. Stevic, Z.; Pamucar, D.; Puska, A.; Chatterjee, P. Sustainable supplier selection in healthcare insudtries using a new MCDM method: Measurement Alternatives and Ranking according to Compromise Solution (MARCOS). *Int. J. Appl. Eng. Res.* **2020**, *140*, 106231.
82. Levine, M.; Ramsey, P.P.; Smidt, R.K. *Applied Statistics for Engineers and Scientists*; Prentice-Hall: Upper Saddle, NJ, USA, 2001.
83. Navidi, W. *Statistics for Engineers and Scientists*, 3rd ed.; McGraw-Hill Science/Engineering: New York, NY, USA, 2010.

Article

Effect of Trace Rare-Earth Element Ce on the Microstructure and Properties of Cold-Rolled Medium Manganese Steel

Qingbo Zhao [1,2], Ruifeng Dong [1,3,*], Yongfa Lu [1], Yang Yang [1], Yanru Wang [1] and Xiong Yang [4]

1   School of Materials Science and Engineering, Inner Mongolia University of Technology, Hohhot 010051, China
2   Beijing BOE Display Technology Co., Ltd., Beijing 102600, China
3   Engineering Research Center of Rare Earth Metals, Inner Mongolia University of Technology, Hohhot 010051, China
4   Technical Center of Inner Mongolia Baotou Steel Union Co., Ltd., Baotou 014010, China
*   Correspondence: drfcsp@163.com

Abstract: Rare-earth elements have been widely used in the field of functional materials, but their effects on the cold-stamping formability of high-strength automotive steels have rarely been studied. In this paper, the effect of the trace rare-earth element Ce on the microstructure and properties of cold-rolled medium manganese steel after ART (austenite-reverted transformation, ART) annealing was studied. The microstructure of the experimental steel was observed using SEM, and the mechanical properties were tested using a universal tensile testing machine. The volume fraction of the retained austenite and the texture of the steel were measured using XRD. The results showed that the original austenite grain size of the experimental steel was smaller after adding the trace rare-earth element Ce. After ART annealing, the grain size distribution of the experimental steel with rare-earth Ce was more uniform, and the comprehensive mechanical properties were better. Under the conditions of quenching at 800 °C for 5 min and annealing at 645 °C for 15 min, the maximum product of tensile strength and elongation was 28.47 GPa·%.

Keywords: trace rare-earth element; cold-rolled medium manganese steel; original austenite; microstructure; properties

Citation: Zhao, Q.; Dong, R.; Lu, Y.; Yang, Y.; Wang, Y.; Yang, X. Effect of Trace Rare-Earth Element Ce on the Microstructure and Properties of Cold-Rolled Medium Manganese Steel. *Metals* **2023**, *13*, 116. https://doi.org/10.3390/met13010116

Academic Editors: Andrey Belyakov, Andrea Di Schino and Claudio Testani

Received: 22 November 2022
Revised: 2 January 2023
Accepted: 4 January 2023
Published: 6 January 2023

1. Introduction

In recent years, the third generation of automobile steel, represented by medium manganese steel (Mn content of 3–10%), has gradually become the main research hotspot at home and abroad due to its excellent high-strength and high-plasticity mechanical properties [1]. The design of alloy composition is an important factor affecting the final properties of materials and has been extensively studied [2] As a strategic resource in China, rare earth (RE) is widely used in the field of functional materials, such as high-performance permanent magnet materials and luminescent materials, but is used less in metal structural materials [3].

Some scholars have studied the effects of rare-earth elements on steel, such as Guo Feng et al. [4], who found that rare earth can improve the morphology of inclusions, purify grain boundaries, improve the strength of grain boundaries, reduce the possibility of crack propagation through defects, and improve impact toughness. Zhao et al. [5] found that the addition of rare earth increased the AC1 and AC3 phase transition temperatures of low-carbon microalloyed high-strength steel, reduced the critical cooling rate of the pearlite transformation, increased the incubation period, and made it easier to obtain a bainite structure. Liu Chengjun et al. [6] pointed out that rare earth can refine the austenite grain boundary to improve impact toughness. QU et al. [7] studied the second phase in rare-earth HSLA steel and found that the average size of precipitates was refined by about 15 nm after adding rare earth, indicating that rare earth can promote the precipitation of fine carbon and nitride, which is beneficial for improving the strength and toughness of materials.

At present, the effect of rare-earth elements on medium manganese steel is not clear, and the role of rare earth in steel is not clear [8–10]. How to reasonably and effectively use rare-earth elements in steel remains to be further studied.

The selection of rare-earth content is also critical to the study of the microstructure and properties of steel. Considering the internal quality of the material, adding a large amount of rare-earth elements produces large inclusions in the structure, which affects the plasticity and performance of the material and deteriorates the forming performance of automobile steel. From the perspective of production practice, automobile steel is mostly plate, and a large number of rare-earth elements are added. During the casting process, large particle inclusions block the casting nozzle and cause casting accidents. This study selected trace rare-earth elements for testing and research.

In this paper, cold-rolled medium manganese experimental steels with 9 ppm rare-earth Ce and without rare earth were selected for comparative study. The two groups of experimental steel were subjected to the ART annealing treatment, and their microstructures, properties, and textures were analyzed to explore the effect of the trace rare-earth element Ce on the microstructure and properties of cold-rolled medium manganese steel, which provided a theoretical basis for the research and development of new third-generation rare-earth automotive steel, which is very economically significant for the development of the rare-earth steel and automotive fields.

2. Experimental Materials and Methods

The experimental steel was melted and cast by a laboratory vacuum induction furnace, and the ingot size was 100 mm × 100 mm × 300 mm. Then, a 1 mm thick experimental steel plate was obtained by rolling. The chemical composition of the test steel is shown in Table 1. The phase transformation temperature of the experimental steel was measured by a NETZSCH STA449 F3 comprehensive thermal analyzer at a rate of 20 °C/min to 950 °C under argon protection. The A_{C1} and A_{C3} values of the experimental steel without rare earth were 596 °C and 734 °C, respectively. The A_{C1} and A_{C3} values of the rare-earth experimental steel were 598 °C and 751 °C, respectively. The heat treatment process was conducted as shown in Figure 1. The experimental steel plate was quenched at 800 °C for 5 min and then annealed at 625 °C, 645 °C, or 665 °C for 15 min.

Table 1. Chemical composition of test steel (%).

Steel	C	Mn	Si	Al	Cu	Ni	Nb	Ti	P	S	Ce
0 RE	0.089	4.873	0.113	0.047	0.28	0.263	0.030	0.028	0.011	0.006	-
9 ppm RE	0.093	4.899	0.128	0.071	0.26	0.246	0.027	0.029	0.011	0.006	0.0009

The original austenite corrosion was carried out on the quenched experimental steel, and the supersaturated picric acid solution was used as the corrosive agent. The metallographic structure after corrosion was observed, and the original austenite grain size distribution was determined using Image Pro Plus software [11]. The small $10 \times 10 \times 1 \ mm^3$ pieces were cut from the annealed steel plate as a microstructure observation sample. The samples were polished and corroded with 4% nitric acid alcohol. The microstructures were observed using a field emission scanning electron microscope. The tensile specimens were prepared according to the GB/T 228.1-2010 standard. The length direction was parallel to the rolling direction. The original gauge was calculated by the proportional gauge ($L_0 = 5.65\sqrt{S_0}$), and the tensile test was carried out on a universal tensile testing machine. Two groups of experimental steels were tested using an X-ray diffractometer (XRD, X Pert PRO MPD). The sample size was 10 mm × 15 mm. The (200), (220), (311) crystal plane diffraction peaks of the FCC phase and the (200) and (211) crystal plane diffraction peaks of the BCC phase were measured, and the reverse-transformation austenite content was calculated using the following formula:

$$Vi = \frac{1}{1 + G(I\alpha / I\gamma)} \tag{1}$$

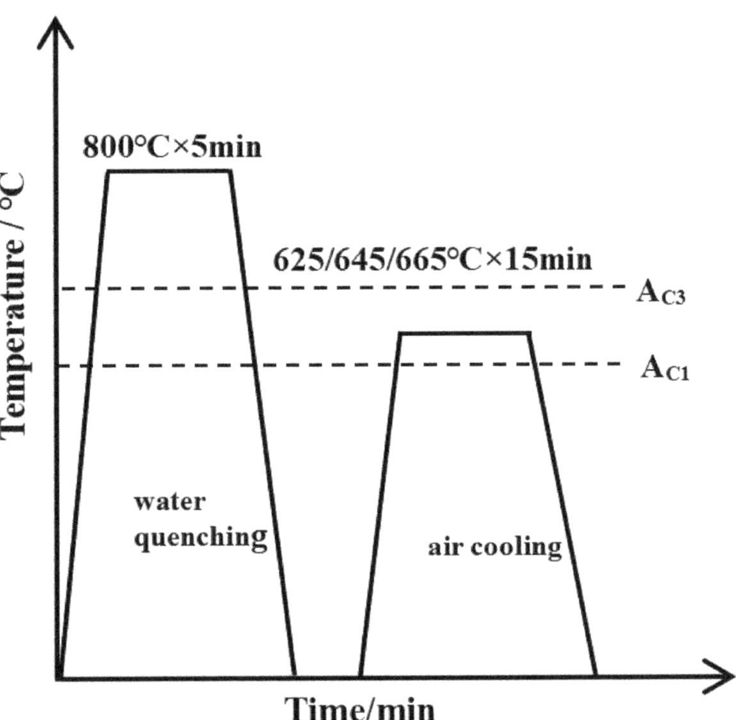

Figure 1. Heat treatment process of test steel.

In the formula, *G* represents the lattice parameter correlation value of the FCC phase and BCC phase [12], and the detailed correspondence is shown in Table 2. *Vi* is the volume fraction of reversed austenite corresponding to different *G* values, *I*γ is the integral intensity of the (200), (220), and (311) crystal plane diffraction peaks of the FCC phase, and *I*α is the integral intensity of the (200) and (211) crystal plane diffraction peaks of the BCC phase. The texture of the samples was also measured using an X-ray diffractometer. The ODF cross section was characterized using the Roe method, and the content of each texture component was calculated using the texture analysis software ResMat-TexTools.

Table 2. G values corresponding to different crystal face indices.

G-Value	(200)γ	(220)γ	(311)γ
(200)α	2.46	1.32	1.78
(211)α	1.21	0.65	0.87

3. Results and Analysis

3.1. Original Austenite Grains

According to the measured phase transition temperature, it was found that the A_{C3} temperature of the experimental steel increased after adding rare-earth elements. To further verify this point, the two groups of experimental steels were subjected to original austenite corrosion to observe the changes in the original austenite grains. Figure 2 shows the original austenite metallographic structure and grain size distribution. It can be seen in panels (a) and (b) that the austenite grains of the experimental steel with rare-earth Ce were smaller. The increase in A_{C3} temperature by 17 °C indicates that the austenite grains nucleated first in the austenitizing process of the experimental steel without RE elements, while the austenite grains nucleated later in the experimental steel with RE elements. Therefore, after the quenching treatment, the original austenite grains in the experimental steel containing rare earth were smaller [11]. The average grain sizes of the original austenite of the two groups of experimental steels were measured using Image-Pro Plus software to be 3.3 μm and 2.5 μm, respectively. The grain size distribution is shown in panels (c) and (d). The grain size of rare-earth experimental steel was small, where the grains between 2 and 4 μm accounted for about 43.91% and the grains less than 2 μm accounted for about 40.92%. The grain size of experimental steel without rare earth was mainly concentrated in 2~4 μm (about 56.23%). It can be seen that with the addition of trace rare-earth elements the austenite grains in the experimental steel were refined and the microstructure became more uniform.

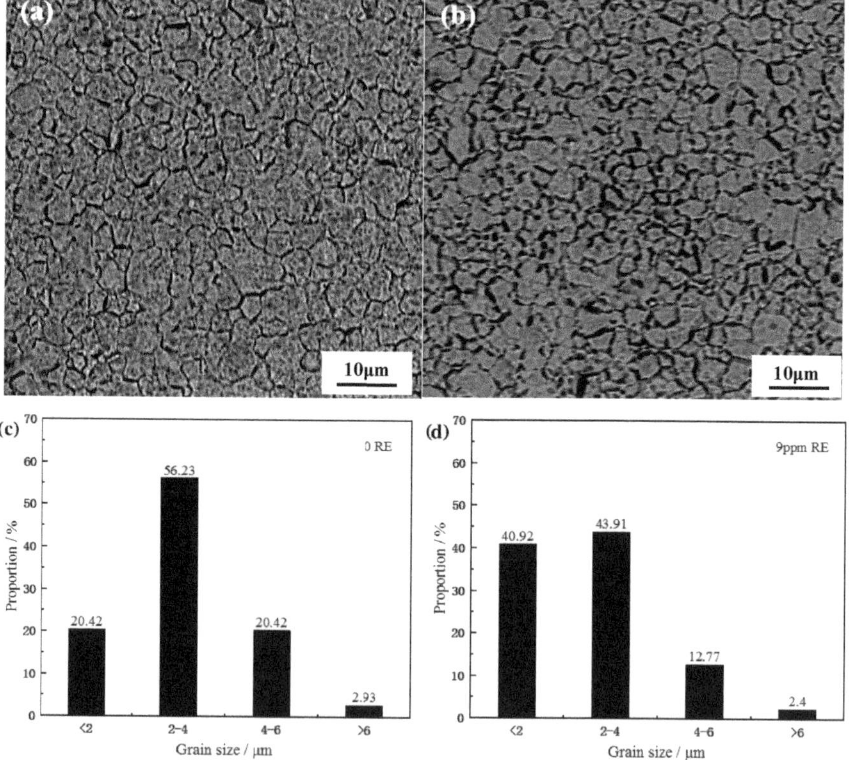

Figure 2. Original austenite microstructure and grain size distribution of two groups of experimental steels: (**a,c**) 0 RE and (**b,d**) 9 ppm RE.

3.2. Microstructural Evolution

Two groups of experimental steels were annealed by ART, and the microstructure after annealing was observed. The results are shown in Figure 3. It can be seen from the diagram that the microstructure after annealing was mainly composed of ferrite and retained austenite (or new martensite), in which the protruding structure was austenite or new martensite and the concave structure was ferrite [13]. Panels (a) and (d) show the microstructure at 625 °C. A large number of white granular carbides were dispersed in the microstructure. Panels (b) and (e) show the microstructure at 645 °C. It can be seen that there were still many white granular carbides on the ferrite [14], while there were almost no white granular carbides on the protruding austenite structure. This was because the annealing temperature was low at this time, the carbides had not been completely dissolved during the growth of austenite, and these carbides were attached to the ferrite. Panels (c) and (f) show the microstructure of the experimental steel at 665 °C. At this time, the carbides had completely disappeared, and the microstructure was composed of ferrite, austenite, and a small amount of new martensite. The austenite grains in the structure grew, and there were two forms, which were the lath and block distributions in the structure. This was also because the austenite grains grew with the increase in temperature, and some larger austenite was transformed into martensite during cooling due to the lower stability of carbon and manganese per unit volume.

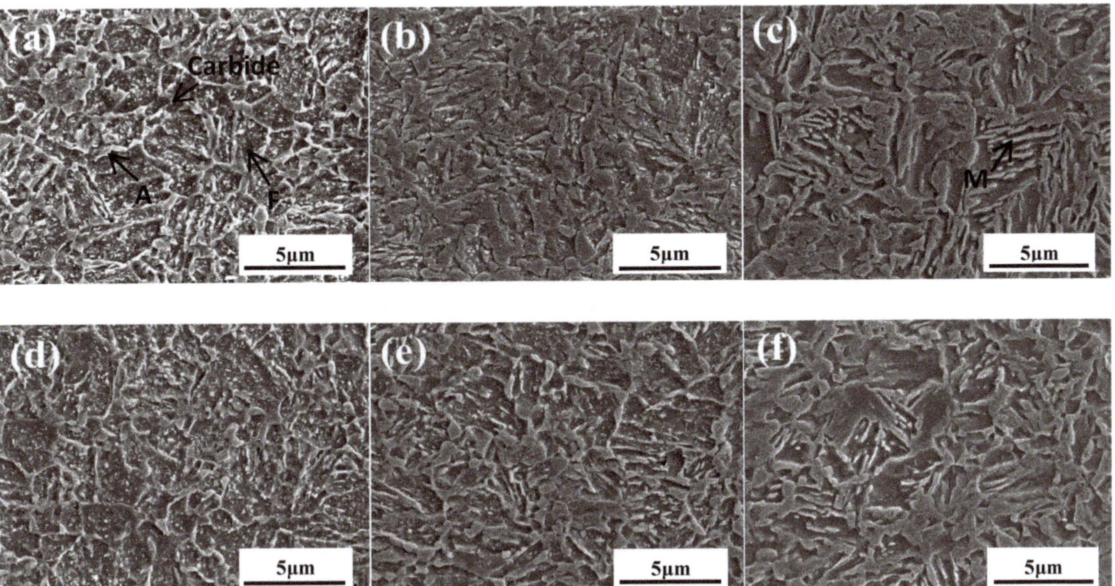

Figure 3. Microstructure of two groups of experimental steels at different annealing temperatures: (**a**,**d**) 625 °C, (**b**,**e**) 645 °C, (**c**,**f**) 665 °C; (**a**–**c**) 0 RE, (**d**–**f**) 9 ppm RE.

Through the comparative analysis of the microstructure of the two groups of experimental steels under different annealing processes, it was found that the grain size of the experimental steel with trace rare-earth elements was smaller. There were two reasons for this. On one hand, rare-earth elements themselves have the effect of grain refinement. The radius of rare-earth atoms is about 1.5 times that of Fe atoms, which can only dissolve into defects in the crystal but cannot dissolve into austenite to form a solid solution. Because there are many defects at the grain boundary, it is easy to accumulate rare-earth atoms near the grain boundary, which hinders the diffusion of atoms, thereby inhibiting the growth of austenite grains and achieving the effect of grain refinement [15]. On the other hand, rare-earth elements have the effect of deoxidation and desulfurization, which can produce fine and stable rare-earth oxides, rare-earth sulfides, or oxygen sulfides. These rare-earth compounds can be used as heterogeneous nuclei to provide excellent conditions for the refinement of crystallization.

3.3. Residual Austenite Content

The experimental steel was tested using XRD after different heat treatment processes, and the results are shown in Figure 4. In the figure, γ represents the FCC phase, α represents the BCC phase, and the volume fraction of austenite was calculated according to the integral strength of the austenite and ferrite diffraction peaks. As can be seen in the figure, with the increase in the annealing temperature, the austenite diffraction peak was enhanced. When the annealing temperature was 645 °C, the austenite content was the highest. When the temperature continued to increase, the diffraction peak of the FCC phase obviously decreased, while that of the BCC phase increased, indicating that the austenite content decreased with the increase in the annealing temperature, and new martensite was formed in the matrix structure. The FCC phase diffraction peak of the experimental steel containing rare earth was significantly higher than that of the experimental steel without rare earth. It can be seen that the addition of trace rare-earth elements increased the volume fraction of the retained austenite in the steel.

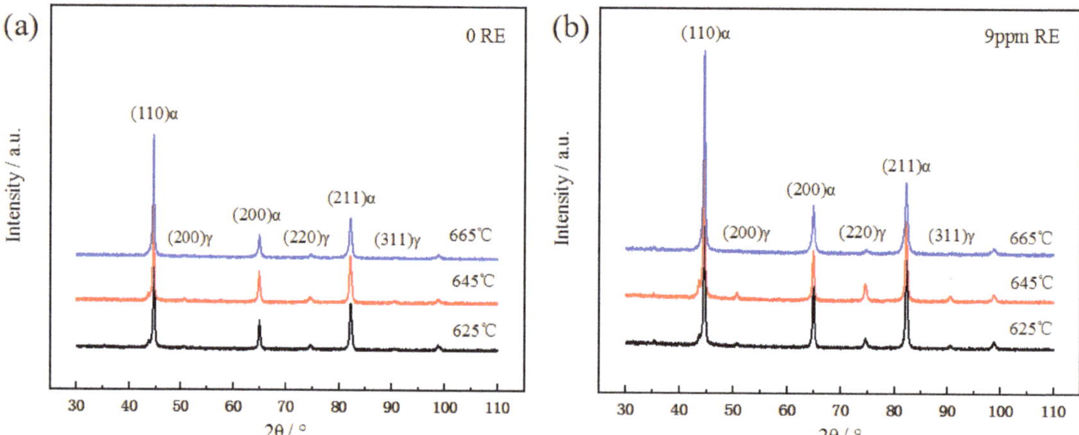

Figure 4. XRD patterns of the experimental steel at different annealing temperatures: (**a**) 0 RE, (**b**) 9 ppm RE.

Figure 5 shows the calculated volume fractions of the retained austenite of the two groups of experimental steels after different annealing processes. It can be seen in the figure that with the increase in temperature, the content of retained austenite increased first and then decreased. This was due to the unstable growth of austenite grains in the microstructure at higher annealing temperatures; the insufficient distribution of C and Mn elements, leading to the low contents of C and Mn elements per unit volume of austenite; and the transformation tendency of martensite to increase during high-temperature annealing, so some austenite transforms into martensite during the cooling process [16]. When the experimental steel was annealed at 645 °C, the austenite content was the largest, and the residual austenite contents of the two experimental steels were 21.1% and 22.8%, respectively. In comparison, it is found that adding trace rare-earth elements to steel was beneficial for retaining more austenite at room temperature.

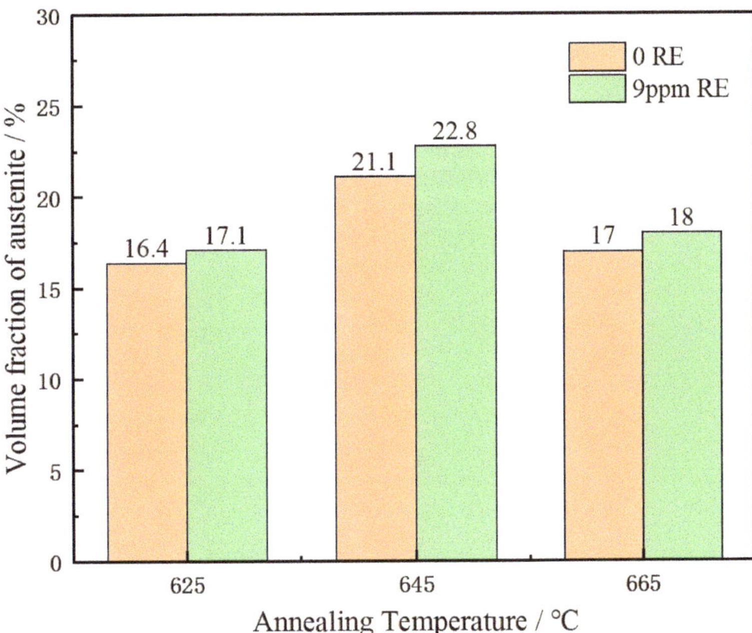

Figure 5. Residual austenite contents of the experimental steels at different annealing temperatures.

3.4. Mechanical Properties

Figure 6 shows a comparison of the mechanical properties of the two groups of experimental steels after the ART annealing treatment. It can be seen in the figure that the yield strength and tensile strength of the two groups of experimental steels after annealing were very similar, but the mechanical properties of the experimental steel with the rare-earth element Ce were improved. The yield strength decreased with the increase in the annealing temperature, and the maximum value was 850 MPa at 625 °C. The tensile strength increased with the increase in the annealing temperature, and the maximum value was 1100 MPa at 665 °C. The elongation increased first and then decreased with the increase in the annealing temperature and reached a maximum at 645 °C. The elongation of the experimental steel without rare earth was 27.82%, while the elongation of the experimental steel with rare earth was 33.89%. The change rule of the product of strength and elongation was consistent with that of elongation, which increased first and then decreased. The maximum value is obtained when the annealing temperature reached 645 °C. At this time, the products of strength and elongation of the experimental steels without rare earth and with rare earth were 24.3 GPa·% and 28.47 GPa·%, respectively. It can be seen that the rare-earth element was beneficial for the improvement of the mechanical properties of the experimental steel, and 800 °C for 5 min and 645 °C for 15 min was the best heat treatment condition for the experimental steel.

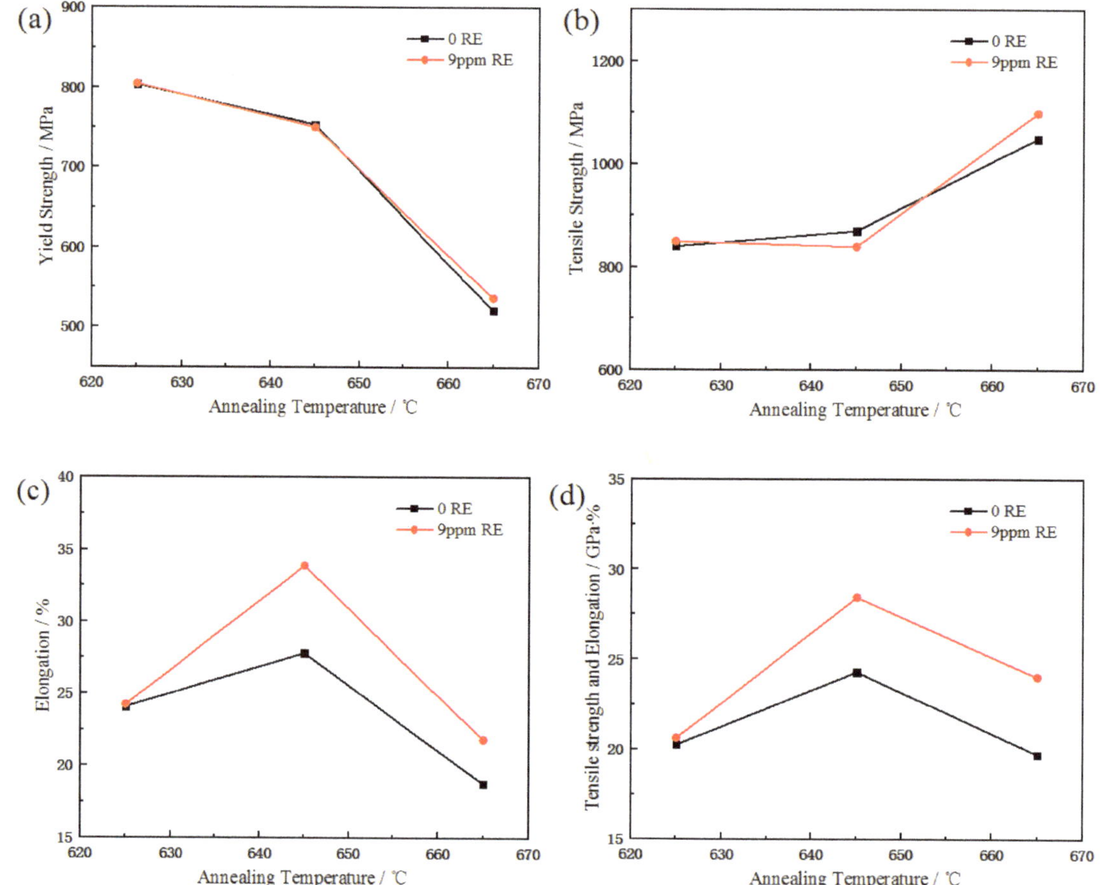

Figure 6. Mechanical properties of two groups of experimental steels at different annealing temperatures: (**a**) yield strength, (**b**) tensile strength, (**c**) elongation, and (**d**) the product of strength and elongation.

The change in the product of strength and elongation was the same as that of the austenite content after annealing at different temperatures, which indicates that the austenite content is an important factor affecting the comprehensive mechanical properties of experimental steel. Combined with the analysis of the XRD results, the reason for this change in the experimental steel may be that at a lower annealing temperature there are carbides in the microstructure and the C element in the matrix does not diffuse on a large scale, that is, it is enriched in the reverse-phase-transformation austenite. At the same time, the volume fraction of austenite is low and the stability is strong at this temperature. The comprehensive effect of the second-phase strengthening of carbides and the solid-solution strengthening of the C element makes the tensile strength and yield strength of the experimental steel remain within a certain range. With the increase in annealing temperature, the carbides in the microstructure continue to dissolve until they disappear, and the second-phase strengthening effect in the alloy also decreases and disappears. When the temperature is high, the microstructure of the experimental steel is transformed into ferrite, austenite, and martensite multiphase structures. The volume fraction of reversed austenite decreases and becomes unstable, and the TRIP (transformation-induced plasticity, TRIP) effect is weakened accordingly. During the cooling process, the hard-phase martensite increases. The higher the temperature, the more the martensite is transformed by cooling, so the tensile strength increases and the elongation decreases. Some scholars have also put forward the same conclusion that increasing the temperature increases the austenite content, but the austenite stability is poor, which is due to the tendency for martensite transformation in high-temperature annealing and the coarsening of austenite, so it is partially transformed into martensite during cooling and the tensile strength increases [17].

3.5. Texture Analysis

Good stamping properties are also an important goal in the field of automobile manufacturing. Deep drawability is related to the composition of a favorable texture in the steel. The larger the {111}/{100} coefficient of the favorable texture, the larger the r-value, indicating that the deep drawability of the steel plate is better [18]. Therefore, the essence of the study of the stamping properties of steel plates is to obtain more favorable textures in the microstructure. The type and density distribution of related textures can be determined with the orientation distribution function (ODF). The denser the orientation lines in the ODF cross section, the greater the orientation density. However, in general, only the main orientation distribution changes in the ODF diagram need to be analyzed. During the rolling process of steel plate, the grain orientation gradually converges to the α orientation line and the γ orientation line. Therefore, we mainly observed the texture type and density ($\varphi_2 = 45°$) on these two orientation lines.

Figure 7 shows the textured ODF cross sections of two groups of experimental steels at the optimal annealing temperature of $\varphi_2 = 45°$. It can be seen that there were {001}<110>, {011}<110>, {111}<110>, and {111}<112> textures in the two experimental steels after holding at 645 °C for 15min, and the texture density levels of the two experimental steels were very close. The strong peak position of the two experimental steels was near the {111}<110> texture on the γ orientation line, but the texture distribution of the experimental steel with the rare-earth element Ce was more uniform, and the texture density was up to 2.6. The contents of each orientation texture of the two groups of experimental steels were compared, as shown in Table 3. It can be seen that the contents of each orientation texture of the experimental steels increased slightly after adding rare-earth elements. Among them, the content of the {111} texture increased more. Combined with the analysis of Figure 7, the trace rare-earth elements may affect the texture of the experimental steel, which is beneficial for improving the forming properties, but the effect is not obvious compared with the effect on the mechanical properties. This requires further analysis of the changes in the favorable texture {111} and unfavorable texture {100} components.

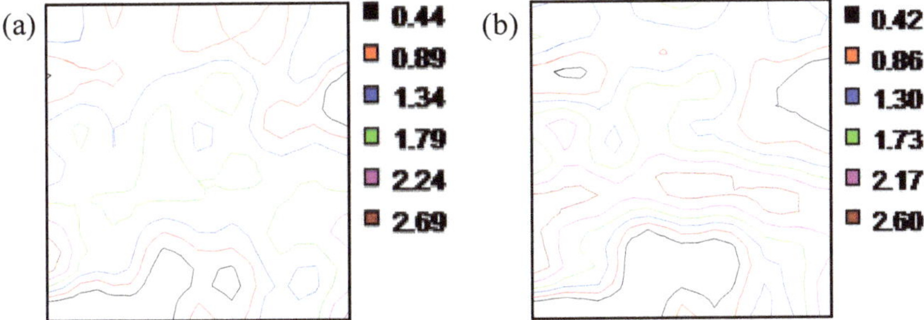

Figure 7. ODF cross sections of $\varphi_2 = 45°$ of two groups of experimental steels under different heat treatment processes: (**a**) 0 RE, (**b**) 9 ppm RE.

Table 3. Texture contents of each orientation after the ART annealing process.

ART Annealing	{001}<110>	{112}<110>	{223}<110>	{111}<110>	{111}<112>
0 RE	2.5	4.8	5.3	4.6	3.2
9 ppm RE	2.9	5.0	5.4	4.9	4.4

The {111} texture of the steel sheet is beneficial for the improvement of stamping performance, the {100} texture is not conducive to the improvement of stamping performance, and the {110} texture is located between the two textures. The texture contents of the {111}, {110}, and {100} planes of the two groups of experimental steels under two heat treatment processes were analyzed and calculated using ResMat-TexTools software, as shown in Figure 8. In the figure, it can be seen that the contents of the favorable textures {111} and {110} in the experimental steel with rare earth were higher than those in the experimental steel without rare earth. The content of the {111} texture was 13.4%, the content of the {110} texture was 19.6%, and the content of the unfavorable texture {100} was only 11.1%. Thus, the addition of trace rare-earth elements is beneficial for improving the formability of steel.

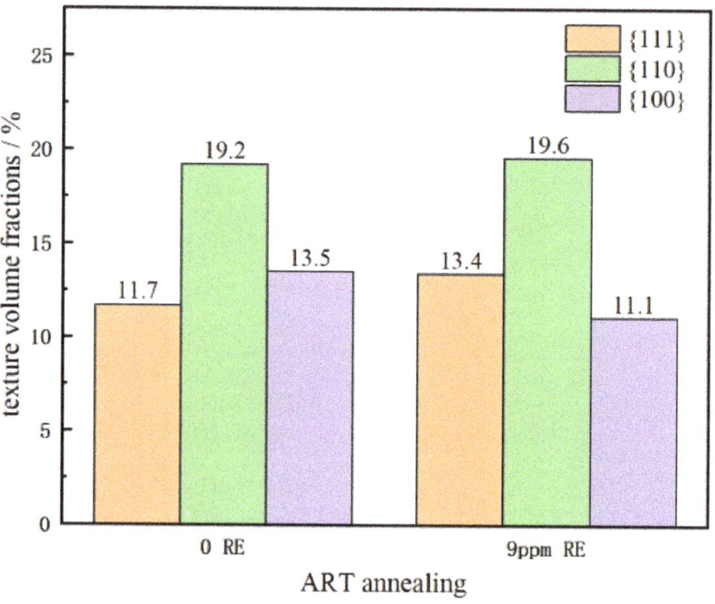

Figure 8. Texture contents of two groups of experimental steels after different annealing processes.

4. Conclusions

(1) After adding trace rare-earth Ce, the A_{C3} temperature of the experimental steel increased, which delayed the nucleation of austenite in the microstructure of the rare-earth experimental steel during quenching, resulting in smaller original austenite grain sizes.

(2) The grain size distribution of the experimental steel with 9 ppm RE was more uniform, the residual austenite content was higher, and the mechanical properties were better than those of RE-free steel. The maximum product of the strength and elongation of the experimental steel with 9 ppm RE was obtained when quenching at 800 °C for 5 min and annealing at 645 °C for 15 min. At this time, the residual austenite content of the experimental steel was 22.8%, the tensile strength was 840 MPa, the elongation was 33.89%, and the product of strength and plasticity was 28.47 GPa·%.

(3) The texture distributions and density levels of the two groups of experimental steel were similar after adding trace rare earth, but the volume fraction of the favorable texture {111} increased and the volume fraction of the unfavorable texture {100} decreased. It can be seen that the addition of trace rare-earth elements can improve the microstructure of steel and improve the comprehensive mechanical properties.

Author Contributions: Methodology, R.D. and X.Y.; validation, Q.Z., Y.L., Y.Y. and Y.W.; writing—original draft preparation, Q.Z.; writing—review and editing, Q.Z. and R.D.; funding acquisition, R.D. All authors have read and agreed to the published version of the manuscript.

Funding: This research was funded by [Advanced Education Research Project in Inner Mongolia Autonomous Region] grant number [NJZZ18076], [Inner Mongolia Autonomous Region Science and Technology Plan Project] grant number [2021GG0238], [2021 undergraduate college-level 'College Students' Innovation and Entrepreneurship Training Program] grant number [2022044059] and [Inner Mongolia University of Technology 2022 Undergraduate Science and Technology Innovation Fund Project] grant number [27].

Data Availability Statement: Data is contained within the article.

Conflicts of Interest: The authors declare no conflict of interest.

References

1. Liu, Q.; Zheng, X.; Zhang, R.; Tian, Y.; Chen, L. Research Status and Development Trend of New High Strength Medium Manganese Steel for Automobile. *Mater. Lett.* **2019**, *33*, 1215–1220.
2. Wang, Z.; Song, C.; Zhang, Y.; Wang, H.; Qi, L.; Yang, B. Effects of yttrium addition on grain boundary character distribution and stacking fault probabilities of 90Cu10Ni alloy. *Mater. Charact.* **2019**, *151*, 112–118. [CrossRef]
3. Gui, W.; Liu, Y.; Zhang, X.; He, L.; Wang, Y.; Wang, Y.; He, E.; Wang, M. Effects of Rare Earth Elements on Microstructure, Mechanical Properties and Nitriding of Medium Carbon Steel. *J. Mater. Res.* **2021**, *35*, 72–80.
4. Guo, F.; Lin, Q. Influence Mechanism of Rare Earth Elements on Impact Toughness of Low Sulfur Oxygen Alloy Structural Steel. *Chin. J. Rare Earths* **2008**, *26*, 97–101.
5. Zhao, Y.; Jiang, Y.; Li, B.; Wang, Y.; Ren, H. Effect of Rare Earth on Microstructure Evolution of Low Carbon Microalloyed High Strength Steel During Cooling. *J. Iron Steel Res.* **2022**, *34*, 169–174.
6. Liu, C.; Jiang, M.; Li, C.; Wang, Y.; Chen, J. Effect Mechanism of Rare Earth on Impact Toughness of Heavy Rail Steel. *J. Proc. Eng.* **2006**, *6*, 135–137.
7. Wei, Q.; Huiping, R.; Zili, J.; Yunping, J.; Binglei, L.; Chaoyi, W. Effect of lanthanum on the microstructure and impact toughness of HSLA steel. *Rare Met. Mater. Eng.* **2018**, *47*, 2087–2092.
8. Zeng, Z.; Reddy, K.M.; Song, S.; Wang, J.; Wang, L.; Wang, X. Microstructure and mechanical properties of Nb and Ti microalloyed lightweight δ-TRIP steel. *Mater. Charact.* **2020**, *164*, 110324. [CrossRef]
9. Wang, Y.L.; Wang, Y.; Xie, X.H.; Chen, L.F.; Chen, H. Research Progress of Rare Earth Cemented Carbides. *Nonferrous Met. Sci. Eng.* **2019**, *10*, 106–112.
10. Song, R.; Zhao, W.; Bao, X.; Chen, L.; Wang, X. Effect of rare earth element Ce and heat treatment on microstructure and mechanical properties of hypereutectoid rail steel. *Met. Heat Treat.* **2022**, *47*, 162–167.
11. Dong, R.; Li, H.; Liu, Z.; Deng, X.; Ren, X. Effect of Rare Earth Elements on Transformation Temperature of Low Carbon Steel. *Rare Earth* **2021**, *42*, 140–146.
12. *YB/T 5338-2006*; Quantitative Determination of Retained Austenite in Steel by X-ray Diffractometer Method. Beijing Iron and Steel Institute: Beijing, China, 2006.
13. Dong, R.; Zhao, Q.; Bi, X.; Deng, X.; Shen, W.; Lu, Y.; Gele, T. Effect of cooling rates after annealing on the microstructure and properties of 1000 MPa grade automobile steel for cold forming. *Mater. Res. Express* **2021**, *8*, 116508. [CrossRef]
14. Chen, J.; Lv, M.Y.; Liu, Z.Y.; Wang, G.D. Influence of Heat Treatments on the Microstructural Evolution and Resultant Mechanical Properties in a Low Carbon Medium Mn Heavy Steel Plate. *Metall. Mater. Trans. A* **2016**, *47*, 2300–2312. [CrossRef]
15. Zhao, Y.; Zhu, R.; Fan, L.; Yue, E. Microstructure evolution of 0.12C-5Mn manganese steel. *Met. Heat Treat.* **2018**, *43*, 82–84.
16. Wang, T.; Xu, H.; Yang, G.; Gao, K.; Ma, P. Study on Strong Plastic Mechanism of Medium Manganese Steel in ART Annealing Process. *Hot Work. Proc.* **2016**, *45*, 41–45.
17. Koyama, M.; Akiyama, E.; Tsuzaki, K. Hydrogen embrittlement in Al-added twinning-induced plasticity steels evaluated by tensile tests during hydrogen charging. *ISIJ Inter.* **2012**, *52*, 2283–2287. [CrossRef]
18. Xu, F.; Xiao, Y.; Chen, Q.; Deng, C.; Sun, X. Effect of cold rolling reduction ratio on microstructure, texture and deep drawing properties of IF steel. *Heat Treat. Met.* **2022**, *47*, 250–255.

Review

Ultrafast Heating Heat Treatment Effect on the Microstructure and Properties of Steels

Matteo Gaggiotti [1], Luciano Albini [2], Paolo Emilio Di Nunzio [2], Andrea Di Schino [1], Giulia Stornelli [3,*] and Giulia Tiracorrendo [2]

[1] Dipartimento di Ingegneria, Università degli Studi di Perugia, Via G. Duranti 93, 06125 Perugia, Italy
[2] RINA Consulting—Centro Sviluppo Materiali SpA, Via Castel Romano 100, 00128 Roma, Italy
[3] Dipartimento di Ingegneria Industriale, Università degli Studi di Roma "Tor Vergata", Via del Politecnico 1, 00133 Roma, Italy
[*] Correspondence: giulia.stornelli@students.uniroma2.eu

Abstract: The adoption of the ultrafast heating (UFH) process has gained much attention in the last few years, as the green energy and minimization of CO_2 emissions are the main aspects of contemporary metal science and thermal treatment. The effect of ultrafast heating (UFH) treatment on carbon steels, non-oriented grain (NGO) electrical steels, and ferritic or austenitic stainless steels is reported in this review. The study highlights the effect of ultrarapid annealing on microstructure and textural evolution in relation to microstructural constituents, recrystallization temperatures, and its effect on mechanical properties. A strong influence of the UFH process was reported on grain size, promoting a refinement in terms of both prior austenite and ferrite grain size. Such an effect is more evident in medium–low carbon and NGO steels than that in ferritic/austenitic stainless steels. A comparison between conventional and ultrafast annealing on stainless steels shows a slight effect on the microstructure. On the other hand, an evident increase in uniform elongation was reported due to UFH. Textural evolution analysis shows the effect of UFH on the occurrence of the Goss component (which promotes magnetic properties), and the opposite with the recrystallization g-fiber. The recovery step during annealing plays an important role in determining textural features; the areas of higher energy content are the most suitable for the nucleation of the Goss component. As expected, the slow annealing process promoted equiaxed grains, whereas rapid heating promoted microstructures with elongated grains as a result of the cold deformation.

Keywords: ultrafast heating; heat treatment; carbon steels; stainless steels; NGO steel; microstructure

Citation: Gaggiotti, M.; Albini, L.; Di Nunzio, P.E.; Di Schino, A.; Stornelli, G.; Tiracorrendo, G. Ultrafast Heating Heat Treatment Effect on the Microstructure and Properties of Steels. *Metals* **2022**, *12*, 1313. https://doi.org/10.3390/met12081313

Academic Editor: Andrey Belyakov

Received: 19 July 2022
Accepted: 3 August 2022
Published: 5 August 2022

Publisher's Note: MDPI stays neutral with regard to jurisdictional claims in published maps and institutional affiliations.

1. Introduction

In recent years, the application of ultrafast heating (UFH) treatment has gained considerable attention from both the academic and industrial communities [1]. Such a heat treatment is enabled by the possibility to electrically heat metallic materials by an electromagnetic induction process based on Joule heating through heat transfer passing through an induction coil, generating an electromagnetic field. The rapidly alternating magnetic field penetrates the object, generating so-called eddy currents inside the metallic component. Such currents flow through the resistance of the material, thus heating it by the Joule effect. UFH could strongly optimize industrial production processes in terms of time and consequently productivity [2–5]. Results very near to industrial applications were reported by Cola Jr. in AISI 8620 and in bainitic steels [6,7], but the most important limit appears to be related to UFH rates with respect to conventional steel-processing lines [8].

UFH is a process based on induction heating (IH) that is widely used to heat-treat specific areas more rapidly, and at a lower cost and lower amount of energy consumption compared to conventional heat treatment methods [8,9]. The inductive heat treatment offers several advantages over conventional heat treatments using a furnace, such as short processing times, high flexibility, and localized annealing [10–12]. The temperature variation over time of the annealing process is represented by the parameter of heating rate (HR). Variation in HR affects the phase transformation kinetics and steel properties; an increase in terms of HR raises the α/γ transformation temperature, thus leading to austenitic grain refinement [13–16]. At the same time, an increase in phase transformation temperatures [17] reduces the carbon amount that can be dissolved in austenite, thus lowering steel hardenability [6]. In addition, during fast heating, the solubility of carbonitrides and pearlite areas is also evidently dependent on heating time and not just temperature, as what usually happens in conventional annealing processes [18,19].

Concerning carbon steels, UFH can be adopted in the austenitization process step (before quenching and tempering) [20–29]. In the case of austenitic stainless steels, UFH is a promising method to be applied after cold rolling to promote grain refinement [30], leading to an increase in terms of mechanical properties [31–33].

In carbon steels, the application of UFH inhibits the dislocation rearrangement that is typical of recovery [33–36]. Hence, recrystallization occurs in direct competition with austenitization [37–39]. The final microstructure of a steel subjected to UFH strongly depends on heating rate and peak temperature [40,41]. In the case of a high HR and inadequate peak temperature, a not fully recrystallized microstructure is achieved [42,43]. When conditions instead allow for complete recrystallization, the microstructure shows a morphology resulting in plastic deformation.

Since recovery is usually negligible in the case of austenitic stainless steels, the immediate start of recrystallization during UFH does not imply significant changes in the mechanism [44]. Ferritic stainless steels, instead, show a final microstructure comparable with that of industrial products [45,46], with a fully recrystallized structure and equiaxed grains for appropriate temperatures [47–51].

The aim of this review paper is to highlight trends in variation in the above-mentioned properties with heating rate, and to predict their behavior when possible for different classes of materials, focusing on the main metallurgical topics (phase transformation, recrystallization, and textural evolution).

2. Effect of UFH on Grain Size

In medium/low carbon steels, the application of UFH leads to a significant decrease in grain size due to the suppression of grain boundary movements, and thus the growth of austenite grains [52–58]. Figure 1 shows the prior austenite grain (PAG) size dependence of heating rate (HR) [40]. In the case of 0.25 wt. % C heat-treated specimens with a peak temperature at 850 °C, HR (between 10 and 1000 °C/s) and cooling rate of 20 °C/s, some authors [40] investigated the effect of HR on the variation in ferritic grain size (Figure 2). Results show an evident average PAG size refinement of about 55% (from 3.8 down to 1.7 µm) (Figure 1) that corresponds to a consequent decrease in ferritic grain size (about 20%, from 2.5 to 2 mm) (Figure 2) as the HR increased from 10 up to 1000 °C/s. A similar steel grade (0.2 wt. % C) was considered by Hernandez-Duran et al. [57], who studied the effect of the same HR as that in [40] on the ferritic grain size, and the results, reported in Figure 2, show a similar refinement of ferritic grain size, with a reduction from 2 to 1.6 mm.

Petrov et al. [59] studied the effect of heating rate increase on a cold-rolled high strength carbon steel grade for automotive applications. They indicated that, when varying the heating rate in the range from 140 to 1500 °C/s, ferritic grain refinement was achieved from 4 to 1 µm. They also showed that excellent ultimate tensile strength (higher than 1200 MPa) and acceptable fracture elongation were achieved. The above results were confirmed by an increase in heating rate of up to 1500 °C/s [40] that, compared to 150 °C/s, led to a more pronounced decrease in ferritic grain size of about 76%, from 2.6 to 0.6 µm (Figure 3).

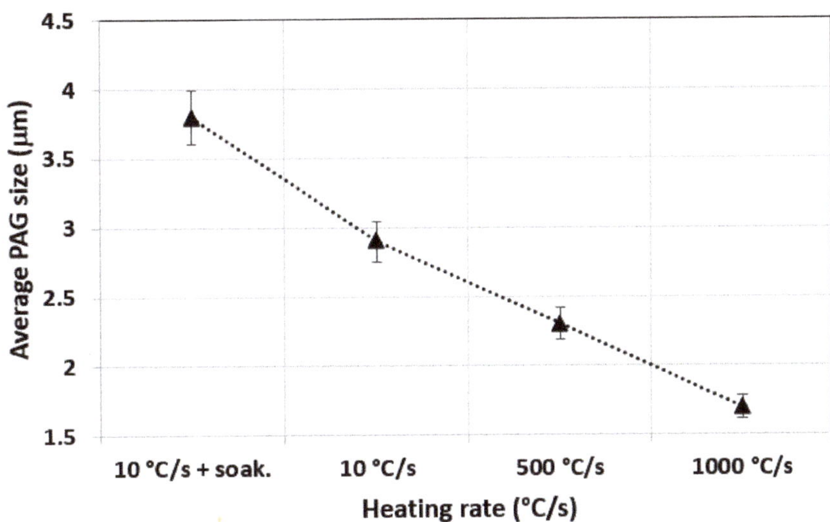

Figure 1. Effect of the heating rate increase on PAG size (data from [40]). Low/medium carbon steel (0.25 wt. % C), peak temperature of 850 °C, and cooling rate of 20 °C/s.

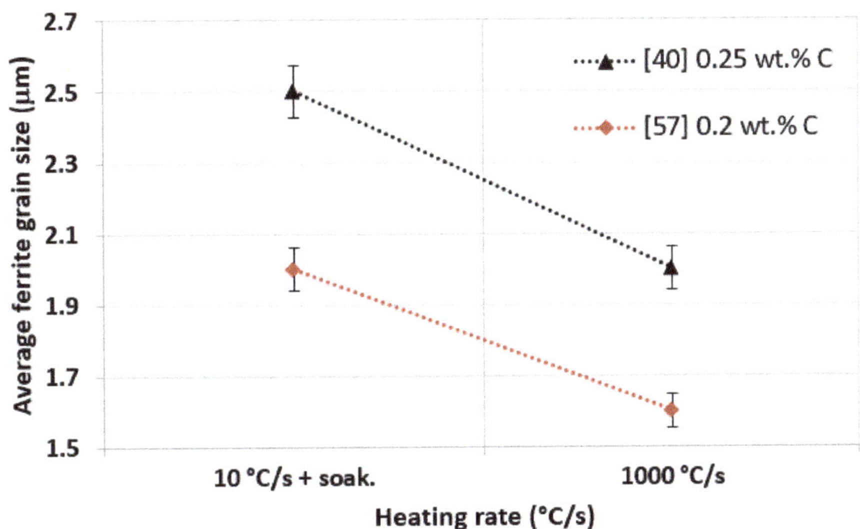

Figure 2. Effect of the heating rate increase on ferritic grain size. Low/medium carbon steel (0.25 wt. % C and 0.2 wt. % C), peak temperature of 850 °C and 950 °C, cooling rate of 20 °C/s and 160 °C/s (data from [40,57]).

An example of microstructures obtained after UFH is reported in Figure 4, referring to the industrial hot-rolled strips of a low-C steel (0.08% C, 0.3% Mn) and of an HSLA steel (0.07% C, 0.5% Mn, 0.03% Nb), cold-rolled from 3 down to 0.5 mm, and processed by imposing different peak temperatures and heating rates to find the optimal processing conditions for achieving full recrystallization. No soaking was applied, and the strips were cooled down just after reaching the peak temperature.

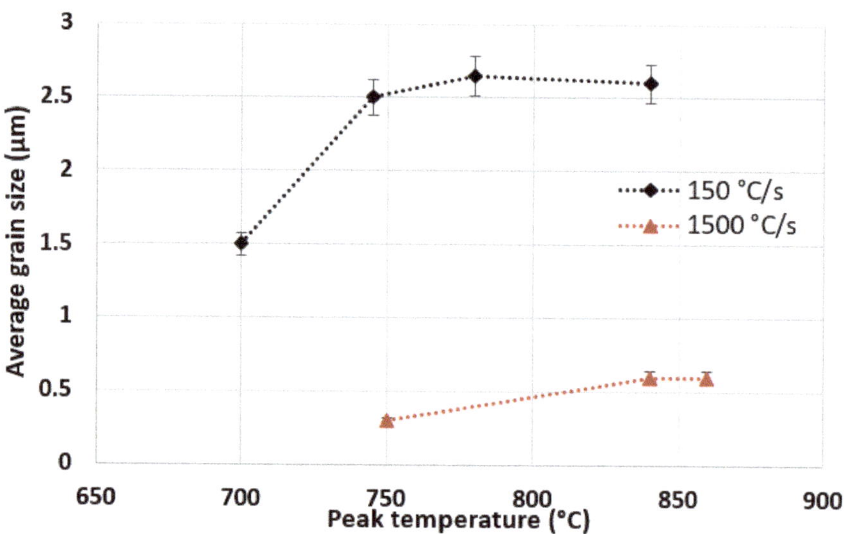

Figure 3. Ferritic grain size variation as function of temperature for the heating rates of 150 and 1500 °C/s (data from [42]).

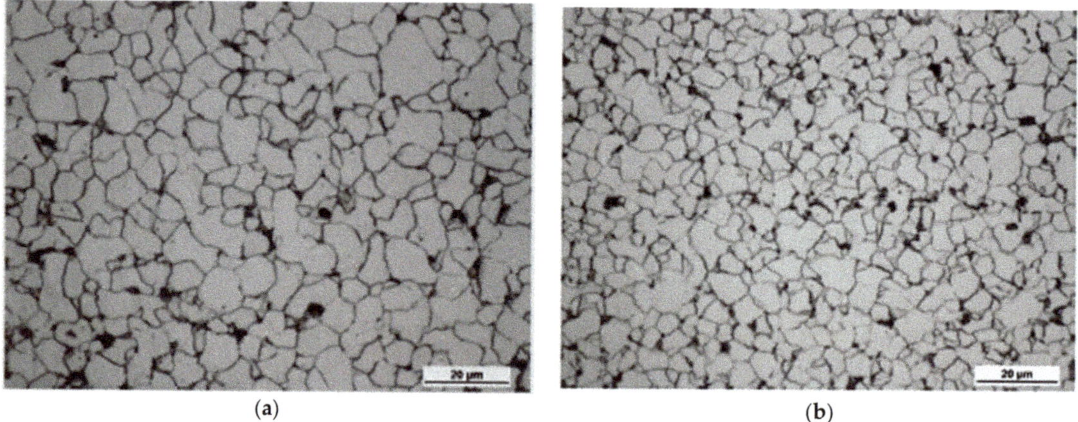

(a)   (b)

Figure 4. Microstructure of fully recrystallized steels subjected to UFH. (**a**) Low-C 0.08% C, 0.3% Mn steel; (**b**) HSLA 0.07% C, 0.5% Mn, 0.03% Nb steel.

The first steel was processed with a peak temperature of 870 °C and a heating rate of 285 °C/s, obtaining an average ferritic grain size of 6.1 μm and a hardness of 136 HV$_{300g}$ (Figure 4a). The second steel was processed with a peak temperature of 820 °C and a heating rate of 325 °C/s, obtaining an average ferritic grain size of 5.4 μm and a hardness of 158 HV$_{300g}$ (Figure 4b). The microstructures appeared to not be very refined, but the resulting tensile properties, nevertheless, complied with grades DX51D and HX340LAD.

If ferritic stainless steels are considered, an increasing heating rate leads to less marked grain refinement, as reported in [45]. Data reported in [45] showed that the average grain size of AISI 430 steel decreased from 7.4 to 6.7 μm for a heating rate of 1000 °C/s. Similarly, Salvatori and Moore [60] analyzed the effect of UFH on AISI 430 ferritic stainless steels, and showed them to be able to reproduce industrial standard products with a heating rate of up 1000 °C/s.

3. Effect of UFH on Ac$_1$–Ac$_3$ Temperatures and on Recrystallization Phenomenon

The quantitative effect of heating rate increase on Ac$_1$ and Ac$_3$ temperatures is related to the steel chemical composition, and several data are found in the literature showing in many cases an increase in both critical temperatures with increasing heating rate.

In a low/medium carbon steel, Valdes-Tabernero et al. [48] (Figure 5) showed a more pronounced increase in Ac$_1$ than that in Ac$_3$ in the range of heating rates from 1 to 200 °C/s. An increase in Si and Mn leads to the opposite effect [38]. In this case, a stronger dependence of Ac$_3$ temperature on heating rate with respect to Ac$_1$ was also reported (Figure 6).

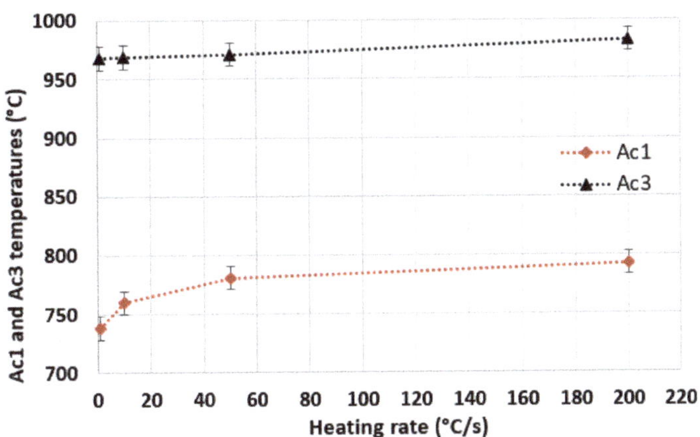

Figure 5. Effect of HV increase on Ac$_1$ and Ac$_3$ temperature variation (steel chemical composition 0.19% C + 0.5% Si + 1.61% Mn + 1.06% Al), peak temperature of 1100 °C, and cooling rate of 300 °C/s (data from [48]).

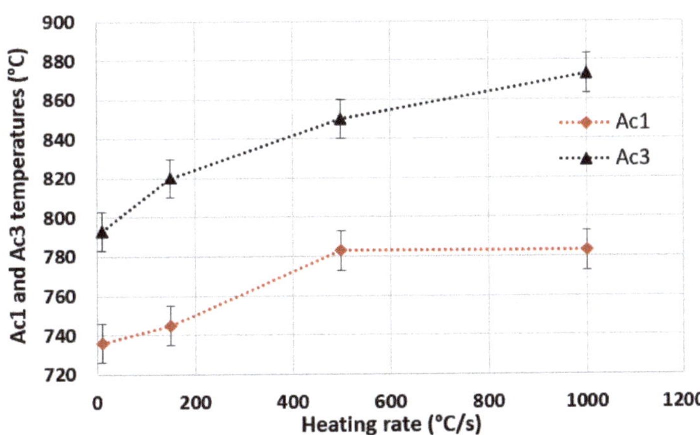

Figure 6. Ac$_1$ and Ac$_3$ temperature variation with an increase in HR (steel chemical composition 0.25% C + 1.5% Si + 3.0% Mn) (data from [40]).

Concerning the recrystallization process, the increase in heating rate led to an increase in the recrystallization temperature, supported by the following evidence:

- The recrystallized fraction decreased with increasing heating rate: in a low carbon steel at 750 °C for a heating rate of 150 °C/s, the recrystallized fraction was ~0.5%. At the same temperature for a heating rate of 1500 °C/s, the recrystallized fraction decreased to 0.1% [42]; in [43], compared to a conventional annealing treatment (10 °C/s and recrystallized fraction of ~77%), during an ultrafast annealing process (778 °C/s), the recrystallized fraction became ~24%.

- In [42], the recrystallized fraction increased with increasing peak temperature: for a peak temperature of 850 °C for two heating rates (150 and 1500 °C/s), the recrystallized fraction was ~0.6% for 150 °C/s and ~0.2% for 1500 °C/s, a value greater than that for 750 °C (Figure 7).
- The recrystallization of ferrite takes place simultaneously with the $\alpha \rightarrow \gamma$ phase transformation [61].

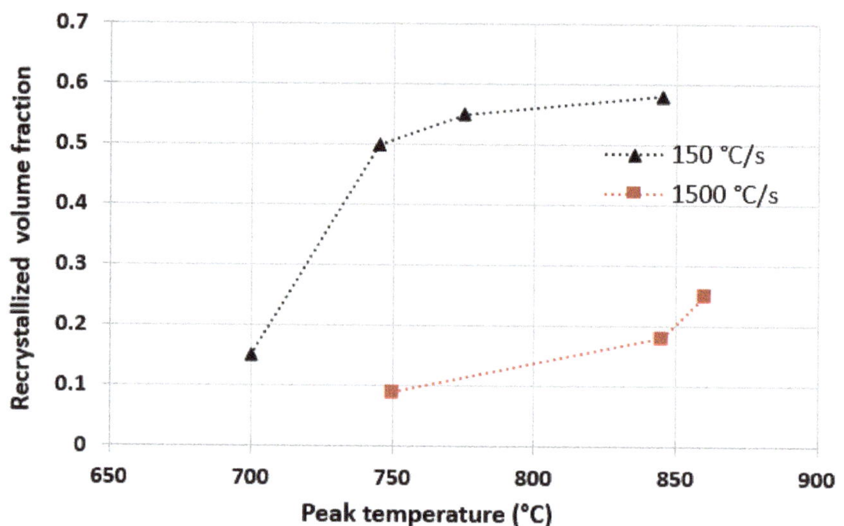

Figure 7. Recrystallized fraction as a function of peak temperature for two different HRs (data from [42]).

4. Effect of UFH on Mechanical Properties

4.1. Carbon Steels

Directly connected to the effect of ultrafast heating on grain size refinement, the mechanical properties of carbon steels subjected to the UFH process showed a clear improvement. Specifically, the evolution of mechanical characteristics was reported for low/medium carbon steel in the case of conventional annealing + soaking (10 °C/s + soak.), conventional annealing (10 °C/s) and ultrafast annealing (500–1000 °C/s) prior to quenching and partitioning (Q and P) heat treatment [40]. Results showed that the Ultimate Tensile Strength (UTS)increased for a higher HR (Figure 8); from conventional annealing (HR of CA = 10 °C/s) to ultrafast heating (HR of UFH = 1000 °C/s), the UTS shifted from 1097 to 1318 MPa (an increase of ~20%). The yield stress (YS), in the same range of heating rate, decreased from 837 to 811 MPa (−3.2%), and uniform elongation ε_c increased from 2.6% to 12.8%.

A comparison of CA (5 °C/s) and UFH (500 °C/s) on dual-phase steel [39] showed the following results: the UTS improved from 625.0 ± 3.6 MPa to 666.0 ± 2.6 MPa (~6.5%), the YS increased from 277.0 ± 8.1 MPa to 372.0 ± 3.0 MPa (~34%), and the uniform elongation and total elongation increased from 16.5 ± 0.2% to 18 ± 0.5% and from 23.3 ± 0.8% to 26.6 ± 0.5%, respectively (Figures 9 and 10).

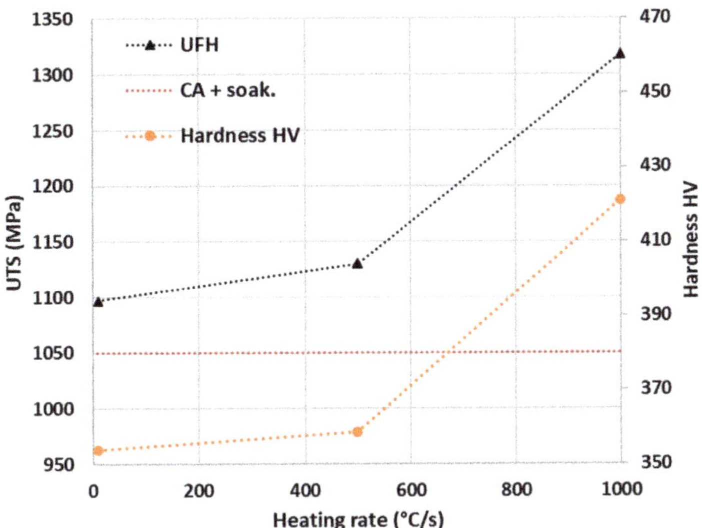

Figure 8. Effect of HR increase on UTS and Vicker hardness value of a low/medium carbon steel undergone the UFH process compared to CA + soaking process (steel chemical composition 0.25% C + 1.5% Si + 3.0% Mn)). Peak temperature of 850 °C, and cooling rate of 20 °C/s (data from [40]).

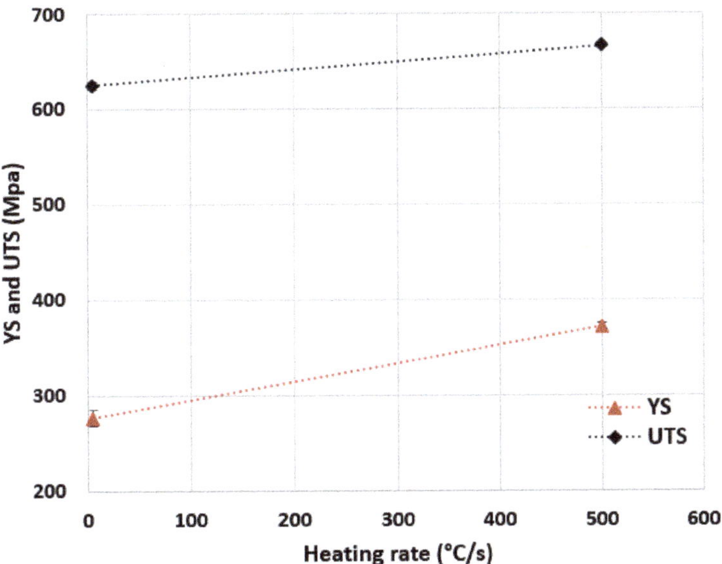

Figure 9. YS and UTS increase with increase in heating rate between CA (5 °C/s) and UFH (500 °C/s), on dual-phase steel (data from [41]).

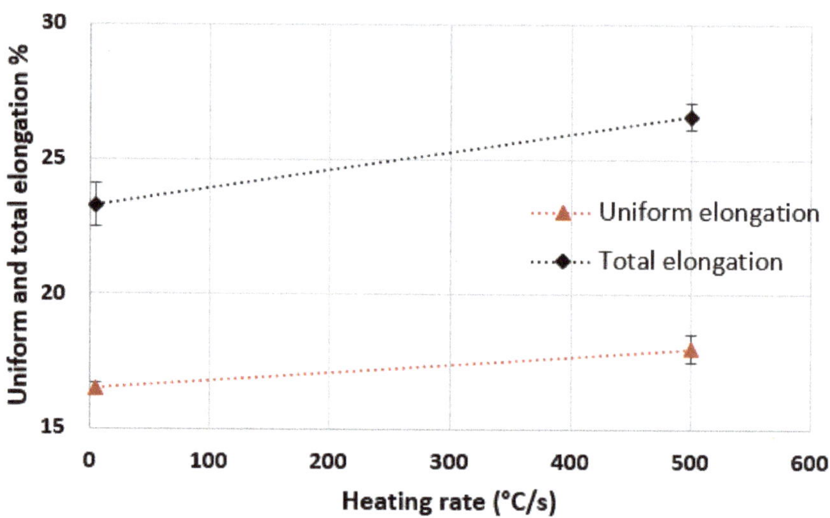

Figure 10. Uniform and total elongation variation with increasing heating rate between CA (5 °C/s) and UFH (500 °C/s) on dual-phase steel (data from [41]).

4.2. Stainless Steels

Regarding the mechanical properties in ferritic stainless steels [60]:

- The values were very similar to those typical of standard industrial production, even if the tensile strength was a little lower. These properties improved as the holding time increased and they did not seem to depend on heating rate. With further increases in holding time, yield stress and tensile strength values remained almost constant, and the total elongation increased.
- A thermal cycle using no holding time in the annealing treatment allowed for obtaining a product that was quite comparable to the industrial standard in the whole range of heating rates (500–1000 °C/s).

Instead, from data reported in [45]:

- The YS, UTS, and ε_c were practically equal for all specimens heated at a peak temperature below 900 °C. At a peak temperature of 900 °C, the yield strength was reduced, even though the grain size and hardness did not change from those at 880 °C.
- As the annealing temperature increased to 900 °C, the yield strength decreased, which is beneficial for formability.

In austenitic stainless steels, data from [44] show that:

- Without any holding time at a peak temperature of 700 °C/s and heating rates of 2 and 20 °C/s, the YS increased to 870 MPa, with an improvement in ductility and uniform elongation (ε_c). Increasing from 2 to 100 °C/s, the UTS and YS did not change too much (UTS from 1013 to 967 MPa and YS from 856 to 778 MPa)), but ductility was improved.
- The uniform elongation was improved from 11% to 37% with the increase in HR from 2 to 100 °C/s.

5. Effect of UFH on Microstructure

The effect of heating rate on the microstructure was clearly reported in [43] for a process involving a heating rate of 778 °C/s in comparison to a standard 10 °C/s heating rate.

The recrystallized volume for a low-carbon steel (0.14 wt. % C), decreased from 77.4% to 27.3%, the fraction of recovered ferrite increased from 8.6% to 54.5%, and the martensite + pearlite fraction increased from 14% to 21.8% (Figure 11).

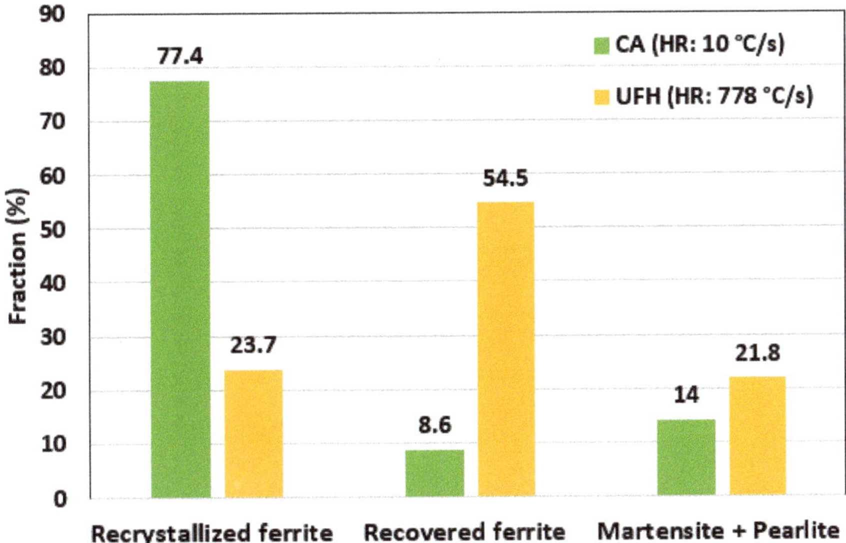

Figure 11. Microstructural evolution for CA (HR: 10 °C/s) and UFH (HR: 778 °C/s), low-carbon steel (0.14% C) (data from [43]).

This result is in contrast with that in [38,39], where an increase in Ac_1 temperature was reported with increasing HR; therefore, less martensite was expected in the microstructure of the UFH sample than that in the CA sample.

The difference can be explained by the lack of ferritic recrystallization in the UFH sample.

A possible explanation of the above results can be that the peak temperature of the specimen treated with CA was lower than that processed by UFH, and that, in the latter treatment, some austenite was formed. In the case of CA, it is reasonable to assume that not only was the peak temperature lower, but also the treatment time at high temperature was longer, so that the recrystallized fraction could be higher than that of the steel treated with UFH. When the steel was cooled, the austenite transformed into martensite.

From the comparison between CA + soaking and UFH prior to Q and P heat treatment [38], for a carbon steel (0.25 wt. % C), the fraction of ferrite increased for higher heating rates from 0% at 10 °C/s + soak. to 25% at 1000 °C/s. This was due to the very short heating time and consequently the absence of complete austenitization. The most important difference from 500 to 1000 °C was the fraction on fresh martensite (15% → 2%) that was replaced by ferrite (5% → 25%) (Figure 12).

Therefore, for this class of steels, UFH heating rates of up to about 500 °C/s are still suitable to achieve a similar microstructure to those of Q and P grades. Instead, for heating rates above 500 °C/s, complex microstructures are produced where ferrite coexists with retained austenite.

The dependence of microconstituent fraction formation as a function of a combination of HR, peak temperature, and steel chemical composition is reported in Figures 13–16 (data from [57]). The comparison is between a low-carbon steel (0.2 wt. % C) and stabilized low-carbon steel (0.2 wt. % C + Mo + Ti + Nb), and the results show that:

- The martensite volume fraction decreased with increasing heating rate (Figures 13 and 14).
- At Ac_3 temperature (Figures 14 and 15), the fraction of martensite showed an opposite trend at 1000 °C/s with respect to lower temperature. This is evidence of the impact of peak temperature on ultrafast heating treatment.
- Concerning the low-C stabilized steel, the effect of alloying elements on hardenability was clear. After UFH at Ac_3 temperature, the fraction of martensite was equal to 100% (Figure 16).

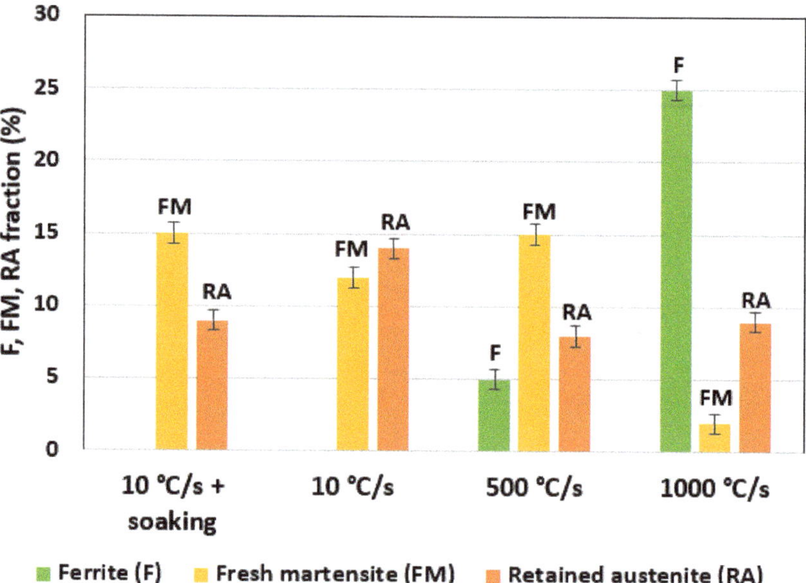

Figure 12. Microstructural evolution for different heating rates (data from [40]; steel chemical composition: 0.25% C + 1.2% Si + 3.0% Mn).

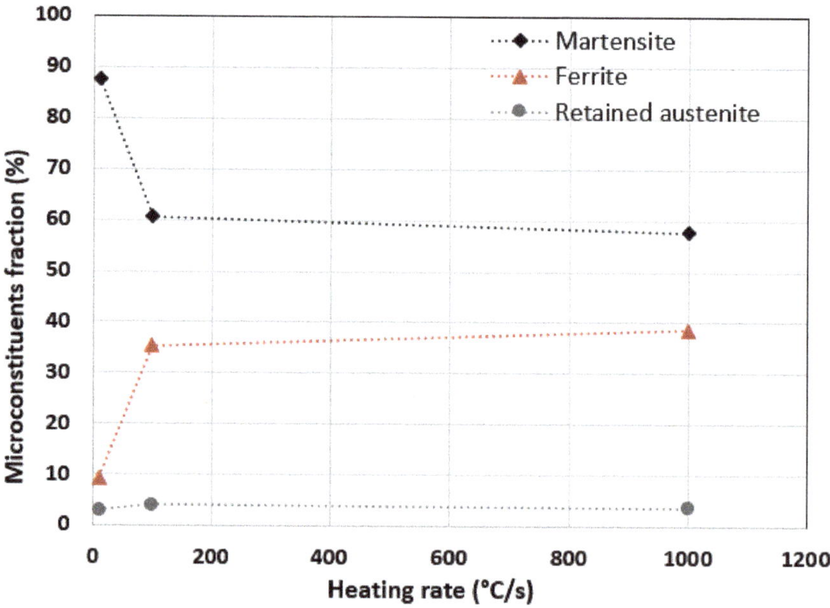

Figure 13. Microstructural evolution at different HRs for peak temperature at 902 °C (data from [57]; steel chemical composition: 0.2% C).

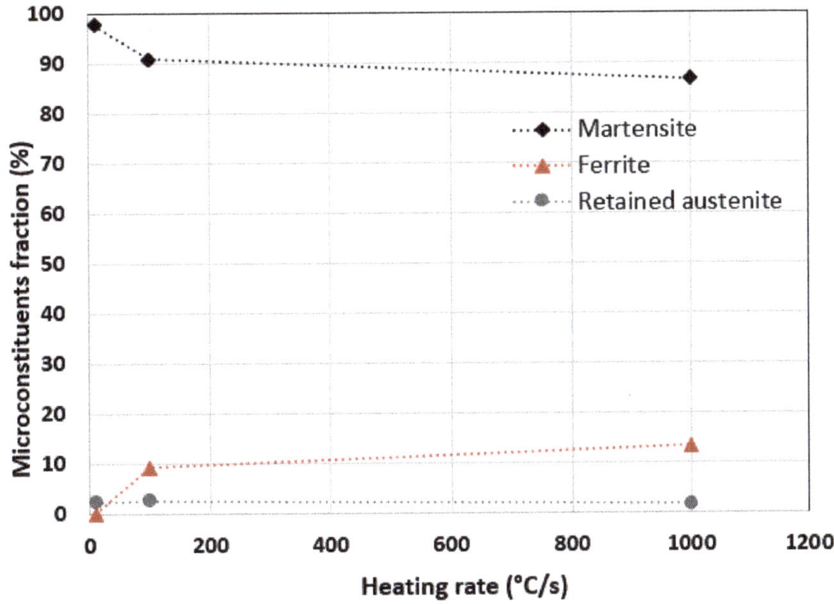

Figure 14. Microstructural evolution at different HRs for peak temperature at 915 °C (data from [57]; steel chemical composition: 0.2% C + Mo + Ti + Nb).

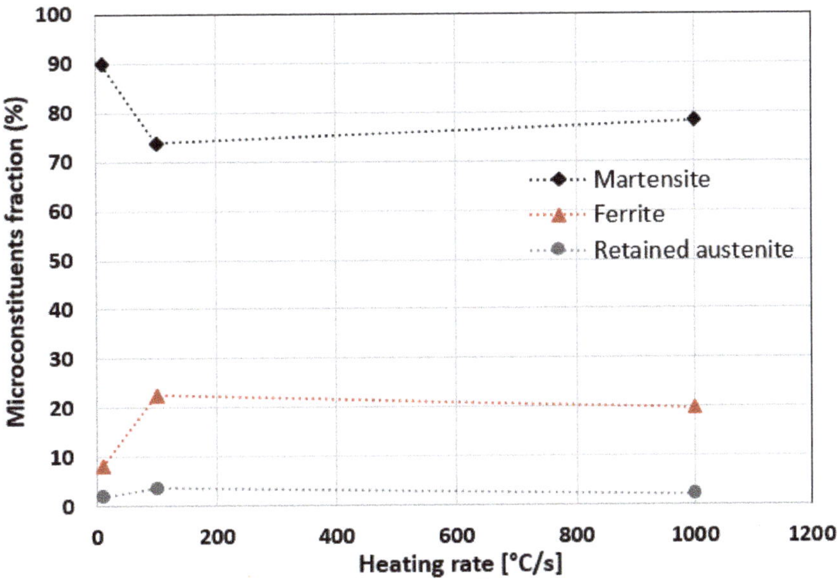

Figure 15. Microstructural evolution at different HRs for peak temperature at Ac3: 950 °C (data from [57]; steel chemical composition: 0.2% C).

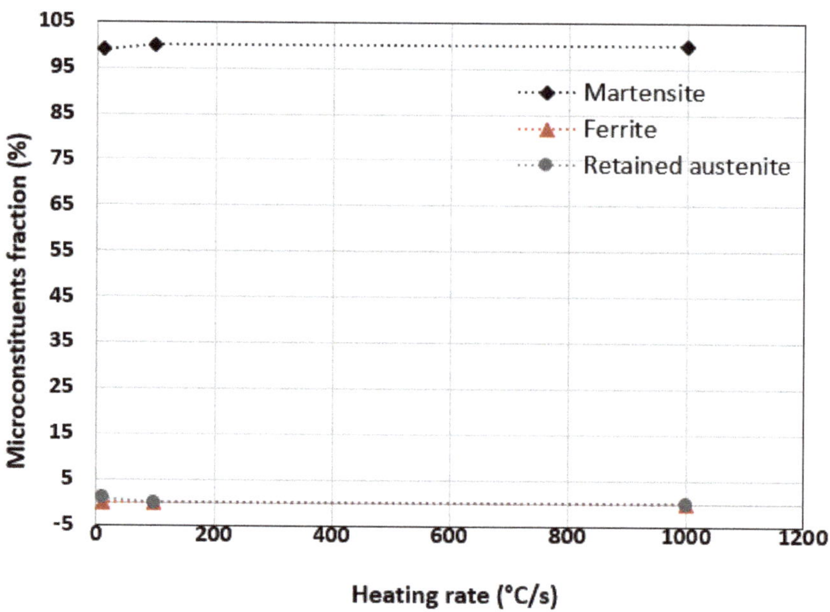

Figure 16. Microstructural evolution at different HRs for peak temperature at Ac3: 950 °C (data from [57]; steel chemical composition: 0.2% C + Mo + Ti + Nb).

The effect of an increased heating rate in reducing the final fraction of martensite at the end of the treatment was apparent at 100 °C/s. Possible reasons to explain why the martensite fraction was generally lower in the low-alloyed steel compared to the alloyed one are, on one hand, the higher critical transformation temperatures and, on the other hand, the lower hardenability. In the former case, the amount of austenite transformed during the process was lower; in the latter case, a lower fraction of the austenite formed at high temperature is converted to martensite during cooling.

6. Effect of UFH on the Textural Evolution and Magnetic Properties of NGO Steels

Nonoriented grain (NGO) electrical steels can be classified as low-C steels, and regarding the microstructural evolution, mechanical properties, and phase transformation temperatures, the effect of UFH can be considered to be similar. In this section, the effect of UFH on the textural evolution and magnetic properties of NGO steels is reported.

For NGO electrical steels, the role of ultrafast annealing on textural evolution is evident, since an increase in the occurrence of Goss component is observed, which leads to an improvement of magnetic properties [61–63].

The magnetic properties of Fe–Si steels strictly depend on textural evolution [64–70], and consequently on the starting microstructure [71–73], annealing conditions [74], and cold-rolling settings [75–80]. Core losses and eddy current losses are strongly dependent on recrystallization and grain size morphology after cold rolling. Grain size and the absence of impurities such as dislocations, grain boundaries, or precipitates are the most important factors determining power losses. The most relevant textural components in electrical steels are the Goss component (110) [001], the rotated Cube component (100) [011], and the γ-fiber [111]//ND [78–82]. After cold rolling, the texture is characterized by very strong Goss and cube component orientations that, in principle, are very positive in terms of magnetic features; in fact, the matrix is highly deformed. However, the effect of the subsequent recrystallization annealing treatment promotes the development of the γ-fiber, the typical recrystallization texture of ferritic steels, and the consequent reduction in deformation components. In conventional industrial lines, the nucleation of textural components starts during the (slow) heating stage and interacts with the concurrent recovery process. Therefore, since the Goss and Cube components nucleate in high-energy zones of the material, the UFH treatment could represent an opportunity to reduce the undesired effect of recovery, and to produce NGO steels with better textural properties.

Wang et al. [64] explored different heat treatment approaches to optimize the magnetic properties of electrical steels. A cold-rolled NGO steel was subjected to an extremely short annealing cycle in the range of 3–30 s, with heating rates from 15 to 300 °C/s and peak temperature from 880 to 980 °C. The presence of silicon (no less than 1%) increases the critical temperatures of the steel, thus permitting the use of peak temperatures higher than those of the corresponding low-carbon steels. In the case of fast heating, a very strong {110}<001> (Goss) texture develops [64,72]. Its intensity increases with increasing heating rate, but decreases as the annealing time increases. Moreover, in coarse-grained specimens, the Goss component is significantly strengthened, and the {111}<112> component is slightly weakened with increasing heating rate. On the other hand, in a fine-grained specimen, the intensity of the Goss component showed only a slight increase, but the {111}<112> component was greatly reduced.

Generally speaking, as the heating rate increases, the typical recrystallization texture of ferrite (γ-fiber) weakens significantly due to the very short heating treatment, whereas the Goss component is extremely favored. Since the nucleation of the Goss component occurs in regions of the deformed matrix with higher stored energy (namely, shear bands), higher heating rates reduce the negative effect of recovery, thus promoting their nucleation and further growth [65–68,71]. This result is in agreement with the work of Hutchinson [83], who reported a schematic ranking of the nucleation rate of different textural components as a function of time during annealing. In fact, from a kinetic viewpoint, the {110} orientations started nucleating earlier than {111} did, thus exploiting a larger kinetic advantage as the heating rate increased.

Regarding the magnetic properties of NGO electrical steels as a function of heating rate, the following results emerged from the experiments [63,64]:

- Magnetic induction was improved as the HR increased from 15 to 300 °C/s. This was due to the optimized recrystallized texture caused by rapid heating.
- Core losses decreased as the heating rate increased from 15 to 100 °C/s. As a matter of fact, this effect was apparent when the heating rate exceeded 100 °C/s since it was associated with the decrease in mean grain size with increasing HR. Incidentally, this induced a parallel decrease in classical eddy current losses, but an increase in hysteresis losses.
- Annealing treatments up to 850 °C after cold rolling improved the magnetic properties (higher permeability and lower energy losses) due to the increase in grain size and the development of a {100} fiber-type texture.

7. Conclusions

The effect of ultrafast heating on different classes of steel, specifically on medium/low-C, NGO, and stainless steels, was analyzed in this paper with respect to the main metallurgical aspects: phase transformation, recrystallization, mechanical properties, and textural evolution. In particular, the variation trend with heating rate was highlighted, focusing on the main metallurgical properties (phase transformation, recrystallization, and textural evolution) for different classes of materials.

The conclusions can be summarized in the following points:

- The UFH treatment had a strong effect on reducing the grain size, in particular on medium/low-C steels, with a PAG size reduction of 55%, shifting from 10 to 1000 °C/s. Ferritic grains showed a maximal reduction of 76%, increasing from 150 to 1500 °C/s. The effect on the stainless steels was similar but much less pronounced: the average grain size reduction was 10% from 25 to 500 °C/s.
- In medium/low-C steels, with increasing heating rate, critical transformation points Ac1 and Ac3 were shifted towards higher temperatures. Furthermore, the recrystal-

lization temperatures were also shifted towards higher values in both carbon and stainless steels.

- The increase in heating rate produces an improvement of almost all mechanical characteristics. In medium/low-C steels, the comparison of conventional annealing and ultrafast heating showed an increase in UTS and ε_c (about 20% and from 2.6% to 12.8%, respectively); the YS remained at the same level. When UFH was applied to ferritic stainless steels, the annealed strip has comparable mechanical characteristics with those from standard industrial production. In austenitic stainless steels, there was instead an improvement of the uniform elongation from 11% to 37% with increasing heating rate from 2 to 100 °C/s.

- Textural evolution was directly connected to magnetic properties of NGO steels, and higher heating rates promoted the nucleation of Cube and Goss components in the high-stored-energy zones. Annealing time had a significant effect on the occurrence of the Goss component that nucleates at the initial stages of recrystallization. Moreover, a slow heating with a longer recovery phase reduced the intensity of the Goss component, and the overall texture is more random. In terms of magnetic properties, the magnetic induction increased from 15 to 300 °C/s, and core losses decreased in a range of heating rate from 15 to 100 °C/s. In the case of a heating rate greater than 100 °C/s, grain size underwent severe refinement, with a consequent decrease in eddy current losses, and an increase in hysteresis losses.

Author Contributions: Conceptualization, M.G., A.D.S.; methodology, M.G.; formal analysis, M.G., G.S: data curation, M.G., G.S.; writing—original draft preparation, M.G.; writing—review and editing, A.D.S., G.S., G.T., P.E.D.N.; supervision, A.D.S., P.E.D.N.; project administration, L.A. All authors have read and agreed to the published version of the manuscript.

Funding: This research was funded by REGIONE UMBRIA based on *Piano Sviluppo e Coesione FSC ex DGR n. 251/2021*, grant number CUP I42C20001280008.

Institutional Review Board Statement: Not applicable.

Informed Consent Statement: Not applicable.

Data Availability Statement: The data presented in this study are available on request from the corresponding author.

Conflicts of Interest: The authors declare no conflict of interest.

References

1. Funatani, K. Heat treatment of automotive components: Current status and future trends. *Trans. Indian Inst. Met.* **2004**, *57*, 381–396.
2. Naar, R.; Bay, F. Numerical optimisation for induction heat treatment processes. *Appl. Math. Model.* **2013**, *37*, 2074–2085. [CrossRef]
3. Yan, P.; Güngör, O.E.; Thibaux, P.; Liebeherr, M.; Bhadeshia, H.K.D.H. Tackling the toughness of steel pipes produced by high frequency induction welding and heat-treatment. *Mater. Sci. Eng. A* **2011**, *528*, 8492–8499. [CrossRef]
4. Sherman, W. Case hardening by megacycle induction heating. *Trans. Electrochem. Soc.* **1944**, *86*, 247. [CrossRef]
5. Kalpande, S.D. Performance improvement, an induction hardening machine, cycle time, and productivity. *Perform. Improv.* **2007**, *46*, 9–16. [CrossRef]
6. Lolla, T.; Cola, G.; Naraynana, B.; Alexandrov, B.; Babu, S.S. Development of rapid heating and cooling (flash processing) process to produce advanced high strength steel microstructures. *Mater. Sci. Technol.* **2022**, *27*, 863. [CrossRef]
7. Cola, G., Jr. Flash bainite nucleated in 80ms by water quenching, AIST Steel Properties and Applications Conference Proceedings, Combined with Ms and T'07. *Mater. Sci. Technol.* **2007**, *6*, 3845.
8. Jászfi, V.; Prevedel, P.; Eggbauer, A.; Godai, Y.; Raninger, P.; Mevec, D.; Panzenböck, M.; Ebner, R. Influence of the parameters of induction heat treatment on the mechanical properties of 50CrMo4. *HTM J. Heat Treat. Mater.* **2019**, *74*, 366–379. [CrossRef]
9. Honda, T.; Santos, E.C.; Kida, K.; Shibukawa, T. Changes in the microstructure of 13Cr-2Ni-2Mo stainless steels through the quenching process by induction heating. *Int. J. Mater. Prod. Technol.* **2012**, *45*, 31–40. [CrossRef]
10. Markovsky, P.E.; Semiatin, S.L. Tailoring of microstructure and mechanical properties of Ti-6Al-4V with local rapid (induction) heat treatment. *Mater. Sci. Eng. A* **2011**, *528*, 3079–3089. [CrossRef]
11. Napoli, G.; Paura, M.; Vela, T.; Di Schino, A. Colouring titanium alloys by anodic oxidation. *Metalurgija* **2018**, *57*, 111–113.

12. Cunningham, J.L.; Medlin, D.J.; Krauss, G. Effects of induction hardening and prior cold work on a microalloyed medium carbon steel. *J. Mater. Eng. Perform.* **1999**, *8*, 401–408. [CrossRef]

13. Vieweg, A.; Ressel, G.; Prevedel, P.; Marsoner, S.; Ebner, R. Effects of the inductive hardening process on the martensitic structure of a 50CrMo4 steel. *HTM J. Heat Treat. Mater.* **2017**, *72*, 3–9. [CrossRef]

14. Clarke, K.D.; Van Tyne, C.J.; Vigil, C.J.; Hackenberg, R.E. Induction hardening 5150 steel: Effects of initial microstructure and heating rate. *J. Mater. Eng. Perform.* **2011**, *20*, 161–168. [CrossRef]

15. Eggbauer, A.; Lukas, M.; Prevedel, P.; Panzenböck, M.; Ressel, G.; Ebner, R. Effect of Initial Microstructure, Heating Rate, and Austenitizing Temperature on the Subsequent Formation of Martensite and Its Microstructural Features in a QT Steel. *Steel Res. Int.* **2019**, *90*, 1800317. [CrossRef]

16. Roumen, P.; Jurij, S.; Wlodzimierz, K.; Leo, K. Grain refinement of a cold rolled TRIP assisted steel after ultra short annealing. *Mater. Sci. Forum* **2012**, *715–716*, 661–666. [CrossRef]

17. Banis, A.; Bouzouni, M.; Petrov, R.H.; Papaefthymiou, S. Simulation and characterisation of the microstructure of ultra-fast heated dual-phase steel. *Mater. Sci. Technol.* **2020**, *36*, 1282–1291. [CrossRef]

18. Ochi, T.; Koyasu, Y. Strengthening of surface induction hardened parts for automotive shafts subject to torsional load. *SAE Tech. Pap.* **1994**, *103*, 570–580. [CrossRef]

19. Sackl, S.; Leitner, H.; Zuber, M.; Clemens, H.; Primig, S. Induction Hardening vs Conventional Hardening of a Heat Treatable Steel. *Metall. Mater. Trans. A Phys. Metall. Mater. Sci.* **2014**, *45*, 5657–5666. [CrossRef]

20. Castro Cerda, F.M.; Sabirov, I.; Goulas, C.; Sietsma, J.; Monsalve, A.; Petrov, R.H. Austenite formation in 0.2% C and 0.45% C steels under conventional and ultrafast heating. *Mater. Des.* **2017**, *116*, 448–460. [CrossRef]

21. Massardier, V.; Ngansop, A.; Fabregue, D.; Cazottes, S.; Merlin, J. Ultra-rapid intercritical annealing to improve deep drawability of low-carbon, al-killed steels. *Metall. Mater. Trans. A Phys. Metall. Mater. Sci.* **2012**, *43*, 2225–2236. [CrossRef]

22. Valdes-Tabernero, M.A.; Vercruysse, F.; Sabirov, I.; Petrov, R.H.; Monclus, M.A.; Molina-Aldareguia, J.M. Effect of Ultrafast Heating on the Properties of the Microconstituents in a Low-Carbon Steel. *Metall. Mater. Trans. A Phys. Metall. Mater. Sci.* **2018**, *49*, 3145–3150. [CrossRef]

23. Castro Cerda, F.M.; Goulas, C.; Sabirov, I.; Kestens, L.A.I.; Petrov, R.H. The effect of the pre-heating stage on the microstructure and texture of a cold rolled FeCMnAlSi steel under conventional and ultrafast heating. *Mater. Charact.* **2017**, *130*, 188–197. [CrossRef]

24. Castro, N.A.; de Campos, M.F.; Landgraf, F.J.G. Effect of deformation and annealing on the microstructure and magnetic properties of grain-oriented electrical steels. *J. Magn. Magn. Mater.* **2006**, *304*, 617–619. [CrossRef]

25. Bacaltchuk, C.M.B.; Castello-Branco, G.A.; Garmestani, H.; Rollett, A.D. Effect of magnetic field during secondary annealing on texture and microstructure of nonoriented silicon steel. *Mater. Manuf. Process.* **2004**, *19*, 611–617. [CrossRef]

26. Di Schino, A.; Testani, C. Corrosion behaviour and mechanical properties of AISI 316 stainless steels clad Q235 plate. *Metals* **2020**, *10*, 552. [CrossRef]

27. Di Schino, A.; Gaggiotti, M.; Testani, C. Heat treatmemnt effect on microstructure evolution in a 7% Cr steel for forging. *Metals* **2020**, *10*, 808. [CrossRef]

28. Stornelli, G.; Montanari, R.; Testani, C.; Pilloni, L.; Napoli, G.; Di Pietro, O.; Di Schino, A. Microstructure refinement effect on EUROFER 97 steel for nuclear fusion application. *Mater. Sci. Forum* **2021**, *1016*, 1392–1397. [CrossRef]

29. Stornelli, G.; Gaggia, D.; Rallini, M.; Di Schino, A. Heat Treatment Effect on Maraging Steel Manufactured by Laser Powder Bed Fusion Technology: Microstructure and Mechanical Properties. *Acta Metall. Slovaca* **2021**, *27*, 122–126. [CrossRef]

30. Stornelli, G.; Gaggiotti, M.; Mancini, S.; Napoli, G.; Rocchi, C.; Tirasso, C.; Di Schino, A. Recrystallization and Grain Growth of AISI 904L Super-Austenitic Stainless Steel: A Multivariate Regression Approach. *Metals* **2022**, *12*, 200. [CrossRef]

31. Masumura, T.; Seto, Y.; Tsuchiyama, T.; Kimura, K. Work-hardening mechanism in high-nitrogen austenitic stainless steel. *Mater. Trans.* **2020**, *61*, 678–684. [CrossRef]

32. Mallick, P.; Tewary, N.K.; Ghosh, S.K.; Chattopadhyay, P.P. Effect of cryogenic deformation on microstructure and mechanical properties of 304 austenitic stainless steel. *Mater. Charact.* **2017**, *133*, 77–86. [CrossRef]

33. Humphreys, F.J.; Hatherly, M. Grain Growth Following Recrystallization. In *Recrystallization and Related Annealing Phenomena*; Elsevier: Oxford, UK, 2004; pp. 333–378. [CrossRef]

34. Di Schino, A.; Kenny, J.M. Effects of the grain size on the corrosion behavior of refined AISI 304 stainless steel. *J. Mater. Sci. Lett.* **2002**, *21*, 1631–1634. [CrossRef]

35. Sinclair, C.W.; Mithieux, J.D.; Schmitt, J.H.; Bréchet, Y. Recrystallization of stabilized ferritic stainless steel sheet. *Metall. Mater. Trans. A Phys. Metall. Mater. Sci.* **2005**, *36*, 3205–3215. [CrossRef]

36. Belyakov, A.; Kimura, Y.; Tsuzaki, K. Recovery and recrystallization in ferritic stainless steel after large strain deformation. *Mater. Sci. Eng. A* **2005**, *403*, 249–259. [CrossRef]

37. Callister, W.D. *Scienza e Ingegneria dei Materiali*; Polytechnic University of Milan: Milan, Italy, 1999.

38. Gao, N.; Baker, T.N. Austenite Grain Growth Behaviour of Microalloyed Al-V-N and Al-V-Ti-N Steels. *ISIJ Int.* **1998**, *38*, 744–751. [CrossRef]

39. Gavard, L.; Montheillet, F.; Coze, J. Le recrystallization and grain growth in high purity austenitic stainless steels. *Scr. Mater.* **1998**, *39*, 1095–1099. [CrossRef]

40. De Knijf, D.; Puype, A.; Föjer, C.; Petrov, R. The influence of ultra-fast annealing prior to quenching and partitioning on the microstructure and mechanical properties. *Mater. Sci. Eng. A* **2015**, *627*, 182–190. [CrossRef]
41. Meng, Q.; Li, J.; Zheng, H. High-efficiency fast-heating annealing of a cold-rolled dual-phase steel. *Mater. Des.* **2014**, *58*, 194–197. [CrossRef]
42. Castro Cerda, F.M.; Kestens, L.A.I.; Monsalve, A.; Petrov, R.H. The effect of ultrafast heating in cold-rolled low carbon steel: Recrystallization and texture evolution. *Metals* **2016**, *6*, 288. [CrossRef]
43. Banis, A.; Duran, E.H.; Bliznuk, V.; Sabirov, I.; Petrov, R.H.; Papaefthymiou, S. The effect of ultra-fast heating on the microstructure, grain size and texture evolution of a commercial low-c, medium-Mn DP steel. *Metals* **2019**, *9*, 877. [CrossRef]
44. Sun, G.S.; Hu, J.; Zhang, B.; Du, L.X. The significant role of heating rate on reverse transformation and coordinated straining behavior in a cold-rolled austenitic stainless steel. *Mater. Sci. Eng. A* **2018**, *732*, 350–358. [CrossRef]
45. Jaskari, M.; Järvenpää, A.; Karjalainen, P. The effect of heating rate and temperature on microstructure and r-value of type 430 ferritic stainless steel. *Mater. Sci. Forum* **2018**, *941*, 364–369. [CrossRef]
46. Di Schino, A. Analysis of phase transformation in high strength low alloyed steels. *Metalurgija* **2017**, *56*, 349–352.
47. Castro Cerda, F.M.; Schulz, B.; Celentano, D.; Monsalve, A.; Sabirov, I.; Petrov, R.H. Exploring the microstructure and tensile properties of cold-rolled low and medium carbon steels after ultrafast heating and quenching. *Mater. Sci. Eng. A* **2019**, *745*, 509–516. [CrossRef]
48. Valdes-Tabernero, M.A.; Celada-Casero, C.; Sabirov, I.; Kumar, A.; Petrov, R.H. The effect of heating rate and soaking time on microstructure of an advanced high strength steel. *Mater. Charact.* **2019**, *155*, 109822. [CrossRef]
49. Senuma, T.; Kawasaki, K.; Takemoto, Y. Recrystallization behavior and texture formation of rapidly annealed cold-rolled extralow carbon steel sheets. *Mater. Trans.* **2006**, *47*, 1769–1775. [CrossRef]
50. Samei, J.; Zhou, L.; Kang, J.; Wilkinson, D.S. Microstructural analysis of ductility and fracture in fine-grained and ultrafine-grained vanadium-added DP1300 steels. *Int. J. Plast.* **2018**, *117*, 58–70. [CrossRef]
51. Speer, J.; Matlock, D.K.; De Cooman, B.C.; Schroth, J.G. Carbon partitioning into austenite after martensite transformation. *Acta Mater.* **2003**, *51*, 2611–2622. [CrossRef]
52. Sharma, D.K.; Filipponi, M.; Di Schino, A.; Rossi, F.; Castaldi, J. Corrosion behaviour of high temperature fuel cells: Issues for materials selection. *Metalurgija* **2019**, *58*, 347–351.
53. Lebedev, A.A.; Kosarchuk, V.V. Influence of phase transformations on the mechanical properties of austenitic stainless steels. *Int. J. Plast.* **2000**, *16*, 749–767. [CrossRef]
54. Jiang, Y.; Tan, H.; Wang, Z.; Hong, J.; Jiang, L.; Li, J. Influence of Creq/Nieq on pitting corrosion resistance and mechanical properties of UNS S32304 duplex stainless steel welded joints. *Corros. Sci.* **2013**, *70*, 252–259. [CrossRef]
55. Nakagawa, H.; Miyazaki, T. Effects of the amount of retained austenite on the microstructures and mechanical properties of a precipitation hardening martensitic stainless steel. *Tetsu-to-Hagane/J. Iron Steel Inst. Jpn.* **1998**, *84*, 43–48. [CrossRef]
56. Stornelli, G.; Schino, A.D.; Mancini, S.; Montanari, R.; Testani, C.; Varone, A. Applied sciences Grain Refinement and Improved Mechanical Properties of EUROFER97 by Thermo-Mechanical Treatments. *Appl. Sci.* **2021**, *11*, 10598. [CrossRef]
57. Hernandez-Duran, E.I.; Corallo, L.; Ros-Yanez, T.; Castro-Cerda, F.M.; Petrov, R.H. Influence of Mo–Nb–Ti additions and peak annealing temperature on the microstructure and mechanical properties of low alloy steels after ultrafast heating process. *Mater. Sci. Eng. A* **2021**, *808*, 140928. [CrossRef]
58. Papaefthymiou, S.; Bouzouni, M.; Petrov, R.H. Study of carbide dissolution and austenite formation during ultra–fast heating in medium carbon chromium molybdenum steel. *Metals* **2018**, *8*, 646. [CrossRef]
59. Roumen, P.; Farideh, H.; Jurij, S.; Jesus, M.; Sietsma, J.; Kestens, L. Ultra-Fast Annealing of High Strength Steel. *Int. Virtual J. Mach. Technol. Mater.* **2012**, *8*, 68–71.
60. Salvatori, I.; Moore, W.B.R. Ultra rapid annealing of cold rolled stainless steels. *ISIJ Int.* **2000**, *40*, 79–83. [CrossRef]
61. Wu, S.Y.; Lin, C.H.; Hsu, W.C.; Chang, L.; Sun, P.L.; Kao, P.W. Effect of heating rate on the development of annealing texture in a 1.09 wt. % Si non-oriented electrical steel. *ISIJ Int.* **2016**, *56*, 326–334. [CrossRef]
62. Iordache, V.E.; Hug, E. Effect of mechanical strains on the magnetic properties of electrical steels. *J. Optoelectron. Adv. Mater.* **2004**, *6*, 1297–1303.
63. Gutiérrez, C.E.; Salinas-Rodríguez, A.; Nava-Vázquez, E. Effect of Fast Annealing on Microstructure and Mechanical Properties of Non-Oriented Al-Si Low C Electrical Steels. *Mater. Sci. Forum* **2007**, *560*, 29–34. [CrossRef]
64. Wang, J.; Li, J.; Mi, X.; Zhang, S.; Volinsky, A.A. Rapid Annealing effects on microstructure, texture, and magnetic properties of non-oriented electrical steel. *Met. Mater. Int.* **2012**, *18*, 531–537. [CrossRef]
65. Gutiérrez-Castañeda, E.J.; Salinas-Rodríguez, A. Effect of annealing prior to cold rolling on magnetic and mechanical properties of low carbon non-oriented electrical steels. *J. Magn. Magn. Mater.* **2011**, *323*, 2524–2530. [CrossRef]
66. Malin, A.S.; Hatherly, M. Shear bands in deformed metals. *Scr. Metall.* **1984**, *13*, 576.
67. Wei, Q.; Jia, D.; Ramesh, K.T.; Ma, E. Evolution and microstructure of shear bands in nanostructured Fe. *Appl. Phys. Lett.* **2002**, *81*, 1240–1242. [CrossRef]
68. Cerda, F.C.; Goulas, C.; Sabirov, I.; Papaefthymiou, S.; Monsalve, A.; Petrov, R.H. Microstructure, texture and mechanical properties in a low carbon steel after ultrafast heating. *Mater. Sci. Eng. A* **2016**, *672*, 108–120. [CrossRef]
69. Stornelli, G.; Faba, A.; Di Schino, A.; Folgarait, P.; Ridolfi, M.R.; Cardelli, E.; Montanari, R. Properties of additively manufactured electric steel powder cores with increased Si content. *Materials* **2021**, *14*, 1489. [CrossRef]

70. Stornelli, G.; Ridolfi, M.R.; Folgarait, P.; De Nisi, J.; Corapi, D.; Repitsch, C.; Di Schino, A. Feasibility assessment of magnetic cores through additive manufacturing techniques. *Metall. Ital.* **2021**, *2*, 50–63.

71. Ray, R.K.; Jonas, J.J.; Hook, R.E. *Cold Rolling and Annealing Textures in Low Carbon and Extra Low Carbon Steels*; Taylor & Francis: Abingdon, UK, 1994; Volume 39.

72. Park, J.T.; Szpunar, J.A. Evolution of recrystallization texture in nonoriented electrical steels. *Acta Mater.* **2003**, *51*, 3037–3051. [CrossRef]

73. Park, J.T.; Szpunar, J.A. Effect of initial grain size on texture evolution and magnetic properties in nonoriented electrical steels. *J. Magn. Magn. Mater.* **2009**, *321*, 1928–1932. [CrossRef]

74. Oldani, C.; Silvetti, S.P. Microstructure and texture evolution during the annealing of a lamination steel. *Scr. Mater.* **2000**, *43*, 129–134. [CrossRef]

75. Wei, X.; Hojda, S.; Dierdorf, J.; Lohmar, J.; Hirt, G. Model for texture evolution in cold rolling of 2.4 wt.-% Si non-oriented electrical steel. In *AIP Conference Proceedings*; AIP Publishing LLC: Melville, NY, USA, 2017; Volume 1896. [CrossRef]

76. Stojakovic, D. Microstructure Evolution in Deformed and Recrystallized Electrical Steel. Ph.D. Thesis, Drexel University, Philadelphia, PA, USA, March 2008.

77. Salih, M.Z.; Weidenfeller, B.; Al-Hamdany, N.; Brokmeier, H.G.; Gan, W.M. Magnetic properties and crystallographic textures of Fe 2.6% Si after 90% cold rolling plus different annealing. *J. Magn. Magn. Mater.* **2014**, *354*, 105–111. [CrossRef]

78. Silva, J.M.; Baêta Júnior, E.S.; Moraes, N.R.D.C.; Botelho, R.A.; Felix, R.A.C.; Brandao, L. Influence of different kinds of rolling on the crystallographic texture and magnetic induction of a NOG 3 wt% Si steel. *J. Magn. Magn. Mater.* **2017**, *421*, 103–107. [CrossRef]

79. Kestens, L.; Aernoudt, E.; Dilewijns, J.; Leuven, K.U.; Metallurgy, P.; Sidmar, N. V The effect of cross rolling on texture and magnetic. *Textures Microstruct.* **1991**, *14*, 921–926.

80. Kvackaj, T.; Bidulska, J.; Biduslky, R. Overview of HSS Steel Grades Development and Study of Reheating Condition Effects on Austenite Grain Size Changes. *Materials* **2021**, *14*, 1988. [CrossRef]

81. Di Pietro, O.; Napoli, G.; Gaggiotti, M.; Marini, R.; Stornelli, G.; Di Schino, A. Analysis of plastic forming paramaters in AISI 441 stainless steels. *Acta Metall. Slov.* **2020**, *26*, 178–183. [CrossRef]

82. Di Schino, A. Precipitation state of warm worked AA7050 alloy: Effect on toughness behavior. *Acta Metall. Slov.* **2021**, *27*, 53–56. [CrossRef]

83. Hutchinson, W.B. Development of textures in recrystallization. *Met. Sci.* **1974**, *8*, 185–196. [CrossRef]

MDPI

St. Alban-Anlage 66

4052 Basel

Switzerland

www.mdpi.com

MDPI Books Editorial Office

E-mail: books@mdpi.com

www.mdpi.com/books